CW01476279

HANDBOOK OF MATHEMATICAL FORMULAS

HANDBOOK OF MATHEMATICAL FORMULAS

By Dipl.-Ing.
HANS-JOCHEN BARTSCH

Translation by
HERBERT LIEBSCHER

With 353 illustrations

ACADEMIC PRESS New York and London
A Subsidiary of Harcourt Brace Jovanovich, Publishers

Translation of the 9th edition of „Mathematische Formeln“
by Herbert Liebscher, Leipzig

Library of Congress Catalog Card Number: 73-2088

ISBN-0-12-080050-0

Edition for Academic Press, Inc. New York and London
Printed and bound in Germany (East)
by Druckhaus Maxim Gorki, Altenburg
114-210/131/74

PREFACE

In our times when technological and scientific progress is made on a large scale it is small wonder that the demands on training and education, especially of the technical and scientific professions, become more and more exacting. Above all this applies to mathematics, the exactest of all sciences. A definite and well-grounded mathematical knowledge only will enable engineers, technicians, designers, and foremen to keep abreast with the rapid technological progress and to meet all requirements with mathematical accuracy. This calls for good courses of instruction in fundamental mathematics at all schools concerned. The purpose of this compilation of formulas is to provide the necessary educational aid. The book starts from secondary school mathematics and is primarily intended for students of technical schools, colleges and universities. It is obvious that not all special subjects could be included. The scope of this collection of formulas covers the whole field from the fundamental rules of arithmetic, via analytic geometry and infinitesimal calculus through to FOURIER's series and the fundamentals of probability calculus. Subjects of topical interest such as matrices, statistics, linear optimization, Boolean algebra, and LAPLACE's transforms are also included.
Great care has been bestowed upon a clear arrangement of the text and a comprehensive index of subjects in order to facilitate the use of the book.
A great number of examples is given to facilitate the grasping of the abstract mathematical formulas. Explanations give the reader an opportunity to consider mathematics not merely formally, but to master it because they induce him to reason about the problems thoroughly and, in addition, they are a contribution to a creative application of mathematics to practice.
Author and publishers present this new edition to the public, hoping that it will meet the requirements; they would be grateful if users of this book would send them suggestions for improving the book.

Author and Publishers

TABLE OF CONTENTS

2. Equations, functions, vectors

3. Geometry

4. Analytical geometry

5. Differential calculus

6. Differential geometry

7. Integral calculus

8. Differential equations

9. Infinite series, Fourier series, Fourier integral, Laplace transforms

10. Theory of probability; statistics; error calculation; mathematical analysis of observations

11. Linear Optimization

12. Algebra of logic (Boolean algebra)

0. Mathematical Signs and Symbols

0.1. Mathematical signs

Signs	Read
1. Ordinal signs	
1.	firstly (or: in the first place)
2.	secondly (or: in the second place)
etc.	etc.
, ·	comma, point Note: In English-speaking countries, the decimal point is placed on the line or raised to the centre or top of the body of the number, e.g. 4.15, 4·15, ·15(= 0·15) In German, the comma is used as decimal sign, e.g. 4,15 Pay attention to the following usage: 50000 (in German) 50,000 (in English)
...	and so on to and so on to infinity
$a_1, a_2, \ldots, a_n$	a sub one, a sub two, a sub n
2. Equality and inequality	
$=$	equals or: is equal to
$\equiv$	is identical with
$\neq$ $\neq$	is not equal to or: does not equal
$\not\equiv$ $\not\equiv$	is not identical with
$\sim$ $\propto$	is (directly) proportional to, varies (directly) as
$\approx$ $\cong$	is approximately equal to or: approximately equals
$\triangleq$	corresponds to
: :	equals (in proportion) or: as
$<$	is less than
$>$	is greater than
$\leqq$	is less than or equal to or: is not greater than
$\geqq$	is greater than or equal to or: is not less than
$\ll$	is much less than
$\gg$	is much greater than
3. Fundamental operations of arithmetic	
$+$	plus
$-$	minus
$\cdot$	multiplied by or: times
$-$ / : $\div$	divided by or: over
:	is to (in proportion)
%	per cent
‰	per mille
( )	parantheses
[ ]	brackets
{ }	braces
⟨ ⟩	angle brackets
4. Geometrical signs	
$\parallel$	is parallel to
$\nparallel$	is not parallel to
$\uparrow\uparrow$	is parallel to, in the same sense
$\uparrow\downarrow$	is parallel to, in the opposite sense
$\perp$	is at right angles to or: is perpendicular to
$\triangle$	triangle
$\cong$	is congruent with
$\sim$	is similar to
$\sphericalangle$	angle
$\overline{AB}$	straight line sement AB
$\overset{\frown}{AB}$	arc AB
5. Algebra and elements of analysis	
sgn	sign
$\lvert z \rvert$	absolute value of z
arc z	arc z
$n!$	n factorial
$\binom{n}{p}$	n over p

Signs	Read
Σ	sum
Π	product
$\sqrt{\ }$; $\sqrt[n]{\ }$	square root of; *n*th root of
i or *j*	imaginary unit
π	pi
( )	matrix
$\vert\ \vert$ or det	determinant
$f(x)$	*f* of *x*
6. Limits	
∞	infinite or: infinity
(a, b)	open interval *ab*
$[a, b]$ or $\langle a, b\rangle$	closed interval *ab*
$\rightarrow$	approaches, approaches as a limit, diverges to, tends to, converges to
lim	limit value or: the limit of
7. Differential calculus	
Δf	delta *f*
$f'(x), f''(x), \ldots, f^{(n)}(x)$	*f* prime *x*, *f* double prime *x*, ..., *f n* prime *x*
$\dot{\varphi}(t), \ddot{\varphi}(t)$	φ dot *t*, φ two dots *t*
$y', y'', \ldots, y^{(n)}$	*y* prime, *y* double prime, ..., *y n* prime
d	differential (sign used in calculus)
$df(x)$	*df* of *x*
$\frac{dy}{dx}$	*dy* by *dx*
$\frac{d^2y}{dx^2}$	*d* two *y* by *dx* squared
$\frac{d^ny}{dx^n}$	*dny* by *dx* to the *n*th
f_x, f_y	*f* with respect to *x*, *f* with respect to *y*
∂	*d* partially
f_{xx}, f_{xy}	*f* with respect to *xx*, *f* with respect to *xy*
f_{yx}, f_{yy},	*f* with respect to *yx*, *f* with respect to *yy*
$df(x, y)$	total differential of $f(x,y)$
8. Integral calculus	
$\int$	integral
$\int f(x)\,dx$	integral of small *f* of *x dx*

Signs	Read
$\int\limits_a^b f(x)\,dx$ in English preferably written as $\int_a^b f(x)\,dx$	integral from *a* to *b* of small *f* of *x dx*
$F(x)\Big\vert_a^b$ in English preferably written as $[F(x)]_a^b$	capital *F* of *x* between the limits *a* and *b*
$\oint$	integral around a closed contour
9. Exponential and logarithmic functions	
exp *x*	exponential function of *x*
log	logarithm (general)
$\log_a$	logarithm to the base *a*
lg	common or Briggsian logarithm (having base 10)
ln	natural logarithm
10. Trigonometrical and hyperbolic functions and their inverses	
sin	sine
cos	cosine
tan	tangent
cot or: ctg	cotangent
arc sin or: $\sin^{-1}$	arc sine, inverse sine
arc cos or: $\cos^{-1}$	arc cosine, inverse cosine
arc tan or: $\tan^{-1}$	arc tangent, inverse tangent
arc cot or: $\cot^{-1}$	arc cotangent, inverse cotangent
sinh	hyperbolic sine
cosh	hyperbolic cosine
tanh	hyperbolic tangent
coth	hyperbolic cotangent
arsinh or: $\sinh^{-1}$	inverse sinh; area sine
arcosh or: $\cosh^{-1}$	inverse cosh; area cosine
artanh or: $\tanh^{-1}$	inverse tanh; area tangent
arcoth or: $\coth^{-1}$	inverse coth; area cotangent

0.2. Symbols used in the theory of sets

$A = \{a_k\}$ $= \{a_1; a_2; a_3; \ldots\}$	Set A consists of the elements $a_1; a_2; a_3; \ldots$	$\cap$	the intersection of
		$\setminus$	difference
		$\rightarrow$	mapped onto
$\{a_k\}_{\text{ordered}}$	ordered set (corresponds to sequence) whose elements correspond to the sequence of the natural numbers	$\times$	sign for Cartesian product (or: cross product set)
		N	set of the natural numbers
		I	set of the whole numbers $N \subset I$
$\in$	is an element of		
$\notin$	is not an element of	K	field of the rational numbers $I \subset K$
$=$	equals		
$\{\,\}; \varnothing$	empty set (or: null set)	R	field of the real numbers $K \subset R$
$\subset$	is a subset of (is contained in)	C	field of the complex numbers $R \subset C$
$\cup$	the union of	S	solution set

0.3. Symbols of logic

$A_1 \Rightarrow A_2$	from A_1 follows A_2	$\vee, +, \cup$	*or*
$A_1 \Leftrightarrow A_2$	the statements A_1 and A_2 are equivalent	$\wedge, \cdot, \cap$	*and* (also without calculating sign)
		$\overline{x}$	logical negation (x cross, not x)

1. Arithmetic

1.1. Set theory

1.1.1. Fundamental notions

When we collect all objects possessing a certain property into an aggregate, the aggregate is called a *set* provided it can be uniquely determined whether or not any one of these objects has the specified property.
The set of the natural numbers

$$N = \{0; 1; 2; \ldots\}$$

The set of the whole numbers

$$I = \{0; 1; -1; 2; -2; \ldots\}$$

If the elements of sets are points of a curve, a plane, or a space, the sets are also called *point sets*. If each element a_i of set A is contained in set B, then A is a *subset* of B:

$$A \subset B$$

The following relations always hold:

$$A \subset A \text{ (reflexive relation)}, \ \emptyset \subset A$$

$$A \subset B \text{ and } B \subset C \Rightarrow A \subset C \text{ (transitive relation)}$$

$$A \subset B \text{ and } B \subset A \Rightarrow A = B$$

$$A = A \text{ (reflexive relation)}$$

$$A = B \Rightarrow B = A \text{ (symmetrical relation)}$$

$$A = B \text{ and } B = C \Rightarrow A = C \text{ (transitive relation)}$$

If $A \subset B$ holds but not $B \subset A$, i.e. if $A \neq B$, then A is called a *proper subset* of B.
A set that contains at least two different elements is called a *field* if its elements can be subjected, without any restriction, to addition,

subtraction, multiplication, and division, the divisor 0 being excluded.
The field K of the rational numbers is the smallest number field that contains the range of the natural numbers.

1.1.2. Set operations

The *union* of two sets $A \cup B$ (read: A united with B) consists of all elements that belong at least to one of the two sets A and B.
The *intersection* of two sets $A \cap B$ (read: A intersecting B) consists of all elements that belong to both set A and set B. The *difference* between two sets $A \setminus B$ (read: difference set between A and B) consists of all elements that belong to A but do not belong to B.
The *Cartesian product* (or the cross product set) of two sets $A \times B$ (read: A cross B) is the set of all ordered pairs of elements (a, b) where $a \in A$, $b \in B$.

Calculating rules

$$A \cup B = B \cup A$$

$$(A \cup B) \cup C = A \cup (B \cup C)$$

$$A \cap B = B \cap A$$

$$(A \cap B) \cap C = A \cap (B \cap C)$$

$$(A \cup B) \cap C = (A \cap C) \cup (B \cap C)$$

$$A \cup \emptyset = A$$

$$A \cap \emptyset = \emptyset$$

$$A \setminus B = A \setminus (A \cap B)$$

$$(A \setminus B) \cap B = \emptyset$$

$$A \setminus A = \emptyset$$

$$A \cup B = (A \setminus B) \cup (B \setminus A) \cup (A \cap B)$$

If one of the three relations

$$A \subset B, \quad A \cup B = B, \quad A \cap B = A$$

holds, then the other two will also hold.

1.1.3. Mappings, cardinality

If each element $x \in X$ is put into correspondence with one or several elements of a set Y, according to a given specification, we speak of a *mapping* of set X onto set Y.
The elements of set Y mapped onto by the element x are called the *images* of x. The element x is also called the *original*.
Set X is called the *original set* or the *domain of definition* D; the set Y is the *image set* or the *range of values* R.
If the mapping is denoted by σ, we write

$$\sigma\colon X \to Y \quad \text{or} \quad x \xrightarrow{\sigma} \sigma(x) = y$$

If in mapping, precisely one element of the image set is mapped onto by one element of the original set, then the mapping is unique (single-valued). If the two sets consist of mathematical objects, we speak of *functions*. If converting a unique mapping also yields a unique mapping, i.e. if each element of the image set corresponds to only one element of the original set, the mapping is *biunique*. Two sets that can be mapped onto each other biuniquely are of the *same cardinality* or *equivalent* to each other.
If set A can be put into one-to-one correspondence with a proper subset B' of a set B but not with set B, then A is of smaller cardinality than the set B.
Set B is said to be of a larger cardinality than set A.

1.2. Real numbers

1.2.1. General

Absolute value of real numbers

$$|a| = |-a|$$

$$|a| \geqq 0$$

$$|a| = \begin{cases} a & \text{for } a \geqq 0 \\ -a & \text{for } a < 0 \end{cases}$$

$$\pm a \leqq |a|$$

$$|a| - |b| \leqq |a + b| \leqq |a| + |b|$$

$$|a| - |b| \leqq |a - b| \leqq |a| + |b|$$

$$|ab| = |a| \cdot |b|$$

$$\left|\frac{a}{b}\right| = \frac{|a|}{|b|} \text{ for } b \neq 0$$

Signs (sgn) before numbers

$$\operatorname{sgn} z = \frac{|z|}{z}$$

$$\operatorname{sgn} z = \begin{cases} +1 & \text{for} \quad z > 0 \\ -1 & \text{for} \quad z < 0 \end{cases}$$

$$\operatorname{sgn} 0 = 0$$

Representation of real numbers

The set of the real numbers can be represented by points on a *straight line.*

1.2.2. Irrational numbers

The nth roots of nth powers of rational numbers are also rational numbers.

$$\sqrt{25} = 5; \quad \sqrt[3]{27} = 3; \quad \sqrt[4]{\frac{1}{256}} = \frac{1}{4}$$

The nth roots of nonnegative real numbers which do not represent nth powers are *irrational* numbers, i.e. nonperiodic decimal numbers with infinitely many places.

Rationalizing of denominators

When operating with fractions having roots in their denominators, the fraction is expressed in higher terms so that the denominator becomes rational.

Examples:

$$\frac{x}{\sqrt{x^3}} = \frac{x\sqrt[4]{x}}{\sqrt[4]{x^3}\,\sqrt[4]{x}} = \frac{x\sqrt[4]{x}}{\sqrt[4]{x^4}} = \frac{x\sqrt[4]{x}}{x} = \sqrt[4]{x}$$

$$\frac{m}{a + \sqrt{b}} = \frac{m\left(a - \sqrt{b}\right)}{\left(a + \sqrt{b}\right)\left(a - \sqrt{b}\right)} = \frac{m\left(a - \sqrt{b}\right)}{a^2 - b}$$

1.2.3. Binomial coefficients, binomial theorem

Binomial coefficient $\binom{n}{k}$ (read: n over k)

$k \in N; \quad n \in K$

$k! = 1 \cdot 2 \cdot 3 \cdot 4 \cdots k$ (k factorial)

$0! = 1$ [definition]

$$\binom{n}{k} = \frac{n(n-1)(n-2)\cdots(n-k+1)}{1 \cdot 2 \cdot 3 \cdots k}$$

$$= \frac{n(n-1)(n-2)\cdots(n-k+1)}{k!}$$

$$\binom{n}{0} = \binom{n}{n} = 1 \text{ [definition]}; \quad \binom{n}{k} = 0 \text{ for } k > n > 0;\ n \in N$$

Examples:

$$\binom{10}{4} = \frac{10 \cdot 9 \cdot 8 \cdot 7}{1 \cdot 2 \cdot 3 \cdot 4} = 210$$

$$\binom{-\frac{1}{2}}{2} = \frac{\left(-\frac{1}{2}\right)\left(-\frac{3}{2}\right)}{1 \cdot 2} = \frac{3}{8}$$

$$\binom{3}{5} = \frac{3 \cdot 2 \cdot 1 \cdot 0 \cdot (-1)}{1 \cdot 2 \cdot 3 \cdot 4 \cdot 5} = 0$$

Pascal's triangle for the determination of the binomial coefficients

$n = 0$						1					
$n = 1$					1		1				
$n = 2$				1		2		1			
$n = 3$			1		3		3		1		
$n = 4$		1		4		6		4		1	
$n = 5$	1		5		10		10		5		1

. . . .

In this triangle $\binom{n}{k}$ is in the $(n + 1)$th row at the $(k + 1)$th place.

Examples;

$\binom{4}{2} = 6$ is in the fifth row at the third place.

Relations between binomial coefficients

$$\binom{n}{k} = \binom{n}{n-k} = \frac{n!}{k!(n-k)!} \qquad n \in N$$

$$\binom{n}{k} + \binom{n}{k+1} = \binom{n+1}{k+1}$$

$$\binom{n}{k+1} = \binom{n}{k} \cdot \frac{n-k}{k+1}$$

$$\binom{n}{k} + \binom{n}{k-1} = \binom{n+1}{k}$$

$$\binom{k}{k} + \binom{k+1}{k} + \binom{k+2}{k} + \cdots$$

$$+ \binom{n}{k} = \binom{n+1}{k+1} \qquad n \in N$$

$$\binom{k}{0} + \binom{k+1}{1} + \binom{k+2}{2} + \cdots$$

$$+ \binom{k+n}{n} = \binom{k+n+1}{n} \qquad n \in N$$

$$\binom{n}{1} = n$$

$$\binom{n}{n-1} = n \qquad n \in N$$

$$\binom{n}{0} + \binom{n}{1} + \binom{n}{2} + \cdots + \binom{n}{n} = 2^n \qquad n \in N$$

$$\binom{n}{0} - \binom{n}{1} + \binom{n}{2} - + \cdots + (-1)^n \binom{n}{n} = 0$$

$$\binom{n}{0} + \binom{n}{2} + \binom{n}{4} + \cdots = 2^{n-1} \qquad n \in N$$

$$\binom{n}{1} + \binom{n}{3} + \binom{n}{5} + \cdots = 2^{n-1} \qquad n \in N$$

$$\binom{n}{0}^2 + \binom{n}{1}^2 + \binom{n}{2}^2 + \cdots + \binom{n}{n}^2 = \binom{2n}{n} \qquad n \in N$$

Binomial theorem in which the exponent is a positive integer

$$(a + b)^n = \binom{n}{0} a^n + \binom{n}{1} a^{n-1}b + \binom{n}{2} a^{n-2}b^2 + \binom{n}{3} a^{n-3}b^3$$

$$+ \cdots + \binom{n}{n-1} ab^{n-1} + \binom{n}{n} b^n = \sum_{k=0}^{n} \binom{n}{k} a^{n-k}b^k$$

$$(a - b)^n = \binom{n}{0} a^n - \binom{n}{1} a^{n-1}b + \binom{n}{2} a^{n-2}b^2 - \binom{n}{3} a^{n-3}b^3$$

$$+ - \cdots + (-1)^{n-1} \binom{n}{n-1} ab^{n-1} + (-1)^n \binom{n}{n} b^n$$

$$= \sum_{k=0}^{n} (-1)^k \binom{n}{k} a^{n-k}b^k$$

General binomial theorem for arbitrary exponents

$$(a + b)^n = \binom{n}{0} a^n + \binom{n}{1} a^{n-1}b + \binom{n}{2} a^{n-2}b^2$$

$$+ \binom{n}{3} a^{n-3}b^3 + \cdots \qquad |a| > |b|$$

$$(a + b)^n = \binom{n}{0} b^n + \binom{n}{1} ab^{n-1} + \binom{n}{2} a^2b^{n-2}$$

$$+ \binom{n}{3} a^3b^{n-3} + \cdots \qquad |a| < |b|$$

Binomial theorem for a few values of n

$n = 2$: $(a \pm b)^2 = a^2 \pm 2ab + b^2$

$n = 3$: $(a \pm b)^3 = a^3 \pm 3a^2b + 3ab^2 \pm b^3$

$n = 4$: $(a \pm b)^4 = a^4 \pm 4a^3b + 6a^2b^2 \pm 4ab^3 + b^4$

$n = 5$: $(a \pm b)^5 = a^5 \pm 5a^4b + 10a^3b^2 \pm 10a^2b^3 + 5ab^4 \pm b^5$

1.3. Imaginary or complex numbers

1.3.1. Imaginary numbers

Imaginary unit

Defining equation: $j^2 = -1$

Powers of the imaginary unit

$$\left.\begin{aligned} j^{4n} &= +1 \\ j^{4n+1} &= j \\ j^{4n+2} &= -1 \\ j^{4n+3} &= -j \end{aligned}\right\} \text{ for } \quad n \in I$$

Square root of a negative number

$$\sqrt{-25} = 5j$$

General: $\sqrt{-a} = j\sqrt{a}$ for $a > 0$

Algebraic sums of imaginary numbers

are either imaginary or 0.

Examples:

$$4j - 7j + 9j = 6j$$
$$4j - 7j + 3j = 0$$

Products of imaginary numbers

are either real or imaginary.

Examples:

$$5j \cdot 7j = 35j^2 = -35$$
$$5j \cdot 7j \cdot 8j = 280j^3 = -280j$$

Quotients of imaginary numbers

are always real.

Example:

$$\frac{14j}{15j} = \frac{14}{15}$$

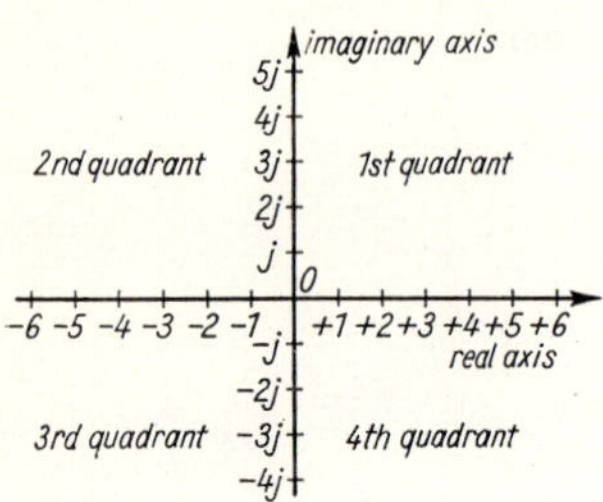

Representation of imaginary numbers

The set of the imaginary numbers can be represented by points on the imaginary axis as multiples of the imaginary unit j (line segment from 0 to j). Real and imaginary axes are perpendicular to each other.

1.3.2. Complex numbers in arithmetical form

$$z = a + bj \qquad a, b \in R$$

$$\left.\begin{aligned} z_1 &= a + bj \\ z_2 &= a - bj \end{aligned}\right\} \textit{conjugate} \text{ complex numbers}$$

Definition

$$a + bj = c + dj \Leftrightarrow a = c; \quad b = d$$

especially holds: $a + bj = 0 \Leftrightarrow a = 0; \quad b = 0$

The absolute value $|z| = \sqrt{a^2 + b^2}$

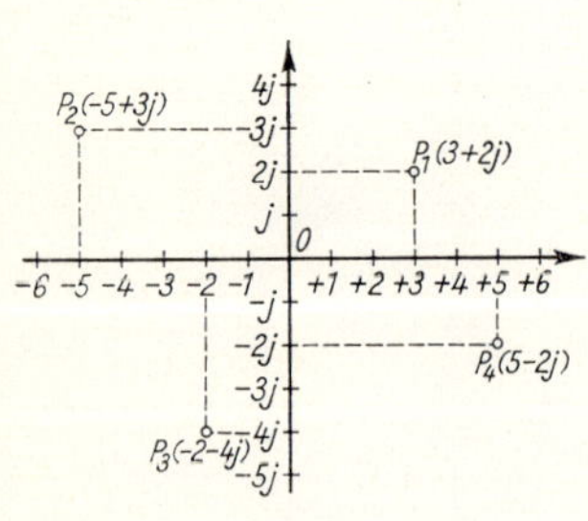

The set of all numbers $a + bj$ forms the **field of the complex numbers** each of which corresponds to an ordered pair (a, b), i.e. to a Cartesian product. All of these Cartesian products are represented by points in the *Gaussian number plane* (one-to-one correspondence established between the set of the complex numbers and the set of points in the Gaussian number plane).

Addition of complex numbers

$$(a + bj) \pm (c + dj) = (a \pm c) + (b \pm d)j$$

$$(a + bj) + (a - bj) = 2a$$

Subtraction of complex numbers

$$(a + bj) - (c + dj) = (a - c) + (b - d)j$$
$$(a + bj) - (a - bj) = 2bj$$

Multiplication of complex numbers

$$(a + bj)(c + dj) = (ac - bd) + (ad + bc)j$$
$$(a + bj)(a - bj) = a^2 + b^2$$

$a^2 + b^2$ is called *norm* of the complex number $a + bj$ or $a - bj$.

Division of complex numbers (rationalization of denominators)

$$\frac{a + bj}{c + dj} = \frac{(a + bj)(c - dj)}{(c + dj)(c - dj)} = \frac{ac + bcj - adj - bdj^2}{c^2 - d^2j^2}$$
$$= \frac{(ac + bd) + (bc - ad)j}{c^2 + d^2} = \frac{ac + bd}{c^2 + d^2} + \frac{bc - ad}{c^2 + d^2} \cdot j$$

When reducing the quotient to higher terms, multiplying it by the complex number conjugate to the divisor, we always obtain a quotient whose denominator is real.

Example:

$$\frac{1 + 2j}{3 - j} = \frac{(1 + 2j)(3 + j)}{(3 - j)(3 + j)} = \frac{1 + 7j}{10} = \frac{1}{10} + \frac{7}{10}j$$

Raising complex numbers to a power

$$(a + bj)^2 = a^2 + 2abj + b^2j^2 = a^2 + 2abj - b^2$$
$$(a + bj)^3 = a^3 + 3a^2bj + 3ab^2j^2 + b^3j^3$$
$$= a^3 + 3a^2bj - 3ab^2 - b^3j$$
$$(a + bj)^4 = a^4 + 4a^3bj + 6a^2b^2j^2 + 4ab^3j^3 + b^4j^4$$
$$= a^4 + 4a^3bj - 6a^2b^2 - 4ab^3j + b^4$$

etc.

Square roots of a complex number

$$z_1 = a + bj; \quad z_2 = \sqrt{z_1} \Rightarrow$$

$$z_{21} = c + dj = \sqrt{\frac{\sqrt{a^2 + b^2} + a}{2}} \pm bj \sqrt{\frac{1}{2\left(a + \sqrt{a^2 + b^2}\right)}}$$

$$z_{22} = e + fj = -\sqrt{\frac{\sqrt{a^2 + b^2} + a}{2}} - bj \sqrt{\frac{1}{2\left(a + \sqrt{a^2 + b^2}\right)}}$$

1.3.3. Complex numbers in a goniometric form

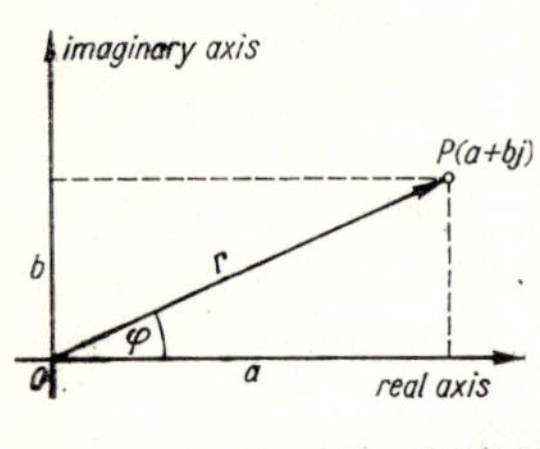

The line segment OP having a definite direction (sense) is called pointer $\mathbf{r}$ (for definition see page 34).

r is the *absolute value* or *modulus* of the complex number

φ is the *argument* or *phase* of the complex number

$$z = a + bj = r(\cos\varphi + j\sin\varphi)$$

$$= r[\cos(\varphi + k\cdot 360°) + j\sin(\varphi + k\cdot 360°)]$$

$$\text{for } k \in G \quad \text{and} \quad 0° \leqq \varphi < 360°$$

In this connection, the following relations hold:

$$\left.\begin{aligned} a &= r\cos\varphi \\ b &= r\sin\varphi \end{aligned}\right\} \tan\varphi = \frac{b}{a}$$

(a and b determine the position of the point in the Gaussian number plane and, hence, φ.)

$$|\mathbf{r}| = r = \sqrt{a^2 + b^2}$$

Example:

Convert $3 - 4j$ into the goniometric form.

$$r = \sqrt{3^3 + (-4)^2} = 5$$

$$\tan\varphi = \frac{-4}{3} = -\frac{4}{3}; \quad \varphi = 306°52'$$

(The second value of φ is eliminated because numbers with $a > 0$ and $b > 0$ lie in the fourth quadrant.)

$$3 - 4j = 5[\cos(306°52' + k\cdot 360°)$$

$$+ j\sin(306°52' + k\cdot 360°)]$$

$$\text{for } k \in G$$

Multiplication of complex numbers

$$z_1 = r_1(\cos\varphi_1 + j\sin\varphi_1)$$

$$z_2 = r_2(\cos\varphi_2 + j\sin\varphi_2)$$

$$z_1 z_2 = r_1 r_2[\cos(\varphi_1 + \varphi_2) + j\sin(\varphi_1 + \varphi_2)]$$

Division of complex numbers

$$z_1 = r_1(\cos\varphi_1 + j\sin\varphi_1)$$

$$z_2 = r_2(\cos\varphi_2 + j\sin\varphi_2)$$

$$\frac{z_1}{z_2} = \frac{r_1}{r_2}[\cos(\varphi_1 - \varphi_2) + j\sin(\varphi_1 - \varphi_2)]$$

Raising complex numbers to a power

$$z = r(\cos\varphi + j\sin\varphi)$$

$$z^n = r^n(\cos n\varphi + j\sin n\varphi) \quad \text{(De Moivre's theorem)} \quad n \in I$$

Extraction of roots of complex numbers

$$z = r(\cos\varphi + j\sin\varphi)$$

$$\sqrt[n]{z} = \sqrt[n]{r}\left(\cos\frac{\varphi + k\cdot 360^\circ}{n} + j\sin\frac{\varphi + k\cdot 360^\circ}{n}\right)$$

$$k \in \{0, 1, 2, \ldots, (n-1)\}$$

This shows that the nth root of a number has n values. $k = 0$ yields the principal value. $n \in N \setminus \{0\}$.

Roots of the positive unit and the negative unit

$$\sqrt[n]{1} = \cos\frac{k\cdot 360^\circ}{n} + j\sin\frac{k\cdot 360^\circ}{n}$$

$$\sqrt[n]{-1} = \cos\frac{180^\circ + k\cdot 360^\circ}{n} + j\sin\frac{180^\circ + k\cdot 360^\circ}{n}$$

for $k \in \{0, 1, 2, \ldots, (n-1)\}$

Examples:

$$\sqrt[3]{1}\begin{cases} e_1 = 1 \\ e_2 = -\dfrac{1}{2} + \dfrac{j}{2}\sqrt{3} \\ e_3 = -\dfrac{1}{2} - \dfrac{j}{2}\sqrt{3} \end{cases} \qquad \sqrt[3]{-1}\begin{cases} e_1 = \dfrac{1}{2} + \dfrac{j}{2}\sqrt{3} \\ e_2 = -1 \\ e_3 = \dfrac{1}{2} - \dfrac{j}{2}\sqrt{3} \end{cases}$$

$$\sqrt[4]{1}\begin{cases} e_1 = 1 \\ e_2 = j \\ e_2 = -1 \\ e_4 = -j \end{cases} \qquad \sqrt[4]{-1}\begin{cases} e_1 = +\dfrac{1}{2}\sqrt{2} + \dfrac{j}{2}\sqrt{2} \\ e_2 = -\dfrac{1}{2}\sqrt{2} + \dfrac{j}{2}\sqrt{2} \\ e_3 = -\dfrac{1}{2}\sqrt{2} - \dfrac{j}{2}\sqrt{2} \\ e_4 = +\dfrac{1}{2}\sqrt{2} - \dfrac{j}{2}\sqrt{2} \end{cases}$$

Solution of the binomial equations

$$\begin{aligned} x^n &= a; & x &= \sqrt[n]{a}\,\sqrt[n]{1} \\ x^n &= -a; & x &= \sqrt{a}\,\sqrt[n]{-1} \end{aligned} \qquad n \in N \setminus \{0\}$$

1.3.4. Complex numbers in exponential form

Definition of $e^{j\varphi}$:

$$\cos\varphi + j\sin\varphi = e^{j\varphi} \qquad \textbf{(Euler's formula)}$$

For $-\varphi$ we have

$$\cos\varphi - j\sin\varphi = e^{-j\varphi}$$

From this follows the exponential form:

$$\begin{aligned} z &= a + bj = r(\cos\varphi + j\sin\varphi) = re^{j\varphi} \\ &= r[\cos(\varphi + 2k\pi) + j\sin(\varphi + 2k\pi)] = re^{j(\varphi+2k\pi)} \quad \text{for } k \in I \end{aligned}$$

Example:

Convert $-2 + 3j$ into the exponential form.

$$r = \sqrt{(-2)^2 + 3^2} = \sqrt{13}; \qquad \tan\varphi = \frac{3}{-2} = -1.5;$$

$$\varphi = 123°41' \triangleq 2.16 \text{ rad}$$

$$-2 + 3j = \underline{\underline{\sqrt{13} \cdot e^{j2.16}}}$$

Abbreviation for practical purposes in engineering calculations:

$$e^{j\varphi} = \angle\underline{\varphi} \text{ (read: versor } \varphi)$$

$$-2 + 3j = \sqrt{13}\, e^{j123°41'} = \sqrt{13} \angle\underline{123°41'}$$

Example:

Convert $17 \cdot e^{j37°22'}$ into the arithmetical form $a + bj$.

$$a + bj = 17(\cos 37°22' + j \sin 37°22')$$

$$= 17(0.795 + j \cdot 0.607) = \underline{\underline{13.5 + 10.3j}}$$

Multiplication of complex numbers

$$\left.\begin{array}{l} z_1 = r_1 e^{j\varphi_1} \\ z_2 = r_2 e^{j\varphi_2} \end{array}\right\} z_1 z_2 = r_1 r_2 e^{j(\varphi_1 + \varphi_2)}$$

Division of complex numbers

$$\left.\begin{array}{l} z_1 = r_1 e^{j\varphi_1} \\ z_2 = r_2 e^{j\varphi_2} \end{array}\right\} \frac{z_1}{z_2} = \frac{r_1}{r_2} e^{j(\varphi_1 - \varphi_2)}$$

Raising complex numbers to a power

$$z = re^{j\varphi} \qquad z^n = r^n e^{jn\varphi} \qquad n \in I$$

Extraction of roots of complex numbers

$$z = re^{j(\varphi + 2k\pi)}$$

$$\sqrt[n]{z} = \sqrt[n]{r}\, e^{j\frac{\varphi + 2k\pi}{n}} \qquad k \in N; \quad n \in N \setminus \{0; 1\}$$

Special values of the factor $e^{j\varphi}$

$$\left.\begin{aligned} e^{j2k\pi} &= 1 \\ e^{j(2k+1)\pi} &= -1 \end{aligned}\right\} \text{ for } k \in I$$

$$e^{j\frac{\pi}{2}} = j; \quad e^{j\frac{2\pi}{3}} = -\frac{1}{2} + \frac{j}{2}\sqrt{3}$$

$$e^{j\frac{3\pi}{2}} = -j; \quad e^{j\frac{4\pi}{3}} = -\frac{1}{2} - \frac{j}{2}\sqrt{3}$$

$$e^{j2k\pi} = 1 \Rightarrow e^{\varphi+2k\pi} = e^{\varphi}$$

Period of the exponential function, hence, $2\pi j$.

1.3.5. Natural logarithms of complex and negative numbers

$$\begin{aligned} \ln(a + bj) &= \ln\left[re^{j(\varphi+2k\pi)}\right] = \ln r + \ln e^{j(\varphi+2k\pi)} \\ &= \ln r + j(\varphi + 2k\pi) \qquad k \in I \end{aligned}$$

For $k = 0$ the *principal value* is obtained.

Example:

Calculate $\ln(3 - 7j)$.

$$r = \sqrt{9 + 49} = \sqrt{58} \qquad \tan\varphi = \frac{-7}{3} = -2.33\ldots$$

$$\varphi = 293°\,12' \mathrel{\hat{=}} 5.12 \text{ rad}$$

$$\begin{aligned} \ln(3 - 7j) &= \ln\sqrt{58} + j(5.12 + 2k\pi) \\ &= 2.03022 + j(5.12 + 2k\pi) \end{aligned}$$

$$\begin{aligned} k = 0&: \ln(3 - 7j) = 2.03022 + 5.12j \\ k = 1&: \qquad\qquad = 2.03022 + j(5.12 + 2\pi) = 2.03022 + 11.4j \\ k = 2&: \qquad\qquad = 2.03022 + j(5.12 + 4\pi) = 2.03022 + 17.68j \end{aligned}$$

etc.

Special cases

Positive imaginary number bj ($a = 0; b > 0$)

$$\ln(bj) = \ln\left[re^{j(\varphi+2k\pi)}\right], \quad \text{where } r = b \text{ and } \varphi = \frac{\pi}{2}$$

$$\ln(bj) = \ln\left[be^{j\left(\frac{\pi}{2}+2k\pi\right)}\right] = \ln b + j\left(\frac{\pi}{2} + 2k\pi\right) \text{ for } k \in I$$

Logarithms of positive imaginary numbers are complex numbers.

Negative imaginary number $-bj$ $(a = 0;\ b > 0)$

$$\ln(-bj) = \ln\left[re^{j(\varphi+2k\pi)}\right], \quad \text{where } r = b \text{ and } \varphi = \frac{3}{2}\pi$$

$$\ln(-bj) = \ln\left[be^{j\left(\frac{3}{2}\pi+2k\pi\right)}\right] = \ln b + j\left(\frac{3}{2}\pi + 2k\pi\right) \text{ for } k \in I$$

Positive real number a $(a > 0;\ b = 0)$

$$\ln a = \ln\left[re^{j(\varphi+2k\pi)}\right], \quad \text{where } r = a \text{ and } \varphi = 0$$

$$\ln a = \ln\left[ae^{j(0+2k\pi)}\right]$$

$$= \ln a + 2k\pi j \quad \text{for } k \in I$$

Negative real number $-a$ $(a > 0;\ b = 0)$

$$\ln(-a) = \ln\left[re^{j(\varphi+2k\pi)}\right], \quad \text{where } r = a \text{ and } \varphi = \pi$$

$$\ln(-a) = \ln\left[ae^{j(\pi+2k\pi)}\right]$$

$$= \ln a + j(\pi + 2k\pi) \quad \text{for } k \in I$$

Similarly, we have:

$$\left.\begin{aligned} \ln 1 &= 2k\pi j \\ \ln(-1) &= (\pi + 2k\pi)j \\ \ln j &= \left(\frac{\pi}{2} + 2k\pi\right)j \\ \ln(-j) &= \left(\frac{3}{2}\pi + 2k\pi\right)j \end{aligned}\right\} \text{ for } k \in I$$

1.3.6. Graphical methods

Graphical addition of complex numbers

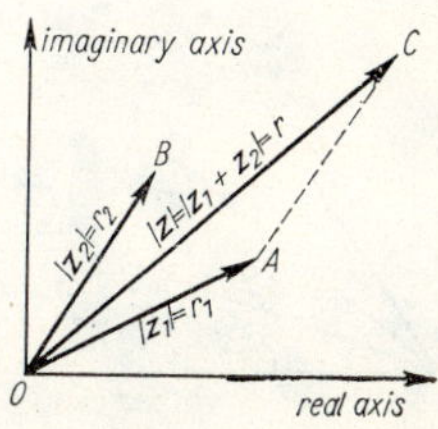

Since complex numbers can be represented by pointers the graphical method of adding complex numbers can be reduced to the addition of these directed line segments. The pointer $OB = \mathbf{z}_2$ is added to the pointer $OA = \mathbf{z}_1$ by displacing the former parallel to itself. The direct segmented $OC = \mathbf{z}$ then is the sum of $\mathbf{z}_1$ and $\mathbf{z}_2$;

$$\mathbf{z} = \mathbf{z}_1 + \mathbf{z}_2$$

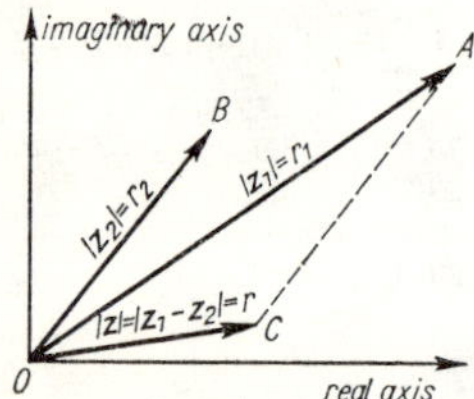

Graphical subtraction of complex numbers

The difference between pointer $\mathbf{z}_1$ and pointer $\mathbf{z}_2$ can be considered the sum of pointer $\mathbf{z}_1$ plus pointer $(-\mathbf{z}_2)$:

$$\mathbf{z}_1 - \mathbf{z}_2 = \mathbf{z}_1 + (-\mathbf{z}_2)$$

To construct the difference, a pointer having a sense opposite to that of the directed segment $\mathbf{z}_2$ is moved parallel to itself till its point of origin coincides with the point of $OA = \mathbf{z}_1$.

The directed segment $OC = \mathbf{z}$ respresents the difference between $\mathbf{z}_1$ and $\mathbf{z}_2$:

$$\mathbf{z} = \mathbf{z}_1 - \mathbf{z}_2$$

Graphical multiplication of complex numbers

Since the multiplication of complex numbers does not correspond to the formation of products in vector algebra, the line segments having a definite direction (sense) which represent the complex numbers are called pointers instead of vectors, a term still frequently used in this connection.
From $z_1 z_2 = r_1 r_2 e^{j(\varphi_1+\varphi_2)} = z = re^{j\varphi}$ we obtain $\varphi = \varphi_1 + \varphi_2$ for the angle and $r = r_1 r_2$ for the modulus. The latter equation can be rearranged to read $r : r_2 = r_1 : 1$.

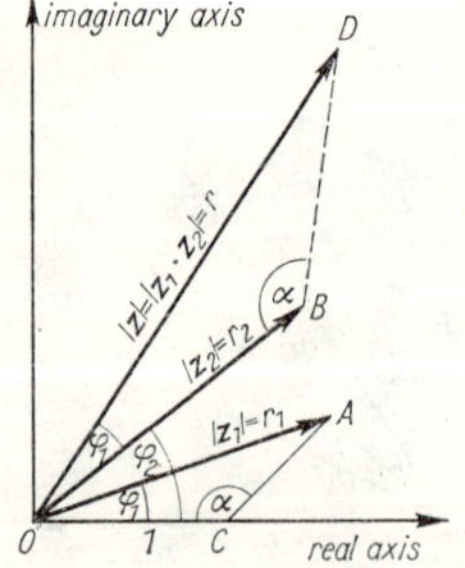

Construction

Step 1: Draw the directed segments OA and OB corresponding to the complex numbers.

Step 2: At the origin lay out the angle φ_1 on the pointer OB in accordance with $\varphi = \varphi_1 + \varphi_2$.

Step 3: Measure off the line segment 1 (point C) on the real axis from O.

Step 4: Connect C with A by a line.

Step 5: Construct the angle $OCA = \alpha$ at point B on the line segment OB. The intersection D of the free arm of this angle and the free arm of angle φ_1 determines the pointer of the result.

Proof: $\triangle OCA \sim \triangle OBD$

(correspondence between the angles) so that the above proportion $r:r_2 = r_1:1$ holds.

Graphical division of complex numbers

From $\frac{z_1}{z_2} = \frac{r_1}{r_2} e^{j(\varphi_1-\varphi_2)} = z = re^{j\varphi}$ we obtain $\varphi = \varphi_1 - \varphi_2$ for the angle and $r = \frac{r_1}{r_2}$ for the modulus. The latter equation can be rearranged to read $r:r_1 = 1:r_2$.

Construction

Step 1: Draw the pointers OA and OB representing the complex numbers.

Step 2: Construct the angle φ_2 on the pointer OA in the negative sense in accordance with $\varphi = \varphi_1 - \varphi_2$.

Step 3: Measure off the line segment 1 (point C) on the real axis from 0.

Step 4: Connect B with C by a line.

Step 5: Construct the angle $OBC = \alpha$ on the line OA at point A.

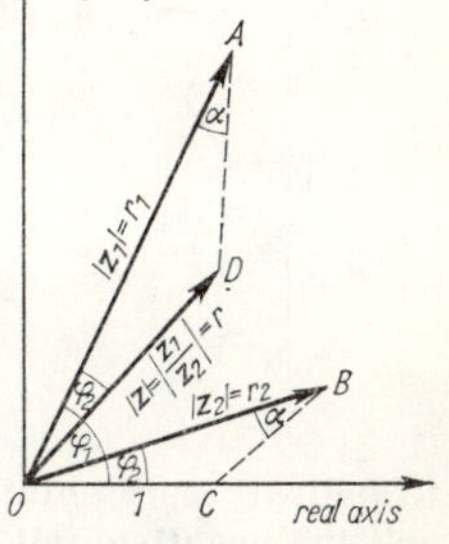

The intersection D of the free arm of this angle and the free arm of the angle φ_2 determines the pointer of the result.

Proof: $\triangle OCB \sim \triangle ODA$

(correspondence between the angles) so that the above proportion $r:r_1 = 1:r_2$ holds.

Graphical raising to a power of complex numbers

This method is based on repetition of the graphical method of multiplication.

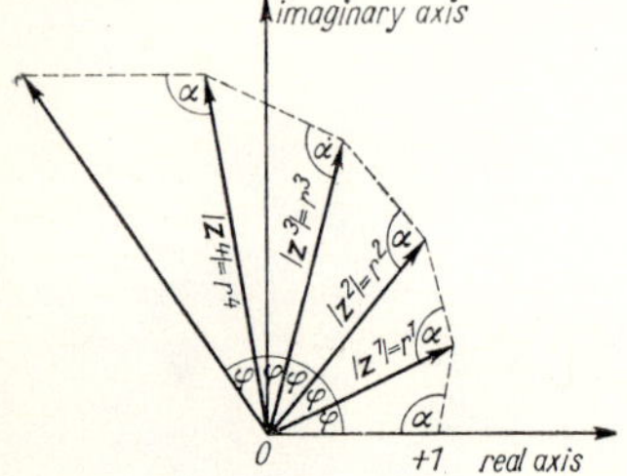

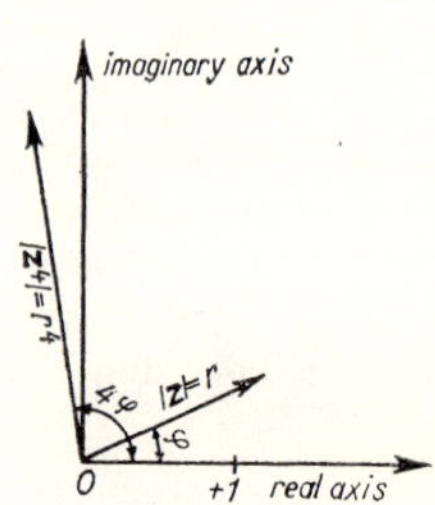

Another method uses the calculated values of $\sphericalangle\, n\varphi$ and r^n. Illustration for $n = 4$.

Graphical extraction of roots of complex numbers

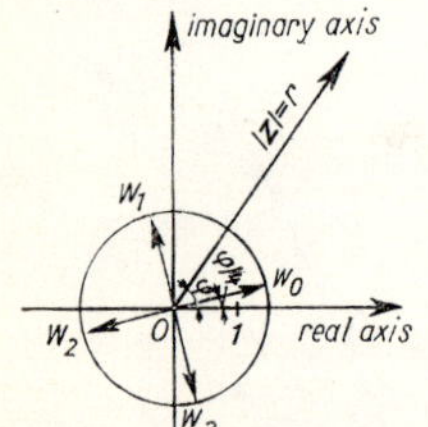

This method uses the calculated values of $\sphericalangle \dfrac{\varphi}{n}$ and $\sqrt[n]{r}$.

The graph for $n = 4$ yields the four root values $w_1, \ldots, w_4$ which are displaced by $\dfrac{2\pi}{n} = \dfrac{2\pi}{4} \mathrel{\hat{=}} 90°$ from one another.

Graphical representation of the roots of the positive and the negative unit

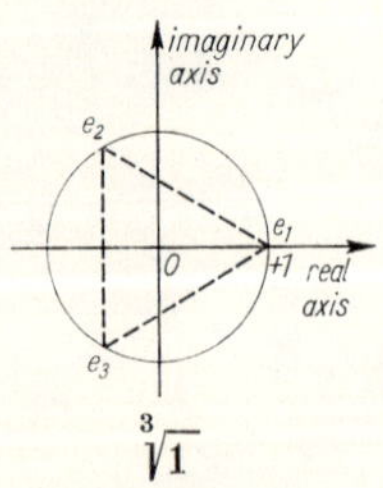

$\sqrt[3]{1}$

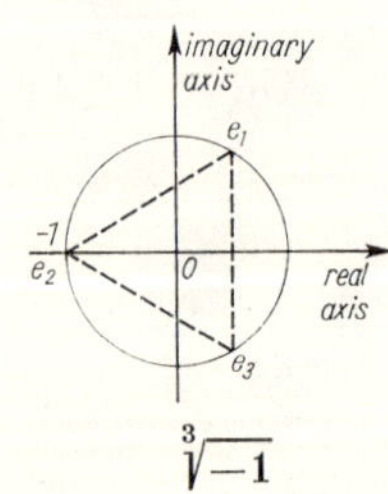

$\sqrt[3]{-1}$

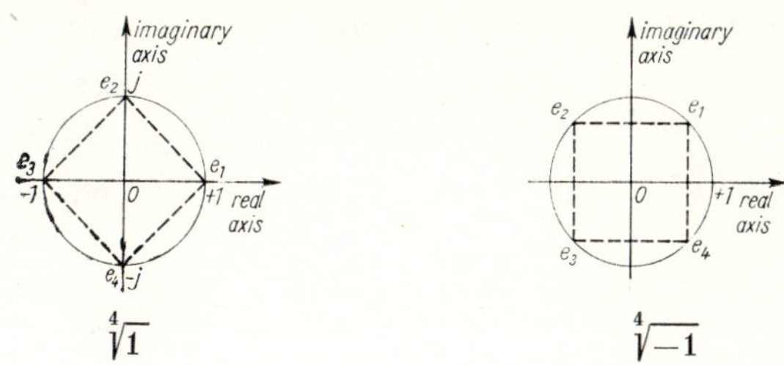

The nth roots of unity divide the unit circle (radius 1) into n equal parts.

1.4. Proportions

Notations

Proportion (ratio equation)

$$a:b = c:d \quad \text{or} \quad \frac{a}{b} = \frac{c}{d}$$

a and d are the outer terms $\quad$ a and c are the antecedents

b and c are the inner terms $\quad$ b and d are the consequents

Reducing terms of a proportion to higher and lower terms:

$$a:b = c:d \Rightarrow \begin{cases} an:bn = c:d \\ an:b \;= cn:d \end{cases}$$

$$a:b = c:d \Rightarrow \begin{cases} \dfrac{a}{n}:\dfrac{b}{n} = c:d \\ \dfrac{a}{n}:b = \dfrac{c}{n}:d \end{cases}$$

(In the place of the sign $=$, the sign :: is sometimes used.)

Product equation

$$a:b = c:d \Leftrightarrow ad = bc$$

(The product of the outer terms is equal to the product of the inner terms.)

Interchanging the terms of a proportion

From $a:b = c:d$ follows: $a:c = b:d$

$$d:b = c:a$$

$$d:c = b:a$$

Derived proportions (corresponding addition and subtraction)

From $a:b = c:d$ follows: $(a + b):a = (c + d):c$

$$(a + b):b = (c + d):d$$

$$(a - b):a = (c - d):c$$

$$(a - b):b = (c - d):d$$

$$(a + b):(a - b) = (c + d):(c - d)$$

$$(pa \pm qb):(ra \pm sb) = (pc \pm qd):(rc \pm sd)$$

Fourth proportional

$a:b = c:x; \quad x = \dfrac{bc}{a}$ (fourth proportional with respect to a, b, c)

For geometrical interpretation, see page 136.

Proportionality factor

$$a:b = c:d \Rightarrow \begin{cases} a = pc \\ b = pd \end{cases}$$

p proportionality factor

Continued proportion

$a:b = b:c$ (proportion with equal inner terms)

Third proportional

$a:b = b:x; \ x = \dfrac{b^2}{a}$ (third proportional to a and b)

Mean proportional

$a:x = x:b; \quad x = \sqrt{ab}$ (mean proportional or *geometric mean* of a and b)

$a, b \geqq 0$

For geometrical interpretation, see page 137.

Continued harmonic proportion

$$(a - x):(x - b) = a:b; \quad x = \frac{2ab}{a + b} \qquad (\textit{harmonic mean} \text{ of } a \text{ and } b)$$

This can also be written as: $\frac{1}{x} = \frac{1}{2}\left(\frac{1}{a} + \frac{1}{b}\right)$

Continuous proportion

$$a:b:c:d: \cdots = a_1:b_1:c_1:d_1: \cdots$$

Any continuous proportion can be rearranged in individual proportions:

$$a:b = a_1:b_1 \qquad b:c = b_1:c_1$$

$$a:c = a_1:c_1 \qquad b:d = b_1:d_1$$

$$a:d = a_1:d_1 \qquad c:d = c_1:d_1 \quad \text{etc.}$$

Mean values of *n* quantities

Arithmetic mean $\mathrm{AM} = \frac{a_1 + a_2 + a_3 + \cdots + a_n}{n}$

Geometric mean $\mathrm{GM} = \sqrt[n]{a_1 \cdot a_2 \cdot a_3 \cdot \cdots \cdot a_n}$ for $a_k > 0$

Harmonic mean $\mathrm{HM} = \frac{n a_1 a_2 a_3 \ldots a_n}{(a_2 a_3 \cdots a_n) + (a_1 a_3 \cdots a_n) + \cdots + (a_1 a_2 \cdots a_{n-1})}$

$$\frac{1}{\mathrm{HM}} = \frac{1}{n}\left(\frac{1}{a_1} + \frac{1}{a_2} + \frac{1}{a_3} + \cdots + \frac{1}{a_n}\right)$$

Root mean square $\mathrm{RMS} = \sqrt{\frac{1}{n}(a_1^2 + a_2^2 + \cdots + a_n^2)}$

Cauchy's theorem

$$\mathrm{AM} > \mathrm{GM} > \mathrm{HM} \quad \text{for } a_i \neq a_k; \quad i, k = 1, 2, \ldots, n$$

1.5. Logarithms

1.5.1. General

Notations

$\log_b a = c$

b base
a antilogarithm
c logarithm

Definition of logarithm

Logarithmic calculation is the second inversion of raising to a power where the exponent is to be found for a given value of the power and a given base.

$$b^x = a \Leftrightarrow x = \log_b a$$

$$(a, b \in R; \quad a, b > 0; \quad b \neq 0; 1)$$

$$b^{\log_b a} = a; \qquad \log_b (b^a) = a; \qquad \log_b (b^{-a}) = -a$$

Common or Briggsian logarithms

Base 10 *written as:* $\log_{10} = \lg$

$$\lg a = x \Leftrightarrow a = 10^x$$

Natural logarithms (Naperian)

Base $e = \lim\limits_{n\to\infty} \left(1 + \dfrac{1}{n}\right)^n = \lim\limits_{h\to 0} (1 + h)^{\frac{1}{h}} = 2.718281828459\ldots$

written as: $\log_e = \ln$

$$\ln a = x \Leftrightarrow e^x = a$$

Relationship between the logarithmic systems

$$\log_b a = \frac{\log_c a}{\log_c b} = \log_c a \frac{1}{\log_c b} = \log_c a \log_b c$$

The conversion factor $\dfrac{1}{\log_c b} = \log_b c$ is called the *modulus.*
Relationship between the common and the natural logarithms

$$\lg a = \frac{\ln a}{\ln 10} = \ln a \lg e$$

$$\frac{1}{\ln 10} = \lg e = \mathrm{M}_{10} = 0.43429\ldots$$

is the *modulus of the common logarithms.*

$$\ln a = \lg a \ln 10$$

$$\ln 10 = \frac{1}{M_{10}} = 2.30259\ldots$$

is the *modulus of the natural logarithms.*

1.5.2. Rules for calculating with logarithms

$\log_b (ac) = \log_b a + \log_b c$	$\lg (ac) = \lg a + \lg c$ $\ln (ac) = \ln a + \ln c$
$\log_b \frac{a}{c} = \log_b a - \log_b c$	$\lg \frac{a}{c} = \lg a - \lg c$ $\ln \frac{a}{c} = \ln a - \ln c$
$\log_b (a^n) = n \log_b a$	$\lg (a^n) = n \lg a$ $\ln (a^n) = n \ln a$
$\log_b \sqrt[n]{a} = \frac{1}{n} \log_b a$	$\lg \sqrt[n]{a} = \frac{1}{n} \lg a$ $\ln \sqrt[n]{a} = \frac{1}{n} \ln a$

Special cases

$$\log_b b = 1; \quad \log_b 1 = 0; \quad \log_b 0 = \begin{cases} -\infty & \text{for } b > 1 \\ +\infty & \text{for } b < 1 \end{cases}$$

$$\lg 0 = -\infty; \quad \ln 0 = -\infty; \quad \log_b \infty = \begin{cases} +\infty & \text{for } b > 1 \\ -\infty & \text{for } b < 1 \end{cases}$$

(for logarithms of complex and negative numbers see page 32).

1.5.3. The use of logarithm tables for finding common logarithms

A common logarithm consists of two parts, the *characteristic* and the *mantissa,* which is the decimal part separated from the integer by a decimal point.

The characteristic of a proper decimal fraction is negative and is equal to the number of zeros before the first significant figure, the zero before the decimal point being included.

Examples:

$$\lg 3745 = 3.\ldots \qquad \lg 75 = 1.\ldots$$

$$\lg 4 = 0.\ldots \qquad \lg 0.0073 = 0.\ldots -3$$

The mantissae have to be read off from logarithm tables. If the antilogarithm has one digit more than the table shows, the mantissa is found by interpolation according to the last digit of the antilogarithm:

$$\text{increase in mantissa} \quad d = \frac{Dn}{10}$$

where D is the tabular difference (difference between the next smallest and the next largest mantissa in the table) and n is the last digit of the antilogarithm which is not in the table.

Examples:

$$\lg 37489 = 4.57391; \qquad \text{since} \quad \lg 37480 = 4.57380$$

$$\lg 37490 = 4.57392$$

Tabular difference $D = 12$, the last digit of the antilogarithm which is not in the table $n = 9$, thus increase in mantissa

$$d = \frac{Dn}{10} = \frac{12 \cdot 9}{10} = 10.8 \approx \underline{\underline{11}}$$

hence

$$\begin{array}{r} \lg 37489 = 4.57380 \\ 11 + \\ \hline \underline{\underline{4.57391}} \end{array}$$

$$\lg 0.12367 = 0.09227 - 1 \qquad D = 35$$

$$n = 7$$

$$d = \frac{35 \cdot 7}{10} \approx \underline{\underline{25}}$$

The increase in mantissae can also be read off from the *proportional parts tables* provided at the edge of many logarithm tables.
Finding the antilogarithm of a given logarithm: The position of the decimal point in the antilogarithm is indicated by the characteristic. If the mantissa happens to be in the table, the figures can be directly read off. Otherwise the figures of the next smallest mantissa have to be read off and the last digit n to be calculated from the increase in mantissa d and the tabular difference D:

$$n = \frac{d \cdot 10}{D}$$

Here again, the use of the proportional parts tables is of advantage.

Examples:

$$\lg x = 3.59814 \qquad d = 1$$

$$x = 3964.1 \qquad D = 11$$

$$n = \frac{1 \cdot 10}{11} \approx \underline{\underline{1}}$$

$$\lg x = 0.99637 - 3 \qquad d = 3$$

$$x = 0.0099168 \qquad D = 4$$

$$n = \frac{3 \cdot 10}{4} \approx \underline{\underline{8}}$$

1.6. Combinatoric analysis

1.6.1. Permutations

Explanation

Permutations of n elements are the arrangements in all possible orderings of all n elements.
If, in a permutation of n different elements, we interchange two elements, then we have performed a transposition. Any permutation of n different elements can be transformed into a different one by successive transpositions.

Number of permutations of n different elements

$$P(n) = 1 \cdot 2 \cdot 3 \cdot \dots \cdot n = n!$$

Permutations of n elements with repetition

Among the elements there are equal elements each occurring r, s, t times:

$$P_{r,s,t,}(n) = \frac{n!}{r!s!t!}$$

Among the elements there are equal elements each occurring r and $n - r$ times:

$$P_{r,n-r}(n) = \frac{n!}{r!(n-r)!} = \binom{n}{r}$$

Examples:

How many different five-digit integers can be formed of the figures 2, 3, 3, 5, 5?

$$P_{2,2}(5) = \frac{5!}{2!2!} = \frac{1 \cdot 2 \cdot 3 \cdot 4 \cdot 5}{1 \cdot 2 \cdot 1 \cdot 2} = 30$$

How many different six-digit integers can be formed of the figures 1, 1, 4, 4, 4, 4?

$$P_{2,4}(6) = \frac{6!}{2!4!} = \frac{1 \cdot 2 \cdot 3 \cdot 4 \cdot 5 \cdot 6}{1 \cdot 2 \cdot 1 \cdot 2 \cdot 3 \cdot 4} = 15 \triangleq \binom{6}{2}$$

Inversion

Explanation

In a certain permutation of elements, two elements of the permutation form an inversion if they are interchanged as compared with their original (natural) order.

Example:

Original arrangement 4, 5, 6, 7, 8
Permuted 8, 4, 6, 5, 7
In this permutation, the elements 8 and 4
8 and 6
8 and 5
8 and 7
6 and 5

form an inversion, hence there are five inversions altogether.
A permutation is called *even* or *odd* depending on whether an even or odd number of inversions was performed.

The number of possible even permutations of n different elements $=$ the number of possible odd permutations $= \dfrac{n!}{2}$.

1.6.2. Variations

Explanation

Arrangements where of n given elements a certain number r of elements occur in all possible orders are called variations.
Written as: $V_r(n)$ means variations of n elements taken r at a time.

Variations of n elements taken r at a time without repetition

Each element occurs only once in a variation.

$$V_r(n) = n(n-1)(n-2)\cdots(n-r+1)$$
$$= \frac{n!}{(n-r)!} = \binom{n}{r} \cdot r!$$

Example:
How many different throws are possible with three dice?

$$V_3(6) = 6 \cdot 5 \cdot 4 = 120$$

Variations of n elements taken r at a time with repetition

Each element may occur several times in a variation.

$$V_r{}^w(n) = n^r$$

Example:
In how many different ways can a pool coupon be filled in?

$n = 3$ (won, lost, drawn)

$r = 12$

$V_{12}^w(3) = 3^{12} = 531441$

1.6.3. Combinations

Explanation

Arrangements that contain a certain number r of elements in an arbitrary order, chosen from n given elements, while no permutations are permissible within this arrangement, are combinations.

Written as: $C_r(n)$ means combinations of n elements taken r at a time.

Combinations of n elements taken r at a time without repetition

Each combination must contain the same element only once.

$$C_r(n) = \binom{n}{r} = \frac{n!}{r!(n-r)!}$$

Example:

In how many different ways can five numbers be marked on a lotto coupon showing 90 numbers?

$$n = 90; \qquad r = 5$$

$$C_5(90) = \binom{90}{5} = 43{,}949{,}268$$

Combinations of n elements taken r at a time with repetition

Each combination may contain the same element several times.

$$C_r{}^w(n) = \left(\frac{n+r-1}{r}\right)$$

Example:

In how many different ways can the five figures 2, 3, 4, 5, 6 be taken three at a time with repetition?

$$C_3{}^w(5) = \binom{5+3-1}{3} = \binom{7}{3} = \frac{7 \cdot 6 \cdot 5}{1 \cdot 2 \cdot 3} = 35$$

1.7. Per cent calculation, interest calculation

1.7.1. Per cent (per mille) calculation

Definition

$$1\% \text{ (per hundred)} = \frac{1}{100} \text{ of the original value}$$

$$p\% \text{ (per hundred)} = \frac{p}{100} \text{ of the original value}$$

Similarly, for per mille calculations we have:

$$1\,{}^0/_{00}\text{ (per thousand)} = \frac{1}{1000}\text{ of the original value}$$

$$p\,{}^0/_{00}\text{ (per thousand)} = \frac{p}{1000}\text{ of the original value}$$

For the per cent value (part of the original value in per cent) P, the rate in per cent or percentage p, and the original value G, the following relation holds:

$$P:p = G:100$$

Per cent "on the hundred" and "within the hundred"

$$p\%\text{ on the hundred} \triangleq \frac{100 \cdot p}{100 + p}\%\text{ per hundred}$$

$$p\%\text{ within the hundred} \triangleq \frac{100 \cdot p}{100 - p}\%\text{ per hundred}$$

Example:

Assume the material lost in a manufacturing process to amount to 23% of the raw material input. Considered from the viewpoint of the finished product, this corresponds to a loss of 23% within the hundred $= \frac{100 \cdot 23}{77}\% = \underline{29.9\%\text{ per hundred}}$.

Example:

The wholesale price of a good is calculated by adding 15% to the price quoted by the manufacturer. This represents 15% on the hundred of the wholesale price or $\frac{100 \cdot 15}{100 + 15}\% = 13\%$ per hundred of the wholesale price.

1.7.2. Interest calculation

Notations

z interest
b principal
p rate of interest
t time

Calculation of interest

$$z = \frac{bpt}{100} \quad \text{or} \quad z = \frac{bpt}{100 \cdot 12} \quad \text{or} \quad z = \frac{bpt}{100 \cdot 360}$$

(t in years or months or days)

Calculation of principal

$$b = \frac{100z}{pt} \quad \text{or} \quad b = \frac{100 \cdot 12z}{pt} \quad \text{or} \quad b = \frac{100 \cdot 360z}{pt}$$

(t in years or months or days)

Calculation of rate of interest

$$p = \frac{100z}{bt} \quad \text{or} \quad p = \frac{100 \cdot 12z}{bt} \quad \text{or} \quad p = \frac{100 \cdot 360z}{bt}$$

(t in years or months or days)

Calculation of time

$$t = \frac{100z}{bp} \quad \text{or} \quad t = \frac{100 \cdot 12z}{bp} \quad \text{or} \quad t = \frac{100 \cdot 360z}{bp}$$

(t in years or months or days)

Interest number and interest divisor

The interest due on a principal lent for any given number of days is calculated as

$$z = \frac{N}{D} \quad \text{where} \quad N = \frac{bt}{100} \quad \text{is the } \textit{interest number}$$

$$\text{and } D = \frac{360}{p} \quad \text{is the } \textit{interest divisor.}$$

1.8. Sequences and series

1.8.1. General

Definition

A *sequence of numbers* is defined as an ordered set of numbers. A sequence of real numbers $\{a_k\}_{\text{ordered}}$ is a single-valued mapping of the set of the natural numbers $k \in N$ upon a subset of the real numbers R where the maintenance of the order is required.

A sequence of numbers is *written as:* $\{a_k\}$ $a_k \in R$ images
$k \in N$ originals

The originals k are also called indices of the terms of the sequence. The images a_k are also called terms of the sequence.

Representations

Representation in words: e.g.: Each positive integer k is put into correspondence with its square.

Analytical representation: $a_k = f(k)$

e.g. $a_k = k^2$ for $k = 1; 2; 3; \ldots; n$ (finite sequence)
$k \in N$ (infinite sequence)

The analytical representation states the general term.
Recursive representation: $a_k = f(a_{k-1})$ with statement of the first term a_1
tabulated representation: $\{a_k\} = 1; 4; 9; 16; \ldots$

Finite sequences terminate in the last term a_n or all $a_k = 0$ for $k > n$, $k \in N$.

Written as: $\{a_k\} = a_1; a_2; \ldots; a_n$ or $a_k = f(k)$ for $k = 1; 2; 3; \ldots; n$

Infinite sequences do not possess a last term, i.e. for each $K \in N$ there is a $k > K$ for which $a_k \neq 0$.

Written as: $\{a_k\} = a_1; a_2; a_3; \ldots$ or $a_k = f(k)$ for $k \in N$

A *subsequence* is obtained if, in an infinite sequence, a finite or infinite number of terms is omitted though an infinite sequence remains,

e.g. $a_k = k$ $k \in N$ $\{a_k\} = 1; 2; 3; \ldots$
subsequence $b_k = a_{2k}$ $\{b_k\} = 2; 4; 6; \ldots$

Difference-sequence: $\{d_k\}$ is the difference-sequence of $\{a_k\}$ if

$$d_k = a_{k+1} - a_k \quad k \in N$$

Quotient-sequence: $\{q_k\}$ is the quotient-sequence of $\{a_k\}$ if

$$q_k = \frac{a_{k+1}}{a_k} \quad k \in N$$

Sequence of partial sums: $\{s_k\}$ is the sequence of partial sums of $\{a_k\}$ if

$$s_k = a_1 + a_2 + a_3 + \cdots + a_k$$

Positive definite sequence: All $a_k > 0$ in $\{a_k\}$
Negative definite sequence: All $a_k < 0$ in $\{a_k\}$
Oscillatory sequence: Alternating signs in $\{a_k\}$; $q_k < 0$; $a_k a_{k+1} < 0$
Monotonic increasing sequence: $a_{k+1} \geqq a_k$
Monotonic decreasing sequence: $a_{k+1} \leqq a_k$
Strictly monotonic increasing sequence: $a_{k+1} > a_k$
Strictly monotonic decreasing sequence: $a_{k+1} < a_k$
Constant sequence: $a_k = a_{k+1} = \text{const.}$
Bounded sequence: The terms of the sequence lie between a lower bound (S) and an upper bound (T) of the sequence. $S, T \in R$

Null sequence:
A null sequence is given if, for sufficiently large k, in $\{a_k\}$ the terms of the sequence tend to zero and never deviate. Then, for any number $\varepsilon > 0$, a $k_0 = k_0(\varepsilon)$ can be given in such a way that the following expession holds for all $k > k_0$:

$$|a_k| < \varepsilon$$

For *limit of a sequence* see page 279.
If the individual terms of a number sequence are connected by addition, a *series* is obtained.

1.8.2. Arithmetic sequences and series

Arithmetic sequence of the first order

$$a_1;\ (a_1 + d);\ (a_1 + 2d);\ (a_1 + 3d);\ \ldots;\ a_n$$

a_1 initial term $\quad$ n number of terms
a_n nth term (final term) $\quad$ d difference

Final term of the arithmetic sequence of the first order

$$a_n = a_1 + (n - 1)d$$

Law of formation of the arithmetic sequence of the first order

$$a_k = a_1 + (k - 1)d \quad \text{for} \quad k = 1; 2; 3; \ldots; n$$

$$a_{k+1} - a_k = d = \text{const.}$$

$$a_k = \frac{1}{2}(a_{k-1} + a_{k+1})$$

Summation formula of the arithmetic series of the first order

$$s_n = \frac{n(a_1 + a_n)}{2} \quad \text{and} \quad s_n = \frac{n}{2}[2a_1 + (n-1)d]$$

Examples:

Sum of the n first natural numbers

$$s_n = \sum_{k=1}^{n} k = \frac{n(n+1)}{2}$$

Sum of the n first even numbers

$$s_n = \sum_{k=1}^{n} 2k = n(n+1)$$

Sum of the n first odd numbers

$$s_n = \sum_{k=1}^{n} (2k-1) = n^2$$

Interpolation of an arithmetic sequence of the first order

If m terms are inserted between each pair of terms of an arithmetic sequence with the difference d, a new arithmetic sequence with the difference $d_1 = \frac{d}{m+1}$ is obtained.

Arithmetic sequences of higher order

An arithmetic sequence of the kth order is given if only the kth difference-sequence has constant terms.

Example:

2; 3; 8; 18; 36; 67; 118; ...	
1 5 10 18 31 51 ...	First difference-sequence
4 5 8 13 20 ...	Second difference-sequence
1 3 5 7 ...	Third difference-sequence
2 2 2 ...	Fourth difference-sequence constant

The primary sequence is a sequence of the fourth order.

Laws of formation of arithmetic sequences of higher order

$$\left.\begin{aligned}
&a_k = b_2(k-1)^2 + b_1(k-1) + b_0\\
&\qquad\text{arithmetic sequence of the second order}\\
&a_k = b_3(k-1)^3 + b_2(k-1)^2 + b_1(k-1) + b_0\\
&\qquad\text{arithmetic sequence of the third order}\\
&\vdots\\
&a_k = b_m(k-1)^m + b_{m-1}(k-1)^{m-1} + \cdots + b_0\\
&\qquad\text{arithmetic sequence of the } m\text{th order}
\end{aligned}\right\}\ \begin{array}{l}\text{for}\\ k \in \{1;2;3;\ldots;n\}\end{array}$$

Examples:

Sum of the first n square-numbers

$$s_n = \sum_{k=1}^{n} k^2 = \frac{n(n+1)(2n+1)}{6}$$

Sum of the first n cubic-numbers

$$s_n = \sum_{k=1}^{n} k^3 = \left[\frac{n(n+1)}{2}\right]^2 = \left[\sum_{k=1}^{n} k\right]^2$$

1.8.3. Geometric sequences and series

Finite geometric sequence

$$a_1;\ a_1q;\ a_1q^2;\ \ldots;\ a_n$$

a_1 initial term
a_n nth term (final term)
n number of terms
q quotient (ratio)

For $q > 1$ the sequence is increasing;
for $0 < q < 1$ the sequence is decreasing;
for $q < 0$ the sequence is oscillatory (alternating signs);
for $q = 1$ the sequence contains equal terms.

Formula for the final term

$$a_n = a_1q^{n-1}$$

Law of formation of the geometric sequence

$$a_k = a_1 q^{k-1} \quad \text{for} \quad k \in \{1; 2; 3; \ldots; n\}$$

$$\frac{a_{k+1}}{a_k} = q = \text{const.}$$

$$a_k = \sqrt{a_{k-1} a_{k+1}}$$

Summation formulas

For $q \neq 1$ the following holds:

$$s_n = \frac{a_1(q^n - 1)}{q - 1} = \frac{a_1(1 - q^n)}{1 - q} \quad \text{for} \quad q > 1$$

$$s_n = \frac{a_n q - a_1}{q - 1} = \frac{a_1 - a_n q}{1 - q}$$

For $q = 1$ we have: $s_n = a_1 n$

Sum of the infinite geometric series

$$\lim_{n \to \infty} s_n = \frac{a_1}{1 - q} = s \quad \text{for} \quad |q| < 1$$

Interpolation of the geometric sequence

If m terms are inserted between each pair of terms of a geometric sequence with the quotient q, a new geometric sequence with the quotient $q_1 = \sqrt[m+1]{q}$ is obtained.

1.8.4. Compound interest calculation

Notations

- b_n final amount
- b_0 original amount or principal, cash value
- n number of conversion periods
- p rate of interest
- $q = 1 + \dfrac{p}{100}$ interest factor
- r annuity, regular payment

Final amount if the interest is added to the principal at the end of each year

$$b_n = b_0 q^n \text{ (} n \text{ denotes the number of years)}$$

Final amount if the interest is added to the principal every six months, every four months, etc.

Every six months: Relative rate of interest $= \dfrac{p}{2}$;

$$\text{thus } q = 1 + \frac{p}{200}$$

Every three months: Relative rate of interest $= \dfrac{p}{4}$;

$$\text{thus } q = 1 + \frac{p}{400} \quad \text{etc.}$$

In this case, n in the formula $b_n = b_0 q^n$ denotes the number of half-years, quarters, etc.
Calculation of the principal (*cash value*)

$$b_0 = \frac{b_n}{q^n}$$

The calculation of the cash value is called *discounting*.
Calculating of rate of interest

$$\textit{rate of interest } q = \sqrt[n]{\frac{b_n}{b_0}}$$

from q results $p = 100(q - 1)$
Calculation of conversion periods, n

$$n = \frac{\lg b_n - \lg b_0}{\lg q}$$

Final amount of regular payments

$$b_n = \frac{r(q^n - 1)}{q - 1}$$ for payments at the end of the year (*subsequently*)

$$b_n = \frac{rq(q^n - 1)}{q - 1}$$ for payments at the beginning of the year (*beforehand*)

1.8.5. Annuities

Final amount of the annuity r to be paid n times *at the beginning of term*

$$b_n = \frac{rq(q^n - 1)}{q - 1}$$

Final amount of the annuity r to be paid n times *at the end of term*

$$b_n = \frac{r(q^n - 1)}{q - 1}$$

Cash value of the annuity r to be paid n times *at the beginning of term*

$$b_0 = \frac{r(q^n - 1)}{q^{n-1}(q - 1)}$$

Cash value of the annuity r to be paid n times *at the end of term*

$$b_0 = \frac{r(q^n - 1)}{q^n(q - 1)}$$

Calculation of amount of annuity from final amount or principal (*end of term*)

$$r = \frac{b_n(q - 1)}{q^n - 1}$$

$$r = \frac{b_0 q^n(q - 1)}{q^n - 1}$$

Calculation of amount of annuity from final amount or principal (*beginning of term*)

$$r = \frac{b_n(q - 1)}{q(q^n - 1)}$$

$$r = \frac{b_0 q^{n-1}(q - 1)}{q^n - 1}$$

Calculation of the time from final amount of principal

$$n = \frac{\lg [b_n(q - 1) + r] - \lg r}{\lg q} \quad \text{for payment at end of term}$$

$$n = \frac{\lg [b_n(q-1) + rq] - \lg (rq)}{\lg q}$$ for payment at beginning of term

$$n = \frac{\lg r - \lg (r + b_0 - b_0 q)}{\lg q}$$ for payment at end of term

$$n = \frac{\lg r + \lg q - \lg (b_0 + rq - b_0 q)}{\lg q}$$ for payment at beginning of term

Cash value of a *perpetual annuity*

$$b_0 = \frac{r}{q-1}$$ (payment at end of term)

$$b_0 = \frac{rq}{q-1}$$ (payment at beginning of term)

Cash value of r paid regularly (in dollars)

$$b_0 = \frac{r(q^n - 1)}{q^n(q-1)}$$ for payments at end of year

$$b_0 = \frac{r(q^n - 1)}{q^{n-1}(q-1)}$$ for payments at beginning of year

Final value of a principal in case of regular deposits and withdrawals of r (in dollars)

$$b_n = b_0 q^n + \frac{r(q^n - 1)}{q-1}$$ for deposits at end of year

$$b_n = b_0 q^n + \frac{rq(q^n - 1)}{q-1}$$ for deposits at beginning of year

$$b_n = b_0 q^n - \frac{r(q^n - 1)}{q-1}$$ for withdrawals at end of year

$$b_n = b_0 q^n - \frac{rq(q^n - 1)}{q-1}$$ for withdrawals at beginning of year

Redemption of a debt (loan)

Annuity $A = \frac{b_0 q^n (q-1)}{q^n - 1}$, b_0 sum of debt

Redemption rate (percentage at which the sum of debt is repaid)

$$i = \frac{p}{q^n - 1}, \quad p \text{ rate of interest}$$

$$\textit{Redemption period} \quad n = \frac{\lg\left(1 + \frac{p}{i}\right)}{\lg q}$$

Continuous yield of interest on a principal b_0 in n years

$$b_n = b_0 e^{\frac{pn}{100}}$$

The same formula applies to any *continuous growth.*
Similarly for a continuous decrease we have

$$b_n = b_0 e^{-\frac{pn}{100}}$$

1.9. Determinants

1.9.1. General

Notations

$$\begin{vmatrix} a_{11} & a_{12} & a_{13} & \cdots & a_{1n} \\ a_{21} & a_{22} & a_{23} & \cdots & a_{2n} \\ a_{31} & a_{32} & a_{33} & \cdots & a_{3n} \\ \vdots & \vdots & \vdots & & \vdots \\ a_{n1} & a_{n2} & a_{n3} & \cdots & a_{nn} \end{vmatrix} = |a_{ik}| = \Delta \quad (n\text{-rowed determinant})$$

The terms a_{ik} (total number n^2) are called *elements* of the determinant (read: *a* one one, *a* one two, etc.).

Horizontal rows rows (the first index i in a_{ik} indicates the row);
vertical rows columns (the second index k in a_{ik} indicates the column);
elements $a_{11}, a_{22}, a_{33}, \ldots, a_{nn}$ form the *principal diagonal*;
elements $a_{1n}, a_{2n-1}, a_{3n-2}, \ldots, a_{n1}$ form the *secondary diagonal*;
product of the elements of the principal diagonal = *principal member*

Definition

An n-rowed determinant is a quantity depending upon any n^2 numbers arranged in the form of a square. The value of an n-rowed determinant is determined by multiplying each term a_{ik} of the first row by its respective minor A_{ik} and adding of the products obtained. A *minor* (*subdeterminant*) A_{ik} is defined as the $(n-1)$-rowed determinant which is obtained when striking out the ith row and the kth column in the given determinant and provided with the sign $(-1)^{i+k}$.

Examples:

2-rowed determinant

$$\Delta = \begin{vmatrix} a_{11} & a_{12} \\ a_{21} & a_{22} \end{vmatrix} = a_{11}a_{22} - a_{12}a_{21}$$

$$\begin{vmatrix} 2 & 4 \\ 6 & 7 \end{vmatrix} = 2 \cdot 7 - 4 \cdot 6 = -10$$

3-rowed determinant

$$\Delta = \begin{vmatrix} a_{11} & a_{12} & a_{13} \\ a_{21} & a_{22} & a_{23} \\ a_{31} & a_{32} & a_{33} \end{vmatrix}$$

$$= a_{11} \begin{vmatrix} a_{22} & a_{23} \\ a_{32} & a_{33} \end{vmatrix} - a_{12} \begin{vmatrix} a_{21} & a_{23} \\ a_{31} & a_{33} \end{vmatrix} + a_{13} \begin{vmatrix} a_{21} & a_{22} \\ a_{31} & a_{32} \end{vmatrix}$$

$$= a_{11}A_{11} + a_{12}A_{12} + a_{13}A_{13}$$

4-rowed determinant

$$\Delta = \begin{vmatrix} 3 & 7 & 4 & 6 \\ 10 & 5 & 9 & 6 \\ 1 & 2 & 7 & 8 \\ 5 & 4 & 2 & 9 \end{vmatrix} = 3 \begin{vmatrix} 5 & 9 & 6 \\ 2 & 7 & 8 \\ 4 & 2 & 9 \end{vmatrix} - 7 \begin{vmatrix} 10 & 9 & 6 \\ 1 & 7 & 8 \\ 5 & 2 & 9 \end{vmatrix} + 4 \begin{vmatrix} 10 & 5 & 6 \\ 1 & 2 & 8 \\ 5 & 4 & 9 \end{vmatrix} - 6 \begin{vmatrix} 10 & 5 & 9 \\ 1 & 2 & 7 \\ 5 & 4 & 2 \end{vmatrix}$$

(expanded about the elements of the first row)

The obtained three-rowed minors provided with signs are to be denoted by A_{11}, A_{12}, A_{13}, and A_{14}.

Sarrus's rule for three-rowed determinants

We write the first two columns again on the right side of the determinant and add up the products parallel to the principal diagonal (with positive sign) and parallel to the secondary diagonal (with negative sign).

$$\Delta = \begin{vmatrix} a_{11} & a_{12} & a_{13} \\ a_{21} & a_{22} & a_{23} \\ a_{31} & a_{32} & a_{33} \end{vmatrix} = a_{11}\,a_{22}\,a_{33} - a_{11}\,a_{23}\,a_{32} + a_{12}\,a_{23}\,a_{31} - a_{12}\,a_{21}\,a_{33} + a_{13}\,a_{21}\,a_{32} - a_{13}\,a_{22}\,a_{31}$$

Example:

$$\begin{vmatrix} 2 & 1 & 9 \\ 1 & -2 & -3 \\ 3 & 5 & 4 \end{vmatrix} \begin{matrix} 2 & 1 \\ 1 & -2 \\ 3 & 5 \end{matrix} = 2\cdot(-2)\cdot 4 + 1\cdot(-3)\cdot 3 + 9\cdot 1\cdot 5 - 3\cdot(-2)\cdot 9 - 5\cdot(-3)\cdot 2 - 4\cdot 1\cdot 1 = 100$$

1.9.2. Theorems on determinants

Transposing the determinant (the rows are written as columns and the columns as rows = reflection in the principal diagonal)

$$\begin{vmatrix} a_{11} & a_{12} & a_{13} & \cdots & a_{1n} \\ a_{21} & a_{22} & a_{23} & \cdots & a_{2n} \\ a_{31} & a_{32} & a_{33} & \cdots & a_{3n} \\ \vdots & \vdots & \vdots & & \vdots \\ a_{n1} & a_{n2} & a_{n3} & \cdots & a_{nn} \end{vmatrix} = \begin{vmatrix} a_{11} & a_{21} & a_{31} & \cdots & a_{n1} \\ a_{12} & a_{22} & a_{32} & \cdots & a_{n2} \\ a_{13} & a_{23} & a_{33} & \cdots & a_{n3} \\ \vdots & \vdots & \vdots & & \vdots \\ a_{1n} & a_{2n} & a_{3n} & \cdots & a_{nn} \end{vmatrix}$$

Interchanging two rows or two columns, e.g.

$$\begin{vmatrix} a_{11} & a_{12} & a_{13} & \cdots & a_{1n} \\ a_{21} & a_{22} & a_{23} & \cdots & a_{2n} \\ a_{31} & a_{32} & a_{33} & \cdots & a_{3n} \\ \vdots & \vdots & \vdots & & \vdots \\ a_{n1} & a_{n2} & a_{n3} & \cdots & a_{nn} \end{vmatrix} = - \begin{vmatrix} a_{11} & a_{12} & a_{13} & \cdots & a_{1n} \\ a_{31} & a_{32} & a_{33} & \cdots & a_{3n} \\ a_{21} & a_{22} & a_{23} & \cdots & a_{2n} \\ \vdots & \vdots & \vdots & & \vdots \\ a_{n1} & a_{n2} & a_{n3} & \cdots & a_{nn} \end{vmatrix}$$

$$\begin{vmatrix} a_{11} & a_{12} & a_{13} & \cdots & a_{1n} \\ a_{21} & a_{22} & a_{23} & \cdots & a_{2n} \\ a_{31} & a_{32} & a_{33} & \cdots & a_{3n} \\ \vdots & \vdots & \vdots & & \vdots \\ a_{n1} & a_{n2} & a_{n3} & \cdots & a_{nn} \end{vmatrix} = - \begin{vmatrix} a_{13} & a_{12} & a_{11} & \cdots & a_{1n} \\ a_{23} & a_{22} & a_{21} & \cdots & a_{2n} \\ a_{33} & a_{32} & a_{31} & \cdots & a_{3n} \\ \vdots & \vdots & \vdots & & \vdots \\ a_{n3} & a_{n2} & a_{n1} & \cdots & a_{nn} \end{vmatrix}$$

Multiplying the determinant by the factor λ, e.g.

$$\lambda\Delta = \begin{vmatrix} \lambda a_{11} & \lambda a_{12} & \lambda a_{13} & \cdots & \lambda a_{1n} \\ a_{21} & a_{22} & a_{23} & \cdots & a_{2n} \\ a_{31} & a_{32} & a_{33} & \cdots & a_{3n} \\ \vdots & \vdots & \vdots & & \vdots \\ a_{n1} & a_{n2} & a_{n3} & \cdots & a_{nn} \end{vmatrix} = \begin{vmatrix} a_{11} & \lambda a_{12} & a_{13} & \cdots & a_{1n} \\ a_{21} & \lambda a_{22} & a_{23} & \cdots & a_{2n} \\ a_{31} & \lambda a_{32} & a_{33} & \cdots & a_{3n} \\ \vdots & \vdots & \vdots & & \vdots \\ a_{n1} & \lambda a_{n2} & a_{n3} & \cdots & a_{nn} \end{vmatrix}$$

the factor λ can be put in any row or column.

Reversing: A factor that is found in all elements of a row or a column can be factored out.

Correspondence of two rows or two columns, e.g.

$$\begin{vmatrix} a_{11} & a_{12} & a_{12} & \cdots & a_{1n} \\ a_{21} & a_{22} & a_{22} & \cdots & a_{2n} \\ a_{31} & a_{32} & a_{32} & \cdots & a_{3n} \\ \vdots & \vdots & \vdots & & \vdots \\ a_{n1} & a_{n2} & a_{n2} & \cdots & a_{nn} \end{vmatrix} = 0$$

The elements of a row or a column are equal to 0, e.g.

$$\begin{vmatrix} a_{11} & a_{12} & a_{13} & \cdots & a_{1n} \\ 0 & 0 & 0 & \cdots & 0 \\ a_{31} & a_{32} & a_{33} & \cdots & a_{3n} \\ \vdots & \vdots & \vdots & & \vdots \\ a_{n1} & a_{n2} & a_{n3} & \cdots & a_{nn} \end{vmatrix} = 0$$

Proportionality among the elements of two rows or two columns, e.g.

$$\begin{vmatrix} a_{11} & a_{12} & k \cdot a_{12} & \cdots & a_{1n} \\ a_{21} & a_{22} & k \cdot a_{22} & \cdots & a_{2n} \\ a_{31} & a_{32} & k \cdot a_{32} & \cdots & a_{3n} \\ \vdots & \vdots & \vdots & & \vdots \\ a_{n1} & a_{n2} & k \cdot a_{n2} & \cdots & a_{nn} \end{vmatrix} = 0 \quad k \text{ proportionality factor}$$

Marginal expansion of the determinant, e.g.

$$\begin{vmatrix} a_{11} & a_{12} & a_{13} & \cdots & a_{1n} \\ a_{21} & a_{22} & a_{23} & \cdots & a_{2n} \\ a_{31} & a_{32} & a_{33} & \cdots & a_{3n} \\ \vdots & \vdots & \vdots & & \vdots \\ a_{n1} & a_{n2} & a_{n3} & \cdots & a_{nn} \end{vmatrix} = \begin{vmatrix} 1 & k_0 & k_1 & k_2 & \cdots & k_n \\ 0 & a_{11} & a_{12} & a_{13} & \cdots & a_{1n} \\ 0 & a_{21} & a_{22} & a_{23} & \cdots & a_{2n} \\ 0 & a_{31} & a_{32} & a_{33} & \cdots & a_{3n} \\ \vdots & \vdots & \vdots & & & \vdots \\ 0 & a_{n1} & a_{n2} & a_{n3} & \cdots & a_{nn} \end{vmatrix}$$

(k_0, k_1, k_2, ..., k_n are arbitrary constants)

Adding a multiple of the elements of a row to the elements of a parallel row, e.g.

$$\begin{vmatrix} a_{11} & a_{12} & a_{13} & \cdots & a_{1n} \\ a_{21} & a_{22} & a_{23} & \cdots & a_{2n} \\ a_{31} & a_{32} & a_{33} & \cdots & a_{3n} \\ \vdots & \vdots & \vdots & & \vdots \\ a_{n1} & a_{n2} & a_{n3} & \cdots & a_{nn} \end{vmatrix} = \begin{vmatrix} a_{11} & a_{12} + ma_{11} & a_{13} & \cdots & a_{1n} \\ a_{21} & a_{22} + ma_{21} & a_{23} & \cdots & a_{2n} \\ a_{31} & a_{32} + ma_{31} & a_{33} & \cdots & a_{3n} \\ \vdots & \vdots & \vdots & & \vdots \\ a_{n1} & a_{n2} + ma_{n1} & a_{n3} & \cdots & a_{nn} \end{vmatrix}$$

Addition of determinants which differ only in one row or one column, e.g.

$$\begin{vmatrix} a_{11} & a_{12} & a_{13} & \cdots & a_{1n} \\ a_{21} & a_{22} & a_{23} & \cdots & a_{2n} \\ a_{31} & a_{32} & a_{33} & \cdots & a_{3n} \\ \vdots & \vdots & \vdots & & \vdots \\ a_{n1} & a_{n2} & a_{n3} & \cdots & a_{nn} \end{vmatrix} \pm \begin{vmatrix} a_{11} & b_1 & a_{13} & \cdots & a_{1n} \\ a_{21} & b_2 & a_{23} & \cdots & a_{2n} \\ a_{31} & b_3 & a_{33} & \cdots & a_{3n} \\ \vdots & \vdots & \vdots & & \vdots \\ a_{n1} & b_n & a_{n3} & \cdots & a_{nn} \end{vmatrix} =$$

$$= \begin{vmatrix} a_{11} & a_{12} \pm b_1 & a_{13} & \dots & a_{1n} \\ a_{21} & a_{22} \pm b_2 & a_{23} & \dots & a_{2n} \\ a_{31} & a_{32} \pm b_3 & a_{33} & \dots & a_{3n} \\ \vdots & \vdots & \vdots & & \vdots \\ a_{n1} & a_{n2} \pm b_n & a_{n3} & \dots & a_{nn} \end{vmatrix}$$

Multiplication theorem of determinants

$$\Delta_1 = \begin{vmatrix} a_{11} & a_{12} & a_{13} \\ a_{21} & a_{22} & a_{23} \\ a_{31} & a_{32} & a_{33} \end{vmatrix}; \qquad \Delta_2 = \begin{vmatrix} b_{11} & b_{12} & b_{13} \\ b_{21} & b_{22} & b_{23} \\ b_{31} & b_{32} & b_{33} \end{vmatrix}$$

$$\Delta = \Delta_1 \Delta_2$$

$$= \begin{vmatrix} a_{11}b_{11} + a_{21}b_{21} + a_{31}b_{31} & a_{11}b_{12} + a_{21}b_{22} + a_{31}b_{32} & a_{11}b_{13} + a_{21}b_{23} + a_{31}b_{33} \\ a_{12}b_{11} + a_{22}b_{21} + a_{32}b_{31} & a_{12}b_{12} + a_{22}b_{22} + a_{32}b_{32} & a_{12}b_{13} + a_{22}b_{23} + a_{32}b_{33} \\ a_{13}b_{11} + a_{23}b_{21} + a_{33}b_{31} & a_{13}b_{12} + a_{23}b_{22} + a_{33}b_{32} & a_{13}b_{13} + a_{23}b_{23} + a_{33}b_{33} \end{vmatrix}$$

Proceed accordingly when multiplying four-row and multirow determinants (linear combination of the columns of Δ_1 with those of Δ_2).

Vandermond's determinant (power determinant)

$$\begin{vmatrix} 1 & a_1 & a_1^2 & a_1^3 & \dots & a_1^{n-1} \\ 1 & a_2 & a_2^2 & a_2^3 & \dots & a_2^{n-1} \\ 1 & a_3 & a_3^2 & a_3^3 & \dots & a_3^{n-1} \\ \vdots & \vdots & \vdots & \vdots & & \vdots \\ 1 & a_n & a_n^2 & a_n^3 & \dots & a_n^{n-1} \end{vmatrix} \begin{aligned} &= (a_2 - a_1)\,(a_3 - a_1)\,\dots\,(a_n - a_1) \\ &\quad \cdot (a_3 - a_2)\,(a_4 - a_2)\,\dots\,(a_n - a_2) \\ &\quad \cdot (a_4 - a_3)\,(a_5 - a_3)\,\dots\,(a_n - a_3) \\ &\quad \dots\dots\dots\dots\dots\dots \\ &\quad (a_n - a_{n-1}) \\ &= \prod (a_r - a_s) \end{aligned}$$

for all $r > s$ from 1 to n.

Explanation: $\prod_{k=1}^{n} a_k = a_1 a_2 a_3 \dots a_n$

Practical calculation of determinants

When using the above theorems on determinants, the following operations are performed:

$$\Delta = \begin{vmatrix} 1 & 7 & 5 & 4 \\ -4 & 4 & 12 & 8 \\ 2 & 6 & 9 & -2 \\ 3 & 1 & 7 & 3 \end{vmatrix} = 4 \begin{vmatrix} 1 & 7 & 5 & 4 \\ -1 & 1 & 3 & 2 \\ 2 & 6 & 9 & -2 \\ 3 & 1 & 7 & 3 \end{vmatrix}$$

common factor of row 2 factored out

$$= 4 \begin{vmatrix} 0 & 8 & 8 & 6 \\ -1 & 1 & 3 & 2 \\ 2 & 6 & 9 & -2 \\ 3 & 1 & 7 & 3 \end{vmatrix} = 4 \begin{vmatrix} 0 & 8 & 0 & 6 \\ -1 & 1 & 2 & 2 \\ 2 & 6 & 3 & -2 \\ 3 & 1 & 6 & 3 \end{vmatrix}$$

row 2 added to row 1 — column 2 subtracted from column 3

$$= 4(-8) \begin{vmatrix} -1 & 2 & 2 \\ 2 & 3 & -2 \\ 3 & 6 & 3 \end{vmatrix} + 4(-6) \begin{vmatrix} -1 & 1 & 2 \\ 2 & 6 & 3 \\ 3 & 1 & 6 \end{vmatrix}$$

developed about the first row

$$= -32 \begin{vmatrix} -1 & 0 & 2 \\ 2 & 5 & -2 \\ 3 & 3 & 3 \end{vmatrix} - 24 \begin{vmatrix} -1 & 0 & 2 \\ 2 & 8 & 3 \\ 3 & 4 & 6 \end{vmatrix}$$

column 3 subtracted from column 2 — column 1 added to column 2

$$= -32 \cdot 3 \begin{vmatrix} -1 & 0 & 2 \\ 2 & 5 & -2 \\ 1 & 1 & 1 \end{vmatrix} - 24 \begin{vmatrix} -1 & 0 & 0 \\ 2 & 8 & 7 \\ 3 & 4 & 12 \end{vmatrix}$$

3 factored out — double column 1 added to column 3

$$= -96(-1) \begin{vmatrix} 5 & -2 \\ 1 & 1 \end{vmatrix} - 96 \cdot 2 \begin{vmatrix} 2 & 5 \\ 1 & 1 \end{vmatrix} - 24(-1) \begin{vmatrix} 8 & 7 \\ 4 & 12 \end{vmatrix}$$

$$= 96(5 + 2) - 192(2 - 5) + 24(96 - 28) = 2880$$

1.9.3. Applications of determinants

Solution of a system of equations consisting of n linear equations with n variables:

$$
\begin{aligned}
a_{11}x_1 + a_{12}x_2 + a_{13}x_3 + \cdots + a_{1n}x_n &= c_1 \\
a_{21}x_1 + a_{22}x_2 + a_{23}x_3 + \cdots + a_{2n}x_n &= c_2 \\
a_{31}x_1 + a_{32}x_2 + a_{33}x_3 + \cdots + a_{3n}x_n &= c_3 \\
\vdots \qquad \vdots \qquad \vdots \qquad\qquad & \vdots \quad \vdots \\
a_{n1}x_1 + a_{n2}x_2 + a_{n3}x_3 + \cdots + a_{nn}x_n &= c_n
\end{aligned}
$$

$$
\Delta = \begin{vmatrix} a_{11} & a_{12} & a_{13} & \cdots & a_{1n} \\ a_{21} & a_{22} & a_{23} & \cdots & a_{2n} \\ a_{31} & a_{32} & a_{33} & \cdots & a_{3n} \\ \vdots & \vdots & \vdots & & \vdots \\ a_{n1} & a_{n2} & a_{n3} & \cdots & a_{nn} \end{vmatrix}
$$

coefficient determinant of the system of equations

Cramer's rule

Normal case

$\Delta \neq 0$ and not all c_i equal 0.

The system of equations possesses exactly one set of solutions (see page 80).

$$
x_1 = \frac{\Delta x_1}{\Delta}, \quad \text{where} \quad \Delta x_1 = \begin{vmatrix} c_1 & a_{12} & a_{13} & \cdots & a_{1n} \\ c_2 & a_{22} & a_{23} & \cdots & a_{2n} \\ c_3 & a_{32} & a_{33} & \cdots & a_{3n} \\ \vdots & \vdots & \vdots & & \vdots \\ c_n & a_{n2} & a_{n3} & \cdots & a_{nn} \end{vmatrix}
$$

$$
x_2 = \frac{\Delta x_2}{\Delta}, \quad \text{where} \ \Delta x_2 = \begin{vmatrix} a_{11} & c_1 & a_{13} & \cdots & a_{1n} \\ a_{21} & c_2 & a_{23} & \cdots & a_{2n} \\ a_{31} & c_3 & a_{33} & \cdots & a_{3n} \\ \vdots & \vdots & \vdots & & \vdots \\ a_{n1} & c_n & a_{n3} & \cdots & a_{nn} \end{vmatrix}
$$

$$x_3 = \frac{\Delta x_3}{\Delta}, \quad \text{where} \quad \Delta x_3 = \begin{vmatrix} a_{11} & a_{12} & c_1 & \dots & a_{1n} \\ a_{21} & a_{22} & c_2 & \dots & a_{2n} \\ a_{31} & a_{32} & c_3 & \dots & a_{3n} \\ \vdots & \vdots & \vdots & & \vdots \\ a_{n1} & a_{n2} & c_n & \dots & a_{nn} \end{vmatrix}$$

etc.

Example:

$$\begin{aligned} x - y + 2z &= 7 \\ 2x - 3y + 5z &= 17 \\ 3x - 2y - z &= 12 \end{aligned}$$

$$\Delta = \begin{vmatrix} 1 & -1 & 2 \\ 2 & -3 & 5 \\ 3 & -2 & -1 \end{vmatrix}$$

$$= \begin{vmatrix} 0 & -1 & 2 \\ -1 & -3 & 5 \\ 1 & -2 & -1 \end{vmatrix} = 1 \begin{vmatrix} -1 & 2 \\ -2 & -1 \end{vmatrix} + 1 \begin{vmatrix} -1 & 2 \\ -3 & 5 \end{vmatrix} = 5 + 1 = 6$$

$$\Delta x = \begin{vmatrix} 7 & -1 & 2 \\ 17 & -3 & 5 \\ 12 & -2 & -1 \end{vmatrix} = \begin{vmatrix} 7 & -1 & 0 \\ 17 & -3 & -1 \\ 12 & -2 & -5 \end{vmatrix}$$

$$= 7 \begin{vmatrix} -3 & -1 \\ -2 & -5 \end{vmatrix} + 1 \begin{vmatrix} 17 & -1 \\ 12 & -5 \end{vmatrix}$$

$$= 7(15 - 2) + 1(-85 + 12) = 18$$

$$\Delta y = \begin{vmatrix} 1 & 7 & 2 \\ 2 & 17 & 5 \\ 3 & 12 & -1 \end{vmatrix} = -12 \quad \Delta z = \begin{vmatrix} 1 & -1 & 7 \\ 2 & -3 & 17 \\ 3 & -2 & 12 \end{vmatrix} = 6$$

$$x = \frac{18}{6} = 3 \quad y = \frac{-12}{6} = -2 \quad z = \frac{6}{6} = 1$$

i.e. $S = \{3; -2; 1\}$

Special cases

(a) *Homogeneous system of equations*

A system of equations is called homogeneous if $c_i = 0$, i.e. if all numerator determinants Δx_i disappear.
If $\Delta \neq 0$, there will be only the trivial solution

$$x_1 = x_2 = x_3 = \cdots = x_n = 0$$

If $\Delta = 0$, there will be systems of solutions where not all variables will disappear.

(b) *Inhomogeneous system of equations*

A system of equations is called inhomogeneous if at least one $c_i \neq 0$.
If $\Delta \neq 0$, the solution will be uniquely determined.
If $\Delta = 0$, there will be either no solution or no uniquely determined solution.
For further applications see relevant Sections.

1.10. Matrices

1.10.1. General

Notations

$$\mathbf{A} = \begin{pmatrix} a_{11} & a_{12} & \cdots & a_{1n} \\ a_{21} & a_{22} & \cdots & a_{2n} \\ \vdots & \vdots & & \vdots \\ a_{m1} & a_{m2} & \cdots & a_{mn} \end{pmatrix} = (a_{ik}) \quad (m,n\text{-matrix})$$

The numbers a_{ik} are called elements of the matrix (read: *a* one one, *a* one two, etc.). The first index indicates the row, the second the column. The elements may be real, complex, or dependent upon a parameter.
A matrix with m rows and n columns is called m,n-matrix and is of type m,n. The requirements $m = n$, as occurs with determinants, is not made necessarily.
$m \neq n$ *rectangular* matrix, $m = n$ *square* matrix.
In a square matrix, the terms $a_{11}, a_{22}, \ldots, a_{mn}$ form the *principal diagonal.*
Matrices with only one row or column are referred to as *vectors.*

Horizontal row = row: *row vector* $\mathbf{a}^i$ (read: a sup i).

$$\mathbf{a}^i = (a_{i1}, a_{i2}, \ldots, a_{in})$$

i in $\mathbf{a}^i$ is an index which in row vectors is usually raised.
Vertical row = column: *column vector* $\mathbf{a}_k$ (read: a sub k)

$$\mathbf{a}_k = \begin{pmatrix} a_{1k} \\ a_{2k} \\ \vdots \\ a_{mk} \end{pmatrix}$$

A matrix with row or column vectors is written as:

$$\mathbf{A} = \begin{pmatrix} a_{11} & a_{12} & \ldots & a_{1n} \\ a_{21} & a_{22} & \ldots & a_{2n} \\ \vdots & \vdots & & \vdots \\ a_{m1} & a_{m2} & \ldots & a_{mn} \end{pmatrix} = \begin{pmatrix} \mathbf{a}^1 \\ \mathbf{a}^2 \\ \vdots \\ \mathbf{a}^m \end{pmatrix} = (\mathbf{a}_1, \mathbf{a}_2, \ldots, \mathbf{a}_n)$$

Definition

A matrix is a rectangular arrangement of $m \cdot n$ numbers (m rows, n columns) which is enclosed in parantheses. It has no numerical value. (In the place of parantheses, two parallel lines can be used.) In a *null matrix*, all elements are zeros, also in null vectors.

$$\mathbf{N} = \begin{pmatrix} 0 & 0 & \ldots & 0 \\ 0 & 0 & \ldots & 0 \\ \vdots & \vdots & & \vdots \\ 0 & 0 & \ldots & 0 \end{pmatrix}$$

Rule: Matrices that are distinguished from each other only by null vectors at the right and lower edges are equal.

$$\begin{pmatrix} 1 & 3 & 5 \\ 2 & 4 & 7 \end{pmatrix} = \begin{pmatrix} 1 & 3 & 5 & 0 & 0 \\ 2 & 4 & 7 & 0 & 0 \\ 0 & 0 & 0 & 0 & 0 \end{pmatrix}$$

In a square *diagonal matrix* all terms away from the principal diagonal disappear.

$$\mathbf{D} = \begin{pmatrix} d_1 & 0 & 0 & \dots & 0 \\ 0 & d_2 & 0 & \dots & 0 \\ 0 & 0 & d_3 & \dots & 0 \\ \vdots & \vdots & \vdots & & \vdots \\ 0 & 0 & 0 & \dots & d_m \end{pmatrix} = (\mathbf{d}_{ik} d_i) = (\mathbf{d}_{ik} d_k)$$

Special case

Unit matrix

$$\mathbf{E} = \begin{pmatrix} 1 & 0 & 0 & \dots & 0 \\ 0 & 1 & 0 & \dots & 0 \\ 0 & 0 & 1 & \dots & 0 \\ \vdots & \vdots & \vdots & & \vdots \\ 0 & 0 & 0 & \dots & 1 \end{pmatrix} = (\mathbf{d}_{ik})$$

Kronecker symbol

$$\mathbf{d}_{ik} = \begin{cases} 0 \text{ for } i \neq k \\ 1 \text{ for } i = k \end{cases}$$

In square triangular matrices,
for an upper triangular matrix $a_{ik} = 0$ for $i > k$
for a lower triangular matrix $a_{ik} = 0$ for $i < k$

Example:

$$\mathbf{A} = \begin{pmatrix} 11 & 2 & 7 \\ 0 & 3 & 4 \\ 0 & 0 & 2 \end{pmatrix}$$ upper triangular matrix

$$\mathbf{B} = \begin{pmatrix} 3 & 0 & 0 \\ 7 & 4 & 0 \\ 9 & 1 & 6 \end{pmatrix}$$ lower triangular matrix

If certain rows and columns of a matrix **A** are eliminated to leave a square matrix with k rows and k columns of which the determinant can be formed, *a k-rowed subdeterminant* of **A** is obtained.

Example:

$$\mathbf{A} = \begin{pmatrix} 1 & 2 & 2 & 3 & 4 \\ 2 & 3 & 6 & 5 & 2 \\ 4 & 0 & 1 & 3 & 1 \\ 1 & 1 & 2 & 4 & 2 \end{pmatrix}$$

4-rowed subdeterminants

$$\begin{vmatrix} 1 & 2 & 2 & 3 \\ 2 & 3 & 6 & 5 \\ 4 & 0 & 1 & 3 \\ 1 & 1 & 2 & 4 \end{vmatrix} \text{ or } \begin{vmatrix} 1 & 2 & 3 & 4 \\ 2 & 3 & 5 & 2 \\ 4 & 0 & 3 & 1 \\ 1 & 1 & 4 & 2 \end{vmatrix} \quad \text{etc.}$$

3-rowed subdeterminants

$$\begin{vmatrix} 1 & 2 & 2 \\ 2 & 3 & 6 \\ 4 & 0 & 1 \end{vmatrix} \text{ or } \begin{vmatrix} 2 & 6 & 5 \\ 4 & 1 & 3 \\ 1 & 2 & 4 \end{vmatrix} \text{ or } \begin{vmatrix} 2 & 3 & 4 \\ 0 & 3 & 1 \\ 1 & 4 & 2 \end{vmatrix} \quad \text{etc.}$$

2-rowed subdeterminants

$$\begin{vmatrix} 1 & 2 \\ 2 & 3 \end{vmatrix} \text{ or } \begin{vmatrix} 3 & 5 \\ 0 & 3 \end{vmatrix} \text{ or } \begin{vmatrix} 2 & 6 \\ 1 & 2 \end{vmatrix} \text{ etc.}$$

If a matrix is square, the determinant can be formed directly of all elements ($\det \mathbf{A} = |\mathbf{A}|$).
The highest k-rowed subdeterminant differing from 0 (of the kth order) determines the *rank* k of the matrix, i.e. all subdeterminants of the $(k+1)$th order with $k+1$ rows and $k+1$ columns are equal to zero.
The rank of a matrix is not changed by taking linear combinations of its rows and columns.

Example:

$$\mathbf{A} = \begin{pmatrix} -1 & 4 & 1 & 3 \\ 2 & -2 & -2 & 0 \\ 0 & 2 & 0 & 2 \end{pmatrix}$$

By a linear combination of the first and the third columns we obtain

$$\begin{pmatrix} 0 & 4 & 1 & 3 \\ 0 & -2 & -2 & 0 \\ 0 & 2 & 0 & 2 \end{pmatrix}$$

All subdeterminants of the third order disappear because three of them only have elements 0 in the first column and the fourth of them shows two equal columns as a result of a linear combination of the first and second columns.

$$\begin{vmatrix} 4 & 1 & 3 \\ -2 & -2 & 0 \\ 2 & 0 & 2 \end{vmatrix} = \begin{vmatrix} 3 & 1 & 3 \\ 0 & -2 & 0 \\ 2 & 0 & 2 \end{vmatrix} = 0$$

Since a subdeterminant of the second order, e.g. $\begin{vmatrix} -1 & 4 \\ 2 & -2 \end{vmatrix}$ is not equal to zero, the matrix has rank $k = 2$.

1.10.2. Theorems on matrices

Given: m,n-matrices $\mathbf{A} = (\mathbf{a}_{ik})$ and $\mathbf{B} = (\mathbf{b}_{ik})$.

Equality of matrices

$$\mathbf{A} = \mathbf{B} \text{ if } (a_{ik}) = (b_{ik}), \text{ i.e. } a_{ik} = b_{ik}.$$

Sum of two matrices

$$\mathbf{A} + \mathbf{B} = (a_{ik} + b_{ik})$$

Terms which correspond to one another are added.

Commutative law $\mathbf{A} + \mathbf{B} = \mathbf{B} + \mathbf{A}$

Associative law $(\mathbf{A} + \mathbf{B}) + \mathbf{C} = \mathbf{A} + (\mathbf{B} + \mathbf{C})$

Multiplication of a matrix by a real number

$$\mu\mathbf{A} = (\mu a_{ik}) = \mathbf{A}\mu$$

Distributive laws $\lambda(\mathbf{A} + \mathbf{B}) = \lambda\mathbf{A} + \lambda\mathbf{B}$

$(\lambda + \mu)\mathbf{A} = \lambda\mathbf{A} + \mu\mathbf{A}$

Limit of a matrix $\mathbf{A}(t)$

If a matrix is dependent upon a parameter t, $\mathbf{A}(t)$, the *limit matrix* is defined as that matrix where the transition $t \to t_0$ has taken place for each member.

$$\lim_{t \to t_0} \mathbf{A}(t) = \left(\lim_{t \to t_0} a_{ik}(t)\right)$$

Differential quotient of a matrix $\mathbf{A}(t)$

The elements are differentiated individually.

$$\frac{d}{dt}\mathbf{A}(t) = \left(\frac{d}{dt}a_{ik}(t)\right) \qquad (a_{ik} \text{ differentiable})$$

The elements are integrated individually.

$$\int_a^b \mathbf{A}(t)\,dt = \left(\int_a^b a_{ik}(t)\,dt\right) \qquad (a_{ik} \text{ can be integrated})$$

Transposed matrix $\mathbf{A}'$

It is obtained by interchanging rows and columns. A row vector is changed into a column vector by transposition.

$$\mathbf{A} = (a_{ik}) \qquad \mathbf{A}' = (a'_{ik}) \qquad a'_{ik} = a_{ki}$$

If $\mathbf{A}$ is of type m, n, then $\mathbf{A}'$ is of type n, m.

$$(\mathbf{A}')' = \mathbf{A}$$

$$(\mathbf{A} + \mathbf{B})' = \mathbf{A}' + \mathbf{B}'$$

$$(k\mathbf{A})' = k\mathbf{A}' \qquad k \text{ scalar factor}$$

Symmetric square matrix

$$\mathbf{A} = \mathbf{A}' \qquad a_{ik} = a_{ki}$$

Example:

$$\mathbf{A} = \mathbf{A}' = \begin{pmatrix} 1 & 5 & 7 \\ 5 & 2 & -6 \\ 7 & -6 & 8 \end{pmatrix}$$

Antisymmetric square matrix

$$\mathbf{A} = -\mathbf{A}' \qquad a_{ik} = -a_{ki}$$

(The elements of the principal diagonal must disappear.)

Example:

$$\mathbf{A} = -\mathbf{A}' = \begin{pmatrix} 0 & 5 & -7 \\ -5 & 0 & 3 \\ 7 & -3 & 0 \end{pmatrix}$$

Conjugate complex matrix $\bar{\mathbf{A}}$

Each element of the original matrix is replaced by its conjugate complex element.

$$\bar{\mathbf{A}} = (\bar{a}_{ik})$$

$$\bar{\bar{\mathbf{A}}} = \mathbf{A}$$

$$\overline{\mathbf{A} + \mathbf{B}} = \bar{\mathbf{A}} + \bar{\mathbf{B}}$$

$$\overline{\mu \mathbf{A}} = \mu \bar{\mathbf{A}}$$

$$(\bar{\mathbf{A}})' = \overline{(\mathbf{A}')}$$

Example:

$$\mathbf{A} = \begin{pmatrix} 1 + 3j & 2 - 5j \\ 5 & 7 + 2j \end{pmatrix} \quad \bar{\mathbf{A}} = \begin{pmatrix} 1 - 3j & 2 + 5j \\ 5 & 7 - 2j \end{pmatrix}$$

$\bar{\mathbf{A}}'$ of a square matrix is called Hermitian. We have:

$$\mathbf{A} = \bar{\mathbf{A}}'; \quad \bar{a}_{ik} = a_{ki}$$

Example:

$$\mathbf{A} = \bar{\mathbf{A}}' = \begin{pmatrix} 2 & 1 - 2j & 3 + 5j \\ 1 + 2j & 3 & 2 - j \\ 3 - 5j & 2 + j & 5 \end{pmatrix}$$

(The elements of the principal diagonal must be real.)
A square matrix is *skew-Hermitian* if:

$$\mathbf{A} = -\bar{\mathbf{A}}'; \quad a_{ik} = -\bar{a}_{ki}$$

Example:

$$\mathbf{A} = -\bar{\mathbf{A}}' = \begin{pmatrix} 2j & 1 - 2j & 3 + 5j \\ -1 - 2j & j & 2 - j \\ -3 + 5j & -2 - j & 5j \end{pmatrix}$$

(The elements of the principal diagonal must be purely imaginary.)

Product of matrices

$$\mathbf{C} = \mathbf{AB} = (a^i b_k) = \left(\sum_{l=1}^{n} a_{il} b_{lk} \right) = (c_{ik})$$

The element c_{ik} is obtained as a scalar product of the row vector a^i (ith row) and the column vector b_k (kth column) (see page 67).
Precondition: number of columns of $\mathbf{A}$ = number of rows of $\mathbf{B}$.
This condition can be established by adding null vectors to the right-hand and lower edges (cf. page 67).

$$a^1 b^1 = (a_1, a_2, \ldots, a_m)\,(b_1, b_2, \ldots, b_n)$$

$$= (a_1, a_2, \ldots, a_m) \begin{pmatrix} b_1 & b_2 & \ldots & b_n \\ 0 & 0 & \ldots & 0 \\ \vdots & \vdots & \vdots & \vdots \\ 0 & 0 & \ldots & 0 \end{pmatrix} = (a_1 b_1, a_1 b_2, \ldots, a_1 b_n)$$

Note: $\mathbf{AB} = \mathbf{N}$ does not necessarily imply $\mathbf{A} = \mathbf{N}$ or $\mathbf{B} = \mathbf{N}$.

but: $\mathbf{A} = \mathbf{N}$ or $\mathbf{B} = \mathbf{N} \Rightarrow \mathbf{AB} = \mathbf{N}$

The transpose of a product

$$(\mathbf{AB})' = \mathbf{B}'\mathbf{A}'$$

$$(\mathbf{ABC})' = \mathbf{C}'\mathbf{B}'\mathbf{A}'$$

Multiplication by the unit matrix E

For square matrices $\mathbf{A}$ and $\mathbf{E}$ of the same order we have

$$\mathbf{AE} = \mathbf{EA} = \mathbf{A}$$

Multiplication by the diagonal matrix D

$$\mathbf{AD} = (a_{ik} d_k) = \begin{pmatrix} a_{11}d_1 & a_{12}d_2 & \ldots & a_{1n}d_n \\ a_{21}d_1 & a_{22}d_2 & \ldots & a_{2n}d_n \\ \vdots\ \ \vdots & \vdots\ \ \vdots & \vdots & \vdots\ \ \vdots \\ a_{m1}d_1 & a_{m2}d_2 & \ldots & a_{mn}d_n \end{pmatrix}$$

Determinant from a square matrix $\det \mathbf{A}' = |\mathbf{A}|$

$$\det \mathbf{A}' = \det \mathbf{A}$$

$$\det (\mathbf{AB}) = \det \mathbf{A} \det \mathbf{B}$$

Singular matrix (A square)

If $\det \mathbf{A} = 0$, $\mathbf{A}$ is called singular.
If $\mathbf{AB} = \mathbf{N}$ and $\mathbf{A} \neq \mathbf{N}$ as well as $\mathbf{B} \neq \mathbf{N}$, then $\mathbf{A}$ and $\mathbf{B}$ are singular. Schematically

$$m \overset{n}{\boxed{\mathbf{A}}} \cdot n \overset{p}{\boxed{\mathbf{B}}} = m \overset{p}{\boxed{\mathbf{C} = \mathbf{AB}}}$$

m,n-matrix $\mathbf{A}$ multiplied by n,p-matrix $\mathbf{B}$ results in the m,p-matrix $\mathbf{C}$.

Example:

$$\mathbf{AB} = \begin{pmatrix} 1 & 3 & 2 \\ 2 & 4 & 1 \end{pmatrix} \begin{pmatrix} 1 & 0 \\ 2 & 3 \\ 4 & 1 \end{pmatrix} = \begin{pmatrix} 1\cdot1+3\cdot2+2\cdot4 & 1\cdot0+3\cdot3+2\cdot1 \\ 2\cdot1+4\cdot2+1\cdot4 & 2\cdot0+4\cdot3+1\cdot1 \end{pmatrix}$$

$$= \begin{pmatrix} 15 & 11 \\ 14 & 13 \end{pmatrix}$$

The **commutative law** does not hold generally

$$\mathbf{AB} \neq \mathbf{BA}$$

This means that, in multiplying a matrix $\mathbf{B}$ by a matrix $\mathbf{A}$ from the left (front), the result obtained in general is different from that obtained when proceeding from the right (rear).

Associative law $(\mathbf{AB})\mathbf{C} = \mathbf{A}(\mathbf{BC}) = \mathbf{ABC}$

Preconditions: m,n-matrix $\mathbf{A}$, n,p-matrix $\mathbf{B}$, p,q-matrix $\mathbf{C}$. The result is a m,q-matrix.

Distributive law: $(\mathbf{A} + \mathbf{B})\mathbf{C} = \mathbf{AC} + \mathbf{BC}$

$$\mathbf{A}(\mathbf{B} + \mathbf{C}) = \mathbf{AB} + \mathbf{AC}$$

Multiplication of row and column vectors

$$\mathbf{a}^1\mathbf{b}_1 = (a_1, a_2, \ldots, a_n)\begin{pmatrix} b_1 \\ b_2 \\ \vdots \\ b_n \end{pmatrix} = \sum_{i=1}^{n} a_i b_i$$

$$\mathbf{a}_1\mathbf{b}^1 = \begin{pmatrix} a_1 \\ a_2 \\ \vdots \\ a_m \end{pmatrix}(b_1, b_2, \ldots, b_n) = \begin{pmatrix} a_1b_1 & a_1b_2 & \ldots & a_1b_n \\ a_2b_1 & a_2b_2 & \ldots & a_2b_n \\ \vdots & \vdots & & \vdots \\ a_mb_1 & a_mb_2 & \ldots & a_mb_n \end{pmatrix}$$

$$\mathbf{a}_1\mathbf{b}_1 = \begin{pmatrix} a_1 \\ a_2 \\ \vdots \\ a_m \end{pmatrix}\begin{pmatrix} b_1 \\ b_2 \\ \vdots \\ b_n \end{pmatrix} = \begin{pmatrix} a_1 & 0 & \ldots & 0 \\ a_2 & 0 & \ldots & 0 \\ \vdots & \vdots & \vdots & \vdots \\ a_m & 0 & \ldots & 0 \end{pmatrix}\begin{pmatrix} b_1 \\ b_2 \\ \vdots \\ b_n \end{pmatrix} = \begin{pmatrix} a_1b_1 \\ a_2b_1 \\ \vdots \\ a_mb_1 \end{pmatrix}$$

Powers of matrices

If $\mathbf{A}$ is a square matrix, we define:

$$\mathbf{A}^n = \mathbf{AAA}\cdots \quad (n \text{ factors})$$

$$\mathbf{A}^{-n} = \mathbf{A}^{-1}\mathbf{A}^{-1}\mathbf{A}^{-1}\cdots \quad (n \text{ factors}) \qquad n \in N$$

$$\mathbf{A}^0 = \mathbf{E}$$

$$\mathbf{A}^n\mathbf{A}^m = \mathbf{A}^{n+m} \qquad n,m \in I$$

Interchanging matrix V

An interchanging matrix has only one 1 in each row and in each column and only zeros elsewhere, and is square. When multiplying a matrix $\mathbf{A}$ by $\mathbf{V}$ from the front (left), a 1 in the rth row and sth column of $\mathbf{V}$ will cause the rth row in the matrix $\mathbf{VA}$ to correspond with the sth row of $\mathbf{A}$.

Example:

$$\overset{r\rightarrow}{\mathbf{VA}} = \begin{pmatrix} 0 & 1 & 0 & 0 \\ 0 & 0 & \boxed{1} & 0 \\ 1 & 0 & 0 & 0 \\ 0 & 0 & \underset{\substack{\uparrow \\ s}}{0} & 1 \end{pmatrix} \left(\begin{array}{cccc} 1 & 3 & 0 & 2 \\ 2 & 4 & 7 & 3 \\ \multicolumn{4}{c}{\boxed{3 \quad 0 \quad 4 \quad 1}} \\ 5 & 2 & 1 & 1 \end{array}\right) \leftarrow s \; = \left(\begin{array}{cccc} 2 & 4 & 7 & 3 \\ \multicolumn{4}{c}{\boxed{3 \quad 0 \quad 4 \quad 1}} \\ 1 & 3 & 0 & 2 \\ 5 & 2 & 1 & 1 \end{array}\right) \leftarrow r$$

When multiplying a matrix **A** by **V** from the rear (right), a 1 in the rth row and sth column of **V** will cause the sth column in the matrix **AV** to correspond with the rth column of **A**.

Example:

$$\mathbf{AV} = \begin{pmatrix} 1 & 2 & 4 & 0 \\ 2 & 4 & 6 & 1 \\ 1 & 3 & 0 & 3 \end{pmatrix} \begin{pmatrix} 0 & 0 & 0 & 1 \\ 1 & 0 & 0 & 0 \\ 0 & 1 & 0 & 0 \\ 0 & 0 & 1 & 0 \end{pmatrix} = \begin{pmatrix} 2 & 4 & 0 & 1 \\ 4 & 6 & 1 & 2 \\ 3 & 0 & 3 & 1 \end{pmatrix}$$

The interchange effected by **V** is undone by $\mathbf{V}^{-1}$. If we replace the 1 in the interchanging matrix by a number p_{ik}, an additional multiplication by p_{ik} of the interchanged row or column is effected.

Reciprocal matrix $\mathbf{A}^{-1}$ (**A** square)

Definition

$$\mathbf{AB} = \mathbf{E} \Leftrightarrow \mathbf{B} = \mathbf{A}^{-1}$$

B is the reciprocal matrix of **A**. The inversion is unique if $\mathbf{A}^{-1}$ exists. Only square nonsingular matrices can be inverted.

$$\mathbf{A} = \begin{pmatrix} a_{11} & a_{12} & \dots & a_{1n} \\ a_{21} & a_{22} & \dots & a_{2n} \\ \vdots & \vdots & \vdots & \vdots \\ a_{n1} & a_{n2} & \dots & a_{nn} \end{pmatrix}; \quad \mathbf{B} = \mathbf{A}^{-1} = \frac{1}{A} \begin{pmatrix} A_{11} & A_{21} & \dots & A_{n1} \\ A_{12} & A_{22} & \dots & A_{n2} \\ \vdots & \vdots & \vdots & \vdots \\ A_{1n} & A_{2n} & \dots & A_{nn} \end{pmatrix}$$

where A_{ki} are subdeterminants (see page 58)

$$A = \det \mathbf{A}$$

Elements of the reciprocal matrix $b_{ik} = \dfrac{A_{ik}}{A}$ (notice the indices)

$$(\mathbf{A}^{-1})^{-1} = \mathbf{A} \qquad \mathbf{AA}^{-1} = \mathbf{A}^{-1}\mathbf{A} = \mathbf{E}$$

$(\mathbf{AC})^{-1} = \mathbf{C}^{-1}\mathbf{A}^{-1}$ **AC** nonsingular

$(\mathbf{A}')^{-1} = (\mathbf{A}^{-1})'$ (*contragradient* matrix of **A**)

Example:

$$\mathbf{A} = \begin{pmatrix} 1 & 0 & 3 \\ 2 & -3 & 1 \\ 1 & 2 & 2 \end{pmatrix} \qquad A = \det \mathbf{A} = \begin{vmatrix} 1 & 0 & 3 \\ 2 & -3 & 1 \\ 1 & 2 & 2 \end{vmatrix} = 13$$

$$A_{11} = (-1)^2 \begin{vmatrix} -3 & 1 \\ 2 & 2 \end{vmatrix} = -8; \quad A_{12} = (-1)^3 \begin{vmatrix} 2 & 1 \\ 1 & 2 \end{vmatrix} = -3;$$

$$A_{13} = (-1)^4 \begin{vmatrix} 2 & -3 \\ 1 & 2 \end{vmatrix} = 7$$

$$A_{21} = (-1)^3 \begin{vmatrix} 0 & 3 \\ 2 & 2 \end{vmatrix} = 6; \quad A_{22} = (-1)^4 \begin{vmatrix} 1 & 3 \\ 1 & 2 \end{vmatrix} = -1;$$

$$A_{23} = (-1)^5 \begin{vmatrix} 1 & 0 \\ 1 & 2 \end{vmatrix} = -2$$

$$A_{31} = (-1)^4 \begin{vmatrix} 0 & 3 \\ -3 & 1 \end{vmatrix} = 9; \quad A_{32} = (-1)^5 \begin{vmatrix} 1 & 3 \\ 2 & 1 \end{vmatrix} = 5;$$

$$A_{33} = (-1)^6 \begin{vmatrix} 1 & 0 \\ 2 & -3 \end{vmatrix} = -3$$

$$\mathbf{A}^{-1} = \frac{1}{13} \begin{pmatrix} -8 & 6 & 9 \\ -3 & -1 & 5 \\ 7 & -2 & -3 \end{pmatrix}$$

1.10.3. Applications

Checking the possibilities of solving systems of linear equations

An *inhomogeneous system of equations of the first degree* of m equations with n unknowns is solvable only if the coefficient matrix and the matrix expanded by the absolute terms have the same rank k.

$k = n$ one unique solution

$k < n$ ∞^{n-k} solutions

Example:

$$3x + 2y + 5z = 10$$

$$6x + 4y + 10z = 30$$

$$12x + 8y + 20z = 40$$

The coefficient matrix $\begin{pmatrix} 3 & 2 & 5 \\ 6 & 4 & 10 \\ 12 & 8 & 20 \end{pmatrix}$ has rank $k = 1$ (see page 69)

The expanded matrix $\begin{pmatrix} 3 & 2 & 5 & 10 \\ 6 & 4 & 10 & 30 \\ 12 & 8 & 20 & 40 \end{pmatrix}$ has rank $k > 1$ since

$$\begin{vmatrix} 5 & 10 \\ 10 & 30 \end{vmatrix} \neq 0$$

There are no solutions of the systems (the equations are inconsistent).

A *homogeneous system of equations of the first degree* of m equations with n unknowns will only give a solution different from the trivial (all unknowns equal to zero) if the rank of the coefficient matrix is smaller than n.

Further applications

Further, matrices are mainly used in quadrupole calculations (resistance matrix, conductivity matrix, iterative matrix, and series parallel matrix), calculations of linear networks (calculation of meshes), solutions of systems of linear differential equations, in planning and solving other economic problems.

2. Equations, functions, vectors

2.1. Equations

2.1.1. General

Expression

An expression is a sequence formed of numbers, letters, and mathematical signs which yields a certain numerical value if the letters are replaced by numbers which can be freely chosen from within the domain of definition.
Notations of expressions: T, S, ...
The set of the expressions forms a field.

Special expressions

Linear expression: The variables are only multiplied by constants and added to constants or expressions of the same kind (linear combination),
e.g. linear expression in the variables x and y

$$L(x, y) = ax + by + c \qquad a, b, c \text{ constants}$$

Whole rational expression (polynomial expression): The variables are only combined by addition, subtraction, and multiplication, e.g. whole rational expression in x and y

$$T(x, y) = a_nx^n + a_{n-1}x^{n-1} + \cdots + a_1x + b_my^m + b_{m-1}y^{m-1} + \cdots + b_1y + a_0 \qquad n, m \in N$$

Rational expression: The variables are combined by the four fundamental operations of arithmetic,
e.g. rational expression

$$T(x) = \frac{a_nx^n + a_{n-1}x^{n-1} + \cdots + a_1x + a_0}{b_mx^m + b_{m-1}x^{m-1} + \cdots + b_1x + b_0} \qquad n, m \in N$$

Definition of equation

When equating two expressions, we obtain an equation

$$T_1 = T_2$$

An equation with at least one variable is a *form of statement* which becomes a true or false *statement* if we coordinate *values of the domain of definition* or *range of variables* and the variables.
An equation without variables represents a true or a false statement, e.g. $7 = 3 + 4$, true statement.
Domain of definition of an equation: $X = X_1 \cap X_2$; $Y = Y_1 \cap Y_2$; ... where $X_1, Y_1, \ldots$ domain of definition for T_1; $X_2, Y_2, \ldots$ domain of definition for T_2.
Roots or solutions of an equation are all elements of the domain of definition for which $T_1(x; y; \ldots) = T_2(x; y; \ldots)$ becomes a true statement.
Solution set, range of validity S is the set of all solutions where S is a subset of the domain of definition.

Written as:

$$S = \{(x; y; \ldots) \mid T_1(x; y; \ldots) = T_2(x; y; \ldots)\}$$

without indication of the range of variables if this is obvious.

or

$$S = \{(x; y; \ldots) \mid x \in X; y \in Y; \ldots \wedge T_1(x; y; \ldots) = T_2(x; y; \ldots)\}$$

with indication of the range of variables.

Special cases

Identity, definition of *equality of expressions:*

$T_1(x; y; \ldots) = T_2(x; y; \ldots)$ for solution set = domain of definition

e.g. equation with one variable

$T_1(x) = T_2(x)$ for $S = X$ or written as

$$S = \{x \mid x \in X \wedge T_1(x) = T_2(x)\} = X$$

Example:

$\sin 2x = 2 \sin x \cos x$ $X = K$, i.e. identity with respect to the field K

or

$$S = \{x \mid x \in K \wedge \sin 2x = 2 \sin x \cos x\} = K$$

Insolvability $S = \emptyset$

Example:

$$3^x = -4 \qquad X = R \qquad S = \emptyset$$

or

$$S = \{x \mid x \in R \wedge 3^x = -4\} = \emptyset$$

Equivalence

Two equations are equivalent with respect to the given range of variables if they possess the same solution set S.

$$T_1 = T_2 \Rightarrow T_1 + T = T_2 + T$$

T a meaningful expression within the range of variables

$T_1 = T_2 \Rightarrow T_1 T = T_2 T \quad T \neq 0$ for all values of the range of variables

2.1.2. Algebraic equations in one variable

Explanation

An equation is called an algebraic equation if it has the following form:

$$a_n x^n + a_{n-1} x^{n-1} + \cdots + a_1 x + a_0 = 0,$$

where $a_n \neq 0; a_{n-1}; a_{n-2}; \ldots; a_0$ rational numbers; $n \in N$.

Numbers obtained as solutions of algebraic equations are called *algebraic.*

2.1.2.1. Linear equations

Standard form: $ax + b = 0 \quad a \neq 0; x \in R$

Solution:

$$x = -\frac{b}{a}$$

or

$$S = \{x \mid x \in R \wedge ax + b = 0 \wedge a \neq 0\} = \left\{-\frac{b}{a}\right\}$$

2.1.2.2. Quadratic equations

General form: $Ax^2 + Bx + C = 0;\ A \neq 0$

Division by A yields

the *standard form* $\{x \mid x^2 + px + q = 0\}$ (*mixed quadratic equation*), where

$$p = \frac{B}{A}, \quad q = \frac{C}{A}$$

Solution:

$$x_{1;2} = -\frac{p}{2} \pm \sqrt{\left(\frac{p}{2}\right)^2 - q}$$

or

$$S = \left\{-\frac{p}{2} + \sqrt{\left(\frac{p}{2}\right)^2 - q};\ -\frac{p}{2} - \sqrt{\left(\frac{p}{2}\right)^2 - q}\right\}$$

Discriminant $D = \left(\frac{p}{2}\right)^2 - q$ (radicand)

$D > 0$ gives two different real solutions.

$D = 0$ gives two equal solutions (double root $x_1 = x_2$).

$D < 0$ gives two conjugate complex solutions.

Special cases

(a) $p = 0$ $\{x \mid x^2 + q = 0\}$ (*pure quadratic equation*)

Solution:

$$x_{1;2} = \pm\sqrt{-q}$$

or

$$S = \{\sqrt{-q};\ -\sqrt{-q}\}$$

(b) $q = 0$ $\{x \mid x^2 + px = 0\}$

Solution:

$$x(x + p) = 0$$

$$x_1 = 0;\ x_2 = -p$$

or

$$S = \{0;\ -p\}$$

(c) Reduction of the *symmetrical equation of the third degree* to a quadratic equation

Explanation

Equations are called *symmetrical* if the coefficients are arranged symmetrically.

$$\{x \mid ax^3 + bx^2 + bx + a = 0\}$$

$$a(x^3 + 1) + bx(x + 1) = 0$$

$$(x + 1)\,[a(x^2 - x + 1) + bx] = 0$$

Hence $x + 1 = 0$; $x_1 = -1$ and

$$a(x^2 - x + 1) + bx = 0 \text{ (quadratic equation)}$$

Example:

$$\{x \mid 6x^3 - 7x^2 - 7x + 6 = 0\}$$

Solution:

$$6(x^3 + 1) - 7x(x + 1) = 0$$

$$(x + 1)\,[6(x^2 - x + 1) - 7x] = 0$$

$$x + 1 = 0; \quad \underline{\underline{x_1 = -1}}$$

$$6x^2 - 13x + 6 = 0$$

$$x^2 - \frac{13}{6}x + 1 = 0 \text{ (standard form)}$$

$$x_{2;3} = \frac{13}{12} \pm \sqrt{\left(\frac{13}{12}\right)^2 - 1} = \frac{13}{12} \pm \frac{5}{12}$$

$$\underline{\underline{x_2 = 1\,\frac{1}{2}}}; \quad \underline{\underline{x_3 = \frac{2}{3}}}$$

or

$$S = \left\{-1;\ 1\,\frac{1}{2};\ \frac{2}{3}\right\}$$

(d) Reduction of the *symmetrical equation of the fourth degree* to a quadratic equation

$$\{x \mid ax^4 + bx^3 + cx^2 + bx + a = 0\}$$

Division by x^2 and factoring yields

$$a\left(x^2 + \frac{1}{x^2}\right) + b\left(x + \frac{1}{x}\right) + c = 0$$

Substitution $y = x + \dfrac{1}{x}$ and $y^2 - 2 = x^2 + \dfrac{1}{x^2}$ yields the quadratic equation

$$a(y^2 - 2) + by + c = 0$$

(e) Reduction of a *biquadratic equation* in which the cubic and linear terms are missing to a quadratic equation

$$\{x \mid ax^4 + cx^2 + e = 0\}$$

Substitution $x^2 = y$ yields the quadratic equation

$$ay^2 + cy + e = 0$$

2.1.2.3. Cubic equations

General form: $Ax^3 + Bx^2 + Cx + D = 0;\ A \neq 0$

Division by A gives

the standard form $\{x \mid x^3 + ax^2 + bx + c = 0\}$

where $a = \dfrac{B}{A},\ b = \dfrac{C}{A},\ c = \dfrac{D}{A}$

Substitution $x = y - \dfrac{a}{3}$

Reduced form $y^3 + py + q = 0$

Cardan's formula for solving the reduced equation

$$y_1 = u + v$$

$$y_2 = -\frac{u+v}{2} + \frac{u-v}{2}\, j\sqrt{3}$$

$$y_3 = -\frac{u+v}{2} - \frac{u-v}{2}\, j\sqrt{3}$$

where

$$u = \sqrt[3]{-\frac{q}{2} + \sqrt{\left(\frac{q}{2}\right)^2 + \left(\frac{p}{3}\right)^3}}$$

$$v = \sqrt[3]{-\frac{q}{2} - \sqrt{\left(\frac{q}{2}\right)^2 + \left(\frac{p}{3}\right)^3}}$$

Discriminant $D = \left(\frac{q}{2}\right)^2 + \left(\frac{p}{3}\right)^3$

$D > 0$ gives one real and two conjugate complex solutions.

$D = 0$ gives three real solutions including a double root.

Casus irreducibilis

$D < 0$ gives three real solutions which can be calculated goniometrically (nonreducible case, *casus irreducibilis*)

$$y_1 = 2\sqrt{\frac{|p|}{3}} \cos\frac{\varphi}{3}$$

$$y_2 = -2\sqrt{\frac{|p|}{3}} \cos\left(\frac{\varphi}{3} - 60°\right)$$

$$y_3 = -2\sqrt{\frac{|p|}{3}} \cos\left(\frac{\varphi}{3} + 60°\right)$$

φ can be calculated from the equation $\cos\varphi = \dfrac{-\frac{q}{2}}{\sqrt{\left(\frac{|p|}{3}\right)^3}}$.

The respective x values are obtained from the above substitution

$$x = y - \frac{a}{3}.$$

First example: $(D > 0)$.

$$\{x \mid x^3 - 3x^2 + 4x - 4 = 0\}$$

substitution $x = y - \frac{-3}{3} = y + 1$

$(y+1)^3 - 3(y+1)^2 + 4(y+1) - 4 = 0$

$y^3 + 3y^2 + 3y + 1 - 3y^2 - 6y - 3 + 4y + 4 - 4 = 0$

reduced form $y^3 + y - 2 = 0 \quad p = 1;\ q = -2$

$$u = \sqrt[3]{-\frac{-2}{2} + \sqrt{\left(\frac{-2}{2}\right)^2 + \left(\frac{1}{3}\right)^3}} = \sqrt[3]{1 + \sqrt{\frac{28}{27}}}$$

$\approx 1.264 \quad (D > 0)$

$$v = \sqrt[3]{1 - \sqrt{\frac{28}{27}}} \approx -0.264$$

$y_1 = 1.264 - 0.264 = 1 \qquad \underline{\underline{x_1 = 2}}$

$$y_2 = -\frac{1}{2} + \frac{1.528}{2} j\sqrt{3} \quad \underline{\underline{x_2 = \frac{1}{2} + 0.764\, j\sqrt{3}}}$$

$$y_3 = -\frac{1}{2} - \frac{1.528}{2} j\sqrt{3} \quad \underline{\underline{x_3 = \frac{1}{2} - 0.764\, j\sqrt{3}}}$$

or

$$S = \left\{2;\ \frac{1}{2} + 0.764\, j\sqrt{3};\ \frac{1}{2} - 0.764\, j\sqrt{3}\right\}$$

Second example: $(D < 0)$

$\{x \mid x^3 - 21x - 20 = 0\}$ (already reduced form)

Discriminant $D = \left(\frac{q}{2}\right)^2 + \left(\frac{p}{3}\right)^3 = (-10)^2 + (-7)^3 < 0$

Goniometrical way of solution:

$$\cos\varphi = \frac{10}{\sqrt{343}};\ \varphi = 57^\circ 19'$$

$x_1 = 2\sqrt{7}\cos 19^\circ 6'; \qquad \underline{\underline{x_1 = 5}}$

$x_2 = -2\sqrt{7}\cos(-40^\circ 54') \qquad \underline{\underline{x_2 = -4}}$

$x_3 = -2\sqrt{7}\cos 79^\circ 6' \qquad \underline{\underline{x_3 = -1}}$

or

$$S = \{5, -4, -1\}$$

Special cases of the cubic equation in reduced form

(a) $p = 0 \quad \{y \mid y^3 + q = 0\}$ (*binomial equation*)

For solution see Section "Complex numbers", page 25

(b) $q = 0 \quad \{y \mid y^3 + py = 0\}$

Solution:

$$y(y^2 + p) = 0 \quad y_1 = 0; \quad y_{2;3} = \pm\sqrt{-p}$$

or

$$S = \left\{0; \ \pm\sqrt{-p}\right\}$$

2.1.2.4. Definitive equation of the *n*th degree

General form: $Ax^n + Bx^{n-1} + Cx^{n-2} + \cdots = 0; \ A \neq 0$
By division by A we obtain the *normal form*

$$x^n + a_{n-1}x^{n-1} + a_{n-2}x^{n-2} + \cdots + a_0 = 0$$

where the coefficients $a_{n-1}, a_{n-2}, \ldots, a_0$ are real or complex numbers. The term a_0 without x is called the *absolute term*.
The solution set S of an algebraic equation with the range of variation $x \in C$ is never empty.

Fundamental theorem of algebra

An algebraic equation of the nth degree possesses exactly n real or complex solutions (roots).

Breaking up into linear factors

If $x_1, x_2, \ldots, x_n$ are the roots of the equation, we have

$$x^n + a_{n-1}x^{n-1} + a_{n-2}x^{n-2} + \cdots + a_0 = 0$$
$$= (x - x_1)(x - x_2)(x - x_3) \cdots (x - x_n)$$

Vieta's theorem of roots

$$x_1 + x_2 + x_3 + \cdots + x_n = -a_{n-1}$$
$$x_1x_2 + x_1x_3 + \cdots + x_2x_3 + x_2x_4 + \cdots + x_{n-1}x_n = a_{n-2}$$
$$x_1x_2x_3 + x_1x_2x_4 + \cdots + x_2x_3x_4 + x_2x_3x_5 + \cdots + x_{n-2}x_{n-1}x_n = -a_{n-3}$$
$$\vdots$$
$$x_1x_2x_3 \ldots x_n = (-1)^n a_0$$

The roots of an equation in its normal form with coefficients that are integers can often be found, by way of trial, using Vieta's theorem of roots if we can presume that they are integers.

Example:

$$\{x \mid x^4 - x^3 - 7x^2 + x + 6 = 0\}$$

$$x_1x_2x_3x_4 = 6 \text{ (absolute term)}$$

Factors of the absolute term are ± 1, ± 2, ± 3, ± 6.
By way of trial we shall find:

$$\underline{\underline{x_1 = 1}}, \quad \underline{\underline{x_2 = -1}}, \quad \underline{\underline{x_3 = -2}} \quad \text{and} \quad \underline{\underline{x_4 = 3}}$$

or

$$S = \{1; -1; -2; 3\}$$

If we find only one solution x_1 by way of trial, the degree of the equation can be reduced by dividing by the respective linear factor $(x - x_1)$.
Horner's array can also be used.

2.1.3. Transcendental equations

Goniometric equations

See page 189.

2.1.3.1. Exponential equations

The variable occurs in the exponent of powers or roots. In a few simple cases, the exponential equations can be reduced to algebraic equations, mostly by a logarithmic procedure on both sides of the equation.

Examples:

1. $\left\{x \,\middle|\, \sqrt[3]{a^{x+2}} = \sqrt{a^{x-5}}\right\}$

Solution:

$$a^{\frac{x+2}{3}} = a^{\frac{x-5}{2}}; \quad \frac{x+2}{3} = \frac{x-5}{2}; \quad 2x + 4 = 3x - 15; \quad x = 19,$$

or

$$S = \{19\}$$

2. $\{x \mid 2^{x+1} + 3^{x-3} = 3^{x-1} - 2^{x-2}\}$

Solution:

$$2^x \cdot 2 + 3^x \cdot 3^{-3} = 3^x \cdot 3^{-1} - 2^x \cdot 2^{-2}$$

$$216 \cdot 2^x + 27 \cdot 2^x = 36 \cdot 3^x - 4 \cdot 3^x; \quad 243 \cdot 2^x = 32 \cdot 3^x;$$

$$\left(\frac{2}{3}\right)^x = \frac{32}{243} = \frac{2^5}{3^5} = \left(\frac{2}{3}\right)^5; \quad \underline{\underline{x = 5}} \text{ or } S = \{5\}$$

3. $\{x \mid 4^{3x} \cdot 5^{2x-3} = 6^x\}$

Solution:

$$3x \lg 4 + (2x - 3) \lg 5 = x \lg 6$$

$$x(3 \lg 4 + 2 \lg 5 - \lg 6) = 3 \lg 5$$

$$\underline{\underline{x = \frac{3 \lg 5}{3 \lg 4 + 2 \lg 5 - \lg 6}}}$$

$$\text{or } S = \left\{\frac{3 \lg 5}{3 \lg 4 + 2 \lg 5 - \lg 6}\right\}$$

Many exponential equations cannot be reduced to algebraic equations and can be solved only by graphical methods.

2.1.3.2. Logarithmic equations

The variable also occurs in the antilogarithm. In general, solution is possible only by graphical methods. Only in special cases, will a reduction to an algebraic equation be successful.

Examples:

1. $\left\{x \mid 3 \ln (2x - 7) + 8 = \sqrt{\ln (2x - 7) + 20}\right\}$

Substitution $\quad y = \ln (2x - 7)$

$$3y + 8 = \sqrt{y + 20}$$

$$9y^2 + 47y + 44 = 0$$

From this one obtains two values for y. By substitution $y = \ln (2x - 7)$ we obtain $e^y = 2x - 7$ from which x can be calculated.

2. $\{x \mid \lg(x^2+1) = 2\lg(3-x)\}$

Solution:

$$\lg(x^2+1) = \lg(3-x)^2$$

$$10^{\lg(x^2+1)} = 10^{\lg(3-x)^2}$$

$$x^2+1 = (3-x)^2$$

$$x = \frac{4}{3} = 1\frac{1}{3} \quad \text{or} \quad S = \left\{1\frac{1}{3}\right\}$$

2.1.4. Approximation methods for determining the roots of an equation

2.1.4.1. Regula falsi (rule of false position) (linear interpolation)

If the equation $x^n + a_{n-1}x^{n-1} + \cdots + a_0 = f(x) = 0$ has a root between the values x_1 and x_2, that is, if $f(x_1)$ and $f(x_2)$ have opposite signs, we obtain an approximate value which is improved as compared with x_1 and x_2, namely

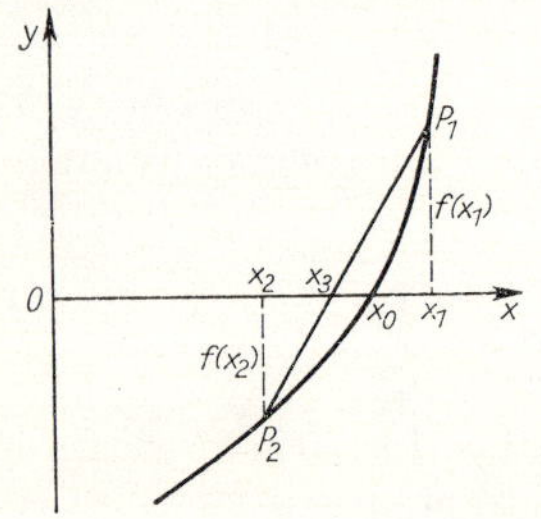

$$x_3 = x_1 - \frac{(x_2 - x_1)\,f(x_1)}{f(x_2) - f(x_1)}$$

Geometrical interpretation: The curve is replaced by the chord through P_1 and P_2 (chord approximation method)

Example:

$$\{x \mid x^3 - x + 7 = 0\}$$

Relevant functional equation

$$f(x) = x^3 - x + 7$$

We read off from the table of values:

$$x_1 = -2; \qquad f(x_1) = 1$$

$$x_2 = -2.5; \qquad f(x_2) = -6.125$$

$$x_3 = -2 + \frac{(-2.5+2)\cdot 1}{1 + 6.125} = -2 + \frac{-0.5}{7.127} = -2.07;$$

$$f(x_3) = 0.200$$

A repetition of the procedure leads to closer approximations,

2.1.4.2. Newton's method of approximation

If the approximate value x_1 of a root is known for the equation $x^n + a_{n-1}x^{n-1} + \cdots + a_0 = f(x) = 0$ a better approximation will be obtained from

$$x_2 = x_1 - \frac{f(x_1)}{f'(x_1)}, \quad f'(x_1) \neq 0$$

The procedure will fail if, at the point of approximation, the curve $f(x)$ takes a course almost parallel to the x-axis, or if an extreme point or a point of inflection with a tangent at the inflection point almost parallel to the x-axis lies between the approximate value and the precise root value.

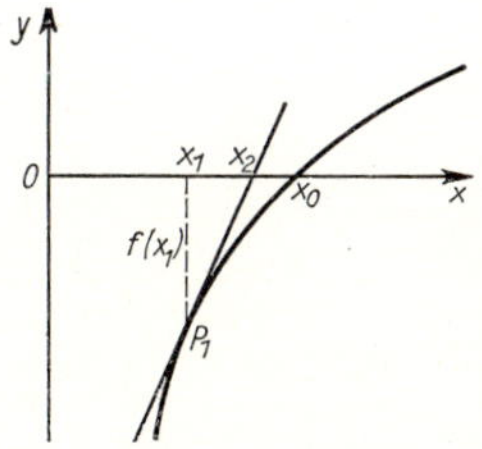

C r i t e r i o n for applicability: In the interval, which contains x_0 and all approximate values, the following must hold:

$$\left| \frac{f(x)f''(x)}{[f'(x)]^2} \right| \leqq m < 1$$

G e o m e t r i c a l interpretation: The curve is replaced by the tangent at the point of approximation P_1 on the curve (*tangential approximation method*).

Example:

$$\{x \mid x^3 + 2x - 1 = 0\}$$

Relevant functional equation $f(x) = x^3 + 2x - 1$.
The table of values gives as approximation $x_1 = 0.5$; $f(x_1) = 0.125$.
Derivatives $f'(x) = 3x^2 + 2$; $f''(x) = 6x$
$f'(x_1) = 2.75$; $f''(x_1) = 3$, hence

$$\left| \frac{f(x_1)f''(x_1)}{[f'(x_1)]^2} \right| \approx 0.05 < 1$$

$$x_2 = 0.5 - \frac{0.125}{2.75} \approx 0.5 - 0.045 \approx 0.455$$

$$f(x_2) \approx 0.004$$

The procedure can be repeated any desired number of times. Both methods can also be applied to nonalgebraic equations.

2.1.4.3. Iterative methods

We bring the equation $f(x) = 0$ to the form $x = \varphi(x)$. If x_1 is an approximate value of a root of an equation, and if the condition $|\varphi'(x_1)| \leqq m < 1$ is satisfied for this value x_1, then $x_2 = \varphi(x_1)$ is a better approximation. Several applications of this procedure will improve the accuracy. For $\varphi'(x_1) < 0$, two successive approximate values lie on different sides of the exact root value, hence, the achieved accuracy can be estimated.
If $|\varphi'(x_1)| > 1$, the inverse function must be introduced.

Example:

$$\{x \mid x^3 + 2x - 6 = 0\}$$

Approximative value $x_1 = 1.45$ for a root, $f(x_1) = -0.051375$

$$x = \frac{6 - x^3}{2} = \varphi(x)$$

$$\varphi'(x) = -\frac{3x^2}{2}; \quad |\varphi'(x_1)| = |\varphi'(1.45)| = 3.14375 > 1$$

Solving the equation for the second term of x:

$$x = \frac{6 - x^3}{2}; \quad x^3 = 6 - 2x; \quad x = \sqrt[3]{6 - 2x} = \psi(x)$$

$$\psi'(x) = \frac{-2}{3 \cdot \sqrt[3]{(6 - 2x)^2}}; \quad |\psi'(1.45)| = \left|\frac{-2}{3 \cdot \sqrt[3]{3.1^2}}\right| < 1$$

From this follows:

$$\underline{\underline{x_1 = 1.45}}$$

$$x_2 = \sqrt[3]{6 - 2 \cdot 1.45} = \underline{\underline{1.4581}}$$

$$x_3 = \sqrt[3]{6 - 2 \cdot 1.4581} = \underline{\underline{1.4556}}$$

etc.

2.1.4.4. Graphical solution of equations

The equation in normal form is transformed into a functional equation by putting the left-hand side equal to y. When this equation is represented graphically, the points of intersection of the curve obtained with the x-axis give the real solutions of this equation ($y = 0$).

Sometimes it is of advantage to write the equation to be solved in the form $\varphi(x) = \psi(x)$ and to represent the two sides as separate functions graphically. Then, the abscissae of the points of intersection of the graphical representation of the two equations correspond to the real roots of the conditional equation.

Example:

$$\{x \mid x^2 - x - 6 = 0\}$$

$$x^2 - x - 6 = y$$

$$\underline{\underline{x_1 = -2}}$$

$$\underline{\underline{x_2 = \quad 3}}$$

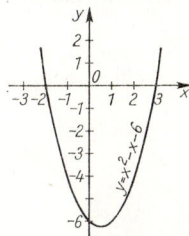

Example:

$$\{x \mid x^3 - 1.5x - 0.5 = 0\}$$

$$x^3 = 1.5x + 0.5$$

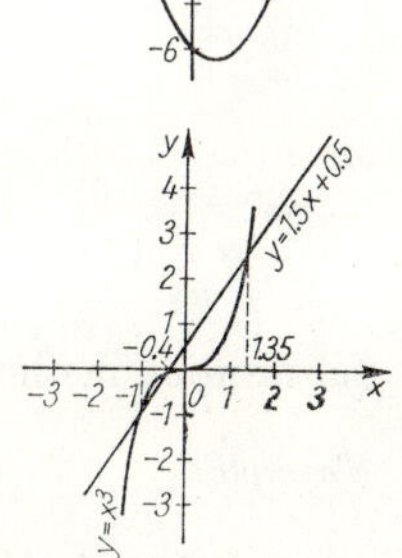

The points of intersection of the curves representing the functions

$$\varphi(x) = y = x^3 \quad \text{and}$$

$$\psi(x) = y = 1.5x + 0.5$$

yield, together with their abscissae, the solutions

$$\underline{\underline{x_1 = -1}}; \quad \underline{\underline{x_2 \approx -0.4}}; \quad \underline{\underline{x_3 \approx 1.35}}$$

2.1.5. Systems of equations

For the unique determination of n variables, we require equations that are independent of each other and simultaneously true. The general procedure for solution is by reducing the n equations with n variables step by step to one equation with one variable.

The value of this one variable calculated from this equation can then be used to determine, step by step, the other variables.
If fewer than n equations, or equations depending one upon another, are available, arbitrary values of the domain of definition are assigned to one or several variables as independent variables and the values pertinent to the variables determined.
The solution set S of a system of equations is equal to the intersection of the solution sets of the individual equations $S_1, S_2, \ldots$

$$S = S_1 \cap S_2 \cap \cdots$$

2.1.5.1. Linear equations in two variables

The solution set is the set of all ordered pairs of values $(x; y) \in R \times R$ for which the forms of statements become true statements.

Written as:

$$S = \{(x; y) \mid (x; y) \in R \times R \wedge a_{11}x + a_{12}y = b_1 \wedge a_{21}x + a_{22}y = b_2\}$$

or, if the domain of definition is obvious, in short

$$S = \{(x; y) \mid a_{11}x + a_{12}y = b_1 \wedge a_{21}x + a_{22}y = b_2\}$$

Substitution method

Example:

$$\{(x; y) \mid 3x + 7y - 7 = 0 \wedge 5x + 3y + 36 = 0\}$$

Solution:

$$\begin{array}{l|} 3x + 7y - 7 = 0 \\ 5x + 3y + 36 = 0 \\ \hline \end{array}$$

$$y = \frac{7 - 3x}{7}$$

$$5x + 3\left(\frac{7 - 3x}{7}\right) + 36 = 0$$

$$x = -10\,\frac{1}{2}; \quad y = \frac{7 + 31\,\frac{1}{2}}{7} = 5\,\frac{1}{2}$$

or

$$S = \left\{\left(-10\,\frac{1}{2};\; 5\,\frac{1}{2}\right)\right\}$$

Equating method

We calculate one of the variables as a function of the other one from both equations and equate the obtained expressions.

Example:

$$\{(x; y) \mid 3x + 7y - 7 = 0 \wedge 5x + 3y + 36 = 0\}$$

Solution:

$$\begin{array}{l|}3x + 7y - 7 = 0 \\ 5x + 3y + 36 = 0\end{array} \qquad \frac{-3x + 7}{7} = \frac{-5x - 36}{3}$$

$$\begin{array}{l|}7y = -3x + 7 \\ 3y = -5x - 36\end{array}$$

$$\begin{array}{l|}y = \dfrac{-3x + 7}{7} \\ y = \dfrac{-5x - 36}{3}\end{array} \qquad \begin{array}{l}x = -10\,\dfrac{1}{2} \\ y = 5\,\dfrac{1}{2}\end{array}$$

or

$$S = \left\{\left(-10\,\frac{1}{2};\; 5\,\frac{1}{2}\right)\right\}$$

Addition method

We multiply both sides of each equation by suitable factors so that one variable shows the same coefficient in both equations. By adding or subtracting the two equations we obtain the value of the remaining variable.

Example:

$$\{(x; y) \mid 3x + 7y - 7 = 0 \wedge 5x + 3y + 36 = 0\}$$

Solution:

$$\left.\begin{aligned} 3x + 7y - 7 &= 0 \\ 5x + 3y + 36 &= 0 \end{aligned}\right| \begin{aligned} &\cdot 3 \\ &\cdot 7 \end{aligned}$$

$$\left.\begin{aligned} 9x + 21y - 21 &= 0 \\ 35x + 21y + 252 &= 0 \end{aligned}\right| \begin{aligned} &- \\ &+ \end{aligned}$$

$$26x + 273 = 0$$

$$\underline{\underline{x = -10\,\frac{1}{2}}}; \quad \underline{\underline{y = 5\,\frac{1}{2}}}$$

or

$$S = \left\{\left(-10\,\frac{1}{2};\ 5\,\frac{1}{2}\right)\right\}$$

2.1.5.2. Linear equations in three variables

Substitution method

Example 1:

$$\left.\begin{aligned} 3x + 3y + z &= 17 \\ 3x + y + 3z &= 15 \\ x + 3y + 3z &= 13 \end{aligned}\right| \begin{aligned} &\text{(I)} \\ &\text{(II)} \\ &\text{(III)} \end{aligned}$$

From (I) $\qquad z = 17 - 3x - 3y$

substituted in (II) and (III)

$$\left.\begin{aligned} 3x + y + 3(17 - 3x - 3y) &= 15 \\ x + 3y + 3(17 - 3x - 3y) &= 13 \end{aligned}\right|$$

$$\left.\begin{aligned} -6x - 8y &= -36 \\ -8x - 6y &= -38 \end{aligned}\right|$$

$$\underline{\underline{x = 3\,\frac{1}{7}}}; \quad \underline{\underline{y = 2\,\frac{1}{7}}}; \quad \underline{\underline{z = 1\,\frac{1}{7}}}$$

or

$$S = \left\{\left(3\frac{1}{7};\ 2\frac{1}{7};\ 1\frac{1}{7}\right)\right\}$$

Equating method

Example 2:

$$\begin{array}{r|}
3x + 3y + z = 17 \\
3x + y + 3z = 15 \\
x + 3y + 3z = 13 \\ \hline
\end{array}$$

$$\begin{array}{l|}
z = 17 - 3x - 3y \\
z = \dfrac{15 - 3x - y}{3} \\
z = \dfrac{13 - x - 3y}{3} \\ \hline
\end{array}$$

$$\begin{array}{l|}
17 - 3x - 3y = \dfrac{15 - 3x - y}{3} \\
\dfrac{15 - 3x - y}{3} = \dfrac{13 - x - 3y}{3} \\ \hline
\end{array}$$

For the result see example 1.

Addition method

Example 3:

$$\begin{array}{r|c|c}
3x + 3y + z = 17 & \cdot 3 + & \\
3x + y + 3z = 15 & - & + \\
x + 3y + 3z = 13 & & - \\ \hline
\end{array}$$

$$\begin{array}{r|}
6x + 8y = 36 \\
2x - 2y = 2 \\ \hline
\end{array}$$

For the result see example 1.

2.1.5.3. Gaussian algorithm

System of equations of n linear equations with n variables

$$\begin{aligned} a_{11}x_1 + a_{12}x_2 + \cdots + a_{1n}x_n &= c_1 \\ a_{21}x_1 + a_{22}x_2 + \cdots + a_{2n}x_n &= c_2 \\ a_{31}x_1 + a_{32}x_2 + \cdots + a_{3n}x_n &= c_3 \\ \vdots \qquad \vdots \qquad\qquad \vdots \quad &\quad \vdots \\ a_{n1}x_1 + a_{n2}x_2 + \cdots + a_{nn}x_n &= c_n \end{aligned} \tag{1}$$

$a_{11} \neq 0$; coefficient determinant $\Delta \neq 0$

We multiply the first equation of (1) by $-\frac{a_{21}}{a_{11}}$ and add to the second equation. If we then multiply the first equation of (1) by $-\frac{a_{31}}{a_{11}}$ and add to the third equation, etc., all terms containing x_1 are eliminated, and the original system of equations (1) is reduced to a system of $n - 1$ equations with $n - 1$ variables.

$$\begin{aligned} a'_{22}x_2 + a'_{23}x_3 + \cdots + a'_{2n}x_n &= c_2' \\ a'_{32}x_2 + a'_{33}x_3 + \cdots + a'_{3n}x_n &= c_3' \\ \vdots \qquad \vdots \qquad\qquad \vdots \quad &\quad \vdots \\ a'_{n2}x_2 + a'_{n3}x_3 + \cdots + a'_{nn}x_n &= c_n' \end{aligned} \tag{2}$$

If we proceed in a similar way with respect to system (2), we arrive at a system of $n - 2$ equations with $n - 2$ variables, etc. Finally, system (1) yields the following so-called staggered system of equations

$$\begin{aligned} a_{11}x_1 + a_{12}x_2 + a_{13}x_3 + \cdots + a_{1n}x_n &= c_1 \\ a'_{22}x_2 + a'_{23}x_3 + \cdots + a'_{2n}x_n &= c_2' \\ a''_{33}x_3 + \cdots + a''_{3n}x_n &= c_3'' \\ \vdots \quad &\quad \vdots \\ a_{nn}^{(n-1)}x_n &= c_n^{(n-1)} \end{aligned} \tag{3}$$

from which, starting with x_n from the last equation, the variables can be calculated one after the other.

Example:

$$\left.\begin{array}{rl} x + 2y - 0.7z = 21 \\ 3x + 0.2y - z = 24 \\ 0.9x + 7y - 2z = 27 \end{array}\right| \cdot(-3) \quad \Big| \cdot(-0.9)$$

$$\left.\begin{array}{rl} x + 2y - 0.7\,z = 21 \\ -5.8y + 1.1\,z = -39 \\ 5.2y - 1.37z = 8.1 \end{array}\right| \cdot\frac{5.2}{5.8}$$

$$\left.\begin{array}{l} x + 2y - 0.7z = 21 \\ -5.8y + 1.1z = -39 \\ -\frac{11.13}{29}z = -\frac{779.1}{29} \end{array}\right|$$

$\underline{\underline{z = 70}}$; $\underline{\underline{y = 20}}$; $\underline{\underline{x = 30}}$, that is, $S = \{30; 20; 70\}$

2.1.5.4. Quadratic equations in two variables

A system of equations in the form

$$\left.\begin{array}{l} a_1x^2 + b_1xy + c_1y^2 + d_1x + e_1y + f_1 = 0 \\ a_2x^2 + b_2xy + c_2y^2 + d_2x + e_2y + f_2 = 0 \end{array}\right|$$

can be solved by the substitution method; however, this procedure is rather troublesome because it leads to an equation of the fourth degree.

Special cases

(a) *One equation is quadratic, one linear*

The substitution method is useful.

Example:

$$\left.\begin{array}{l} x^2 + y^2 = 25 \\ x - y = 4 \end{array}\right|$$

$y = x - 4$ substituted in the first equation leads to $x^2 + (x - 4)^2 = 25$ (quadratic equation).

(b) *Pure quadratic equations* (without any term with xy)
The addition method is useful.

Example:

$$\begin{array}{l|c|c} 9x^2 - 2y^2 = 18 & \cdot 3 & + \\ 5x^2 + 3y^2 = 47 & \cdot 2 & + \end{array}$$

$$37x^2 = 148$$

$$x^2 = 4$$

$$x_1 = 2; \quad x_2 = 2; \quad x_3 = -2; \quad x_4 = -2$$

$$y_1 = 3; \quad y_2 = -3; \quad y_3 = 3; \quad y_4 = -3$$

or

$$S = \{(2; 3); \quad (2; -3); \quad (-2; 3); \quad (-2; -3)\}$$

(c) *Equations where only xy occur as a quadratic term*
Both the addition method and the substitution method are useful.

Example:

$$\begin{array}{l|l} 5x + y + 3 = 2xy & \text{(I)} \\ xy = 2x - y + 9 & \text{(II)} \end{array}$$

$$\begin{array}{r|c|c} -2xy + 5x + y + 3 = 0 & & + \\ xy - 2x + y - 9 = 0 & \cdot 2 & + \end{array}$$

$$x + 3y - 15 = 0$$

$$x = 15 - 3y$$

substituted in equation (II) yields

$$y(15 - 3y) = 2(15 - 3y) - y + 9$$

a quadratic equation for y.

(d) *Two homogeneous quadratic equations*
Equations are called *homogeneous* if their left-hand sides are homogeneous functions of the variable. A quadratic function in x and y is called homogeneous if it contains only terms of the same degree, that is,

$$f(x, y) = ax^2 + bxy + cy^2$$

The substitution $y = xz$ leads to a quadratic equation for z.

Example:

$$\left.\begin{aligned} x^2 - xy + y^2 &= 39 \\ 2x^2 - 3xy + 2y^2 &= 43 \end{aligned}\right| \quad y = xz$$

$$\left.\begin{aligned} x^2 - x^2z + x^2z^2 &= 39 \\ 2x^2 - 3x^2z + 2x^2z^2 &= 43 \end{aligned}\right|$$

$$\left.\begin{aligned} x^2(1 - z + z^2) &= 39 \\ x^2(2 - 3z + 2z^2) &= 43 \end{aligned}\right|$$

$$\frac{1 - z + z^2}{2 - 3z + 2z^2} = \frac{39}{43} \quad \text{(quadratic equation for } z\text{)}$$

$$z_1 = \frac{7}{5}; \quad z_2 = \frac{5}{7}$$

By substitution, we obtain quadratic equations for x_1 or x_2 from this.

2.1.5.5. Graphical solution of systems of equations in two variables

Systems of equations of the first degree

Each equation of the first degree in x and y can be considered a functional equation and, when graphically represented, yields a straight line. These lines intersect in a point whose coordinates represent the real solutions of the system of equations.

Example:

$$\left.\begin{aligned} x + 3y &= 3 \\ x - 3y &= 9 \end{aligned}\right|$$

$$y = -\frac{1}{3}x + 1$$

$$y = \frac{1}{3}x - 3$$

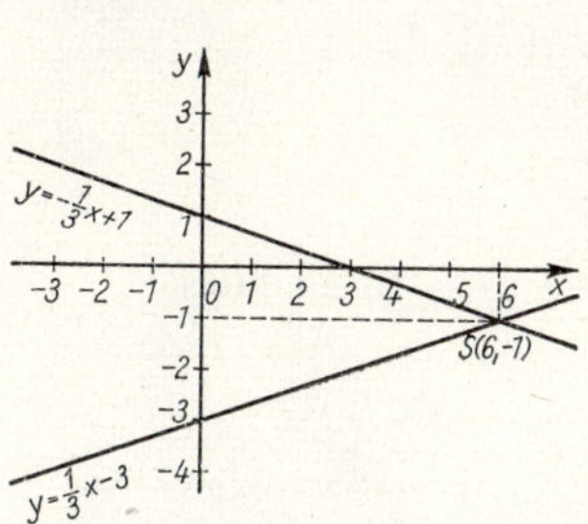

Solution:

$$x_s = x = 6; \quad y_s = y = -1 \quad \text{or} \quad S = \{(6; -1)\}$$

Note: If the two lines are parallel to each other or if they coincide, the equations of the system contradict each other or are no longer independent of each other.

Systems of equations of the second degree

An equation of the second degree in x and y, taken as a functional equation, yields a conic section (parabola, ellipse, hyperbola) in a graphical representation.
If the second equation of the system is linear, it is represented by a straight line intersecting the conic section; the coordinates of these points of intersection give the required solution of the system.
If both equations are quadratic, their graphical representation yields two conic sections. Then the coordinates of the points of intersection, of which there may be two or four, are the real pairs of roots of the system of equations.

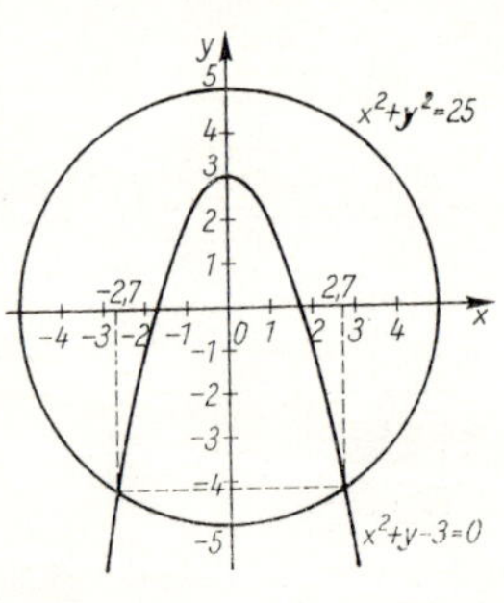

Example:

$$\begin{array}{rcl|c} x^2 + y^2 &=& 25 & \text{(I)} \\ x^2 + y &=& 3 & \text{(II)} \end{array}$$

Equation (I) represents a circle with radius $r = 5$ described about the origin as center; equation (II) represents a parabola with the vertex $(0; 3)$.

The coordinates of the points of intersection give the following pairs of roots:

$$\underline{\underline{x_1 \approx 2.7}}; \quad \underline{\underline{y_1 \approx -4.2}}; \quad \underline{\underline{x_2 \approx -2.7}}; \quad \underline{\underline{y_2 \approx -4.2}}$$

or

$$S = \{(2.7; -4.2); (-2.7; -4.2)\}$$

2.2. Inequalities

Definition

If mathematical expressions are connected by one of the ordering symbols $<$, $\leq$, $>$, $\geq$, the resulting proposition is referred to as an *inequality*.
If expressions without variables are in this way connected with

each other, an inequality represents a true or false statement, e.g. $7 < 5$ is a false statement.
An inequality with at least one variable represents a form of statement which is transformed into a true or false statement if the variables are assigned values of the domain of definition of the variables.
The solution set is the set of all values for which the inequality represents a true statement.

Equivalence:

$$T_1 < T_2 \Rightarrow T_1 + T < T_2 + T \qquad T \text{ meaningful expression in the domain of variables}$$

$$T_1 < T_2 \Rightarrow T_1T < T_2T \qquad T > 0$$

Calculating with inequalities

$$T_1 < T_2 \Rightarrow T_1T > T_2T \qquad T < 0$$

$$0 < T_1 < T_2 \Rightarrow \frac{1}{T_1} > \frac{1}{T_2}$$

$$T_1 < T_2 \quad \text{and} \quad T_3 < T_4 \Rightarrow T_1 + T_3 < T_2 + T_4$$

$$T_1 < T_2 \quad \text{and} \quad T_3 < T_4 \Rightarrow T_1T_3 < T_2T_4 \qquad T_2, T_3 > 0$$

$$T_1 < T_2 \Rightarrow -T_1 > -T_2$$

$$T_1^n + T_2^n \leqq (T_1 + T_2)^n \qquad T_1, T_2 > 0; \quad n \neq 0; \quad n \in N$$

$$(1 + T)^n \geqq 1 + nT \qquad T \geqq 0, \quad n \in N$$

$$2^x > x \qquad x \in N$$

$$\sqrt[n]{1 + T} < 1 + \frac{T}{n} \qquad T > 0, \qquad n \in N \quad \{0; 1\}$$

2.3. Functions

2.3.1. General

Definition

If the mapping of the set X onto the set Y is unique, that is, if exactly one $y \in Y$ is made to correspond to each $x \in X$, the mapping is called a *function*.

It represents the set f of ordered pairs (x, y), which is a subset of the Cartesian product $R \times R$.

Set X = *domain of definition, original set,* set of the *argument values*

Set Y = *range of values, image set,* set of the *function values*

x independent variable

y dependent variable

If both the originals ($x \in R$) and the images ($y \in R$) are real numbers, we speak of a *real function.*

Written as:

$$y = f(x) \quad \text{for the mapping} \quad x \to y$$

Representations

Analytical representation in *explicit* form	$y = f(x)$
in *implicit* form	$f(x, y) = 0$
in *parametric* form	$x = \varphi(t)$
	$y = \psi(t)$

Exactest way of writing

$$f = \{(x, y) \mid y = f(x), \quad x \in X, \quad y \in Y\}$$

or

$$f = \{[x, f(x)] \mid x \in X, y \in Y\}$$

Tabulated representation
Graphical representation

Classification of functions

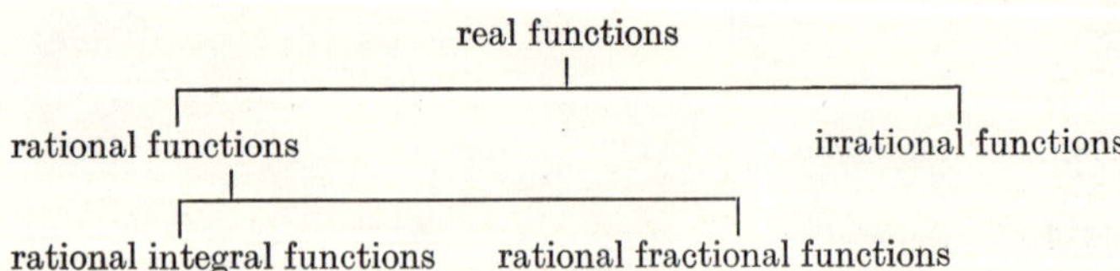

Rational integral function of the nth degree

$$y = a_n x^n + a_{n-1} x^{n-1} + \cdots + a_1 x + a_0$$

for $n \in N$; $a_k \in R$; $a \neq 0$

Rational fractional function

$$y = \frac{a_n x^n + a_{n-1} x^{n-1} + \cdots + a_1 x + a_0}{b_m x^m + b_{m-1} x^{m-1} + \cdots + b_1 x + b_0}$$

for $\quad n, m \in N, \quad a_k, b_k \in R, \quad a_n, b_m \neq 0$

Proper fractional function for $n < m$
Improper fractional function for $n \geqq m$
Linear fractional function for $n = m = 1$

Algebraic function

$$a_n y^n + a_{n-1} y^{n-1} + \cdots + a_1 y + a_0 = 0$$

for $n \in N$, $a_n \neq 0$, a_k rational integral functions of x of the kth degree

Transcendental function

A nonalgebraic function is called transcendental.

Identical functions

If the domain of definition and range of values of two functions are in agreement and if, by both functions, each $x \in X$ is mapped onto the same function value $y \in Y$, the two functions are identical:

$$f(x) \equiv g(x)$$

However, $f(x) = g(x)$ means agreement only with respect to certain argument and function values.

Even and odd functions

Even function: $\quad f(-x) = f(x) \qquad x \in X$

Odd function: $\quad f(-x) = -f(x) \qquad x \in X$

Monotonic functions

A function $y = f(x)$ is referred to as *monotonically increasing* if within the domain of definition $f(x_1) \leqq f(x_2)$ holds for $x_1 < x_2$. It is called *monotonically decreasing* if within the domain of definition $f(x_1) \geqq f(x_2)$ holds for $x_1 < x_2$.

If, in the whole domain of definition, $f(x_1) < f(x_2)$ or $f(x_1) > f(x_2)$ holds for $x_1 < x_2$, the function is called *strictly monotonically increasing* and *strictly monotonically decreasing*, respectively.

Homogeneous functions

A function

$$f(x_1, x_2, x_3, \ldots, x_n)$$

is called *homogeneous* if it satisfies the condition

$$f(tx_1, tx_2, tx_3, \ldots, tx_n) = t^k f(x_1, x_2, x_3, \ldots, x_n)$$

k degree of homogeneity

Zeros, poles, gaps

Zero x_0 of a function is the value of the independent variable x where the function value disappears. (In the case of rational fractional functions, numerator $= 0$ or denominator $\xrightarrow[x \to x_0]{} \infty$.)
Pole x_p of a function is the value of the independent variable x where the function value $y \to \infty$.
(In the case of rational fractional functions, denominator $= 0$ or numerator $\xrightarrow[x \to x_p]{} \infty$.)
Gap of a rational fractional function is the value of the independent variable x where numerator and denominator disappear.

Continuous functions

The function $y = f(x)$ is said to be continuous at point a if it is defined in an arbitrary neighborhood of a and if for each $\varepsilon > 0$ there is a $\delta = \delta(a, \varepsilon)$ so that, in the given neighborhood of a, for all x for which $|x - a| < \delta$, the inequality $|f(x) - f(a)| < \varepsilon$ holds.
A function $y = f(x)$ defined in the interval $[a, b]$ is said to be *uniformly continuous* if, for each $\varepsilon > 0$ there is a number $\delta = \delta(\varepsilon) > 0$ independent of x so that, for all x and $x + h$ in the interval $[a, b]$, the inequality $|f(x + h) - f(x)| < \varepsilon$ holds if $|h| < \delta = \delta(\varepsilon)$.
If the functions $f(x)$ and $g(x)$ are uniformly continuous in the interval $[a, b]$, then the functions $f(x) \pm g(x)$, $f(x) \cdot g(x)$ and, if the inequality $g(x) \neq 0$ holds in $[a, b]$, $\dfrac{f(x)}{g(x)}$ are also uniformly continuous in $[a, b]$.

All points for which the conditions of continuity are not satisfied, are referred to as *discontinuities*.

Examples:

$y = \frac{1}{x}$ is discontinuous at point $x = 0$ (pole)

$y = \frac{3}{1 - e^{\frac{1}{x}}}$ is discontinuous at point $x = 0$ (the function jumps from one value to another one)

$y = \frac{x^3 - 1}{x^2 - 1}$ is discontinuous at points $x = \pm 1$ (gap at $x = +1$ and pole at $x = -1$)

If $\lim_{x \to a} f(x)$ exists, the discontinuities can be reduced; the limit $\lim_{x \to a} f(x)$ is assigned to the point $x = a$ at which the function is not defined.

Periodic functions

$$f(x) = f(x + kT_0) \qquad T_0 \text{ period}$$

$$k \in G$$

Inverse functions

If we map the images onto their primitive elements, we obtain the inverted mapping of the original mapping.
In case this mapping is unique, the function has an *inverse function*, that is, exactly one image is mapped onto each original and one original onto each image.
We write: $f^{-1}(x)$ is called the inverse function of $f(x)$. We obtain the inverse function of $y = f(x)$ by interchanging the letters x and y and solving the new equation also for y, if possible.

Example:

$$y = \frac{x}{1 + x} = f(x); \qquad \text{interchange: } x = \frac{y}{1 + y}$$

$$y = f^{-1}(x) = \frac{x}{1 - x}$$

Each strictly monotonic function has an inverse function.

2.3.2. Further methods of analytic representation

Horner's array

It is useful for rapidly calculating a function value of a rational integral function.

$$y = a_n x^n + a_{n-1} x^{n-1} + \cdots + a_0 \quad \text{for} \quad x = x_1$$

$$\begin{array}{lllll}
a_n & a_{n-1} & a_{n-2} & a_{n-3} & \cdots a_0 \\
0 & a_n x_1 & a_n x_1^2 + a_{n-1} x_1 & a_n x_1^3 + a_{n-1} x_1^2 + a_{n-2} x_1 & \\
\hline
a_n & a_n x_1 + a_{n-1} & a_n x_1^2 + a_{n-1} x_1 + a_{n-2} & & \\
 & & & a_n x_1^3 + a_{n-1} x_1^2 + a_{n-2} x_1 + a_{n-3} & \\
\cdots & & a_0 & & \\
\cdots & & a_n x_1^n + a_{n-1} x_1^{n-1} + a_{n-2} x_1^{n-2} + \cdots & & \\
\hline
\end{array}$$

$$a_n x_1^n + a_{n-1} x_1^{n-1} + a_{n-2} x_1^{n-2} + \cdots + a_0 = f(x_1)$$

First line: Coefficients of the function. It should be noted that, if one term $a_k x^{n-k}$ is missing, zero must be put in the respective place in the first line of Horner's array.

Step 1: multiply a_n by x_1 and add to a_{n-1}

Step 2: multiply the sum again by x_1 and add to a_{n-2}

Step 3: multiply the sum again by x_1 and add to a_{n-3}

etc.

The last sum is the function value to be found.

Example:

$$f(x) = x^4 + 2x^3 - 5x + 7$$

at the point $x_1 = 2.3$

to be calculated.

1	2	0	−5	7	(coefficient line)
0	2.3	9.89	22.747	40.8181	
1	4.3	9.89	17.747	47.8181	$= f(2.3)$

$$f(2.3) \approx 47.82$$

Approximate representation of functions by interpolation formulas

Lagrange's formuia

$$y = f(x) = y_1 \frac{(x - x_2)(x - x_3)(x - x_4) \cdots (x - x_n)}{(x_1 - x_2)(x_1 - x_3)(x_1 - x_4) \cdots (x_1 - x_n)}$$

$$+ y_2 \frac{(x - x_1)(x - x_3)(x - x_4) \cdots (x - x_n)}{(x_2 - x_1)(x_2 - x_3)(x_2 - x_4) \cdots (x_2 - x_n)}$$

$$\vdots$$

$$+ y_n \frac{(x - x_1)(x - x_2)(x - x_3) \cdots (x - x_{n-1})}{(x_n - x_1)(x_n - x_2)(x_n - x_3)(x_n \cdots - x_{n-1})}$$

where $y_1, y_2, y_3, \ldots, y_n$ are the function values belonging to $x_1, x_2, x_3, \ldots, x_n$ and which are known or given.

Example:

Find the function in analytical representation for the following table of values.

x	1	4	6	9
y	2	5	3	6

$$y = 2 \cdot \frac{(x - 4)(x - 6)(x - 9)}{(1 - 4)(1 - 6)(1 - 9)} + 5 \cdot \frac{(x - 1)(x - 6)(x - 9)}{(4 - 1)(4 - 6)(4 - 9)}$$

$$+ 3 \cdot \frac{(x - 1)(x - 4)(x - 9)}{(6 - 1)(6 - 4)(6 - 9)} + 6 \cdot \frac{(x - 1)(x - 4)(x - 6)}{(9 - 1)(9 - 4)(9 - 6)}$$

$$y = -\frac{1}{60}(x^3 - 19x^2 + 114x - 216)$$

$$+ \frac{1}{6}(x^3 - 16x^2 + 69x - 54)$$

$$- \frac{1}{10}(x^3 - 14x^2 + 49x - 36)$$

$$+ \frac{1}{20}(x^3 - 11x^2 + 34x - 24)$$

$$\underline{\underline{y = \frac{1}{10}x^3 - \frac{3}{2}x^2 + \frac{32}{5}x - 3}}$$

Newton's formula

$$y = y_1 + A_1(x - x_1) + A_2(x - x_1)(x - x_2)$$
$$+ A_3(x - x_1)(x - x_2)(x - x_3) + \cdots$$
$$+ A_{n-1}(x - x_1)(x - x_2)(x - x_3) \cdots (x - x_{n-1})$$

where

$$A_1 = \frac{y_2 - y_1}{x_2 - x_1}; \quad A_2 = \frac{(y_3 - y_1) - A_1(x_3 - x_1)}{(x_3 - x_1)(x_3 - x_2)};$$

$$A_3 = \frac{(y_4 - y_1) - A_1(x_4 - x_1) - A_2(x_4 - x_1)(x_4 - x_2)}{(x_4 - x_1)(x_4 - x_2)(x_4 - x_3)}$$

etc.

This formula becomes much simpler if the x values are equally spaced, that is, if

$$x_2 - x_1 = x_3 - x_2 = x_4 - x_3 = \cdots = x_n - x_{n-1} = h$$

By putting

$$y_2 - y_1 = \Delta y_1,$$
$$y_3 - y_2 = \Delta y_2,$$
$$y_4 - y_3 = \Delta y_3, \ldots$$
$$\Delta y_2 - \Delta y_1 = \Delta^2 y_1,$$
$$\Delta y_3 - \Delta y_2 = \Delta^2 y_2,$$
$$\Delta y_4 - \Delta y_3 = \Delta^2 y_3, \ldots$$
$$\Delta^2 y_2 - \Delta^2 y_1 = \Delta^3 y_1,$$
$$\Delta^2 y_3 - \Delta^2 y_2 = \Delta^3 y_2,$$
$$\Delta^2 y_4 - \Delta^2 y_3 = \Delta^3 y_3, \ldots$$
$$\cdot \quad \cdot \quad \cdot \quad \cdot \quad \cdot \quad \cdot \quad \cdot \quad \cdot \quad \cdot \quad \cdot \quad \cdot \quad \cdot$$
$$\Delta^{n-2} y_2 - \Delta^{n-2} y_1 = \Delta^{n-1} y_1,$$
$$\Delta^{n-2} y_3 - \Delta^{n-2} y_2 = \Delta^{n-1} y_2,$$
$$\Delta^{n-2} y_4 - \Delta^{n-2} y_3 = \Delta^{n-1} y_3, \ldots,$$

we obtain

$$y = y_1 + \frac{\Delta y_1(x - x_1)}{1!\,h} + \frac{\Delta^2 y_1(x - x_1)(x - x_2)}{2!\,h^2}$$

$$+ \frac{\Delta^3 y_1(x - x_1)(x - x_2)(x - x_3)}{3!\,h^3} + \cdots$$

$$+ \frac{\Delta^{n-1} y_1 (x - x_1)(x - x_2)(x - x_3) \cdots (x - x_{n-1})}{(n-1)!\,h^{n-1}}$$

Example:

Find the function in analytical representation for the following table of values.

x	2	3	4	5	6
y	3	5	4	2	7

Step 1: determination of the differences

$$\Delta y_1 = 5 - 3 = 2$$
$$\Delta y_2 = 4 - 5 = -1$$
$$\Delta y_3 = 2 - 4 = -2$$
$$\Delta y_4 = 7 - 2 = 5$$

$$\Delta^2 y_1 = -1 - 2 = -3$$
$$\Delta^2 y_2 = -2 + 1 = -1$$
$$\Delta^2 y_3 = 5 + 2 = 7$$

$$\Delta^3 y_1 = -1 + 3 = 2$$
$$\Delta^3 y_2 = 7 + 1 = 8$$

$$\Delta^4 y_1 = 8 - 2 = 6$$

Step 2: putting in the formula

$$y = 3 + \frac{2(x-2)}{1! \cdot 1} + \frac{-3(x-2)(x-3)}{2! \cdot 1^2}$$

$$+ \frac{2(x-2)(x-3)(x-4)}{3! \cdot 1^3}$$

$$+ \frac{6(x-2)(x-3)(x-4)(x-5)}{4! \cdot 1^4}$$

$$\underline{\underline{y = \frac{1}{4}x^4 - \frac{19}{6}x^3 + \frac{53}{4}x^2 - \frac{61}{3}x + 12}}$$

2.3.3. Graphical representation of functions

Algebraic functions

Rational integral function of the first degree (linear function)

General form: $y = a_1x + a_0$

Form of curve: straight line

$a_1 = \tan\alpha$ *direction factor* where α is the angle which the straight line makes in a rectangular coordinate system with the positive x-axis

$a_1 > 0$ increasing straight line

$a_1 < 0$ decreasing straight line

a_0 segment on the y-axis

$a_0 = 0$ straight line through the zero point

Example:

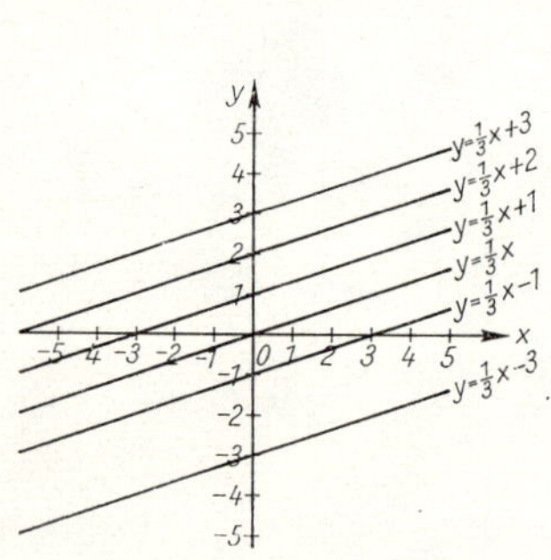

$a_1 =$ const., parallel straight lines

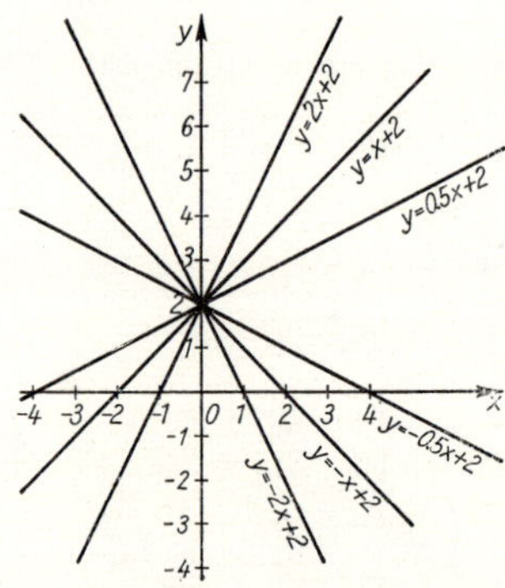

$a_0 =$ const., straight lines through the same point on the y-axis

Rational integral function of the second degree (quadratic function)

General form: $y = a_2x^2 + a_1x + a_0$

Form of curve: quadratic parabola, axis parallel to ordinate axis

$a_2 > 0$ opening above

$a_2 < 0$ opening below

Parabola vertex $V\left(\frac{-a_1}{2a_2},\ -\frac{a_1^2}{4a_2} + a_0\right)$

Normal form: $y = x^2 + px + q$ with

$$V\left(-\frac{p}{2},\ -\left[\left(\frac{p}{2}\right)^2 - q\right]\right)$$

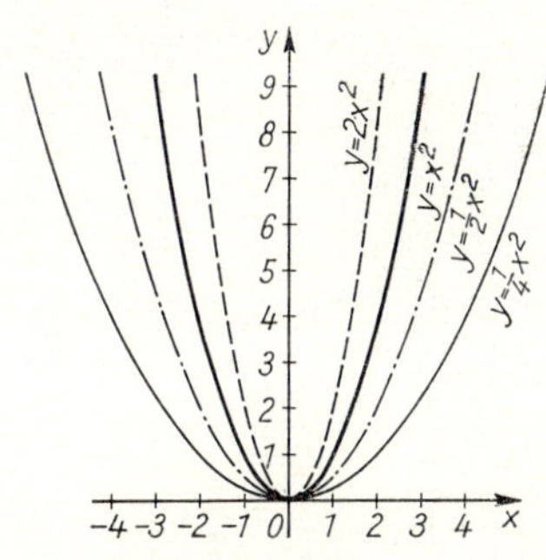

Form of curve: normal parabola, axis parallel to ordinate axis

Special cases

$y = a_2x^2$	parabola with vertex at origin
$a_2 = 1$	*normal parabola*
$\lvert a_2\rvert < 1$	wider parabola
$\lvert a_2\rvert > 1$	narrower parabola

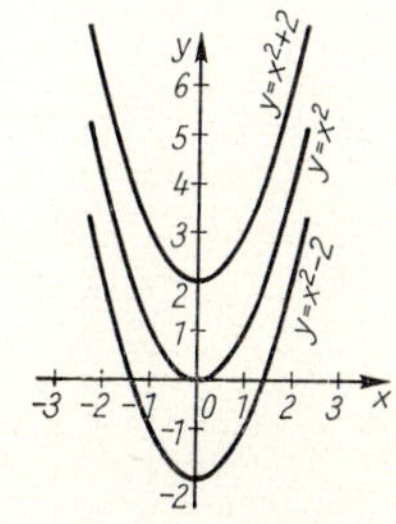

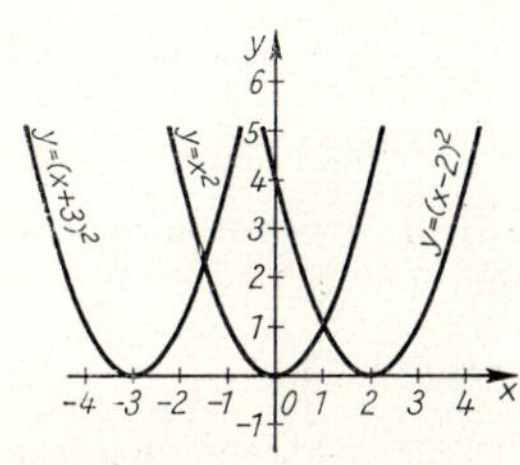

$y = x^2 + a_0$	normal parabola with vertex $V(0, a_0)$
$y = (x + b)^2$	normal parabola with vertex $V(-b, 0)$

Rational integral function of the third degree

$$y = a_3x^3 + a_2x^2 + a_1x + a_0$$

$y = x^3 - 6x^2 + 11x - 5$

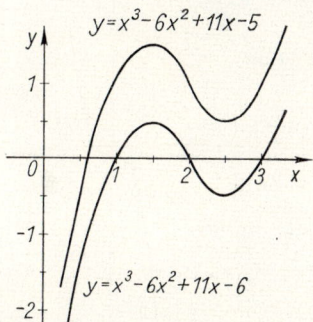

$y = x^3 - 6x^2 + 11x - 6$

Form of curve:

cubic parabola

$a_3 > 0$	The parabola extends from the lower half-plane to the upper half-plane
$a_3 < 0$	inverse behavior
a_0	displaces the curve in the direction of the ordinate

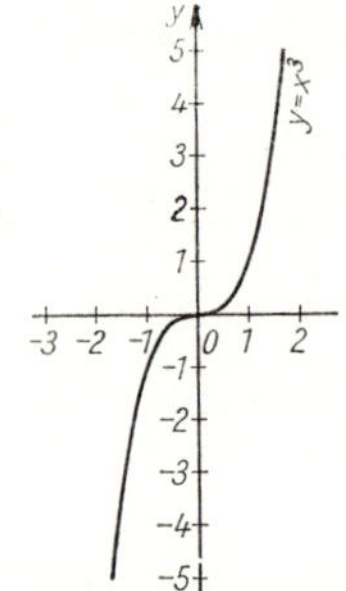

Special case:

$y = x^3$ *cubic normal parabola*

Even power functions

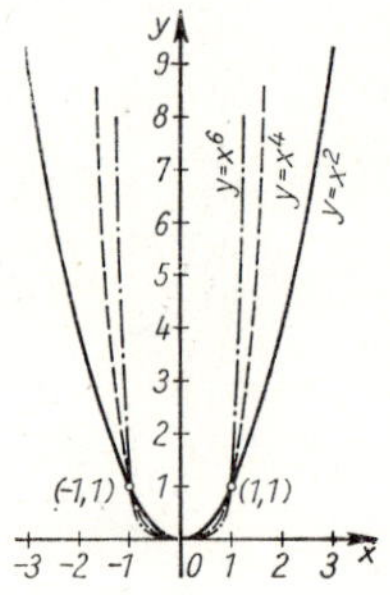

$y = x^{2n}$ for $n \in N$

Form of curve:

Family of curves of normal parabolas of the second, fourth, ... degree

Vertex $V(0, 0)$,

$y = -x^{2n}$ yields the same parabolas opening below.

Odd power functions

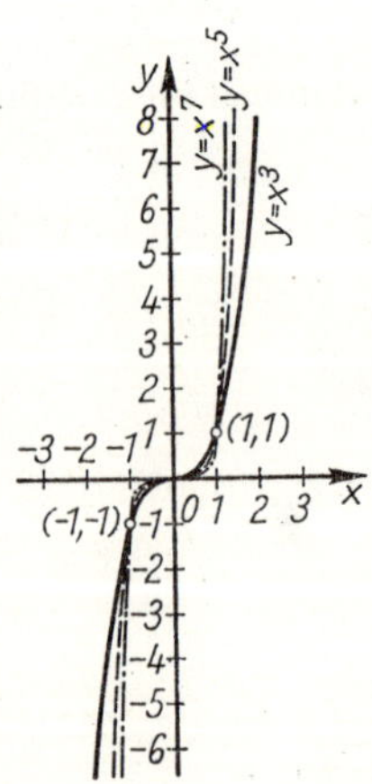

$y = x^{2n+1}$ for $n \in N$

Form of curve:

Family of curves of normal parabolas of the third, fifth, ... degree in the first and third quadrants; origin as center of symmetry $y = -x^{2n+1}$ yields the respective family of curves in the second and fourth quadrants which is obtained by reflection from the x-axis.

Even power function with negative exponent

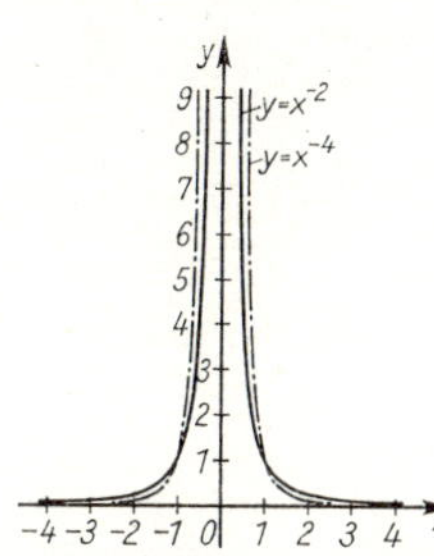

$$y = x^{-2n} \quad \text{for} \quad n \in N$$

Form of curve:

Family of curves with hyperbolas in the first and second quadrants $y = -x^{-2n}$ yields the corresponding family of curves in the third and fourth quadrants which are obtained by reflection from the x-axis.

Odd power functions with negative exponent

$$y = x^{-(2n+1)} \quad \text{for} \quad n \in N$$

Form of curve:

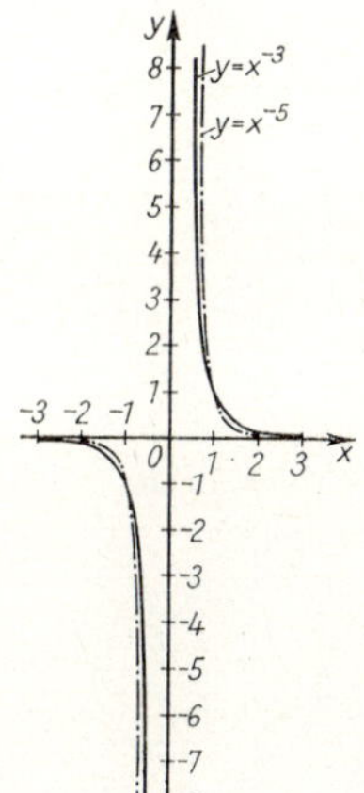

Family of curves of hyperbolas in the first and third quadrants, origin as center of symmetry

Special case: $n = 0$; $y = x^{-1} = \dfrac{1}{x}$
equilateral hyperbola

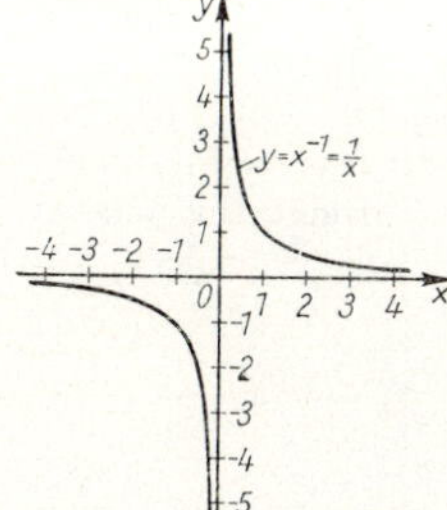

$y = -x^{-(2n+1)}$ yields the corresponding curves in the second and fourth quadrants which are obtained by reflection in the x-axis.

Inverse function

The image of the inverse function is found by reversing the image of the original function (original curve) with reference to the straight line $y = x$.

Root functions

$y = \sqrt[n]{x}$ is the inverse function of $y = x^n$

$n \in N \setminus \{0; 1\}$

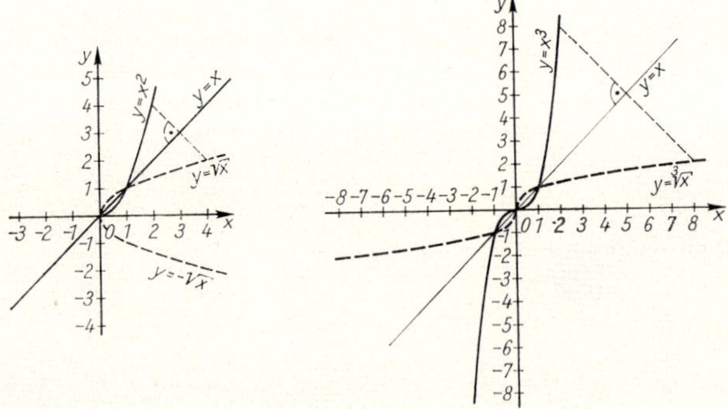

For periodic functions see Section 9.

Transcendental functions

Exponential functions

$y = a^x$ yields exponential curves for positive values of a which always pass through the point (0, 1).

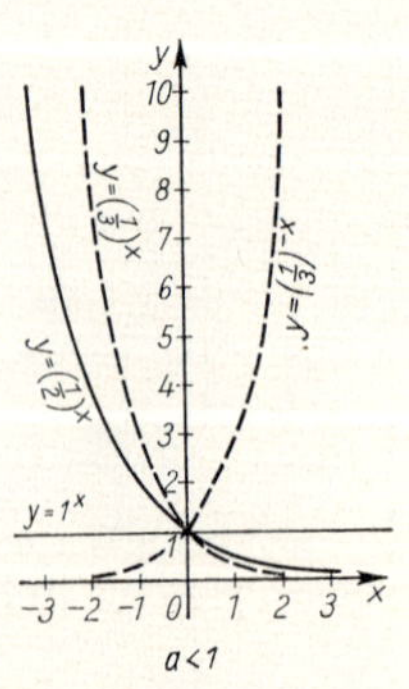

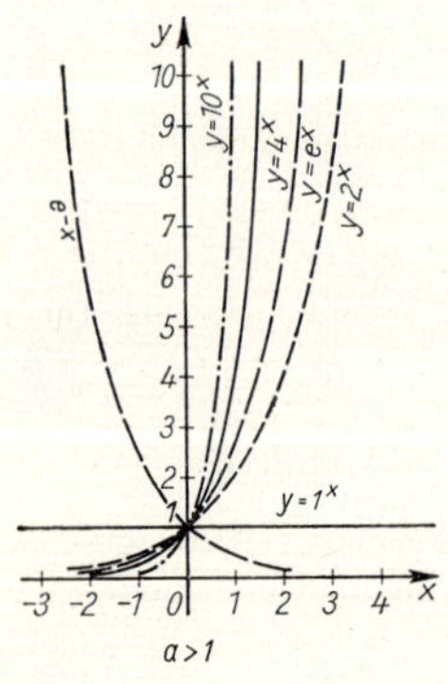

Logarithmic functions

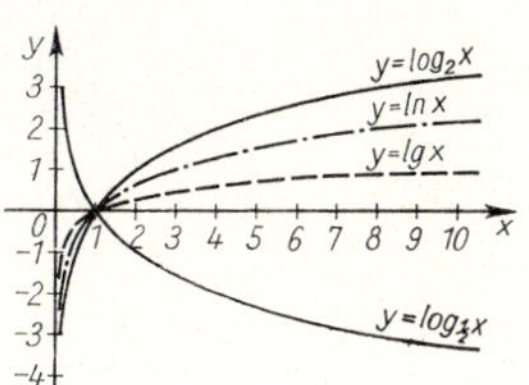

$y = \log_a x$ yields logarithmic curves for $x > 0$; $a > 0$; $a \neq 1$ which always pass through the point (1, 0). The logarithmic functions are the inverse of the corresponding exponential functions.
For trigonometric functions, antitrigonometric functions, hyperbolic functions and their inverse see the respective Sections.

2.4. Vector calculus

2.4.1. General

The definition of the vector as a single-rowed matrix, given on page 66, admits of a geometric interpretation which is used especially in physics.
The essential properties of a vector are a numerical value or magnitude (a length), a direction and a directional sense. It is graphically represented by an arrow whose direction and sense correspond to those of the vector and whose length is proportional to a numerical value.
For example, a parallel translation in a plane or in space can be represented by a vector, the direction of displacement being specified by the direction of the vector and the length of the displacement-segment by the magnitude of the vector.

Definitions

Vectors are denoted by $\mathbf{a}$, $\mathbf{b}$, ... or $\overrightarrow{AB}$, $\overrightarrow{BC}$, ...

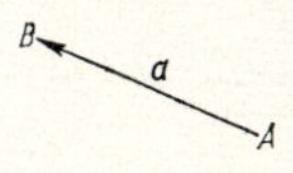

The length of a vector is called its *magnitude* and is written as $|\mathbf{a}| = a$. *Nullvectors* have length zero and are regarded as having no definite direction.

Position vectors are vectors with a common point of origin.
Radius vectors $\mathbf{r}$ are position vectors whose points of origin are at the origin of a coordinate system.

Unit vectors are of length 1. They are denoted by $\mathbf{a}^0$, $\mathbf{c}^0$, ...

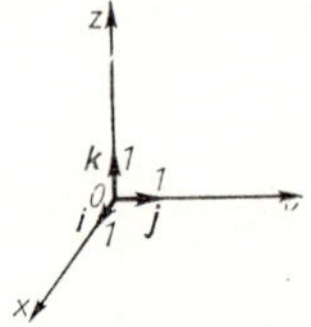

Unit vectors with a common point of origin O whose directions coincide with the axes of a rectangular coordinate system in space are denoted by **i**, **j**, **k** (unit coordinate vectors or *basis vectors*).
The following considerations are based on a *right-handed system* of coordinates which is characterized by the fact that a rotation from the positive x-axis toward the positive y-axis with a simultaneous displacement in the direction of the positive z-axis yields a right-handed screw.

Collinear vectors are parallel to the same straight line.

Coplanar vectors are parallel to the same plane.

Opposite vectors have the same magnitude but opposite directions:

$$\left.\begin{aligned}\overrightarrow{AB} &= \mathbf{a}\\ \overrightarrow{BA} &= -\mathbf{a}\end{aligned}\right\} \quad |\mathbf{a}| = |-\mathbf{a}|$$

Two vectors **a** and **b** are equal if they are of the same magnitude, have the same direction, and the same sense. They can be arbitrarily translated in space (*free vectors*). *Bound vectors* such as occurring in physics (e.g. field vectors) are vectors whose point of origin is fixed at a definite point in space.
Quantities that are defined only by a numerical value are called *scalars* to distinguish them from vectors.

Representation of vectors in terms of their components

The projections of a vector **a** onto the three coordinate axes x, y, z yield the *components of the vector* **a**:

$$\mathbf{a}_x, \mathbf{a}_y, \mathbf{a}_z$$

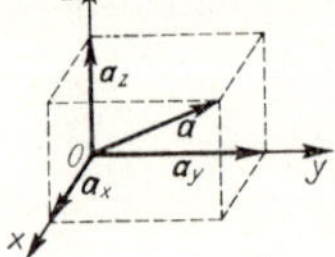

Scalar components of **a**:

$$|\mathbf{a}_x| = a_x; \quad |\mathbf{a}_y| = a_y; \quad |\mathbf{a}_z| = a_z$$

If point of origin and terminus of **a** have the coordinates x_1, y_1, z_1 and x_2, y_2, z_2, respectively, it follows that

$$a_x = x_2 - x_1; \quad a_y = y_2 - y_1; \quad a_z = z_2 - z_1$$

A vector is **written as**

$$\mathbf{a} = (a_x, a_y, a_z) = a_x\mathbf{i} + a_y\mathbf{j} + a_z\mathbf{k} = \mathbf{a}_x + \mathbf{a}_y + \mathbf{a}_z$$

$$= a\mathbf{a}^0 = a[\mathbf{i}\cos(\mathbf{a}, \mathbf{i}) + \mathbf{j}\cos(\mathbf{a}, \mathbf{j}) + \mathbf{k}\cos(\mathbf{a}, \mathbf{k})]$$

Magnitude (absolute value)

$$|\mathbf{a}| = \sqrt{a_x^2 + a_y^2 + a_z^2} = a = \sqrt{\mathbf{a}^2}$$

(Pythagoras's theorem)

Similarly, we have

$$\mathbf{i} = (1, 0, 0); \qquad \mathbf{j} = (0, 1, 0); \qquad \mathbf{k} = (0, 0, 1)$$

Example:

$$\mathbf{a} = 5\mathbf{i} + 2\mathbf{j} - 6\mathbf{k}$$

$$a = |\mathbf{a}| = \sqrt{5^2 + 2^2 + (-6)^2} = \sqrt{65} \approx 8.062$$

Direction cosine

The angles which a vector **a** makes with the positive coordinate axes result from the so-called direction cosines of **a**:

$$\cos(\mathbf{a}, \mathbf{i}) = \frac{a_x}{a}; \quad \cos(\mathbf{a}, \mathbf{j}) = \frac{a_y}{a}; \quad \cos(\mathbf{a}, \mathbf{k}) = \frac{a_z}{a}$$

$$\cos^2(\mathbf{a}, \mathbf{i}) + \cos^2(\mathbf{a}, \mathbf{j}) + \cos^2(\mathbf{a}, \mathbf{k}) = 1$$

$$a_x\cos(\mathbf{a}, \mathbf{i}) + a_y\cos(\mathbf{a}, \mathbf{j}) + a_z\cos(\mathbf{a}, \mathbf{k}) = a$$

Example:

Direction cosine for $\mathbf{a} = 5\mathbf{i} + 2\mathbf{j} - 6\mathbf{k}$ (see above example)

$$\cos(\mathbf{a}, \mathbf{i}) = \frac{5}{8.062} = 0.6202; \quad \sphericalangle(\mathbf{a}, \mathbf{i}) = 51°40'$$

$$\cos(\mathbf{a}, \mathbf{j}) = \frac{2}{8.062} = 0.2481; \quad \sphericalangle(\mathbf{a}, \mathbf{j}) = 75°38'$$

$$\cos(\mathbf{a}, \mathbf{k}) = \frac{-6}{8.062} = -0.7442; \quad \sphericalangle(\mathbf{a}, \mathbf{k}) = 138°6'$$

Addition and subtraction of vectors

Vector $\mathbf{b}$ is added to vector $\mathbf{a}$ by parallel translation. The connection of the point of origin of $\mathbf{a}$ with the terminus of $\mathbf{b}$ results in the vector sum or *resultant vector* $\mathbf{s} = \mathbf{a} + \mathbf{b}$.

Subtraction of one or several vectors is reduced to an addition of opposite vectors (see illustration):

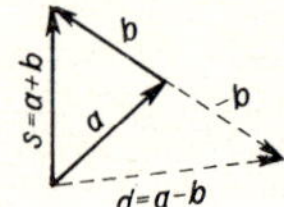

$$\mathbf{d} = \mathbf{a} - \mathbf{b} = \mathbf{a} + (-\mathbf{b})$$

In this way, an addition of vectors leads to an open *polygon* which is closed by the resultant vector.

Commutative law of addition

$$\mathbf{a} + \mathbf{b} = \mathbf{b} + \mathbf{a}$$

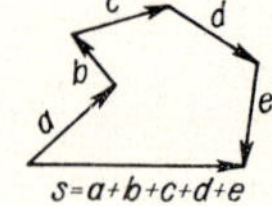

Associative law of addition

$$\mathbf{a} + (\mathbf{b} + \mathbf{c}) = (\mathbf{a} + \mathbf{b}) + \mathbf{c} \quad \text{etc.}$$

For the absolute value we have

$$|\mathbf{a} + \mathbf{b}| \leqq |\mathbf{a}| + |\mathbf{b}|; \quad |\mathbf{a} - \mathbf{b}| \geqq |\mathbf{a}| - |\mathbf{b}|$$

Formation of the sum or difference of vectors in terms of their components:

$$\mathbf{a} \pm \mathbf{b} = \mathbf{a}_x + \mathbf{a}_y + \mathbf{a}_z \pm (\mathbf{b}_x + \mathbf{b}_y + \mathbf{b}_z)$$

$$= (a_x \pm b_x)\mathbf{i} + (a_y \pm b_y)\mathbf{j} + (a_z \pm b_z)\mathbf{k}$$

Example:

$$\mathbf{a} = -5\mathbf{i} + 12\mathbf{j} + 7\mathbf{k} \qquad \mathbf{b} = 3\mathbf{i} - 6\mathbf{j} - 7\mathbf{k}$$

$$\mathbf{s} = \mathbf{a} + \mathbf{b} = (-5 + 3)\mathbf{i} + (12 - 6)\mathbf{j} + (7 - 7)\mathbf{k}$$

$$= -2\mathbf{i} + 6\mathbf{j}$$

2.4.2. Multiplication of vectors

Multiplication of one vector by one scalar

$n\mathbf{a} = \mathbf{a}n$ represents a vector which is collinear with the vector **a**.

$$n \in R$$

$$n\mathbf{a} = n\mathbf{a}_x + n\mathbf{a}_y + n\mathbf{a}_z$$

$$|n\mathbf{a}| = |n| \cdot |\mathbf{a}|$$

$n > 0$ yields $n\mathbf{a} \uparrow\uparrow \mathbf{a}$ (n**a** has the same direction as **a**)

$n = 0$ yields $n\mathbf{a} = 0$ (nullvector)

$n < 0$ yields $n\mathbf{a} \uparrow\downarrow \mathbf{a}$ (n**a** is opposed to **a**)

$$n_1(n_2\mathbf{a}) = (n_1 n_2)\mathbf{a}$$

Scalar product (inner product)

We write: **ab** or $\mathbf{a} \cdot \mathbf{b}$ (read: **ab** or **a** dot **b**)

Definition:

$$\mathbf{ab} = |\mathbf{a}| \cdot |\mathbf{b}| \cos (\mathbf{a}, \mathbf{b})$$

where $\sphericalangle (\mathbf{a}, \mathbf{b})$ is the smallest angle through which one of the vectors must be rotated in order to become parallel to the other.
The scalar product also is a scalar.

Commutative law: $\mathbf{ab} = \mathbf{ba}$

Distributive law: $(\mathbf{a} + \mathbf{b})\mathbf{c} = \mathbf{ac} + \mathbf{bc}$

Associative law:

$$(n\mathbf{a})\mathbf{b} = \mathbf{a}(n\mathbf{b}) = n(\mathbf{ab})$$

but: $(\mathbf{ab})\mathbf{c} \neq \mathbf{a}(\mathbf{bc})$

For $\mathbf{a} \uparrow\uparrow \mathbf{b}$ holds: $\mathbf{ab} = |\mathbf{a}| \cdot |\mathbf{b}|$

For $\mathbf{a} \uparrow\downarrow \mathbf{b}$ holds: $\mathbf{ab} = -|\mathbf{a}| \cdot |\mathbf{b}|$

For $\mathbf{a} \perp \mathbf{b}$ holds: $\mathbf{ab} = 0$

For the unit vectors, it follows that

$$\mathbf{i}^2 = 1 \qquad \mathbf{j}^2 = 1 \qquad \mathbf{k}^2 = 1$$

$$\mathbf{ij} = 0 \qquad \mathbf{jk} = 0 \qquad \mathbf{ki} = 0$$

From this we have

$$\mathbf{ab} = a_x b_x + a_y b_y + a_z b_z$$

$$\mathbf{a}^2 = a_x{}^2 + a_y{}^2 + a_z{}^2$$

Intersecting angle of two straight lines:

$$\cos(\mathbf{a}, \mathbf{b}) = \frac{a_x b_x + a_y b_y + a_z b_z}{\sqrt{(a_x{}^2 + a_y{}^2 + a_z{}^2)(b_x{}^2 + b_y{}^2 + b_z{}^2)}} = \frac{\mathbf{ab}}{|\mathbf{a}| \cdot |\mathbf{b}|}$$

Orthogonality of two vectors ($\mathbf{a} \perp \mathbf{b}$):

$$a_x b_x + a_y b_y + a_z b_z = 0$$

Cosine theorem:

$$(\mathbf{a} \pm \mathbf{b})^2 = \mathbf{a}^2 \pm 2\mathbf{ab} + \mathbf{b}^2$$

$$|\mathbf{a} \pm \mathbf{b}| = \sqrt{\mathbf{a}^2 \pm 2\mathbf{ab} + \mathbf{b}^2}$$

$$(\mathbf{a} + \mathbf{b})^2 - (\mathbf{a} - \mathbf{b})^2 = 4\mathbf{ab}$$

Example:

$$\mathbf{a} = 16\mathbf{i} + 4\mathbf{j} - 7\mathbf{k} \qquad \mathbf{b} = 3\mathbf{i} - 9\mathbf{j} - 4\mathbf{k}$$

$$\mathbf{ab} = 16 \cdot 3 + 4 \cdot (-9) + (-7)(-4) = 40$$

$$|\mathbf{a}| = \sqrt{16^2 + 4^2 + 7^2} = \sqrt{321} \approx 17.9$$

$$|\mathbf{b}| = \sqrt{3^2 + 9^2 + 4^2} = \sqrt{106} \approx 10.3$$

$$\cos(\mathbf{a}, \mathbf{b}) = \frac{40}{17.9 \cdot 10.3} \approx 0.2168; \quad \sphericalangle(\mathbf{a}, \mathbf{b}) \approx 77°29'$$

Vector product (outer product)

We write: $\mathbf{a} \times \mathbf{b}$ (read: **a** cross **b**)

Definition

$$\mathbf{a} \times \mathbf{b} = \mathbf{c}$$

$$|\mathbf{c}| = |\mathbf{a}| \cdot |\mathbf{b}| \sin(\mathbf{a}, \mathbf{b})$$

a, **b**, **c** form an orthogonal right-handed system.

Geometrical interpretation: The vector $\mathbf{c}$ is perpendicular to the vectors $\mathbf{a}$ and $\mathbf{b}$. The absolute value $|\mathbf{c}|$ is equal to the dimensions of the area of the parallelogram specified by $\mathbf{a}$ and $\mathbf{b}$.

$$\mathbf{a} \times \mathbf{b} = -\mathbf{b} \times \mathbf{a} \text{ (The commutative law is not applicable!)}$$

Associative law: $n(\mathbf{a} \times \mathbf{b}) = (n\mathbf{a}) \times \mathbf{b} = \mathbf{a} \times (n\mathbf{b})$

Distributive law: $(\mathbf{a} + \mathbf{b}) \times \mathbf{c} = \mathbf{a} \times \mathbf{c} + \mathbf{b} \times \mathbf{c}$

$$\mathbf{a} \uparrow\uparrow \mathbf{b} \quad \text{and } \mathbf{a} \uparrow\downarrow \mathbf{b} \Rightarrow \mathbf{a} \times \mathbf{b} = 0$$

$$\mathbf{a} \perp \mathbf{b} \qquad \Rightarrow |\mathbf{a} \times \mathbf{b}| = |\mathbf{a}| \cdot |\mathbf{b}|$$

For the unit vectors, it follows that

$$\mathbf{i} \times \mathbf{i} = 0 \qquad \mathbf{j} \times \mathbf{j} = 0 \qquad \mathbf{k} \times \mathbf{k} = 0$$

$$\mathbf{i} \times \mathbf{j} = \mathbf{k} \qquad \mathbf{j} \times \mathbf{k} = \mathbf{i} \qquad \mathbf{k} \times \mathbf{i} = \mathbf{j}$$

$$\mathbf{j} \times \mathbf{i} = -\mathbf{k} \qquad \mathbf{k} \times \mathbf{j} = -\mathbf{i} \qquad \mathbf{i} \times \mathbf{k} = -\mathbf{j}$$

The vector product represented in terms of vector components

$$\mathbf{a} \times \mathbf{b} = (a_y b_z - b_y a_z)\mathbf{i} + (a_z b_x - b_z a_x)\mathbf{j} + (a_x b_y - b_x a_y)\mathbf{k}$$

$$\mathbf{a} \times \mathbf{b} = \begin{vmatrix} \mathbf{i} & \mathbf{j} & \mathbf{k} \\ a_x & a_y & a_z \\ b_x & b_y & b_z \end{vmatrix}$$

Example:

$$\mathbf{a} = 16\mathbf{i} + 4\mathbf{j} - 7\mathbf{k} \qquad \mathbf{b} = 3\mathbf{i} - 9\mathbf{j} - 4\mathbf{k}$$

$$\mathbf{a} \times \mathbf{b} = \begin{vmatrix} \mathbf{i} & \mathbf{j} & \mathbf{k} \\ 16 & 4 & -7 \\ 3 & -9 & -4 \end{vmatrix} = -79\mathbf{i} + 43\mathbf{j} - 156\mathbf{k}$$

Multiple products of vectors

There are no laws for the combination of three and more vectors, neither for the scalar product nor for the vector product. In one calculating operation, only two vectors can be combined with respect to their scalar or vector products.

Scalar product of three vectors:

$$(\mathbf{a}\times\mathbf{b})\mathbf{c} = \begin{vmatrix} a_x & a_y & a_z \\ b_x & b_y & b_z \\ c_x & c_y & c_z \end{vmatrix} = \mathbf{a}(\mathbf{b}\times\mathbf{c}) > 0 \quad \text{for} \quad \mathbf{a}, \mathbf{b}, \mathbf{c} \text{ right-handed system}$$

Geometrical representation: The scalar product of three vectors in terms of magnitude is equal to the volume of the prism specified by the three vectors **a**, **b**, **c**.

$$(\mathbf{a}\times\mathbf{b})\mathbf{c} = (\mathbf{b}\times\mathbf{c})\mathbf{a} = (\mathbf{c}\times\mathbf{a})\mathbf{b} = -(\mathbf{b}\times\mathbf{a})\mathbf{c}$$
$$= -(\mathbf{c}\times\mathbf{b})\mathbf{a} = -(\mathbf{a}\times\mathbf{c})\mathbf{b}$$

Example:

Determine the volume of the prism specified by the following vectors

$$\mathbf{a} = 3\mathbf{i} + 6\mathbf{j} - 2\mathbf{k}$$
$$\mathbf{b} = 5\mathbf{i} - \mathbf{j} + 7\mathbf{k}$$
$$\mathbf{c} = 6\mathbf{i} - 3\mathbf{j} + 8\mathbf{k}$$

$$\mathbf{abc} = \begin{vmatrix} 3 & 6 & -2 \\ 5 & -1 & 7 \\ 6 & -3 & 8 \end{vmatrix} = 69$$

$(\mathbf{a}\times\mathbf{b})\mathbf{c} = 0$ if the three vectors are contained in one plane (coplanar vectors) or if one vector is the nullvector.
Double vector product:

$$\mathbf{a}\times(\mathbf{b}\times\mathbf{c}) = (\mathbf{ac})\mathbf{b} - (\mathbf{ab})\mathbf{c} \quad (\textit{theorem of expansion})$$

The double vector product represents a vector which lies in the plane of the two vectors **b** and **c**.
Products with four factors:

$$(\mathbf{a}\times\mathbf{b})(\mathbf{c}\times\mathbf{d}) = \begin{vmatrix} \mathbf{ac} & \mathbf{bc} \\ \mathbf{ad} & \mathbf{bd} \end{vmatrix}$$

$$(\mathbf{a}\times\mathbf{b})^2 = \begin{vmatrix} \mathbf{aa} & \mathbf{ab} \\ \mathbf{ab} & \mathbf{bb} \end{vmatrix}$$

$$(\mathbf{a}\times\mathbf{b})\times(\mathbf{c}\times\mathbf{d}) = \mathbf{c}[\mathbf{abd}] - \mathbf{d}[\mathbf{abc}]$$
$$= \mathbf{b}[\mathbf{acd}] - \mathbf{a}[\mathbf{bcd}]$$

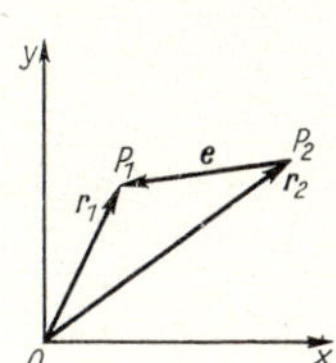

2.4.3. Geometrical applications of vector calculus

Distance between two points P_1 and P_2

$$|\mathbf{e}| = |\mathbf{r}_1 - \mathbf{r}_2|$$

Division of a line segment AB in the ratio λ

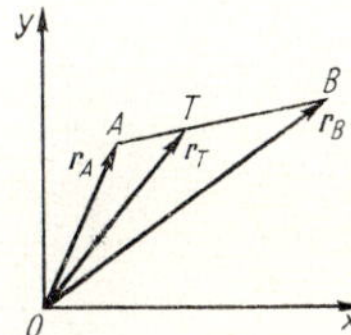

$$\mathbf{r}_T = \frac{\mathbf{r}_A + \lambda \mathbf{r}_B}{1 + \lambda}$$

$\lambda > 0$ inner point of division; $\lambda < 0$ outer point of division $\lambda = +1$ midpoint of the line segment

$$\mathbf{r}_M = \frac{\mathbf{r}_A + \mathbf{r}_B}{2}$$

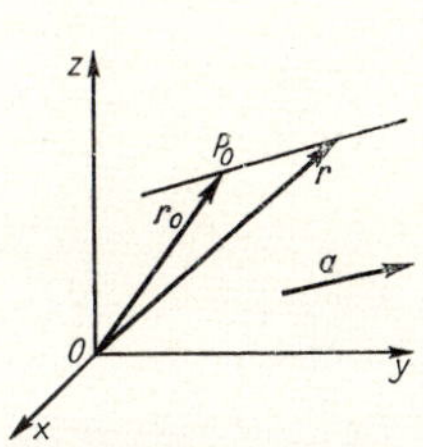

Pointslope equation of the straight line

$$\mathbf{r} = \mathbf{r}_0 + t\mathbf{a}, \quad t \in R$$

Example:

$$P_0(3, -4, 6); \quad \mathbf{a} = 2\mathbf{i} + 4\mathbf{j} - 5\mathbf{k}$$

$$\mathbf{r} = 3\mathbf{i} - 4\mathbf{j} + 6\mathbf{k} + t(2\mathbf{i} + 4\mathbf{j} - 5\mathbf{k})$$

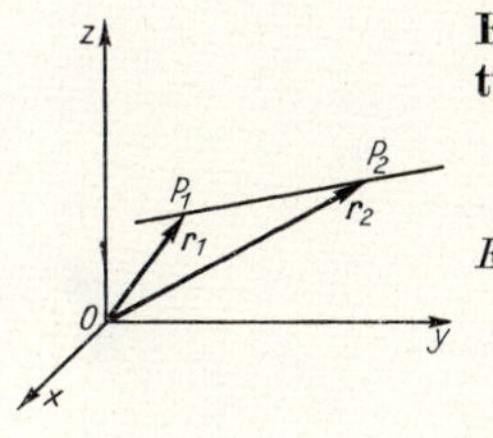

Equation of the straight line through two points

$$\mathbf{r} = \mathbf{r}_1 + t(\mathbf{r}_2 - \mathbf{r}_1), \qquad t \in R$$

Example:

$$P_1(-1, 5, 7)$$

$$P_2(3, -4, 2)$$

$$\begin{aligned}\mathbf{r} &= (-1)\mathbf{i} + 5\mathbf{j} + 7\mathbf{k} \\ &\quad + t[3\mathbf{i} - 4\mathbf{j} + 2\mathbf{k} - (-1)\mathbf{i} - 5\mathbf{j} - 7\mathbf{k}] \\ &= -\mathbf{i} + 5\mathbf{j} + 7\mathbf{k} + t(4\mathbf{i} - 9\mathbf{j} - 5\mathbf{k})\end{aligned}$$

Angle of intersection of two straight lines

$$\sphericalangle(\mathbf{a}, \mathbf{b}) = \arccos \frac{\mathbf{a}\mathbf{b}}{|\mathbf{a}| \cdot |\mathbf{b}|}$$

Distance between two skew straight lines

$$\mathbf{d} = \left| \frac{([\mathbf{a} \times \mathbf{b}](\mathbf{r}_1 - \mathbf{r}_2))}{[\mathbf{a} \times \mathbf{b}]} \right| = \left| \frac{\Delta}{[\mathbf{a} \times \mathbf{b}]} \right|$$

$\mathbf{a}$, $\mathbf{b}$ direction vectors of the two straight lines
$\mathbf{r}_1$, $\mathbf{r}_2$ radius vectors to one arbitrary point on each of the two straight lines

$$\mathbf{r}_1 = x_1\mathbf{i} + y_1\mathbf{j} + z_1\mathbf{k}; \quad \mathbf{r}_2 = x_2\mathbf{i} + y_2\mathbf{j} + z_2\mathbf{k}$$

$$\Delta = \begin{vmatrix} a_x & a_y & a_z \\ b_x & b_y & b_z \\ x_1 - x_2 & y_1 - y_2 & z_1 - z_2 \end{vmatrix}$$

The straight lines intersect if $\Delta = 0$.

Example:

$$\mathbf{g}_1 \equiv (2\mathbf{i} + 4\mathbf{j} + 3\mathbf{k})t + 3\mathbf{i} - \mathbf{j} + 2\mathbf{k}$$

$$\mathbf{g}_2 \equiv (-4\mathbf{i} + 4\mathbf{j} + 6\mathbf{k})\tau - \mathbf{i} + 5\mathbf{j} + 10\mathbf{k}$$

$$\Delta = \begin{vmatrix} 2 & 4 & 3 \\ -4 & 4 & 6 \\ 4 & -6 & -8 \end{vmatrix} = 0 \quad \text{The straight lines intersect.}$$

For the last row of the determinant, the radius vectors of the straight line points for $t = 0$ or $\tau = 0$ were taken as a basis.
The point of intersection is obtained by equating the equations of the straight lines and comparison of coefficients.

$$(2\mathbf{i} + 4\mathbf{j} + 3\mathbf{k})t + 3\mathbf{i} - \mathbf{j} + 2\mathbf{k}$$

$$= (-4\mathbf{i} + 4\mathbf{j} + 6\mathbf{k})\tau - \mathbf{i} + 5\mathbf{j} + 10\mathbf{k}$$

$$2t + 3 = -4\tau - 1$$

$$4t - 1 = 4\tau + 5$$

$$3t + 2 = 6\tau + 10$$

From two of the above equations we have

$$t = \frac{1}{3}; \quad \tau = -\frac{7}{6}$$

These values must also satisfy the third equation, otherwise there will be no point of intersection.
Radius vector of the point of intersection

$$\mathbf{r}_s = (2\mathbf{i} + 4\mathbf{j} + 3\mathbf{k}) \cdot \frac{1}{3} + 3\mathbf{i} - \mathbf{j} + 2\mathbf{k}$$
$$= \frac{11}{3}\mathbf{i} + \frac{1}{3}\mathbf{j} + 3\mathbf{k}$$

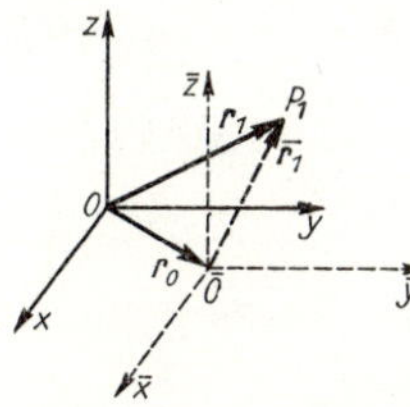

Parallel translation of the system of coordinates

$$\overline{\mathbf{r}}_1 = \mathbf{r}_1 - \mathbf{r}_0$$

Equation of a plane through three points which do not lie on a straight line

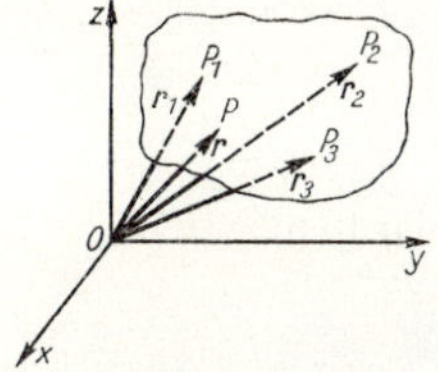

$$(\mathbf{r} - \mathbf{r}_1)(\mathbf{r} - \mathbf{r}_2)(\mathbf{r} - \mathbf{r}_3) = 0$$

This scalar product from three vectors in the form of determinants

$$\begin{vmatrix} x - x_1 & y - y_1 & z - z_1 \\ x - x_2 & y - y_2 & z - z_2 \\ x - x_3 & y - y_3 & z - z_3 \end{vmatrix} = 0$$

From this we obtain by marginal expansion

$$\begin{vmatrix} x & y & z & 1 \\ x_1 & y_1 & z_1 & 1 \\ x_2 & y_2 & z_2 & 1 \\ x_3 & y_3 & z_3 & 1 \end{vmatrix} = 0$$

$$Ax + By + Cz + D = 0$$

A, B, C, D are subdeterminants of the elements of the first row of the above determinant

Form of intercepts: a, b, c intercepts on x, y, z-axes

$$\begin{vmatrix} x & y & z & 1 \\ a & 0 & 0 & 1 \\ 0 & b & 0 & 1 \\ 0 & 0 & c & 1 \end{vmatrix} = 0 \quad \text{or} \quad \frac{x}{a} + \frac{y}{b} + \frac{z}{c} = 1$$

Parametric representation of the plane

$$\mathbf{r} = \mathbf{r}_1 + s(\mathbf{r}_2 - \mathbf{r}_1) = t(\mathbf{r}_2 - \mathbf{r}_3)$$

s, t parameters

$\mathbf{r}_1$, $\mathbf{r}_2$, $\mathbf{r}_3$ radius vectors of three given points on the plane

$$\mathbf{r} = \mathbf{r}_0 + s\mathbf{a} + t\mathbf{b}$$

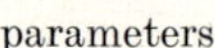

s, t parameters

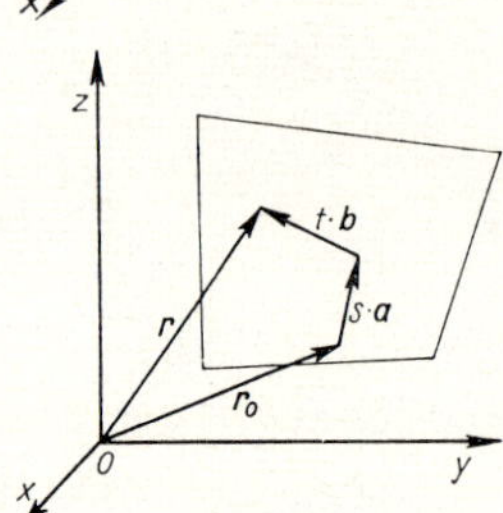

$\mathbf{r}_0$ radius vector of the given point

$\mathbf{a}$, $\mathbf{b}$ direction vectors

Pointdirection equation of the plane

(Equation of the plane through P_0, perpendicular to $\mathbf{a}$)

Position vector $\mathbf{n}_0 = \dfrac{\mathbf{n}}{|\mathbf{n}|}$

$$\mathbf{n}(\mathbf{r} - \mathbf{r}_0) = 0$$

$$\mathbf{n}\mathbf{r} = \mathbf{n}\mathbf{r}_0$$

$$\mathbf{n}\mathbf{r} + D = 0$$

Hessian normal form of the equation of a plane

From the above follows that

$$\mathbf{n}^0\mathbf{r} - p = 0$$

$$p = -\frac{D}{|\mathbf{n}|}$$ distance of the origin from the plane

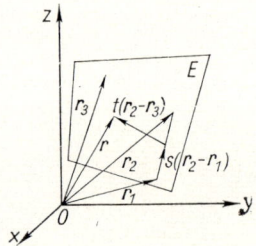

Distance of a point P_0 from the plane

$$d = \mathbf{n}^0\mathbf{r}_0 - p \begin{cases} < 0 \text{ for } P_0 \text{ and origin on the same side of the plane} \\ > 0 \text{ for } P_0 \text{ and origin on different sides of the plane} \end{cases}$$

$d = \mathbf{n}^0(\mathbf{r}_1 - \mathbf{r}_0)$ $\mathbf{r}_1$ radius vector of a point of the plane

$$d = \frac{Ax_0 + By_0 + Cz_0 + D}{\sqrt{A^2 + B^2 + C^2}}$$

Angle between two planes

$$\cos(E_1, E_2) = \frac{\mathbf{n}_1\mathbf{n}_2}{|\mathbf{n}_1| \cdot |\mathbf{n}_2|}$$

Area of the plane triangle

$$2A = \mathbf{r}_1 \times \mathbf{r}_2 + \mathbf{r}_2 \times \mathbf{r}_3 + \mathbf{r}_3 \times \mathbf{r}_1$$

$\mathbf{r}_1$, $\mathbf{r}_2$, $\mathbf{r}_3$ radius vectors to the corners

Volume of a three-sided pyramid with vertex at the origin

$$V = \left|\frac{1}{6}(\mathbf{r}_1 \times \mathbf{r}_2)\mathbf{r}_3\right|$$

Volume of a three-sided pyramid in arbitrary position

$$V = \left|\frac{1}{6}(\mathbf{r}_1 - \mathbf{r}_4)(\mathbf{r}_2 - \mathbf{r}_4)(\mathbf{r}_3 - \mathbf{r}_4)\right| \quad P_4 \text{ vertex}$$

$$V = \frac{1}{6}\begin{vmatrix} x_1 & y_1 & z_1 & 1 \\ x_2 & y_2 & z_2 & 1 \\ x_3 & y_3 & z_3 & 1 \\ x_4 & y_4 & z_4 & 1 \end{vmatrix}$$

2.5. Reflection in a circle, inversion

Inversion:

$$\mathbf{A} = |\mathbf{A}|\, e^{j\varphi} \Leftrightarrow \mathbf{B} = \frac{1}{|\mathbf{A}|}\, e^{-j\varphi}$$

Application: Inversion of transfer loci, conversion of resistance into conductivity, etc.

Inversion of a pointer (directed segment) according to the above equation by graphical methods based on the unit circle:

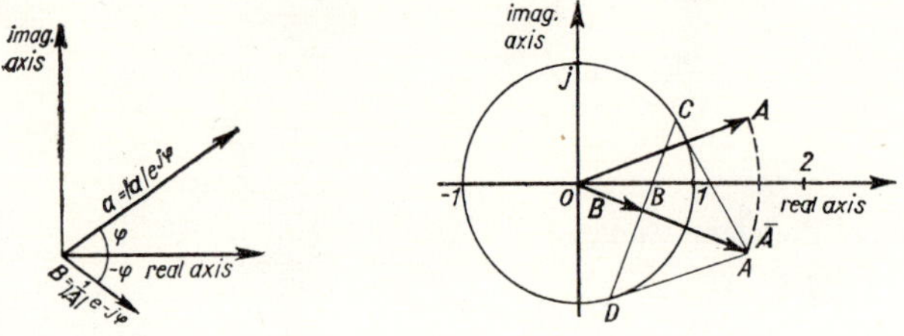

1. Drawing of the conjugate complex pointer $\bar{\mathbf{A}} = |\mathbf{A}|\, e^{-j\varphi}$
2. Drawing the tangents from the terminus of pointer $\bar{\mathbf{A}}$ to the circle
3. The connecting line between the points of contact of the tangents intersect pointer $\bar{\mathbf{A}}$ in B, the terminus of the inverse pointer $\mathbf{B} = \mathbf{A}'$

Proof:

$$\triangle\, OBD \sim \triangle\, ODA; \quad \frac{|\mathbf{A}|}{1} = \frac{1}{|\mathbf{B}|}$$

Taking the scale into consideration, we have:

$$\mathbf{r} = 1 \mathrel{\hat{=}} a \text{ units of } \mathbf{A} \Leftrightarrow \frac{1}{a} \text{ units of } \mathbf{B}$$

e.g. diameter of circle $r = 1 \mathrel{\hat{=}} 5\,\Omega \Leftrightarrow \frac{1}{5}\,\mathrm{S}$

when converting resistance and conductivity.

Inversion of curves

First theorem on inversion: A straight line through the origin by inversion again gives a straight line through the origin.

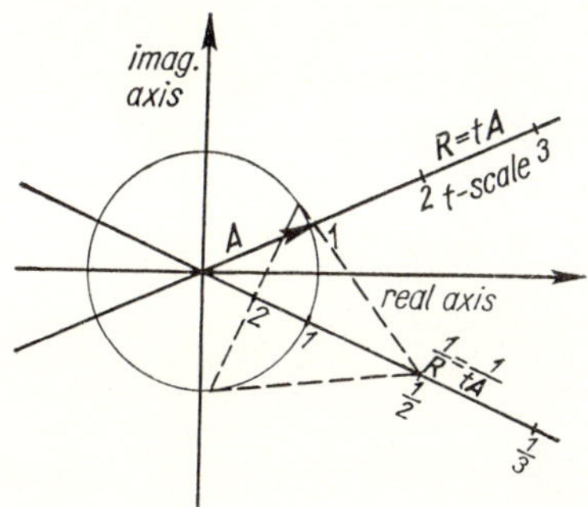

Given:

Straight line $\mathbf{R}(t) = t\mathbf{A}$

Inverted:

Straight line $\dfrac{1}{\mathbf{R}} = \dfrac{1}{t\mathbf{A}} = \mathbf{R}'$

1. Drawing the conjugate direction through $\mathbf{A} \Rightarrow \bar{\mathbf{A}}$
2. Plotting the parameter scale t for

$$\mathbf{R}' = \frac{1}{t\mathbf{A}}$$

Second theorem on inversion: A straight line not through the origin by inversion yields a circle through the origin.

Given:

Straight line $\mathbf{R}(t) = \mathbf{R}_0 + t\mathbf{A}$

Inverted:

Circle $\dfrac{1}{\mathbf{R}} = \mathbf{R}' = \dfrac{1}{\mathbf{R}_0 + t\mathbf{A}}$

1. Drawing the conjugate straight line $\overline{\mathbf{R}}$
2. Drawing the normal to $\overline{\mathbf{R}} \Rightarrow \overline{\mathbf{R}}_{\min}$
3. Reflection of $\overline{\mathbf{R}}_{\min}$ in the unit circle $\Rightarrow \mathbf{R}'_{\max}$

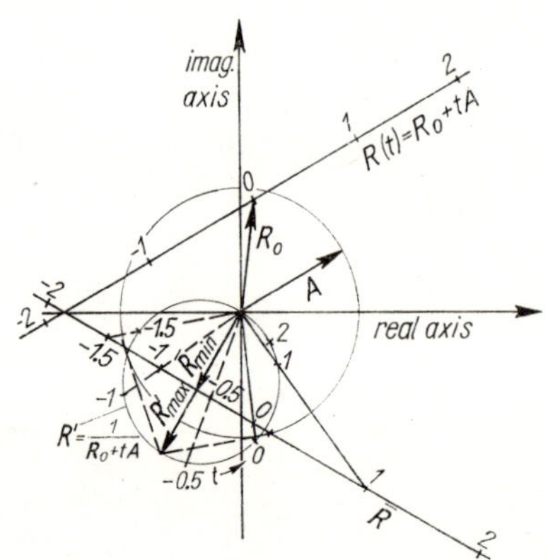

The tangent points T_1 and T_2 are points of intersection of $\overline{\mathbf{R}}$ with the unit circle and, at the same time, points of the circle obtained by inversion.

Third theorem on inversion: A circle not through the origin by inversion again results in a circle not through the origin.

Given:

$$\text{Circle } \mathbf{R}(t) = \mathbf{R}_0 + \frac{1}{\mathbf{A} + t\mathbf{B}} = \frac{\mathbf{C} + t\mathbf{D}}{\mathbf{A} + t\mathbf{B}}$$

Inverted:

$$\text{Circle } \mathbf{R}' = \frac{\mathbf{A} + t\mathbf{B}}{\mathbf{C} + t\mathbf{D}} \text{ of the same kind}$$

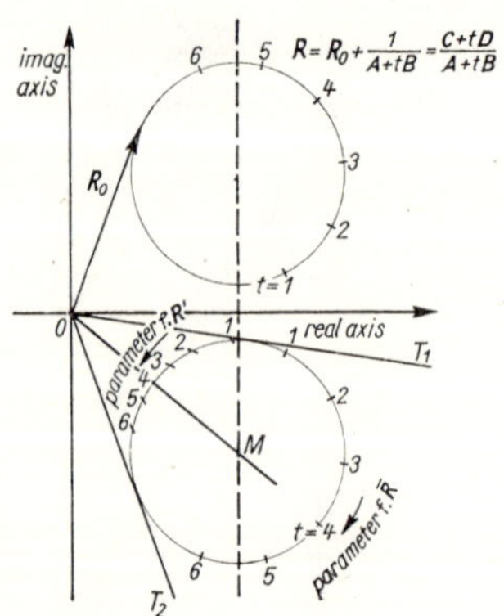

1. Drawing the conjugate complex circle $\overline{\mathbf{R}}$.
2. It is advisable to choose the scale for the inverted circle so that conjugate and inverted circles coincide. At this scale plot the parameter scale for the inverted circle $\mathbf{R}'$ as points of intersection of the straight lines through the parameter points on the conjugate circle to the origin with this circle.

When choosing a different scale for the inverted circle it should be noted that the tangents from the origin to the conjugate complex circle are also tangents to the inverted circle whereby the center will be always on the straight line through M and O.

3. Geometry

3.1. General

Angles

Adjacent supplementary angles add up to 180°.
Vertically opposite angles are equal.
Complementary angles add up to 90°.
Supplementary angles add up to 180°.

Angular measures

Angles are measured in *degrees*, *gons*, or in *radians* (circular measure). The degree (sexagesimal measure) is divided in minutes ($1° = 60'$) and seconds ($1' = 60''$).
The gon (decimal division) is divided in minutes ($1^g = 100^c$) and seconds ($1^c = 100^{cc}$).

$$360° \triangleq 400^g \qquad 90° \triangleq 100^g$$

The measure of an angle in radian (circular) measure is the length of arc cut off by the angle α on the unit circle with center at the vertex of the angle.

$$\operatorname{arc} \alpha = \frac{\pi\alpha}{180°} = \hat{\alpha}$$

e.g.

$$\operatorname{arc} 360° = 2\pi \qquad \operatorname{arc} 60° = \frac{\pi}{3}$$

$$\operatorname{arc} 270° = \frac{3}{2}\pi \qquad \operatorname{arc} 45° = \frac{\pi}{4}$$

$$\operatorname{arc} 180° = \pi \qquad \operatorname{arc} 30° = \frac{\pi}{6}$$

$$\operatorname{arc} 90° = \frac{\pi}{2} \qquad \operatorname{arc} 1° = 0.01745$$

$$\operatorname{arc} 1' = 0.00029$$

An angle of one radian measure, $1 = 57°17'45''$ (1 radian = 1 rad)
1 rad is the plane angle for which the ratio of the length of the associated circular arc to its radius is equal to 1.

Angles formed by transversal to parallel lines

Pairs of step angles are equal.

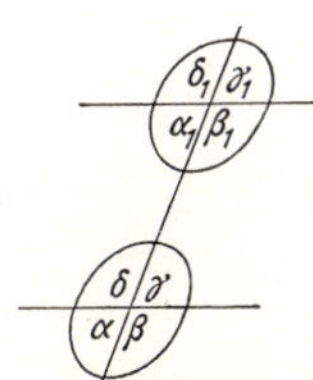

$$\alpha = \alpha_1 \qquad \gamma = \gamma_1$$

$$\beta = \beta_1 \qquad \delta = \delta_1$$

Pairs of alternate angles are equal.

$$\alpha = \gamma_1 \qquad \gamma = \alpha_1$$

$$\beta = \delta_1 \qquad \delta = \beta_1$$

Pairs of opposed angles are supplementary (add up to 180°).

$$\alpha + \delta_1 = 180° \qquad \gamma + \beta_1 = 180°$$

$$\beta + \gamma_1 = 180° \qquad \delta + \alpha_1 = 180°$$

Symmetry

A plane figure is called *axially symmetric* if a straight line exists (the axis of symmetry) such that the figure can be superposed on itself by rotation about the line through 180°.
A plane figure is said to be *centrally symmetric* with respect to a point (the center of symmetry), if when rotated through 180° about the center of symmetry, the figure comes into coincidence with itself.

Ray theorems

First theorem on rays: If the rays of a pencil of rays are cut by paralle- lines, the ratio of segments on one ray is equal to the ratio of correl sponding segments on each other ray.
Second theorem on rays: If the rays of a pencil of rays are cut by parallel lines, the ratio of the segment cut off on the transversals equals the ratio of the segments cut off on any ray.

First theorem on rays:

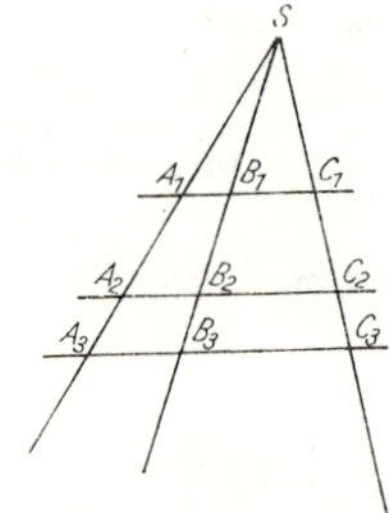

$$\overline{SA_1}:\overline{SA_2}:\overline{SA_3} = \overline{SB_1}:\overline{SB_2}:\overline{SB_3}$$

$$= \overline{SC_1}:\overline{SC_2}:\overline{SC_3}$$

or

$$\overline{SA_1}:\overline{A_1A_2}:\overline{A_2A_3}$$

$$= \overline{SB_1}:\overline{B_1B_2}:\overline{B_2B_3}$$

$$= \overline{SC_1}:\overline{C_1C_2}:\overline{C_2C_3}$$

etc.

Second theorem on rays:

$$\overline{A_1B_1}:\overline{A_2B_2}:\overline{A_3B_3} = \overline{SA_1}:\overline{SA_2}:\overline{SA_3}$$

or

$$\overline{B_1C_1}:\overline{B_2C_2}:\overline{B_3C_3} = \overline{SC_1}:\overline{SC_2}:\overline{SC_3}$$

etc.

From this follows that

$$\overline{A_1B_1}:\overline{B_1C_1} = \overline{A_2B_2}:\overline{B_2C_2} = \overline{A_3B_3}:\overline{B_3C_3}$$

Fourth proportional

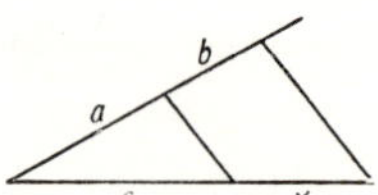

a, b, c given line segments

$a:b = c:x$

Division of a line segment in a given ratio

To divide a segment AB in the ratio $m:n$.

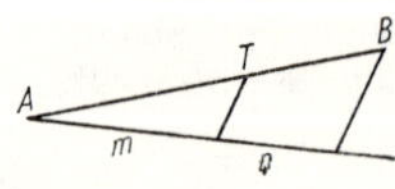

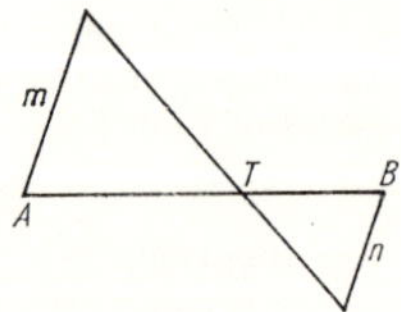

The point of division obtained is more precisely called the *inner* point of division.

Harmonic division of a line segment

If a segment is divided internally and externally in the same ratios, we say that the segment is harmonically divided.

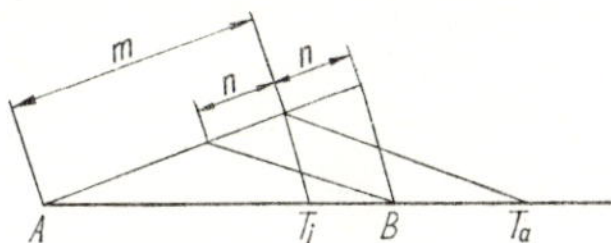

Application of the first theorem on rays

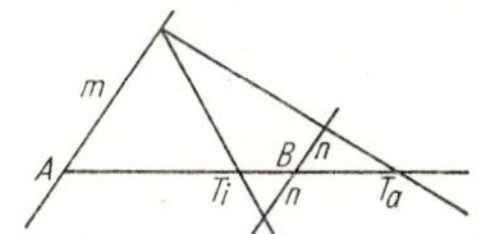

Application of the second theorem on rays

T_i inner point of division
T_0 outer point of division

In a triangle, the bisectors of its interior angle and exterior angle divide the opposite side harmonically, in the ratio of the two other sides.

$$\overline{AD}:\overline{BD} = \overline{AE}:\overline{BE} = b:a$$

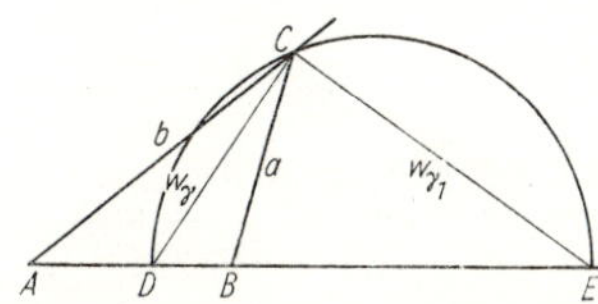

The circle on $\overline{DE}$ as diameter is the locus of the vertices of all triangles for which one side ($\overline{AB}$ in the illustration) is given and the ratio of the other two sides is specified (the circle of APOLLONIUS).

Mean proportional (geometric mean)

$$x^2 = a \cdot b \Rightarrow a:x = x:b$$

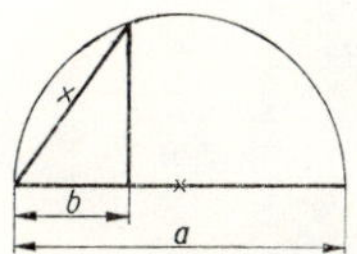

Euclid's theorem

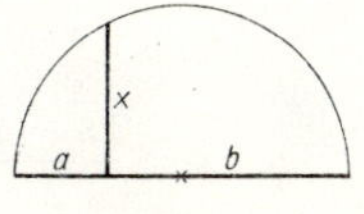

Altitude theorem

See also tangent secant theorem.

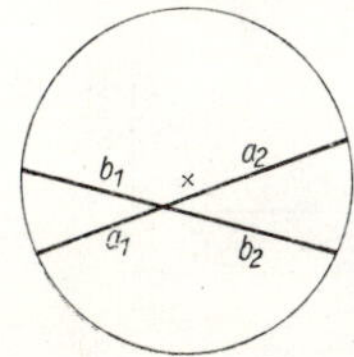

Circle and straight line

Chord theorem: If two chords are drawn through one point in the interior of a circle, the product of their segments is constant:

$$a_1 a_2 = b_1 b_2$$

Secant theorem: If two secants are drawn from a point outside a circle, the product of each segment and its outer segment is constant:

$$aa_1 = bb_1$$

Tangent secant theorem: If a secant and a tangent are drawn from a point outside a circle, the product of the secant and its outer segment is equal to the square of the length of the tangent.

$$aa_1 = x^2 \quad \text{or} \quad a:x = x:a_1$$

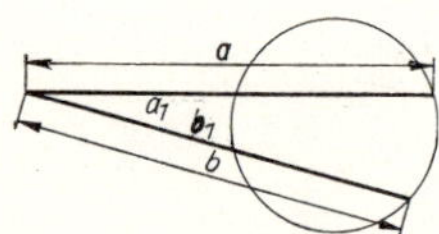

Secant theorem

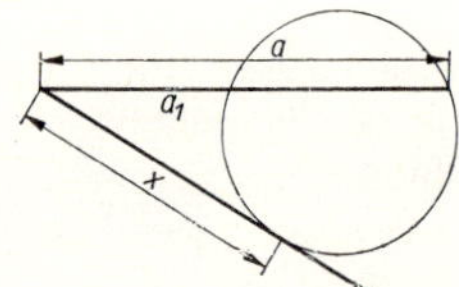

Tangent secant theorem

The length of the tangent is the mean proportional of the secant and its outer segment (construction of the mean proportional).

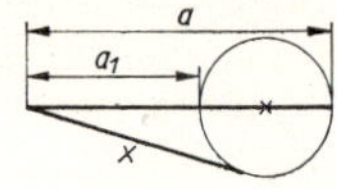

Golden section

A line segment is said to be divided in golden section if the larger segment is the mean proportional between the complete segment and the smaller segment.

$$a:x = x:(a - x)$$

$$x = \frac{\sqrt{5} - 1}{2} a \approx 0.618a$$

Construction

Erect the perpendicular $BC = \frac{1}{2} a$ from B on the segment $AB = a$, draw the join from A to C, with center C and radius $\frac{1}{2} a$ draw a circle which intersects $\overline{AC}$ in D and mark off $\overline{AD}$ on $\overline{AB}$ from A. E divides $\overline{AB}$ in golden section.

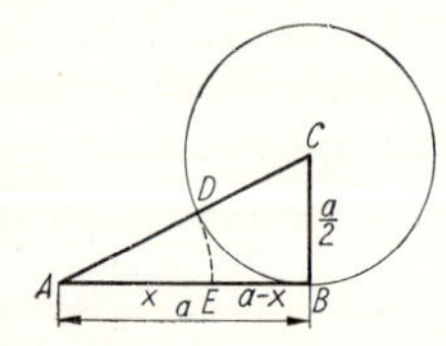

Application: The side of the regular decagon is the larger segment of the radius divided in golden section of its circumcircle.

Similarity

Plane polygons which agree in respect of the main attributes of shape are called similar ($\sim$).

Theorems on similarity

Two triangles are similar if they coincide

in respect of two angles or

in respect of the ratio of two sides and the angle included by these sides or

in respect of the ratio of two sides and the angle opposite to the larger of these sides or

in respect of the ratios of their three sides.

Similar triangles are divided into similar triangles by altitudes or angle-bisectors or side-bisectors.

In similar triangles the ratio of any two corresponding altitudes, angle-bisectors, and side-bisectors are equal to the ratio of two corresponding sides.

The perimeters of two similar triangles are to each other as two corresponding line segments (sides, altitudes, side-bisectors, etc.).

$$u_1 : u_2 = a_1 : a_2 = b_1 : b_2 = c_1 : c_2 = k$$

(*ratio of similarity, linear increase*)

The areas of two similar triangles are to each other as the squares of two corresponding segments (sides, altitudes, side-bisectors, etc.).

$$A_1 : A_2 = a_1^2 : a_2^2 = b_1^2 : b_2^2 = c_1^2 : c_2^2 = k^2 \qquad (k \text{ see above})$$

Polygons are called similar if they agree in respect of all angles and ratios of pairs of sides.

The perimeters of similar polygons are to each other as a pair of corresponding segments (sides, altitudes, diagonals, etc.).

$$u_1 : u_2 = a_1 : a_2 = b_1 : b_2 = \ldots = k$$

The areas of similar polygons are to each other as the squares of corresponding segments (sides, altitudes, diagonals, etc.).

$$A_1 : A_2 = a_1^2 : a_2^2 = b_1^2 : b_2^2 = \ldots = k^2$$

The *increase of the area* of polygons is equal to the square of the linear increase.

Position of similarity

Similar polygons are in a similar position if corresponding sides are parallel and corresponding points lie on rays of a pencil of rays. The vertex V of the pencil of rays is called *point of similarity*.

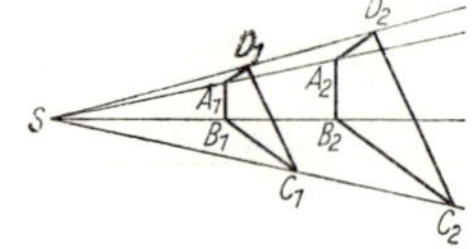

Congruence

Polygons that agree not only in respect of shape but also in respect of the size of homologous pieces are called congruent ($\cong$).

Congruence theorem

Triangles are congruent

if one side and its two adjacent angles coincide or
if two corresponding sides and their included angle coincide or
if two corresponding sides and the angle that lies opposite the larger of the two sides coincide or
if all three pairs of corresponding sides coincide.

3.2. Planimetry

3.2.1. Triangle *ABC*

Notations

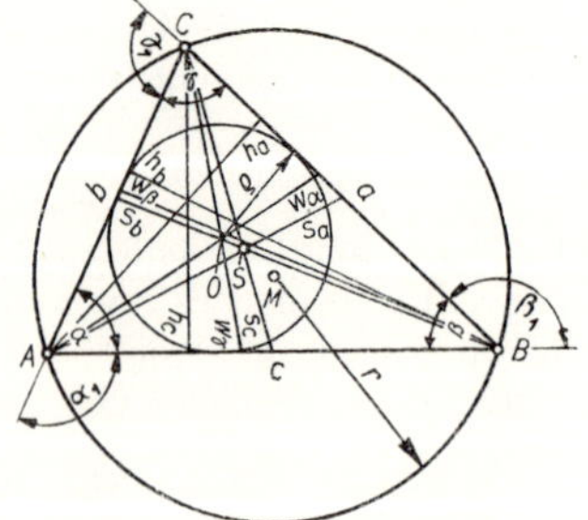

α, β, γ	interior angles
$\alpha_1, \beta_1, \gamma_1$	exterior angles
a	side opposite the vertex A
b	side opposite the vertex B
c	side opposite the vertex C
A	area

$$s = \frac{a + b + c}{2}$$

h_a altitude associated with side a
h_b altitude associated with side b
h_c altitude associated with side c

s_a side-bisector from A
s_b side-bisector from B
s_c side-bisector from C

$w_\alpha, w_\beta, w_\gamma$ angle-bisectors of interior angles
$w_{\alpha_1}, w_{\beta_1}, w_{\gamma_1}$ angle-bisectors of exterior angles

r radius of the circumcircle
ϱ radius of the incircle
$\varrho_a, \varrho_b, \varrho_c$ radius of the escribed circle (the circle touches on the outside of the side given as index)

Angle theorems

$$\alpha + \beta + \gamma = 180° \qquad \alpha_1 + \beta_1 + \gamma_1 = 360°$$

$$\alpha_1 = \beta + \gamma; \quad \beta_1 = \alpha + \gamma; \quad \gamma_1 = \alpha + \beta$$

Side theorems

$$a + b > c \qquad |a - b| < c$$

$$b + c > a \qquad |b - c| < a$$

$$a + c > b \qquad |a - c| < b$$

Side-bisectors

Point of intersection of the side-bisectors with each other = the *centroid* S of the triangle. The distance of the centroid from one side is equal to one third of the associated altitude.
The centroid divides the side-bisectors in the ratio 2:1 (from the vertex)

$$s_a = \frac{1}{2}\sqrt{2(b^2 + c^2) - a^2}$$

$$s_b = \frac{1}{2}\sqrt{2(a^2 + c^2) - b^2}$$

$$s_c = \frac{1}{2}\sqrt{2(a^2 + b^2) - c^2}$$

Altitudes

$$h_a : h_b : h_c = \frac{1}{a} : \frac{1}{b} : \frac{1}{c}$$

Angle-bisectors

Each angle-bisector w_α, w_β, w_γ divides the opposite side *internally* in the ratio of the sides adjacent to the angle bisected.

Each angle-bisector w_{α_1}, w_{β_1}, w_{γ_1} divides the opposite side *externally* in the ratio of the adjacent sides.

$$\overline{AD}:\overline{BD} = b:a$$ (cf. graph on page 137)

$$\overline{AE}:\overline{BE} = b:a$$

$$w_\alpha = \frac{1}{b+c}\sqrt{bc[(b+c)^2 - a^2]}$$

$$w_\beta = \frac{1}{a+c}\sqrt{ac[(a+c)^2 - b^2]}$$

$$w_\gamma = \frac{1}{a+b}\sqrt{ab[(a+b)^2 - c^2]}$$

Circumcircle, incircle, ecircles

Center of circumcircle = point of intersection of the three midperpendiculars (perpendiculars to the sides in their midpoints)

Center of incircle = point of intersection of the three angle-bisectors w_α, w_β, w_γ

Centers of escribed circles = points of intersection of w_α, w_{β_1}, w_{γ_1} or of w_β, w_{α_1}, w_{γ_1} or of w_γ, w_{α_1}, w_{β_1}

$$r = \frac{bc}{2h_a} = \frac{ac}{2h_b} = \frac{ab}{2h_c} = \frac{abc}{4A}$$

$$\varrho = \frac{A}{s} = \sqrt{\frac{(s-a)(s-b)(s-c)}{s}}$$

Distance of the center of the circumcircle from the center of the incircle $= \sqrt{r^2 - 2r\varrho}$

$$\varrho_a = \frac{A}{s-a};\ \varrho_b = \frac{A}{s-b};\ \varrho_b = \frac{A}{s-c}$$

$$\frac{1}{\varrho_a} = \frac{1}{h_b} + \frac{1}{h_c} - \frac{1}{h_a} \qquad \frac{1}{\varrho} = \frac{1}{h_a} + \frac{1}{h_b} + \frac{1}{h_c}$$

$$\frac{1}{\varrho_b} = \frac{1}{h_a} + \frac{1}{h_c} - \frac{1}{h_b} \qquad \frac{1}{\varrho} = \frac{1}{\varrho_a} + \frac{1}{\varrho_b} + \frac{1}{\varrho_c}$$

$$\frac{1}{\varrho_c} = \frac{1}{h_a} + \frac{1}{h_b} - \frac{1}{h_c} \qquad \varrho_a + \varrho_b + \varrho_c = 4r + \varrho$$

Area

$$A = \frac{abc}{4r} = \sqrt{\varrho\varrho_a\varrho_t\varrho_c}$$

$$A = \frac{ah_a}{2} = \frac{bh_b}{2} = \frac{ch_c}{2}$$

$$A = \sqrt{s(s-a)(s-b)(s-c)} \quad \text{(Heron's formula)}$$

$$A = \varrho s = \varrho_a(s-a) = \varrho_b(s-b) = \varrho_c(s-c)$$

$$A = \frac{1}{2}\,ab \sin\gamma = \frac{1}{2}\,bc \sin\alpha = \frac{1}{2}\,ac \sin\beta$$

$$A = \frac{a^2 \sin\beta \sin\gamma}{2 \sin\alpha} = \frac{b^2 \sin\alpha \sin\gamma}{2 \sin\beta} = \frac{c^2 \sin\alpha \sin\beta}{2 \sin\gamma}$$

$$A = 2r^2 \sin\alpha \sin\beta \sin\gamma$$

$$A = s^2 \tan\frac{\alpha}{2} \tan\frac{\beta}{2} \tan\frac{\gamma}{2}$$

$$A = \varrho^2 \cot\frac{\alpha}{2} \cot\frac{\beta}{2} \cot\frac{\gamma}{2}$$

If the vertices of a triangle are given in a rectangular system of coordinates, $A(x_1, y_1, z_1)$, $B(x_2, y_2, z_2)$, $C(x_3, y_3, z_3)$, the following holds:

$$A = \frac{1}{2}\,[x_1(y_2 - y_3) + x_2(y_3 - y_1) + x_3(y_1 - y_2)]$$

The indices of the three summands in square brackets result from one another by *interchanging in cyclic order*.

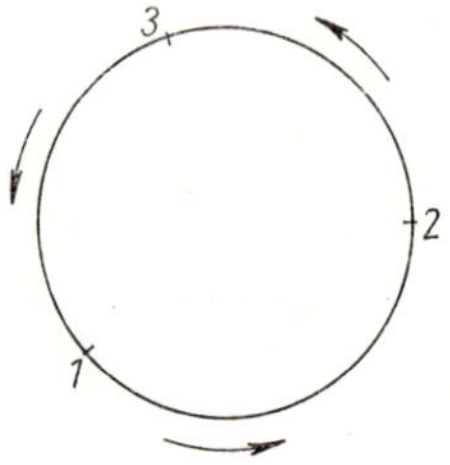

$$A = \frac{1}{2}\left[(x_1 - x_2)(y_1 + y_2) + (x_2 - x_3)(y_2 + y_3) + (x_3 - x_1)(y_3 + y_1)\right]$$

$$= \frac{1}{2} \cdot \begin{vmatrix} x_1 & y_1 & 1 \\ x_2 & y_2 & 1 \\ x_3 & y_3 & 1 \end{vmatrix} = \frac{1}{2} \cdot \begin{vmatrix} x_2 - x_1 & y_2 - 1 \\ x_3 - x_1 & y_3 - 1 \end{vmatrix}$$

The sign of the area is positive if the vertices of the triangle succeed one another in the counterclockwise sence (*mathematical sense*), otherwise the sign is negative. In the latter case, the absolute value must be taken.

If the vertices of a triangle are given in polar coordinates, i.e. $A(r_1, \varphi_1)$, $B(r_2, \varphi_2)$, $C(r_3, \varphi_3)$, we have:

$$A = \frac{1}{2}\left[r_1 r_2 \sin(\varphi_1 - \varphi_2) + r_2 r_3 \sin(\varphi_2 - \varphi_3) + r_3 r_1 \sin(\varphi_3 - \varphi_1)\right]$$

Generalized theorem of Pythagoras

$$a^2 = b^2 + c^2 \pm 2bp \qquad \alpha \gtrless 90^\circ \quad p \text{ projection of } c \text{ on } b$$

$$b^2 = c^2 + a^2 \pm 2cq \quad \text{for } \beta \gtrless 90^\circ \quad q \text{ projection of } a \text{ on } c$$

$$c^2 = a^2 + b^2 \pm 2ar \qquad \gamma \gtrless 90^\circ \quad r \text{ projection of } b \text{ on } a$$

Basic problems on triangles

A triangle can be uniquely determined if there are given
one side and two angles or
two sides and the included angle or
two sides and the angle opposite the larger side or
three sides.

The right-angled triangle

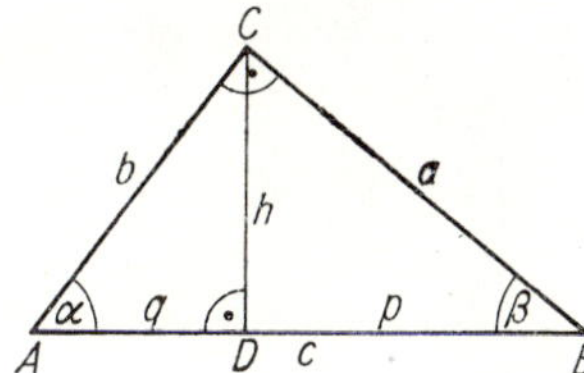

Notations

a and b legs ($\gamma = 90°$)
c hypotenuse
h altitude
p projection of a on c
q projection of b on c

Theorem of Pythagoras

$$a^2 + b^2 = c^2 \qquad (\gamma = 90°)$$

Pythagorean numbers

Pythagorean numbers satisfy the equation $a^2 + b^2 = c^2$, $a, b, c \in I$. Pythagorean numbers are obtained by putting $a = 2pq$, $b = p^2 - q^2$, $c = p^2 + q^2$ $(p, q \in I)$.

Table of Pythagorean numbers:

p	q	a	b	c
2	1	4	3	5
3	1	6	8	10
4	1	8	15	17
5	1	10	24	26
3	2	12	5	13
4	2	16	12	20
5	2	20	21	29
4	3	24	7	25
5	3	30	16	34
5	4	40	9	41 etc.

We obtain further Pythagorean numbers when we replace the a, b, c values belonging together in the above table by λa, λb, λc ($\lambda \in N$). If the dimensions of a triangle are Pythagorean numbers, the triangle is right-angled.

Euclid's theorem

$$a^2 = cp; \quad b^2 = cq$$

Altitude theorem (Euclid)

$$h^2 = pq$$

Area

$$A = \frac{ch}{2} = \frac{ab}{2}$$

Distance of the *centroid* S from the hypotenuse $= \frac{1}{3}\,h$

Distance of the *centroid* S from the side $a = \frac{1}{3}\,b$

Distance of the *centroid* S from the side $b = \frac{1}{3}\,a$

Equilateral triangle

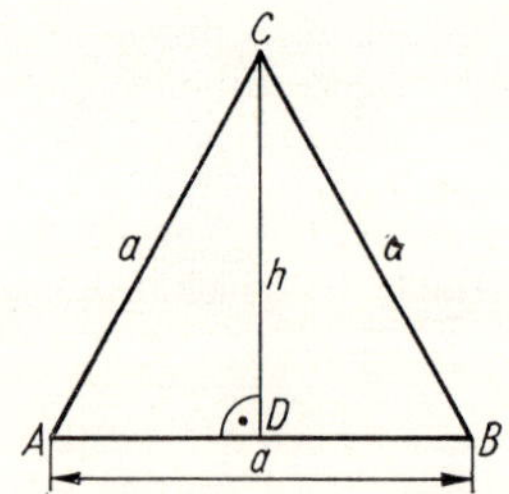

$$h = \frac{a\sqrt{3}}{2}$$

$$A = \frac{a^2\sqrt{3}}{4}$$

Distance of the *centroid* S from one side $= \frac{a}{6}\sqrt{3}$

3.2.2. Quadrilaterals

Notations

$\alpha, \beta, \gamma, \delta$	interior angles
$\alpha_1, \beta_1, \gamma_1, \delta_1$	exterior angles
a, b, c, d	sides
e, f	diagonals
ϱ	radius of the incircle
h_a	altitude associated with side a
h_b	altitude associated with side b

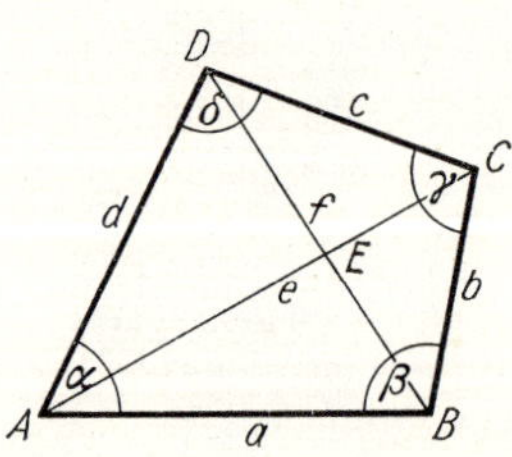

Sums of angles

$$\alpha + \beta + \gamma + \delta = 360°$$

$$\alpha_1 + \beta_1 + \gamma_1 + \delta_1 = 360°$$

Parallelogram

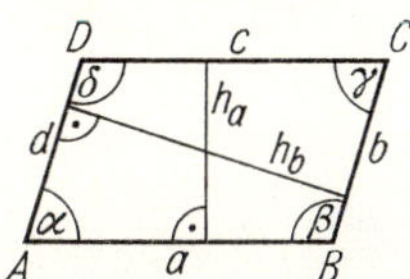

$a \parallel c; \quad b \parallel d; \quad a = c; \quad b = d$

$\alpha = \gamma; \quad \beta = \delta$

$$\alpha + \beta = \beta + \gamma = \gamma + \delta = \delta + \alpha = 180°$$

The diagonals bisect each other.
Centroid S = point of intersection of the diagonals.

$A = ah_a = bh_b = gh$ — g base line, h associated altitude

Rectangle

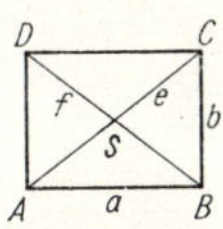

$$e = f = \sqrt{a^2 + b^2}$$

The diagonal bisect each other.

$$A = ab$$

Centroid S = point of intersection of the diagonals.

Square

The diagonals e and f bisect each other and are perpendicular to each other.

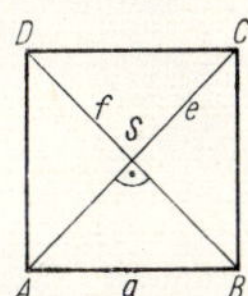

$$e = f = a\sqrt{2}$$

$$A = a^2 = \frac{1}{2} e^2$$

Centroid S = point of intersection of the diagonals.

Trapezium

$$a \parallel c$$

$$A = \frac{a + c}{2} h = mh$$

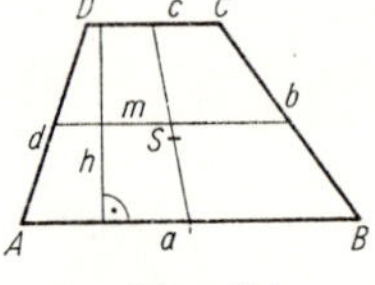

m midparallel

The *centroid* S lies on the line joining the midpoints on the parallel base sides at a distance $\frac{h}{3} \cdot \frac{a + 2c}{a + c}$ from the base a.

Rhombus

$$a = b = c = d$$

The diagonals e and f are perpendicular to each other, bisect each other and the rhombus angles.

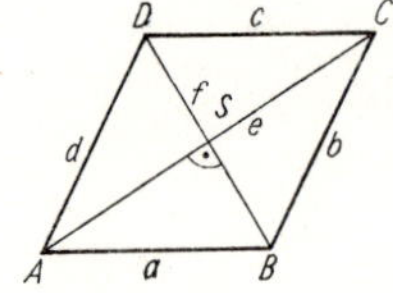

$$A = \frac{ef}{2}$$

Centroid S = point of intersection of the diagonals.

Quadrilateral with circumcircle

The sides of such a quadrilateral are chords of the circumcircle.

$$\alpha + \gamma = 180°; \quad \beta + \delta = 180°$$

$$A = \sqrt{(s - a)(s - b)(s - c)(s - d)}; \quad s = \frac{a + b + c + d}{2}$$

$$ac + bd = ef \quad \text{(Ptolemy's theorem)}$$

$$e = \sqrt{\frac{(ac + bd)(bc + ad)}{ab + cd}}; \quad f = \sqrt{\frac{(ac + bd)(ab + cd)}{bc + ad}}$$

$$r = \frac{1}{4} \sqrt{\frac{(ab + cd)(ac + bd)(bc + ad)}{(s - a)(s - b)(s - c)(s - d)}}$$

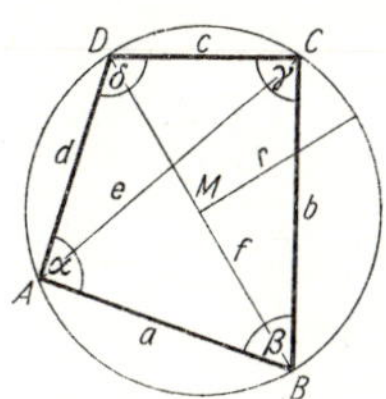

Quadrilateral with circumcircle

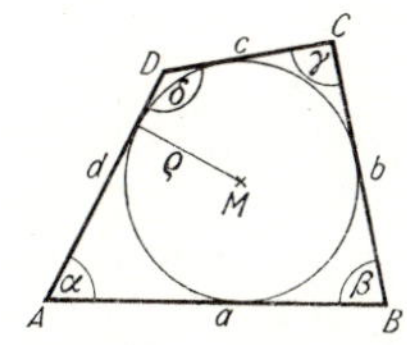

Quadrilateral with incircle

Quadrilateral with incircle

The sides of such a quadrilateral are tangents of the incircle.

$$a + c = b + d$$

$A = \varrho s$ For s see "quadrilateral with circumcircle"

The isosceles kite-shaped quadrilateral or deltoid

$$A = \frac{ef}{2}$$

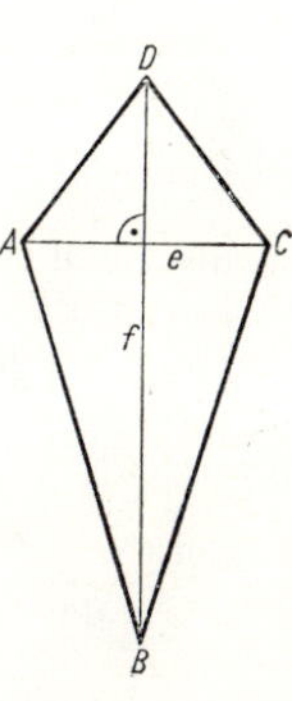

deltoid

3.2.3. Polygons (n-sided polygons)

Sum of angles of the n-sided polygon

Sum of interior angles $= (2n - 4) \cdot 90°$
Sum of exterior angles $= 360°$

Number of diagonals of the n-sided polygon

$$\frac{n(n-3)}{2}$$

Area of the n-sided polygon

The n-sided polygon is divided either into triangles by diagonals drawn from a vertex or into right-angled triangles and trapeziums by perpendiculars drawn from the vertices to a base line.

$$A = \sum_{1}^{k} A_i$$

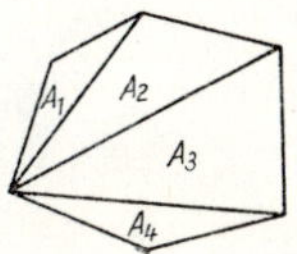

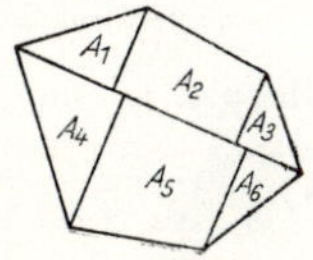

Regular n-sided polygon

The sides are all of equal length and the angles are all equal.

Notations

s_n side of the regular n-gon
s_{2n} side of the regular $2n$-gon
r radius of the circumcircle
ϱ_n radius of the incircle
α interior angle
α_1 exterior angle

Centroid S = center of the circumcircle

Angles of the regular n-gon:

$$\alpha = \frac{2n-4}{n} \cdot 90^\circ \qquad \alpha_1 = \frac{360^\circ}{n}$$

Relations between s_n, r, and ϱ_n

$$\varrho_n = \frac{1}{2}\sqrt{4r^2 - s_n^{\,2}}$$

Calculation of the side s_{2n} from s_n:

$$s_{2n} = \sqrt{2r^2 - r\sqrt{4r^2 - s_n^{\,2}}}$$

$$s_{2n} = \sqrt{2r^2 - 2r\varrho_n}$$

Area of the n-gon:

$$A_n = \frac{ns_n\varrho_n}{2} = \frac{ns_n\sqrt{4r^2 - s_n^{\,2}}}{4}$$

$$A_n = \frac{nr^2}{2}\sin\gamma \qquad \gamma \text{ center angle}$$

Simple regular polygons

Given: circumcircle radius r

Regular triangle (equilateral triangle)

$$s_3 = r\sqrt{3} \qquad \varrho_3 = \frac{r}{2} \qquad A = \frac{3r^2\sqrt{3}}{4}$$

Regular quadrilateral (square)

$$s_4 = r\sqrt{2} \quad \varrho_4 = \frac{r}{2}\sqrt{2} \qquad A = 2r^2$$

Regular pentagon

$$s_5 = \frac{r}{2}\sqrt{10 - 2\sqrt{5}} \qquad \varrho_5 = \frac{r}{4}\left(\sqrt{5} + 1\right)$$

$$A = \frac{5}{8}r^2\sqrt{10 + 2\sqrt{5}}$$

Regular hexagon

$$s_6 = r \quad \varrho_6 = \frac{r}{2}\sqrt{3} \qquad A = \frac{3}{2}r^2\sqrt{3}$$

Regular octagon

$$s_8 = r\sqrt{2 - \sqrt{2}} \qquad \varrho_8 = \frac{r}{2}\sqrt{2 + \sqrt{2}} \qquad A = 2r^2\sqrt{2}$$

Regular decagon

$$s_{10} = \frac{r}{2}\left(\sqrt{5} - 1\right) \qquad \varrho_{10} = \frac{r}{4}\sqrt{10 + 2\sqrt{5}}$$

$$A = \frac{5}{4}r^2\sqrt{10 - 2\sqrt{5}}$$

Given: side s_n of the polygon

Regular triangle (equilateral triangle)

$$r = \frac{s_3}{3}\sqrt{3} \quad \varrho_3 = \frac{s_3}{6}\sqrt{3} \quad A = \frac{s_3^2}{4}\sqrt{3}$$

Regular quadrilateral (square)

$$r = \frac{s_4}{2}\sqrt{2} \quad \varrho_4 = \frac{s_4}{2} \quad A = s_4^2$$

Regular pentagon

$$r = \frac{s_5}{10}\sqrt{50 + 10\sqrt{5}} \qquad \varrho_5 = \frac{s_5}{10}\sqrt{25 + 10\sqrt{5}}$$

$$A = \frac{s_5^{\,2}}{4}\sqrt{25 + 10\sqrt{5}}$$

Regular hexagon

$$r = s_6 \qquad \varrho_6 = \frac{s_6}{2}\sqrt{3} \qquad A = \frac{3}{2}s_6^{\,2}\sqrt{3}$$

Regular octagon

$$r = \frac{s_8}{2}\sqrt{4 + 2\sqrt{2}} \qquad \varrho_8 = \frac{s_8}{2}\left(\sqrt{2} + 1\right)$$

$$A = 2s_8^{\,2}\left(\sqrt{2} + 1\right)$$

Approximate value for the side of the regular nonagon ($r = 1$)

$$s_9 \approx \frac{2\sqrt{5} + 1}{8} \qquad \text{(after H. Kissel, Ludwigshafen/Rh.)}$$

Regular decagon

$$r = \frac{s_{10}}{2}\left(\sqrt{5} + 1\right) \qquad \varrho_{10} = \frac{s_{10}}{2}\sqrt{5 + 2\sqrt{5}}$$

$$A = \frac{5}{2}s_{10}^{\,2}\sqrt{5 + 2\sqrt{5}}$$

Construction of the simple regular polygons

Given: radius r of the circumcircle
For *regular triangle* see regular hexagon.

Regular quadrilateral and octagon

Two mutually perpendicular diameters are drawn within a circle of radius r and their endpoints are joined. Bisecting the angles that the diagonals make with one another and joining the points of intersection of the bisectors with the circle, the vertices of a regular octagon are obtained. The procedure can be continued to obtain the regular 2^n-gon ($n \in N$).

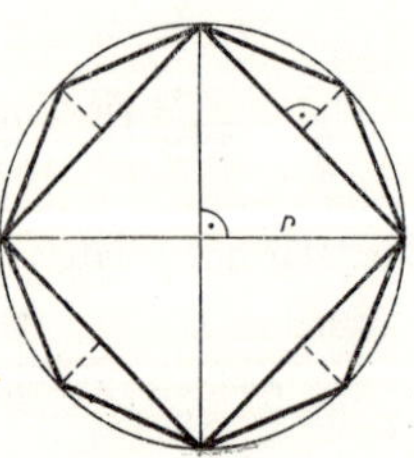

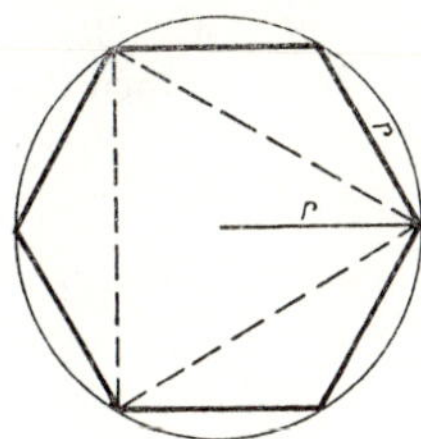

Regular hexagon and triangle

If chords of length r are successively laid out on the circumference of a circle of radius r, the six vertices of a regular hexagon are obtained. Omit alternate vertices, and the three vertices of a regular triangle inscribed in the same circle are obtained.

Regular dodecagon

By erection of the midperpendiculars on the sides of the regular hexagon and extending them so as to intersect the circle, the vertices of the regular dodecagon are obtained. This procedure can be repeated to construct the regular polygon with 24 vertices, with 48 vertices, ..., in general the regular polygon with $3 \cdot 2^n$ vertices ($n \in N$).

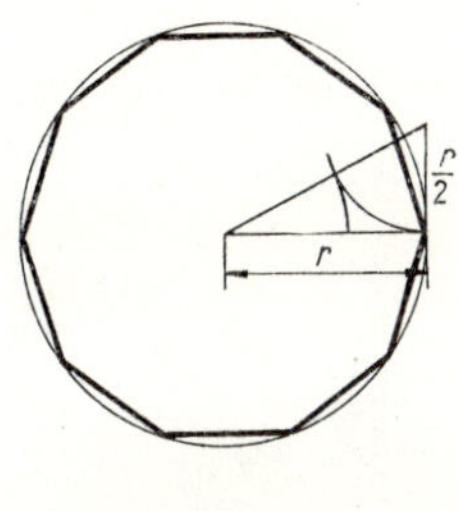

Regular decagon and pentagon

By dividing the given radius r in continuous proportion (golden section) and drawing the larger segment of the radius as chord of the circle ten times in succession, we obtain the regular decagon.
The regular pentagon is obtained from the regular decagon by omitting alternate vertices. The regular 20-gon, regular 40-gon, ..., in general the regular $5 \cdot 2^n$-gon ($n \in N$) are obtained from the decagon by successive bisections of the central angles.

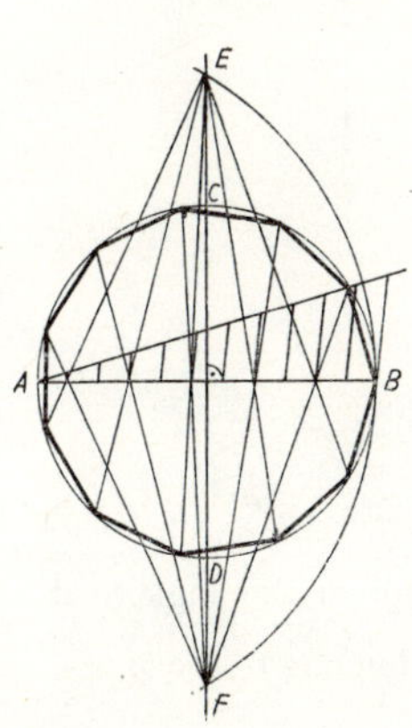

Approximate constructions of any regular n-gon

Given: Radius r of the circumcircle

Two mutually perpendicular diameters AB and CD are drawn within the circle of the given radius r. Then one diameter (in the figure $\overline{AB}$) is divided into n equal parts (in the figure $n = 11$) and a circle of radius $2r$ is drawn about one endpoint (in the figure endpoint A). The circle intersects the extensions of the other diameter in E and F. Rays are drawn from E and F through the points of division; then the points of intersection of the rays with the original circle are the vertices of the n-gon.

3.2.4. Circle

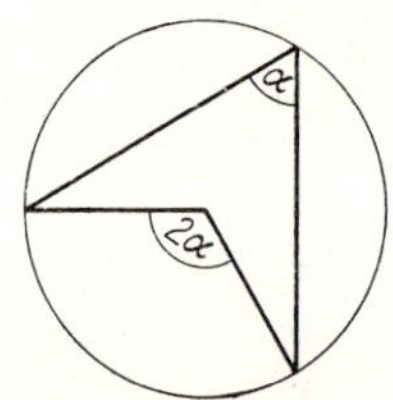

Notations

r radius
d diameter

The *peripheral angle* is equal to one half of the *central angle* above the same arc.

Theorem of Thales

The peripheral angle in the semicircle is 90°.
The *chord-tangent angle* is equal to half the central angle over the same arc and, hence, equal to the peripheral angle over the same arc.
(Chord theorem, secant theorem, tangent-secant theorem see page 137, 138)

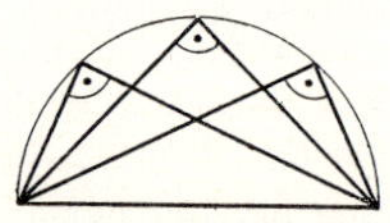

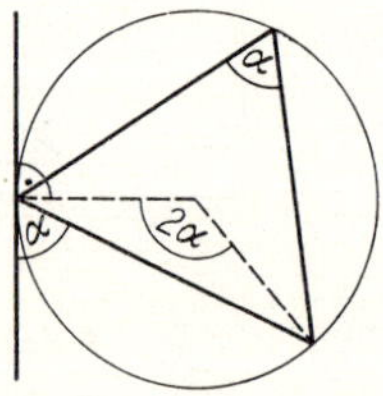

Circumference of a circle

$$u = 2\pi r = \pi d$$

Arc of a circle

$$b = \frac{\pi r \alpha}{180°} = \frac{\pi d \alpha}{360°} = r \operatorname{arc} \alpha = r\hat{\alpha}$$

α associated central angle

The *centroid* S of the arc b lies on the bisector of the angle α at a distance $\frac{rs}{b}$ from the center, s is the chord belonging to arc b.

Area of a circle

$$A = \pi r^2 = \frac{\pi d^2}{4}$$

Centroid S = center of the circle

Area of a sector of a circle

$$A = \frac{\pi r^2 \alpha}{360^\circ} = \frac{br}{2}$$

$$A = \frac{r^2}{2} \operatorname{arc} \alpha = \frac{r^2}{2} \hat{\alpha}$$

Centroid S lies on the axis of symmetry at a distance $\frac{2}{3} \cdot \frac{rs}{b}$ from the center M.

Area of a segment of a circle

$$s = 2\sqrt{2hr - h^2}$$

$$h = r - \frac{1}{2}\sqrt{4r^2 - s^2} \quad \text{for} \quad h < r$$

$$A = \text{sector} - \text{triangle } AMB$$

$$A = \frac{1}{2}[br - s(r - h)]$$

$$A \approx \frac{2}{3} hs$$

$$A = \frac{r^2}{2}\left(\frac{\pi\alpha}{180^\circ} - \sin\alpha\right)$$

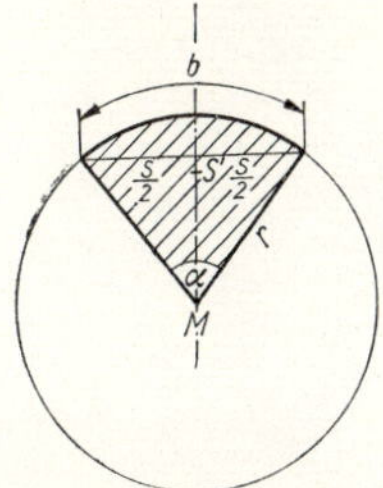

Sector of a circle

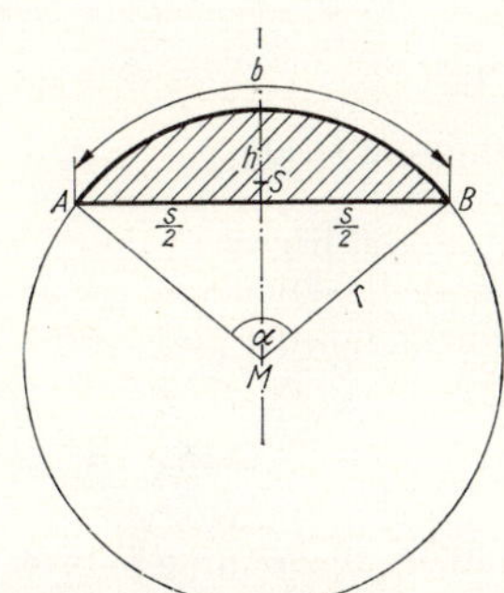

Segment of a circle

Centroid S is on the axis of symmetry at a distance $\frac{s^3}{12A}$ from the center M.

Circular ring

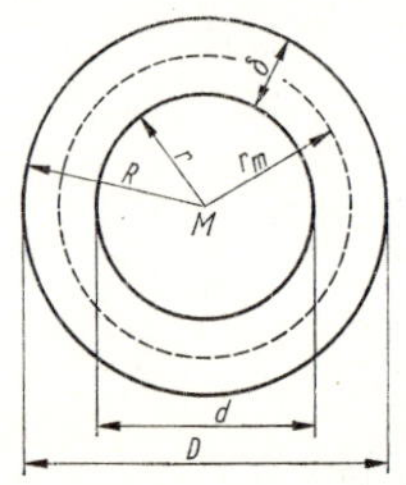

$$A = \pi(R^2 - r^2)$$

$$= \frac{\pi}{4}(D^2 - d^2) = 2\pi r_m \delta$$

$$\delta = R - r$$

Centroid S = center M

3.3. Stereometry

Notations

V	volume	A_B	base area
A_S	surface	A_T	area of upper surface
h	height	A_C	area of curved surface

3.3.1. General theorems

Cavalieri's theorem

If solid bodies are of equal height and have bases of equal area, and if all plane sections parallel to the bases and at the same distances from the corresponding bases are equal in area, then the solids are equal in volume.

Simpson's rule

If a solid has two parallel bases A_B and A_T and if, at height x, each parallel section has an area which at most is a rational integral function of the third degree of x, then

$$V = \frac{h}{6}(A_B + A_T + 4A_m) \qquad A_m \text{ median section}$$

Euler's theorem on polyhedra

Polyhedra are solid bodies that are bounded only by plane polygons.

On the condition that a polyhedron has no reentrant corner, the following holds:

$$e + f - k = 2$$

e number of corners
f number of faces
k number of edges

Guldin's rule

The volume of a solid of revolution is equal to the product of the area of the generatrix and the circumference of the circle described by its centroid:

$$V = 2\pi y_0 A = 2\pi M_x$$

A generatrix

y_0 distance of its centroid from the axis of revolution
M_x static moment

The surface area of a solid of revolution is equal to the product of the length of the generatrix and the circumference of the circle described by its centroid:

$$A_T = 2\pi y_0 s$$

s rotating generatrix
y_0 distance of its centroid from the axis of revolution

Solid angles

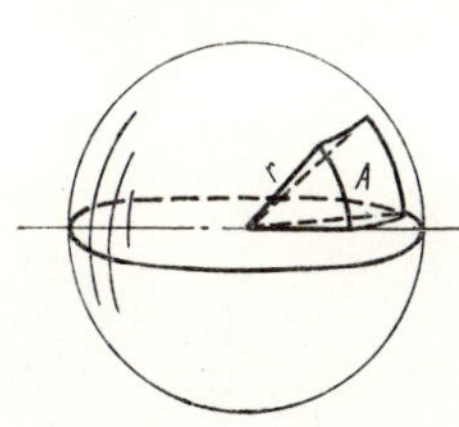

A solid angle can be measured by the ratio of a plane A cut out from a sphere (about the vertex of the angle) to the square of the radius of the sphere. That solid angle is taken as the unit for which the ratio of the spherical surface to the square of the radius of the sphere has the numerical value 1. This unit is called steradian or spheradian (symbol sr).

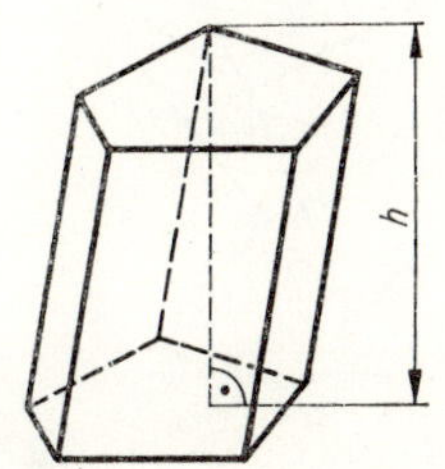

3.3.2. Solids bounded by plane surfaces

Notation

d body diagonal

Prism (right and oblique)

$$V = A_B h$$

$$A_S = A_C + 2A_B$$

Centroid S = point of bisection of the join between the centroids of the base and upper surfaces.

Right parallelepiped cuboid

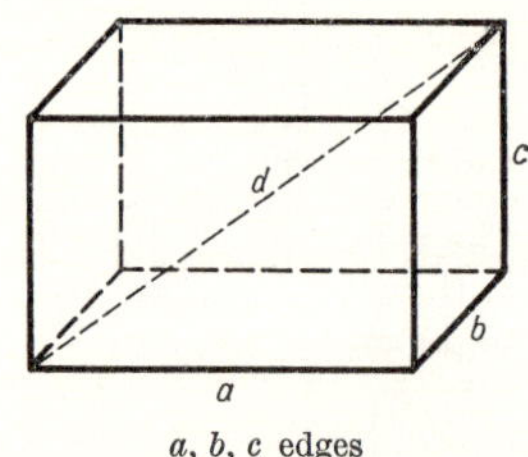

a, b, c edges

$$V = abc$$

$$A_S = 2(ab + ac + bc)$$

$$d = \sqrt{a^2 + b^2 + c^2}$$

Centroid S = point of intersection of the body diagonals

Cube

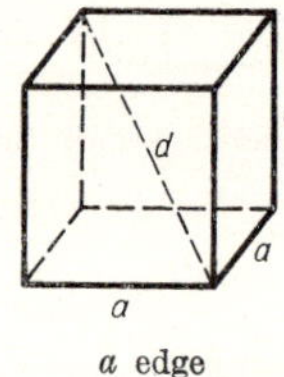

a edge

$$V = a^3$$

$$A_S = 6a^2$$

$$d = a\sqrt{3}$$

Centroid S = point of intersection of the body diagonals

Obliquely cut right three-sided prism

$$V = A_B \frac{a + b + c}{3} \quad \text{(figure to the left)}$$

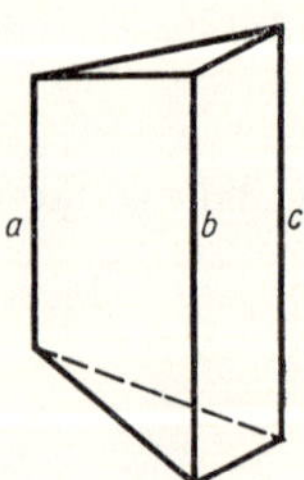

a, b, c the three parallel edges

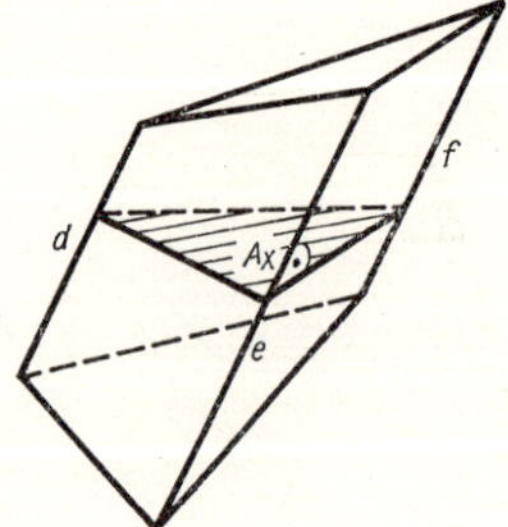

A_X section perpendicular to the edges;
d, e, f the three parallel edges

Obliquely cut oblique three-sided prism

$$V = A_X \frac{d + e + f}{3}$$ (figure to the right, page 158, bottom)

Obliquely cut n-sided prism

$$V = A_X s$$

s segment joining the centroids of the base and upper surfaces
A_X sections perpendicular to s

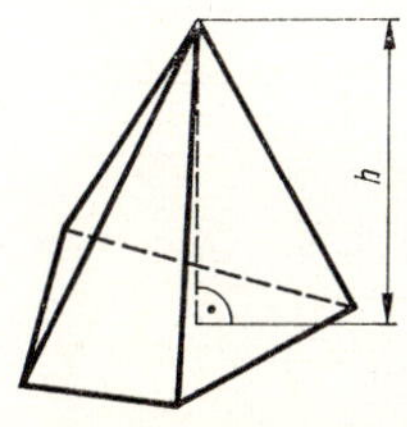

Pyramid (right and oblique)

$$V = \frac{1}{3} A_B h \qquad A_S = A_B + A_C$$

Centroid S lies on the line joining the vertex with the centroid of the base surface at a distance $\frac{h}{4}$ from the base surface.

Three-sided pyramid

Vertices of the base

$$P_1(x_1, y_1, z_1), \qquad P_2(x_2, y_2, z_2), \qquad P_3(x_3, y_3, z_3)$$

Vertex at the origin

$$V = \frac{1}{6} \cdot \begin{vmatrix} x_1 & y_1 & z_1 \\ x_2 & y_2 & z_2 \\ x_3 & y_3 & z_3 \end{vmatrix}$$

V has the positive sign if the vertices P_1, P_2, P_3 succeed one another in the mathematically positive sense (counterclockwise sense).

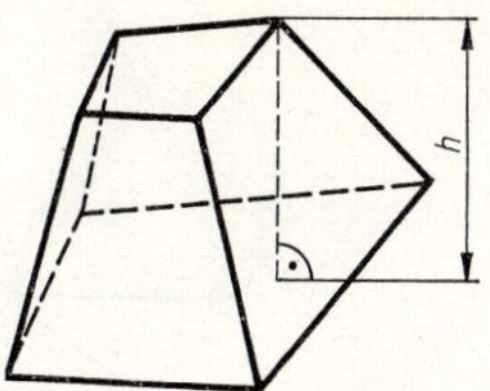

Truncated pyramid (frustrum)

$$V = \frac{h}{3}\left(A_B + \sqrt{A_B A_T} + A_T\right)$$

$$A_S = A_B + A_T + A_C$$

Centroid S lies on the line joining the centroids of base and upper surfaces. Distance from the

$$\text{base} = \frac{h}{4} \cdot \frac{A_B + 2\sqrt{A_B A_T} + 3A_T}{A_B + \sqrt{A_B A_T} + A_T}$$

Formula of approximation for the truncated pyramid

$$V \approx \frac{A_B + A_T}{2} h$$

The formula yields good approximate values if A_B differs but slightly from A_T.

The five regular polyhedra

Notations

a side
r radius of the circumscribed sphere
ϱ radius of the inscribed sphere

Tetrahedron (bounded by four equilateral triangles)

$$V = \frac{a^3 \sqrt{2}}{12} = \frac{1}{6} \cdot \begin{vmatrix} x_1 & y_1 & z_1 \\ x_2 & y_2 & z_2 \\ x_3 & y_3 & z_3 \end{vmatrix}$$ (see also "three-sided pyramid")

$$A_S = a^2 \sqrt{3}$$

$$r = \frac{a}{4} \sqrt{6}\,; \quad \varrho = \frac{a}{12} \sqrt{6}$$

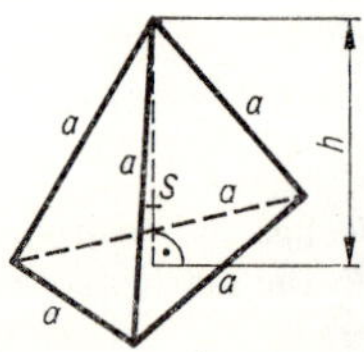

Centroid S lies on the altitude line at a distance $\frac{h}{4}$ from the base surface.

Octahedron (bounded by eight equilateral triangles)

$$V = \frac{a^3 \sqrt{2}}{3}\,; \quad A_S = 2a^2 \sqrt{3}$$

$$r = \frac{a}{2} \sqrt{2}\,; \quad \varrho = \frac{a}{6} \sqrt{6}$$

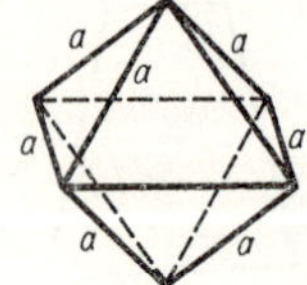

Centroid S = point of intersection of the diagonals of the common base square

Icosahedron (bounded by 20 equilateral triangles)

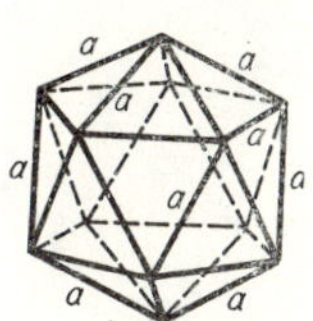

$$V = \frac{5a^3(3 + \sqrt{5})}{12}$$

$$A_S = 5a^2 \sqrt{3}$$

$$r = \frac{a}{4} \sqrt{2(5 + \sqrt{5})}; \qquad \varrho = \frac{a\sqrt{3}\,(3 + \sqrt{5})}{12}$$

Hexahedron (cube) (bounded by six squares)

$$V = a^3; \qquad A_S = 6a^2;$$

$$r = \frac{a}{2}\sqrt{3}; \qquad \varrho = \frac{a}{2}$$

Centroid S = point of intersection of the body diagonals

Dodecahedron (bounded by 12 regular pentagons)

$$V = \frac{a^3(15 + 7\sqrt{5})}{4}; \qquad A_S = 3a^2 \sqrt{5\,(5 + 2\sqrt{5})}$$

$$r = \frac{a\sqrt{3}\,(1 + \sqrt{5})}{4}; \qquad \varrho = \frac{a\sqrt{10(25 + 11\sqrt{5})}}{20}$$

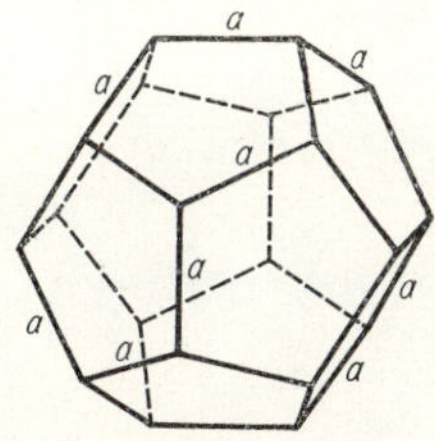

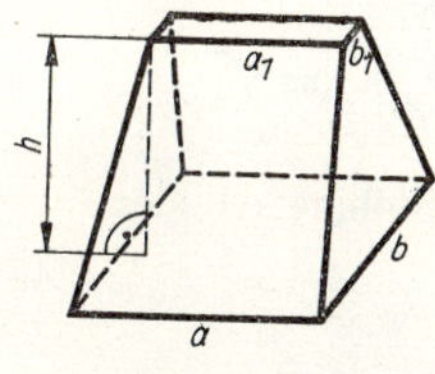

Obelisk

Obelisk

Base and upper surfaces are nonsimilar parallel rectangles, the lateral surfaces are trapeziums.

$$V = \frac{h}{6}\,[(2a + a_1)b + (2a_1 + a)b_1]$$

$$= \frac{h}{6}\,[ab + (a + a_1)(b + b_1) + a_1 b_1]$$

Distance of the *centroid* S from the base surface ab is

$$\frac{h}{2} \cdot \frac{ab + ab_1 + a_1b + 3a_1b_1}{2ab + ab_1 + a_1b + 2a_1b_1}$$

Wedge

Base surface is rectangular, lateral surfaces are isosceles triangles and isosceles trapeziums.

$$V = \frac{bh}{6}(2a + a_1)$$

Distance of the *centroid* S from the base surface ab is

$$\frac{h}{2} \cdot \frac{a + a_1}{2a + a_1}$$

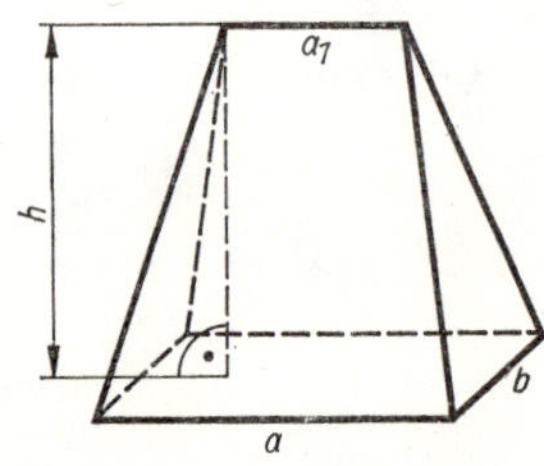

Prismatoid

Prismatoids are solid bodies having only straight edges and plane and curved boundary surfaces and whose vertices or base surfaces lie in two parallel planes (calculation with the help of SIMPSON's rule, see page 156).

3.3.3. Solids bounded by curved surfaces

Notations

r radius
d diameter

Oblique cylinder

$$V = A_B h; \quad A_C = us$$

u circumference of the section perpendicular to the axis
s generatrix

Right circular cylinder

$$V = \pi r^2 h = \frac{\pi d^2}{4} h$$

$$A_C = 2\pi r h = \pi d h$$

$$A_S = 2\pi r(r + h) = \pi d \left(\frac{d}{2} + h\right)$$

Centroid S lies on the axis at distance $\frac{h}{2}$ from the base.

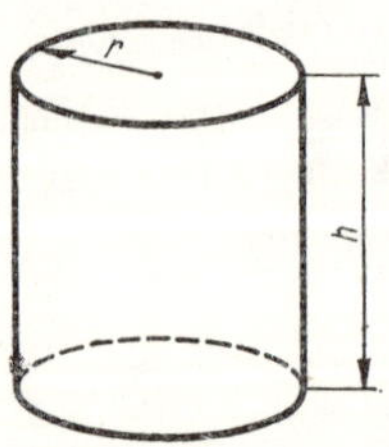

Obliquely cut right circular cylinder

$$V = \frac{\pi r^2}{2}(s_1 + s_2) = \frac{\pi d^2}{8}(s_1 + s_2)$$

$$A_C = \pi r(s_1 + s_2) = \frac{\pi d}{2}(s_1 + s_2)$$

$$A_S = \pi r\left[s_1 + s_2 + r + \sqrt{r^2 + \left(\frac{s_1 - s_2}{2}\right)^2}\right]$$

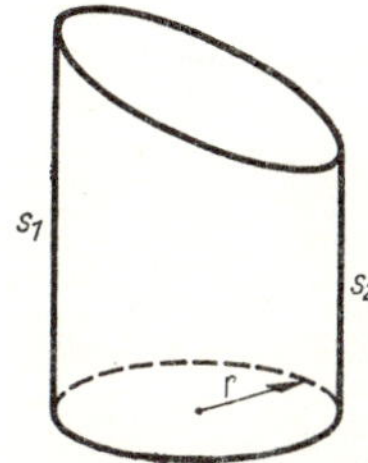

Centroid S lies on the axis at distance

$$\frac{s_1 + s_2}{4} + \frac{1}{4} \cdot \frac{r^2 \tan^2 \alpha}{s_1 + s_2}$$

from the base.
α angle of inclination which the upper surface makes with the base.

s_1, s_2 longest and shortest elements

Portion of cylinder having smaller base area than base circle

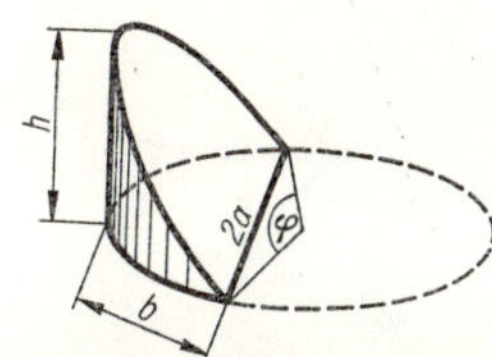

φ central angle of the base circle in circular measure, $2a$ edge of cylinder portion, h longest element, b perpendicular from the foot of h to the edge of the cylinder portion

$$V = \frac{h}{3b}\left[a(3r^2 - a^2) + 3r^2(b - r)\frac{\varphi}{2}\right]$$

$$A_C = \frac{2rh}{b}\left[(b - r)\frac{\varphi}{2} + a\right]$$

For $a = b = r$ we get

$$V = \frac{2}{3}r^2 h = \frac{d^2}{6}h$$

$$M = 2\,rh = dh$$

Hollow cylinder (tube)

$$\begin{aligned} V &= \pi h(r_1^2 - r_2^2) = \frac{\pi h}{4}(d_1^2 - d_2^2) \\ &= \pi a h(2r_1 - a) = \pi a h(2r_2 + a) \\ A_C &= 2\pi h(r_1 + r_2) = \pi h(d_1 + d_2) \\ A_S &= 2\pi(r_1 + r_2)(h + r_1 - r_2) \\ &= \pi(d_1 + d_2)\left(h + \frac{d_1 - d_2}{2}\right) \end{aligned}$$

Centroid S lies on the axis at distance $\frac{h}{2}$ from the base.

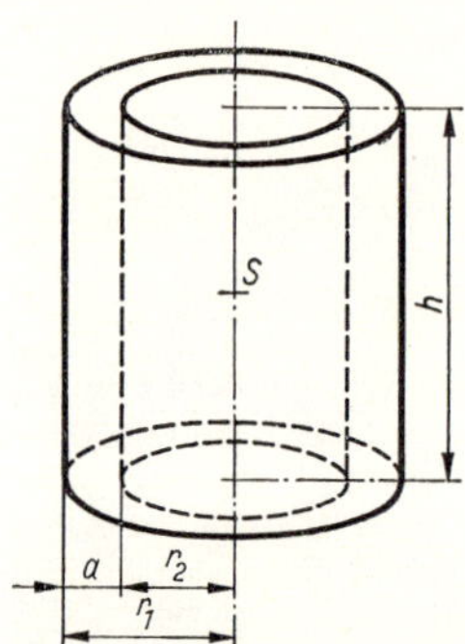

Cone (right and oblique)

$$V = \frac{1}{3} A_B h$$

$r_1(d_1)$ outer radius (diameter)
$r_2(d_2)$ inner radius (diameter)
$a = r_1 - r_2$ thickness of the wall

Right circular cone

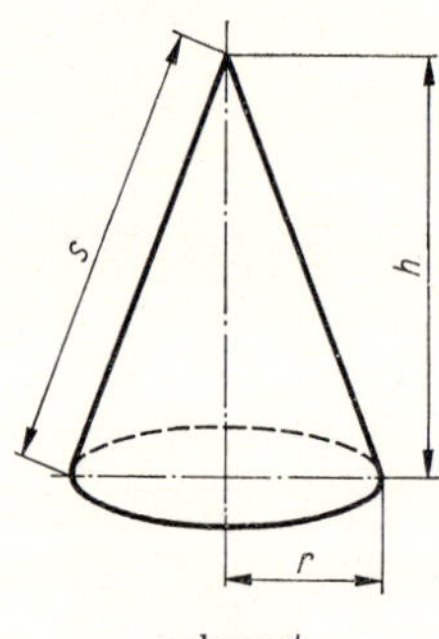

$$V = \frac{1}{3} \pi r^2 h = \frac{\pi}{12} d^2 h$$

$$A_C = \pi r s = \frac{\pi}{2} d s$$

$$A_S = \pi r (r + s) = \pi \frac{d}{2} \left(\frac{d}{2} + s \right)$$

Centroid S lies on the axis at distance $\frac{h}{4}$ from the base.

s element

Frustrum of right circular cone

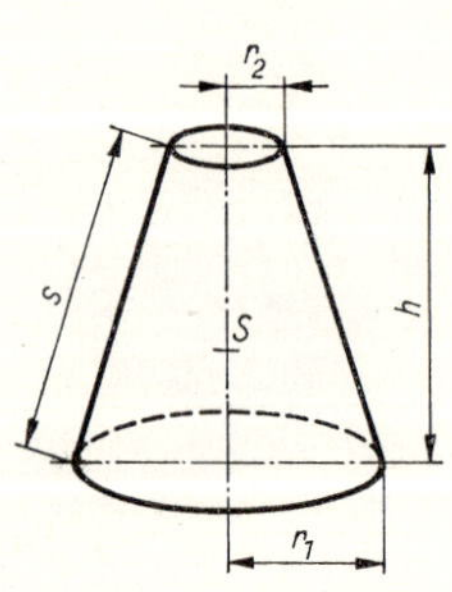

$$V = \frac{1}{3} \pi h (r_1^2 + r_1 r_2 + r_2^2)$$

$$= \frac{1}{12} \pi h (d_1^2 + d_1 d_2 + d_2^2)$$

$$A_C = \pi s (r_1 + r_2) = \frac{\pi s}{2} (d_1 + d_2)$$

$$s = \sqrt{h^2 + (r_1 - r_2)^2}$$

$$= \frac{1}{2} \sqrt{4h^2 + (d_1 - d_2)^2}$$

$$A_S = \pi [r_1^2 + r_2^2 + s(r_1 + r_2)] = \frac{\pi}{4} [d_1^2 + d_2^2 + 2s(d_1 + d_2)]$$

Centroid S lies on the axis at distance

$$\frac{h}{4} \cdot \frac{r_1^2 + 2r_1r_2 + 3r_2^2}{r_1^2 + r_1r_2 + r_2^2}$$

from the base.

Approximation formula for the volume of the frustrum of a cone

$$V \approx \frac{\pi}{2} h(r_1^2 + r_2^2) = \frac{\pi}{8} h(d_1^2 + d_2^2)$$

The formula yields good approximations if r_1 deviates but slightly from r_2.

$$V \approx \frac{\pi}{4} h(r_1 + r_2)^2 = \frac{\pi}{16} h(d_1 + d_2)^2$$

$$V \approx \pi h r_m^2 = \frac{\pi}{4} h d_m^2$$

$$r_m = \frac{r_1 + r_2}{2} \quad \text{radius of median section}$$

$$d_m = \frac{d_1 + d_2}{2} \quad \text{diameter of median section}$$

Sphere

$$V = \frac{4}{3} \pi r^3 = \frac{\pi}{6} d^3 = \frac{1}{6} \sqrt{\frac{A_S^3}{\pi}}$$

$$A_S = 4\pi r^2 = \pi d^2 = \sqrt[3]{36\pi V^2}$$

$$r = \frac{1}{2} \sqrt{\frac{A_S}{\pi}} = \sqrt[3]{\frac{3V}{4\pi}}$$

$$d = \sqrt{\frac{A_S}{\pi}} = 2 \sqrt[3]{\frac{3V}{4\pi}}$$

Centroid S = center of sphere

Spherical segment (of one base)

$$V = \frac{1}{6}\pi h(3\varrho^2 + h^2) = \frac{1}{3}\pi h^2(3r - h) = \frac{1}{6}\pi h^2(3d - 2h)$$

$$A_C = 2\pi rh = \pi dh = \pi(\varrho^2 + h^2)$$

$$A_S = \pi(2rh + \varrho^2)$$
$$= \pi(h^2 + 2\varrho^2) = \pi h(4r - h)$$

$$\varrho = \sqrt{h(2r - h)}$$

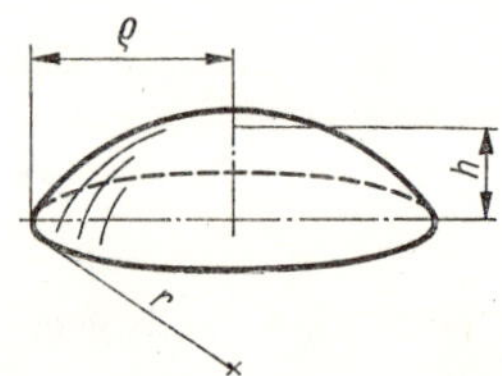

ϱ radius of the base of the segment
h altitude of the segment

Centroid S lies on the axis of symmetry of the segment at distance $\frac{3}{4} \cdot \frac{(2r - h)^2}{3r - h}$ from the center of sphere.

Spherical sector

$$V = \frac{2\pi r^2 h}{3} = \frac{\pi}{6} d^2 h$$

$$A_S = \pi r(2h + \varrho)$$

Centroid S lies on the axis of symmetry of the sector at distance $\frac{3}{8}(2r - h)$ from the center of sphere.

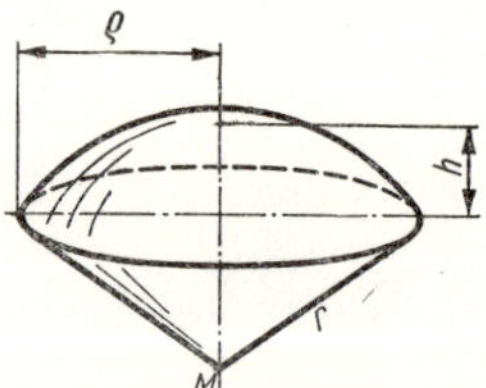

h altitude of the associated sector
ϱ radius of the base circle of the associated sector

Spherical segment of two bases

$$V = \frac{1}{6}\pi h(3\varrho_1^2 + 3\varrho_2^2 + h^2)$$

$$A_C = 2\pi rh = \pi dh$$

(spherical zone)

$$A_S = \pi(2rh + \varrho_1^2 + \varrho_2^2)$$
$$= \pi(dh + \varrho_1^2 + \varrho_2^2)$$

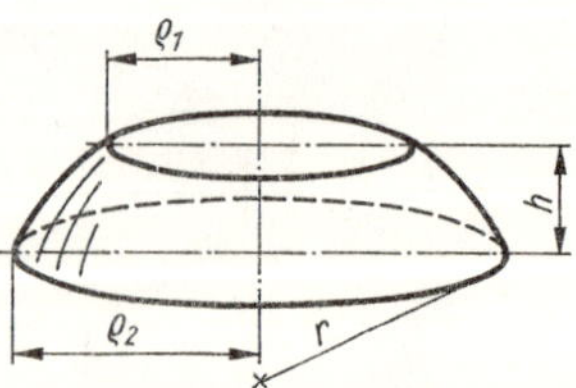

ϱ_1, ϱ_2 radii of the boundary circles
h altitude of the segment

Spherical cap

The curved part of the surface of a spherical segment with two bases is called a spherical cap (graph see page 166).

$$A = 2\pi rh = \pi dh$$

A area
h height of the associated segment

Spherical lune

$$A = \frac{\pi r^2 \alpha}{90°}$$

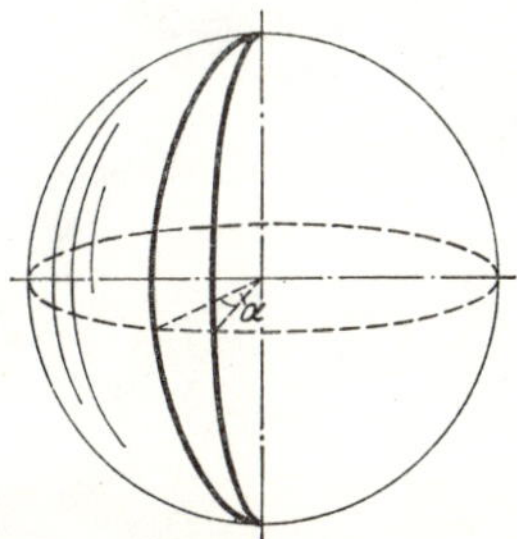

α angle between the bounding great circles

Spherical triangle

$$A = \frac{\pi r^2 \varepsilon}{180°}$$

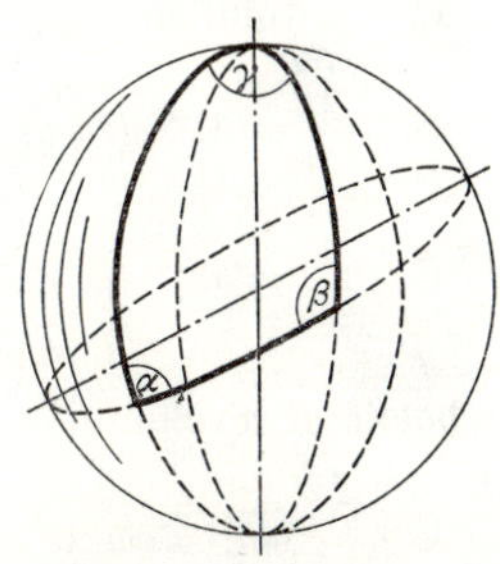

ε spherical excess
$\varepsilon = \alpha + \beta + \gamma - 180°$
α, β, γ angles of the triangle

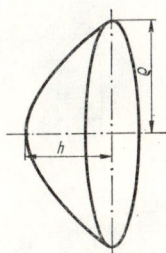

Paraboloid of revolution

$$V = \frac{1}{2}\pi\varrho^2 h$$

Centroid S lies on the axis at distance $\frac{2}{3}h$ from the vertex.

ϱ radius of the base circle bounding the body
h height

Truncated paraboloid of revolution

$$V = \frac{1}{2}\pi({\varrho_1}^2 + {\varrho_2}^2)h$$

Ellipsoid

$$V = \frac{4}{3}\pi abc$$

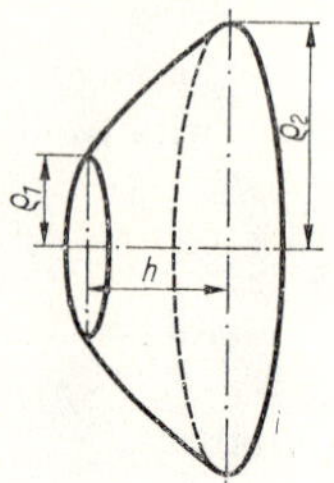

ϱ_1, ϱ_2 radii of the parallel base circles

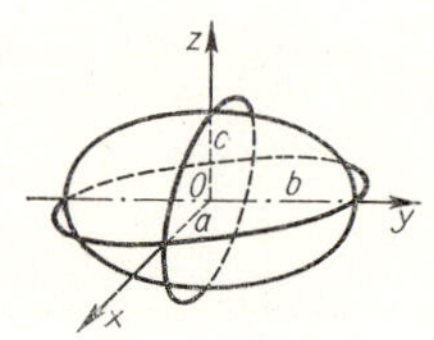

a, b, c semiaxes

Ellipsoid of revolution

$$V = \frac{4}{3}\pi ab^2 \qquad (2a \text{ axis of rotation})$$

$$V = \frac{4}{3}\pi a^2 b \qquad (2b \text{ axis of rotation})$$

Hyperboloid of revolution of one sheet

$$V = \frac{\pi h}{3}(2a^2 + \varrho^2)$$

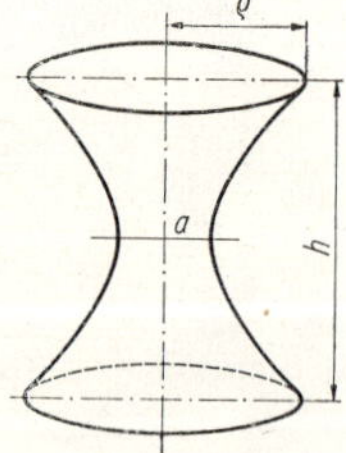

h height
a transverse semiaxis
ϱ radius of the bounding base circles

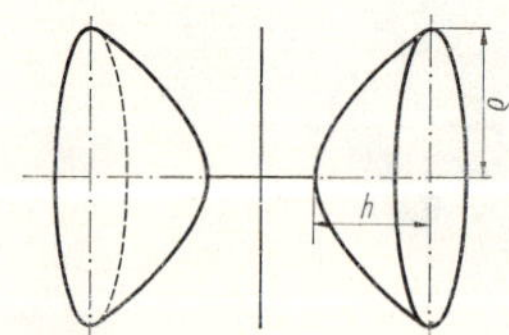

h height of one sheet
ϱ radius of the parallel base circles

Hyperboloid of revolution of two sheets

$$V = \frac{\pi h}{3}\left(3\varrho^2 - \frac{b^2h^2}{a^2}\right) \qquad \text{(figure above on the right)}$$

Barrel

Spherical and elliptic curvature

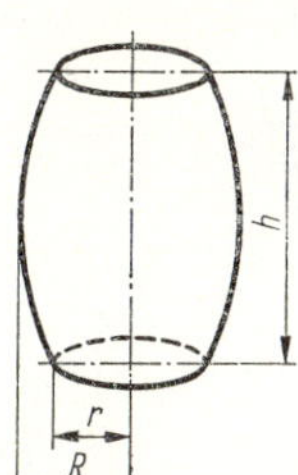

$$V = \frac{1}{3}\pi h(2R^2 + r^2) = \frac{1}{12}\pi h(2D^2 + d^2)$$

Parabolic curvature

$$V = \frac{1}{15}\pi h(8R^2 + 4Rr + 3r^2)$$

$$= \frac{1}{60}\pi h(8D^2 + 4dD + 3d^2)$$

For other curvatures, the above formulas yield approximate values.

Ring with circular section (torus)

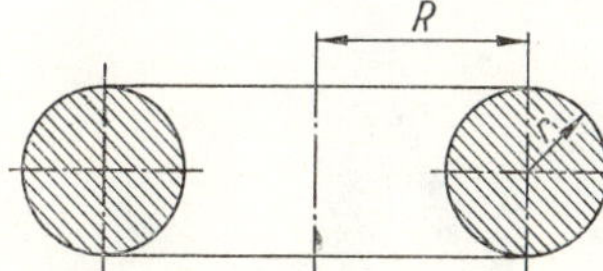

$$A_S = 4\pi^2 rR$$

$$V = 2\pi^2 r^2 R$$

r radius of the circular section
R mean radius of the ring

3.4. Goniometry, plane trigonometry, hyperbolic functions

3.4.1. Goniometry

Definition of trigonometric functions

In a right-angled triangle ABC with angles α, β, γ, $\overline{AC}$ is called the side adjacent to α, $\overline{BC}$ is called the side opposite to α, $\overline{AB}$ is the hypotenuse.

We define:

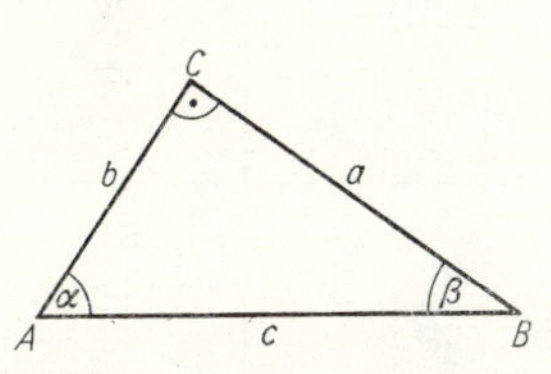

$$\sin\alpha = \frac{a}{c} = \frac{\text{opposite side}}{\text{hypotenuse}}$$

$$\cos\alpha = \frac{b}{c} = \frac{\text{adjacent side}}{\text{hypotenuse}}$$

$$\tan\alpha = \frac{a}{b} = \frac{\text{opposite side}}{\text{adjacent side}}$$

$$\cot\alpha = \frac{b}{a} = \frac{\text{adjacent side}}{\text{opposite side}}$$

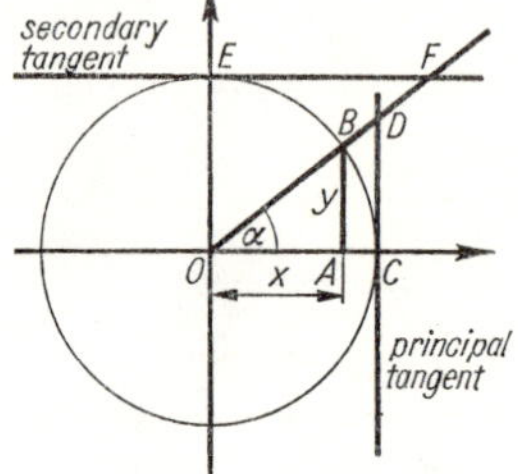

In the *unit circle* ($r = 1$) we have:

The sine of the angle is equal to dimension of the ordinate.

$$\sin \alpha = y = \overline{AB}$$

The cosine of the angle is equal to the dimension of the abscissa.

$$\cos \alpha = x = \overline{OA}$$

The tangent of the angle is equal to the dimension of the intercept of the principal tangent.

$$\tan \alpha = \overline{CD}$$

The cotangent of the angle is equal to the dimension of the intercept of the secondary tangent.

$$\cot \alpha = \overline{EF}$$

Note: $\dfrac{1}{\sin \alpha} = \dfrac{c}{a} = \operatorname{cosec} \alpha$; $\dfrac{1}{\cos \alpha} = \dfrac{c}{b} = \sec \alpha \equiv \csc \alpha$

Complementary relations

$$\sin \alpha = \cos (90° - \alpha)$$
$$\cos \alpha = \sin (90° - \alpha)$$
$$\tan \alpha = \cot (90° - \alpha)$$
$$\cot \alpha = \tan (90° - \alpha)$$

Signs of the trigonometric functions corresponding to angles in the four quadrants

Quadrant	sin	cos	tan	cot
I	+	+	+	+
II	+	−	−	−
III	−	−	+	+
IV	−	+	−	−

Reduction formulas for arbitrary angles

	$-\alpha$	$90° \pm \alpha$	$180° \pm \alpha$	$270° \pm \alpha$	$360° - \alpha$
sin	$-\sin \alpha$	$+\cos \alpha$	$\mp\sin \alpha$	$-\cos \alpha$	$-\sin \alpha$
cos	$+\cos \alpha$	$\mp\sin \alpha$	$-\cos \alpha$	$\pm\sin \alpha$	$+\cos \alpha$
tan	$-\tan \alpha$	$\mp\cot \alpha$	$\pm\tan \alpha$	$\mp\cot \alpha$	$-\tan \alpha$
cot	$-\cot \alpha$	$\mp\tan \alpha$	$\pm\cot \alpha$	$\mp\tan \alpha$	$-\cot \alpha$

Periodicity of the trigonometric functions

$$\left.\begin{aligned} \sin \alpha &= \sin (\alpha + k \cdot 360°) \\ \cos \alpha &= \cos (\alpha + k \cdot 360°) \\ \tan \alpha &= \tan (\alpha + k \cdot 180°) \\ \cot \alpha &= \cot (\alpha + k \cdot 180°) \end{aligned}\right\} \text{for} \quad k \in I$$

Behavior of the trigonometric functions

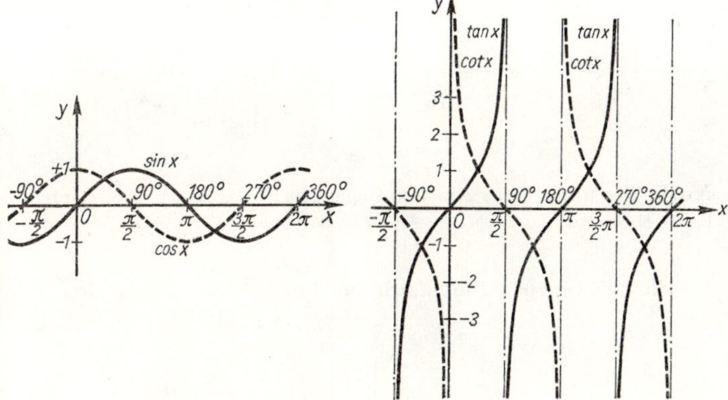

$$\sin \alpha, \; \cos \alpha \in [-1, 1]$$

$$\tan \alpha, \cot \alpha \in (-\infty, +\infty)$$

Examples:

$$\sin(-500^\circ) = -\sin 500^\circ = -\sin 140^\circ$$
$$= -\sin 40^\circ = -\cos 50^\circ$$

$$\cos 1000^\circ = \cos 280^\circ = \cos 80^\circ = \sin 10^\circ$$

$$\tan 1500^\circ = \tan 60^\circ = \cot 30^\circ$$

$$\cot(-2000^\circ) = -\cot 2000^\circ = -\cot 20^\circ = -\tan 70^\circ$$

Special values of the trigonometric functions

	0° 180°	30° 150°	45° 135°	60° 120°	90° 270°
sin	0	$\frac{1}{2}$	$\frac{1}{2}\sqrt{2}$	$\frac{1}{2}\sqrt{3}$	± 1
cos	± 1	$\pm\frac{1}{2}\sqrt{3}$	$\pm\frac{1}{2}\sqrt{2}$	$\pm\frac{1}{2}$	0
tan	0	$\pm\frac{1}{3}\sqrt{3}$	± 1	$\pm\sqrt{3}$	$\pm\infty$
cot	$\pm\infty$	$\pm\sqrt{3}$	± 1	$\pm\frac{1}{3}\sqrt{3}$	0

Note: The upper signs apply to the angles of the first row of the box head, the lower signs for the second row.

Relations between trigonometric functions of the same angle

$$\sin^2\alpha + \cos^2\alpha = 1$$

$$\tan\alpha = \frac{\sin\alpha}{\cos\alpha} = \frac{1}{\cot\alpha}$$

$$\cot\alpha = \frac{\cos\alpha}{\sin\alpha} = \frac{1}{\tan\alpha}$$

$$\tan\alpha\cot\alpha = 1$$

$$1 + \tan^2\alpha = \frac{1}{\cos^2\alpha}$$

$$1 + \cot^2\alpha = \frac{1}{\sin^2\alpha}$$

Calculation of a function from another function of the same angle

	sin	cos	tan	cot
$\sin\alpha =$	—	$\pm\sqrt{1-\cos^2\alpha}$	$\pm\dfrac{\tan\alpha}{\sqrt{1+\tan^2\alpha}}$	$\pm\dfrac{1}{\sqrt{1+\cot^2\alpha}}$
$\cos\alpha =$	$\pm\sqrt{1-\sin^2\alpha}$	—	$\pm\dfrac{1}{\sqrt{1+\tan^2\alpha}}$	$\pm\dfrac{\cot\alpha}{\sqrt{1+\cot^2\alpha}}$
$\tan\alpha =$	$\pm\dfrac{\sin\alpha}{\sqrt{1-\sin^2\alpha}}$	$\pm\dfrac{\sqrt{1-\cos^2\alpha}}{\cos\alpha}$	—	$\dfrac{1}{\cot\alpha}$
$\cot\alpha =$	$\pm\dfrac{\sqrt{1-\sin^2\alpha}}{\sin\alpha}$	$\pm\dfrac{\cos\alpha}{\sqrt{1-\cos^2\alpha}}$	$\dfrac{1}{\tan\alpha}$	—

For arbitrary angles α, the signs of the root are given by the quadrant of the angle.

ADDITION THEOREMS

Functions of the sum and difference of two angles

$$\sin(\alpha\pm\beta)=\sin\alpha\cos\beta\pm\cos\alpha\sin\beta$$

$$\cos(\alpha\pm\beta)=\cos\alpha\cos\beta\mp\sin\alpha\sin\beta$$

$$\tan(\alpha\pm\beta)=\frac{\tan\alpha\pm\tan\beta}{1\mp\tan\alpha\tan\beta}$$

$$\cot(\alpha\pm\beta)=\frac{\cot\alpha\cot\beta\mp 1}{\cot\beta\pm\cot\alpha}$$

$$\sin(\alpha+\beta)\sin(\alpha-\beta)=\cos^2\beta-\cos^2\alpha$$

$$\cos(\alpha+\beta)\cos(\alpha-\beta)=\cos^2\beta-\sin^2\alpha$$

Functions of double argument and half argument

$$\sin 2\alpha = 2\sin\alpha\cos\alpha$$

$$\cos 2\alpha = \cos^2\alpha - \sin^2\alpha = 1 - 2\sin^2\alpha = 2\cos^2\alpha - 1$$

$$\tan 2\alpha = \frac{2\tan\alpha}{1-\tan^2\alpha} = \frac{2}{\cot\alpha - \tan\alpha}$$

$$\cot 2\alpha = \frac{\cot^2\alpha - 1}{2\cot\alpha} = \frac{\cot\alpha - \tan\alpha}{2}$$

$$\sin\alpha = 2\sin\frac{\alpha}{2}\cos\frac{\alpha}{2}$$

$$\cos\alpha = \cos^2\frac{\alpha}{2} - \sin^2\frac{\alpha}{2} = 1 - 2\sin^2\frac{\alpha}{2} = 2\cos^2\frac{\alpha}{2} - 1$$

$$\tan\alpha = \frac{2\tan\frac{\alpha}{2}}{1-\tan^2\frac{\alpha}{2}} = \frac{2}{\cot\frac{\alpha}{2} - \tan\frac{\alpha}{2}}$$

$$\cot\alpha = \frac{\cot^2\frac{\alpha}{2} - 1}{2\cot\frac{\alpha}{2}} = \frac{\cot\frac{\alpha}{2} - \tan\frac{\alpha}{2}}{2}$$

$$\sin\frac{\alpha}{2} = \sqrt{\frac{1-\cos\alpha}{2}}$$

$$\cos\frac{\alpha}{2} = \sqrt{\frac{1+\cos\alpha}{2}}$$

$$\tan\frac{\alpha}{2} = \sqrt{\frac{1-\cos\alpha}{1+\cos\alpha}} = \frac{1-\cos\alpha}{\sin\alpha} = \frac{\sin\alpha}{1+\cos\alpha}$$

$$\cot\frac{\alpha}{2} = \sqrt{\frac{1+\cos\alpha}{1-\cos\alpha}} = \frac{1+\cos\alpha}{\sin\alpha} = \frac{\sin\alpha}{1-\cos\alpha}$$

$$\sin\alpha = \sqrt{\frac{1-\cos 2\alpha}{2}}$$

$$\cos\alpha = \sqrt{\frac{1+\cos 2\alpha}{2}}$$

$$\tan\alpha = \sqrt{\frac{1-\cos 2\alpha}{1+\cos 2\alpha}} = \frac{\sin 2\alpha}{1+\cos 2\alpha} = \frac{1-\cos 2\alpha}{\sin 2\alpha}$$

$$\cot\alpha = \sqrt{\frac{1+\cos 2\alpha}{1-\cos 2\alpha}} = \frac{\sin 2\alpha}{1-\cos 2\alpha} = \frac{1+\cos 2\alpha}{\sin 2\alpha}$$

Functions of further multiples of an angle

$$\sin 3\alpha = 3 \sin \alpha - 4 \sin^3 \alpha$$

$$\sin 4\alpha = 8 \sin \alpha \cos^3 \alpha - 4 \sin \alpha \cos \alpha$$

$$\sin 5\alpha = 16 \sin \alpha \cos^4 \alpha - 12 \sin \alpha \cos^2 \alpha + \sin \alpha$$

$$\cos 3\alpha = 4 \cos^3 \alpha - 3 \cos \alpha$$

$$\cos 4\alpha = 8 \cos^4 \alpha - 8 \cos^2 \alpha + 1$$

$$\cos 5\alpha = 16 \cos^5 \alpha - 20 \cos^3 \alpha + 5 \cos \alpha$$

$$\sin n\alpha = n \sin \alpha \cos^{n-1} \alpha - \binom{n}{3} \sin^3 \alpha \cos^{n-3} \alpha + \binom{n}{5} \sin^5 \alpha \cos^{n-5} \alpha - + \cdots$$

$$\cos n\alpha = \cos^n \alpha - \binom{n}{2} \sin^2 \alpha \cos^{n-2} \alpha + \binom{n}{4} \sin^4 \alpha \cos^{n-4} \alpha - + \cdots$$

$$\tan 3\alpha = \frac{3 \tan \alpha - \tan^3 \alpha}{1 - 3 \tan^2 \alpha}$$

$$\tan 4\alpha = \frac{4 \tan \alpha - 4 \tan^3 \alpha}{1 - 6 \tan^2 \alpha + \tan^4 \alpha}$$

$$\cot 3\alpha = \frac{\cot^3 \alpha - 3 \cot \alpha}{3 \cot^2 \alpha - 1}$$

$$\cot 4\alpha = \frac{\cot^4 \alpha - 6 \cot^2 \alpha + 1}{4 \cot^3 \alpha - 4 \cot \alpha}$$

Sums and differences of trigonometric functions

$$\sin \alpha + \sin \beta = 2 \sin \frac{\alpha + \beta}{2} \cos \frac{\alpha - \beta}{2}$$

$$\sin \alpha - \sin \beta = 2 \cos \frac{\alpha + \beta}{2} \sin \frac{\alpha - \beta}{2}$$

$$\cos\alpha + \cos\beta = 2\cos\frac{\alpha+\beta}{2}\cos\frac{\alpha-\beta}{2}$$

$$\cos\alpha - \cos\beta = -2\sin\frac{\alpha+\beta}{2}\sin\frac{\alpha-\beta}{2}$$

$$\cos\alpha + \sin\alpha = \sqrt{2}\sin(45^\circ + \alpha) = \sqrt{2}\cos(45^\circ - \alpha)$$

$$\cos\alpha - \sin\alpha = \sqrt{2}\cos(45^\circ + \alpha) = \sqrt{2}\sin(45^\circ - \alpha)$$

Products of trigonometric functions

$$\sin\alpha\sin\beta = \frac{1}{2}[\cos(\alpha-\beta) - \cos(\alpha+\beta)]$$

$$\cos\alpha\cos\beta = \frac{1}{2}[\cos(\alpha-\beta) + \cos(\alpha+\beta)]$$

$$\sin\alpha\cos\beta = \frac{1}{2}[\sin(\alpha+\beta) + \sin(\alpha-\beta)]$$

$$\cos\alpha\sin\beta = \frac{1}{2}[\sin(\alpha+\beta) - \sin(\alpha-\beta)]$$

$$\tan\alpha\tan\beta = \frac{\tan\alpha + \tan\beta}{\cot\alpha + \cot\beta} = -\frac{\tan\alpha - \tan\beta}{\cot\alpha - \cot\beta}$$

$$\cot\alpha\cot\beta = \frac{\cot\alpha + \cot\beta}{\tan\alpha + \tan\beta} = -\frac{\cot\alpha - \cot\beta}{\tan\alpha - \tan\beta}$$

$$\tan\alpha\cot\beta = \frac{\tan\alpha + \cot\beta}{\cot\alpha + \tan\beta} = -\frac{\tan\alpha - \cot\beta}{\cot\alpha - \tan\beta}$$

$$\cot\alpha\tan\beta = \frac{\cot\alpha + \tan\beta}{\tan\alpha + \cot\beta} = -\frac{\cot\alpha - \tan\beta}{\tan\alpha - \cot\beta}$$

$$\sin\alpha\sin\beta\sin\gamma = \frac{1}{4}[\sin(\alpha+\beta-\gamma) + \sin(\beta+\gamma-\alpha) + \sin(\gamma+\alpha-\beta) - \sin(\alpha+\beta+\gamma)]$$

$$\cos\alpha\cos\beta\cos\gamma = \frac{1}{4}[\cos(\alpha+\beta-\gamma) + \cos(\beta+\gamma-\alpha) + \cos(\gamma+\alpha-\beta) + \cos(\alpha+\beta+\gamma)]$$

$$\sin\alpha \sin\beta \cos\gamma = \frac{1}{4}[-\cos(\alpha+\beta-\gamma) + \cos(\beta+\gamma-\alpha) + \cos(\gamma+\alpha-\beta) - \cos(\alpha+\beta+\gamma)]$$

$$\sin\alpha \cos\beta \cos\gamma = \frac{1}{4}[\sin(\alpha+\beta-\gamma) - \sin(\beta+\gamma-\alpha) + \sin(\gamma+\alpha-\beta) + \sin(\alpha+\beta+\gamma)]$$

Powers of trigonometric functions

$$\sin^2\alpha = \frac{1}{2}(1 - \cos 2\alpha)$$

$$\cos^2\alpha = \frac{1}{2}(1 + \cos 2\alpha)$$

$$\sin^3\alpha = \frac{1}{4}(3\sin\alpha - \sin 3\alpha)$$

$$\cos^3\alpha = \frac{1}{4}(3\cos\alpha + \cos 3\alpha)$$

$$\sin^4\alpha = \frac{1}{8}(\cos 4\alpha - 4\cos 2\alpha + 3)$$

$$\cos^4\alpha = \frac{1}{8}(\cos 4\alpha + 4\cos 2\alpha + 3)$$

$$\sin^5\alpha = \frac{1}{16}(10\sin\alpha - 5\sin 3\alpha + \sin 5\alpha)$$

$$\cos^5\alpha = \frac{1}{16}(10\cos\alpha + 5\cos 3\alpha + \cos 5\alpha)$$

$$\sin^6\alpha = \frac{1}{32}(10 - 15\cos 2\alpha + 6\cos 4\alpha - \cos 6\alpha)$$

$$\cos^6\alpha = \frac{1}{32}(10 + 15\cos 2\alpha + 6\cos 4\alpha + \cos 6\alpha)$$

Relations between the trigonometric functions and the exponential function (Euler's formulas)

$$e^{j\varphi} = \cos\varphi + j\sin\varphi$$

$$e^{-j\varphi} = \cos\varphi - j\sin\varphi$$

From this follows:

$$\sin\varphi = \frac{e^{j\varphi} - e^{-j\varphi}}{2j} \qquad \cos\varphi = \frac{e^{j\varphi} + e^{-j\varphi}}{2}$$

$$\tan\varphi = -\frac{j(e^{j\varphi} - e^{-j\varphi})}{e^{j\varphi} + e^{-j\varphi}} \qquad \cot\varphi = \frac{j(e^{j\varphi} + e^{-j\varphi})}{e^{j\varphi} - e^{-j\varphi}}$$

Trigonometric functions of imaginary arguments

$$\sin j\varphi = j\sinh\varphi \qquad \tan j\varphi = j\tanh\varphi$$

$$\cos j\varphi = \cosh\varphi \qquad \cot j\varphi = -j\coth\varphi$$

φ circular measure (radian)

Graphical representation of various trigonometric functions

Amplitude variation:

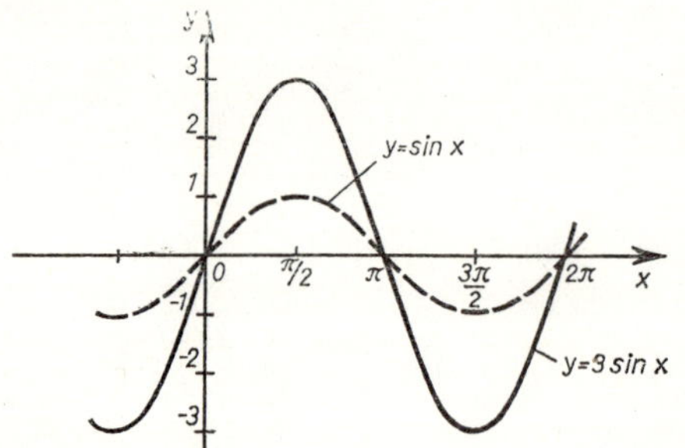

$$y = a\sin x$$

Image for $a = 3 \Rightarrow y = 3\sin x$

Reference curve $y = \sin x$

Frequency variation:

$$y = \sin ax$$

Image for $a = \frac{1}{4} \Rightarrow$

$$y = \sin\frac{x}{4}$$

for $a = 2 \Rightarrow y = \sin 2x$

Reference curve $y = \sin x$

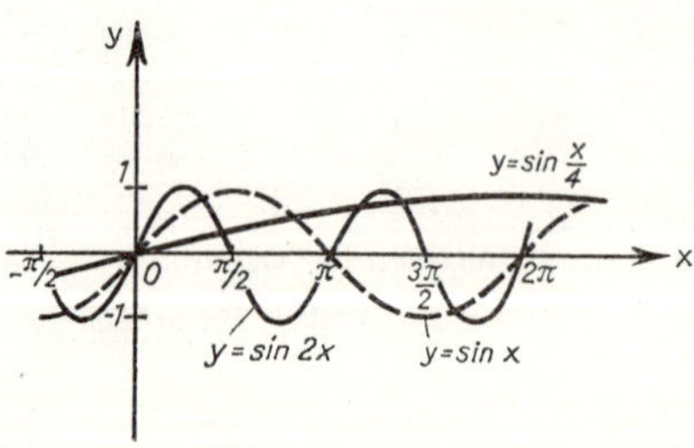

Superposition of two trigonometrical curves:

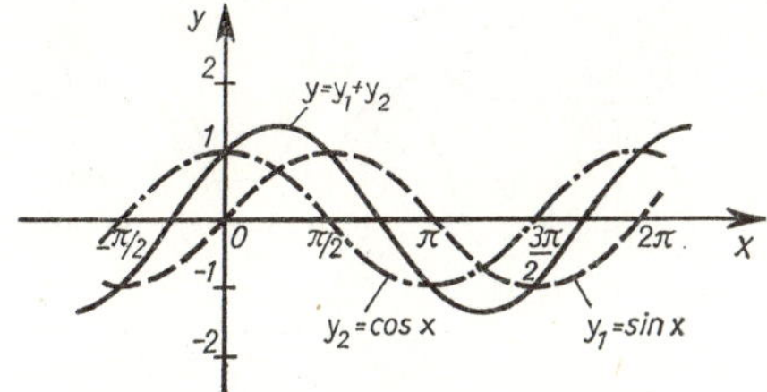

$$y = \sin x + \cos x$$

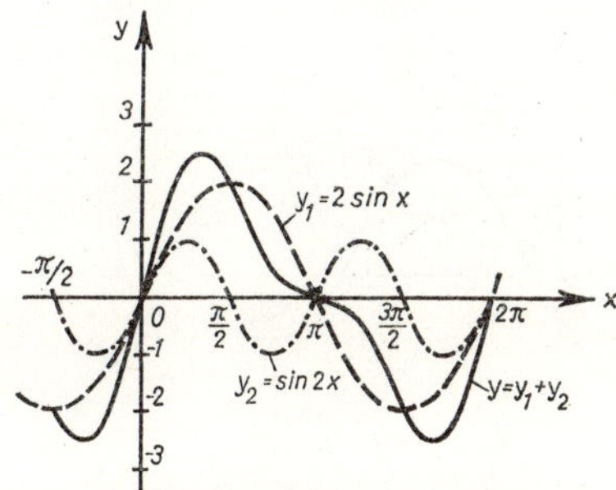

$$y = 2 \sin x + \sin 2x$$

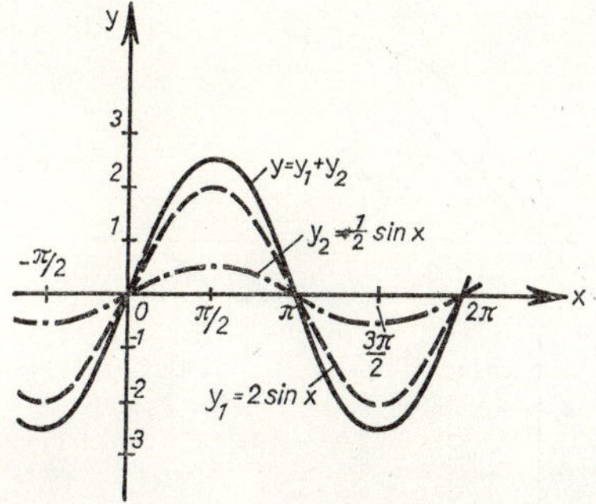

$$y = 2 \sin x + \frac{1}{2} \sin x$$

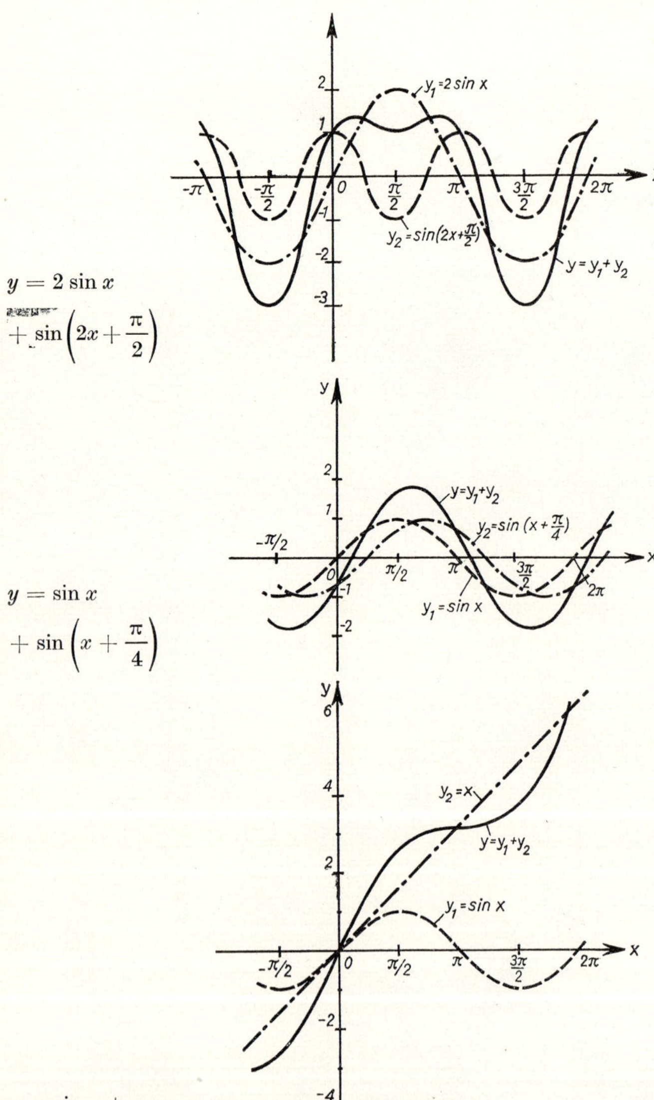

$y = 2\sin x + \sin\left(2x + \frac{\pi}{2}\right)$

$y = \sin x + \sin\left(x + \frac{\pi}{4}\right)$

$y = \sin x + x$

Products of trigonometric functions:

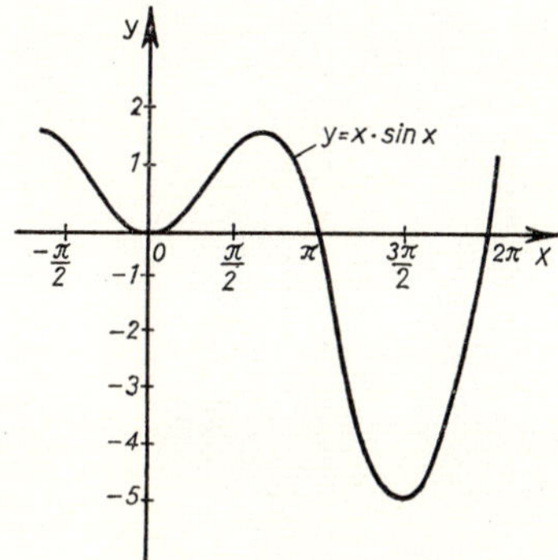

$y = x \sin x$

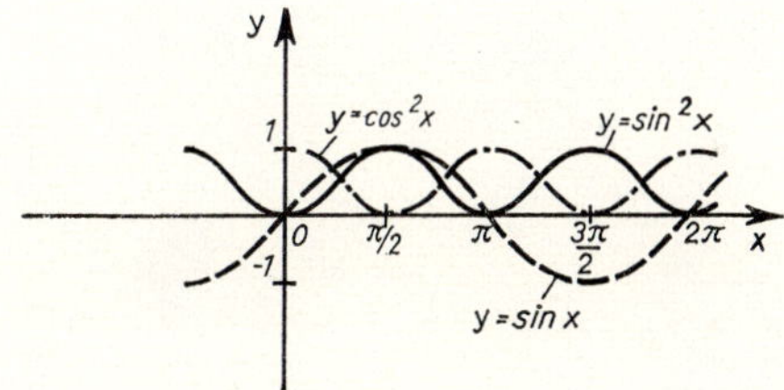

$y = \sin^2 x$

Reference curve $y = \sin x$

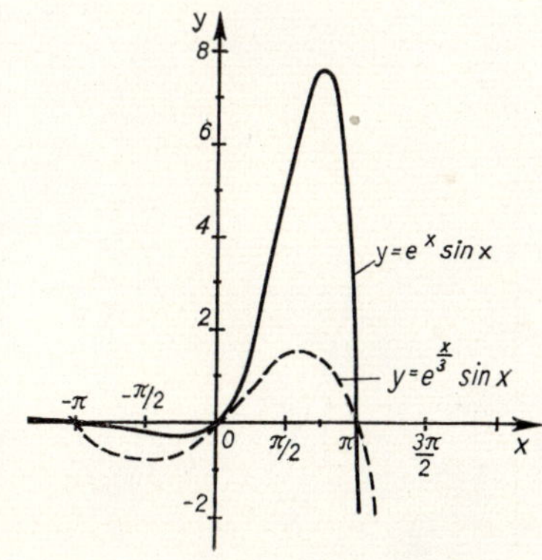

$y = e^x \sin x$

$y = e^{\frac{x}{3}} \sin x$

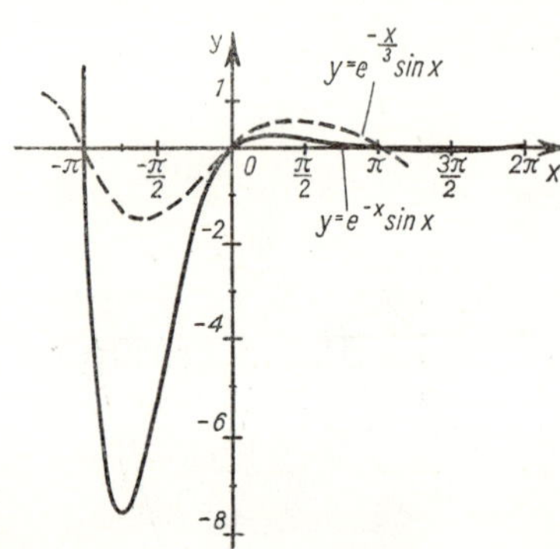

$y = e^{-x} \sin x \quad y = e^{-\frac{x}{3}} \sin x$

Vector diagram of sine functions

Sine functions $f = \{x, y \mid y = A \sin (x + \varphi)\}$ can be symbolized by rotating vectors with the determining quantities

A length of the vector

$\frac{x}{t}$ angular velocity

φ displacement with respect to the axis of projection at $x = 0$ (or $t = 0$)

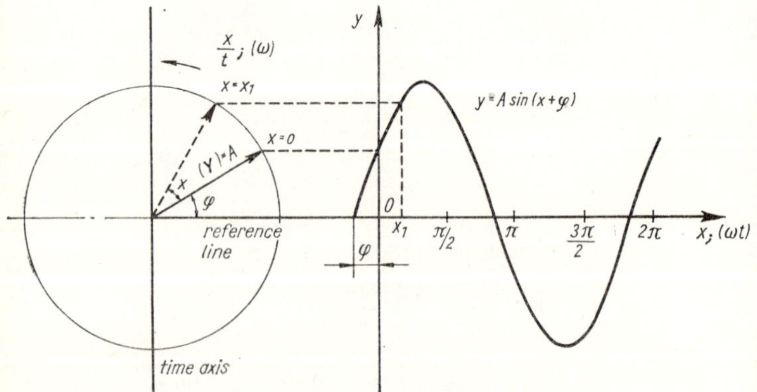

In electrical engineering: $x = \omega t = \frac{t}{T}$ where ω angular frequency, T cycle duration, and, φ phase position.
The physically real instantaneous value is obtained by projection of the vector onto a plane perpendicular to the axis of projection.

Since, in a vector diagram, a sinusoidal variation with $\frac{x}{t} = \omega = \text{const}$ is assumed, A and φ characterize the function so that resting vectors will suffice; they are obtained by allowing the sine axis, which is perpendicular to the axis of projection, to rotate in the opposite direction. Though these resting vectors can betreated as vectors, they differ from the usual concept of a vector by their definition.
In a vector diagram, two vectors can be added according to the vector calculus. The vector sum mirrors the physical or geometrical reality of the addition of the instantaneous values $y = y_1 = y_2$ in accordance with the scalar addition of the components of vectors. Its magnitude and position with reference to the axis of projection (phase position) can be read off from the vector diagram. Its angular velocity is equal to that of the two individual vectors.

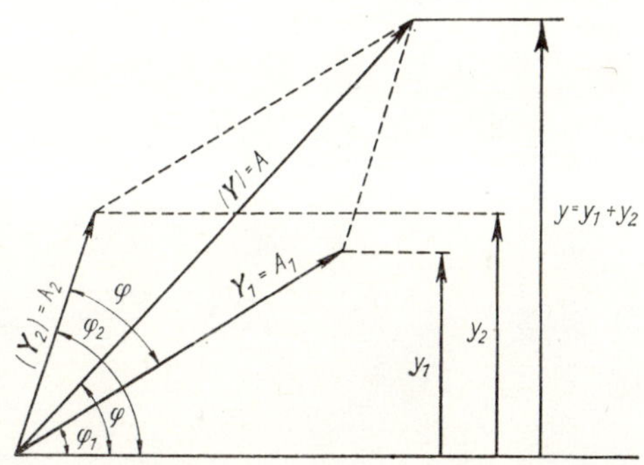

$$y_1 = A_1 \sin (x + \varphi_1)$$

$$y_2 = A_2 \sin (x + \varphi_2)$$

$$y = y_1 + y_2 = A \sin (x + \varphi)$$

where

$$A = A_1^2 + A_2^2 + 2A_1A_2 \cos (\varphi_2 - \varphi_1)$$

$$\tan \varphi = \frac{A_1 \sin \varphi_1 + A_2 \sin \varphi_2}{A_1 \cos \varphi_1 + A_2 \cos \varphi_2}$$

It is obvious that the vector addition is simpler than the analytical procedure. The diagrammatic inaccuracy included in the result can frequently be neglected.

3.4.2. Trigonometric formulas for oblique-angled triangles

Relations between angles

$$\sin\alpha + \sin\beta + \sin\gamma = 4\cos\frac{\alpha}{2}\cos\frac{\beta}{2}\cos\frac{\gamma}{2}$$

$$\cos\alpha + \cos\beta + \cos\gamma = 1 + 4\sin\frac{\alpha}{2}\sin\frac{\beta}{2}\sin\frac{\gamma}{2}$$

$$\sin 2\alpha + \sin 2\beta + \sin 2\gamma = 4\sin\alpha\sin\beta\sin\gamma$$

$$\cos 2\alpha + \cos 2\beta + \cos 2\gamma = -(4\cos\alpha\cos\beta\cos\gamma + 1)$$

$$\tan\alpha + \tan\beta + \tan\gamma = \tan\alpha\tan\beta\tan\gamma$$

$$\sin^2\alpha + \sin^2\beta + \sin^2\gamma = 2(1 + \cos\alpha\cos\beta\cos\gamma)$$

$$\cos^2\alpha + \cos^2\beta + \cos^2\gamma = 1 - 2\cos\alpha\cos\beta\cos\gamma$$

$$\cot\alpha\cot\beta + \cot\alpha\cot\gamma + \cot\beta\cot\gamma = 1$$

$$\cot\frac{\alpha}{2} + \cot\frac{\beta}{2} + \cot\frac{\gamma}{2} = \cot\frac{\alpha}{2}\cot\frac{\beta}{2}\cot\frac{\gamma}{2}$$

$$(\sin\alpha + \sin\beta + \sin\gamma)(\sin\alpha + \sin\beta - \sin\gamma)$$
$$\times (\sin\alpha - \sin\beta + \sin\gamma)(-\sin\alpha + \sin\beta + \sin\gamma)$$
$$= 4\sin^2\alpha\sin^2\beta\sin^2\gamma$$

Sine theorem

$$a:b:c = \sin\alpha:\sin\beta:\sin\gamma$$

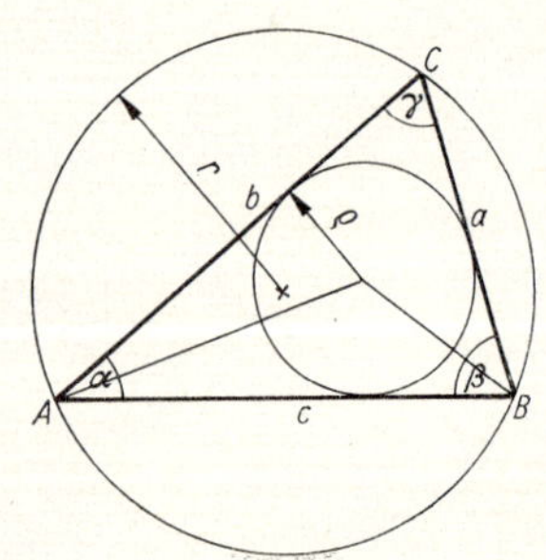

Cosine theorem

$$a^2 = b^2 + c^2 - 2bc\cos\alpha$$

$$b^2 = a^2 + c^2 - 2ac\cos\beta$$

$$c^2 = a^2 + b^2 - 2ab\cos\gamma$$

Mollweide's formulas

$$\frac{a+b}{c} = \frac{\cos\dfrac{\alpha-\beta}{2}}{\sin\dfrac{\gamma}{2}} \qquad \frac{a-b}{c} = \frac{\sin\dfrac{\alpha-\beta}{2}}{\cos\dfrac{\gamma}{2}}$$

$$\frac{b+c}{a} = \frac{\cos\dfrac{\beta-\gamma}{2}}{\sin\dfrac{\alpha}{2}} \qquad \frac{b-c}{a} = \frac{\sin\dfrac{\beta-\gamma}{2}}{\cos\dfrac{\alpha}{2}}$$

$$\frac{a+c}{b} = \frac{\cos\dfrac{\alpha-\gamma}{2}}{\sin\dfrac{\beta}{2}} \qquad \frac{a-c}{b} = \frac{\sin\dfrac{\alpha-\gamma}{2}}{\cos\dfrac{\beta}{2}}$$

Tangent theorem

$$\frac{a+b}{a-b} = \frac{\tan\dfrac{\alpha+\beta}{2}}{\tan\dfrac{\alpha-\beta}{2}} = \frac{\cot\dfrac{\gamma}{2}}{\tan\dfrac{\alpha-\beta}{2}}$$

$$\frac{b+c}{b-c} = \frac{\tan\dfrac{\beta+\gamma}{2}}{\tan\dfrac{\beta-\gamma}{2}} = \frac{\cot\dfrac{\alpha}{2}}{\tan\dfrac{\beta-\gamma}{2}}$$

$$\frac{a+c}{a-c} = \frac{\tan\dfrac{\alpha+\gamma}{2}}{\tan\dfrac{\alpha-\gamma}{2}} = \frac{\cot\dfrac{\beta}{2}}{\tan\dfrac{\alpha-\gamma}{2}}$$

Half-angle-theorems

$$\sin\frac{\alpha}{2} = \sqrt{\frac{(s-b)(s-c)}{bc}}; \quad \sin\frac{\beta}{2} = \sqrt{\frac{(s-a)(s-c)}{ac}}$$

$$\sin\frac{\gamma}{2} = \sqrt{\frac{(s-a)(s-b)}{ab}} \qquad s = \frac{a+b+c}{2}$$

$$\cos\frac{\alpha}{2} = \sqrt{\frac{s(s-a)}{bc}}; \qquad \cos\frac{\beta}{2} = \sqrt{\frac{s(s-b)}{ac}}$$

$$\cos\frac{\gamma}{2} = \sqrt{\frac{s(s-c)}{ab}}$$

$$\tan\frac{\alpha}{2} = \sqrt{\frac{(s-b)(s-c)}{s(s-a)}}; \quad \tan\frac{\beta}{2} = \sqrt{\frac{(s-a)(s-c)}{s(s-b)}}$$

$$\tan\frac{\gamma}{2} = \sqrt{\frac{(s-a)(s-b)}{s(s-c)}}; \quad \text{where} \quad s = \frac{a+b+c}{2}$$

Circumcircle radius r

$$r = \frac{a}{2\sin\alpha} = \frac{b}{2\sin\beta} = \frac{c}{2\sin\gamma}$$ Figure see page 184

$$s = 4r\cos\frac{\alpha}{2}\cos\frac{\beta}{2}\cos\frac{\gamma}{2} \qquad s = \frac{a+b+c}{2}$$

Incircle radius ϱ

$$\varrho = (s-a)\tan\frac{\alpha}{2} = (s-b)\tan\frac{\beta}{2} = (s-c)\tan\frac{\gamma}{2}$$

$$\varrho = s\tan\frac{\alpha}{2}\tan\frac{\beta}{2}\tan\frac{\gamma}{2}$$

$$\varrho = 4r\sin\frac{\alpha}{2}\sin\frac{\beta}{2}\sin\frac{\gamma}{2}$$

Ecircle radii $\varrho_a, \varrho_b, \varrho_c$

$$\varrho_a = s\tan\frac{\alpha}{2} = \frac{\alpha\cos\frac{\beta}{2}\cos\frac{\gamma}{2}}{\cos\frac{\alpha}{2}}$$

$$\varrho_b = s\tan\frac{\beta}{2} = \frac{b\cos\frac{\alpha}{2}\cos\frac{\gamma}{2}}{\cos\frac{\beta}{2}}$$

$$\varrho_c = s\tan\frac{\gamma}{2} = \frac{c\cos\frac{\alpha}{2}\cos\frac{\beta}{2}}{\cos\frac{\gamma}{2}}$$

Medians s_a, s_b, s_c

$$s_a = \frac{1}{2}\sqrt{b^2 + c^2 + 2bc\cos\alpha}$$

$$s_b = \frac{1}{2}\sqrt{a^2 + c^2 + 2ac\cos\beta}$$

$$s_c = \frac{1}{2}\sqrt{a^2 + b^2 + 2ab\cos\gamma}$$

Angle-bisectors w_α, w_β, w_γ

$$w_\alpha = \frac{2bc\cos\frac{\alpha}{2}}{b + c} \qquad w_\beta = \frac{2ac\cos\frac{\beta}{2}}{a + c}$$

$$w_\gamma = \frac{2ab\cos\frac{\gamma}{2}}{a + b}$$

Altitudes h_a, h_b, h_c

$$\begin{array}{l|l|l} h_a = b\sin\gamma & h_b = a\sin\gamma & h_c = a\sin\beta \\ \quad\;\, = c\sin\beta & \quad\;\, = c\sin\alpha & \quad\;\, = b\sin\alpha \end{array}$$

Calculation of obtuse-angled triangles by trigonometrical procedures

Basic problem 1:

Given one side and two angles (e.g. a, α, γ).

Solution:

$$\beta = 180° - (\alpha + \gamma)$$

$$b = \frac{a\sin\beta}{\sin\alpha} \quad \text{(sine theorem)}$$

$$c = \frac{a\sin\gamma}{\sin\alpha} \quad \text{(sine theorem)}$$

Basic problem 2:

Given two sides and the included angle (e.g. a, c, β).

Solution 1:

$$\tan\frac{\alpha-\gamma}{2}=\frac{a-c}{a+c}\cot\frac{\beta}{2} \qquad \text{(tangent theorem)}$$

From this $\frac{\alpha-\gamma}{2}$

By addition and subtraction of

$$\frac{\alpha-\gamma}{2} \quad \text{and} \quad \frac{\alpha+\gamma}{2}=90^\circ-\frac{\beta}{2}$$

we obtain α and γ.

$$b=\frac{a\sin\beta}{\sin\alpha} \qquad \text{(sine theorem)}$$

Solution 2:

$$b=\sqrt{a^2+c^2-2ac\cos\beta} \qquad \text{(cosine theorem)}$$

$$\sin\alpha=\frac{a\sin\beta}{b} \qquad \text{(sine theorem)}$$

$$\gamma=180^\circ-(\alpha+\beta)$$

Basic problem 3:

Given two sides and the angle opposite to the larger one of them (e.g. a, b, α; $a>b$).

Solution:

$$\sin\beta=\frac{b\sin\alpha}{a} \qquad \text{(sine theorem)}$$

$$\gamma=180^\circ-(\alpha+\beta)$$

$$c=\frac{a\sin\gamma}{\sin\alpha} \qquad \text{(sine theorem)}$$

Basic problem 4:

Given the three sides a, b, c.

Solution 1:

$$\tan\frac{\alpha}{2} = \sqrt{\frac{(s-b)(s-c)}{s(s-a)}} \quad \text{(half-angle theorem)}$$

$$\sin\beta = \frac{b\sin\alpha}{a} \quad \text{(sine theorem)}$$

$$\gamma = 180^\circ - (\alpha + \beta)$$

Solution 2:

$$\cos\alpha = \frac{b^2 + c^2 - a^2}{2bc} \quad \text{(cosine theorem)}$$

$$\sin\beta = \frac{b\sin\alpha}{a} \quad \text{(sine theorem)}$$

$$\gamma = 180^\circ - (\alpha + \beta)$$

3.4.3. Goniometric equations

Explanation

Goniometric equations contain functions of unknown angles besides known quantities.

Technique of solution (calculatory):

If different functions of angles occur in an equation, then with the help of goniometric formulas we attempt to modify then equation so that only one angular function of the same angle occurs in it. If different arguments of an angular function occur, they must also be reduced to the same argument. Because of the periodicity of angular functions, a goniometric equation may have infinitely many solutions. Mostly the solutions are confined to values between 0° and 360° (0 and 2π) (principal values).
Each solution must be checked for validity by substitution in the initial equation. For example, we may obtain too many solutions if, during the solution of the equation, the degree of the equation is increased by squaring.

Example 1:

$$\{x \mid \sin 2x = \sin x\} \quad \text{(use formula for } \sin 2x\text{)}$$

$$2 \sin x \cos x = \sin x$$

$$2 \sin x \cos x - \sin x = 0$$

(Factor out! Do not divide by $\sin x$ otherwise solutions will be lost)

$$\sin x(2 \cos x - 1) = 0 \quad \text{yields}$$

$$\left.\begin{array}{l} \sin x = 0 \text{ yields } x_1 = 0° \\ \qquad x_2 = 180° \\ \qquad x_3 = 360° \\ 2 \cos x - 1 = 0 \\ \cos x = \frac{1}{2} \text{ yields } x_4 = 60° \\ \qquad x_5 = 300° \end{array}\right\} \text{(principal values)}$$

The proof shows that all solutions are valid.
Hence $S = \{0°; 60°; 180°; 300°; 360°\}$

Example 2:

$$\{x \mid 17 \sin^2 x - 3 \cos^2 x = 2\}$$

$$17 \sin^2 x - 3(1 - \sin^2 x) = 2$$

$$20 \sin^2 x = 5$$

$$(\sin x)_{1;2} = \pm \frac{1}{2}$$

$$(\sin x)_1 = + \frac{1}{2} \quad \text{gives as main values } x_1 = 30° \text{ and } x_2 = 150°$$

$$(\sin x)_2 = - \frac{1}{2} \quad \text{gives as main values } x_3 = 210° \text{ and } x_4 = 330°$$

All solutions are valid.
Hence $S = \{30°; 150°; 210°; 330°\}$

Example 3:

$$\{x \mid 2 \sin x + \cos x = 2\}$$

Solution 1:

$\cos x$ to be replaced by $\sqrt{1 - \sin^2 x}$

$$2 \sin x + \sqrt{1 - \sin^2 x} = 2 \text{ (root on the left to be isolated)}$$

$$\sqrt{1 - \sin^2 x} = 2 - 2 \sin x \text{ (square the equation)}$$

$$1 - \sin^2 x = 4 - 8 \sin x + 4 \sin^2 x$$

$$(\sin x)_{1;2} = \frac{4}{5} \pm \sqrt{\frac{16}{25} - \frac{15}{25}} = \frac{4}{5} \pm \frac{1}{5}$$

$(\sin x)_1 = 1$ yields $x_1 = 90°$ as main value

$(\sin x)_2 = \dfrac{3}{5}$ yields $x_2 = 36°52'$

$$[x_3 = 143°8']$$

Checking shows that x_3 does not satisfy the equation.
Hence $S = \{90°; 36°52'\}$

Solution 2:

$\sin x + \dfrac{1}{2} \cos x = 1$ (introduction of an auxiliary angle by the substitution $\dfrac{1}{2} = \tan z$ from which $z = 26°34'$ results.)

$$\sin x + \tan z \cos x = 1$$

$$\sin x + \frac{\sin z}{\cos z} \cos x = 1$$

$$\sin x \cos z + \sin z \cos x = \cos z$$

$$\sin (x + z) = \cos z = \cos 26°34' = 0.8944$$

$$(x + z)_1 = \ \ 63°26' \text{ yields } x_1 = 36°52'$$

$$(x + z)_2 = 116°34' \text{ yields } x_2 = 90°$$

Example 4:

$$\{x \mid \sin 2x = \tan x\}$$

$$2 \sin x \cos x = \frac{\sin x}{\cos x}$$

$$2 \sin x \cos^2 x - \sin x = 0$$

$$\sin x\,(2 \cos^2 x - 1) = 0$$

$$\sin x = 0 \text{ yields } \left.\begin{aligned} x_1 &= 0° \\ x_2 &= 180° \\ x_3 &= 360° \end{aligned}\right\} \text{as principal values}$$

$$2 \cos^2 x - 1 = 0$$

$$\cos^2 x = \frac{1}{2}$$

$$(\cos x)_{1;2} = \pm \frac{1}{2}\sqrt{2} \text{ yields } \begin{aligned} x_4 &= 45° \\ x_5 &= 135° \\ x_6 &= 225° \\ x_7 &= 315° \end{aligned}$$

All values are possible.
Hence $S = \{0°; 45°; 135°; 180°; 225°; 315°; 360°\}$

Technique of solution (graphically):

We treat both the left-hand side and the right-hand side of the equation as functions of x, represent them graphically, and find the solutions as the abscissae of the points of intersection of the two curves.

Example:

$\{x \mid 2 \sin x + \cos x = 2\}$ (calculatory solution see above)

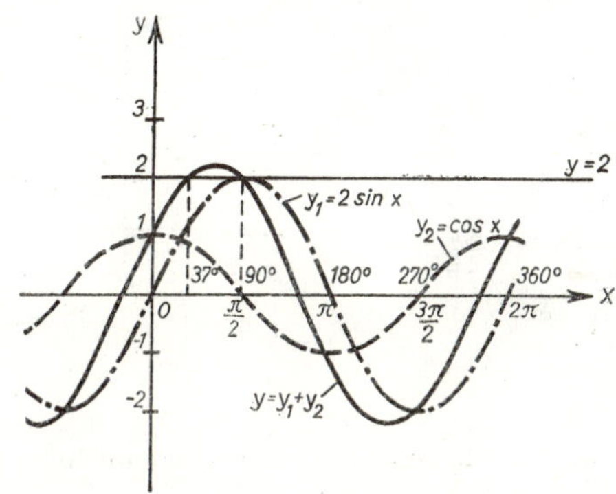

3.4.4. Inverse trigonometric functions

$y = \sin^{-1} x$ is the inverse function of $y = \sin x$

$y = \cos^{-1} x$ is the inverse function of $y = \cos x$

$y = \tan^{-1} x$ is the inverse function of $y = \tan x$

$y = \cot^{-1} x$ is the inverse function of $y = \cot x$

$\sin^{-1} x$ is also written as arc sin x (i.e. the angle, in radian measure, whose sine has the value x), etc.
Because of the periodicity of the trigonometric functions, inverse trigonometric functions have infinitely many solutions.
Principal values of the inverse trigonometric functions

$y = \text{arc sin } x$ within the domain $y \in \left[-\frac{\pi}{2}, \frac{\pi}{2}\right]$

$y = \text{arc cos } x$ within the domain $y \in [0, \pi]$

$y = \text{arc tan } x$ within the domain $y \in \left(-\frac{\pi}{2}, \frac{\pi}{2}\right)$

$y = \text{arc cot } x$ within the domain $y \in (0, \pi)$

Graphically, the inverse trigonometric functions are the mirror-image of the trigonometric functions in the straight line $y = x$.

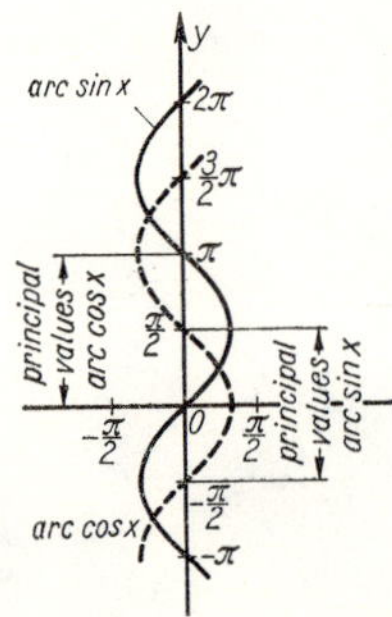

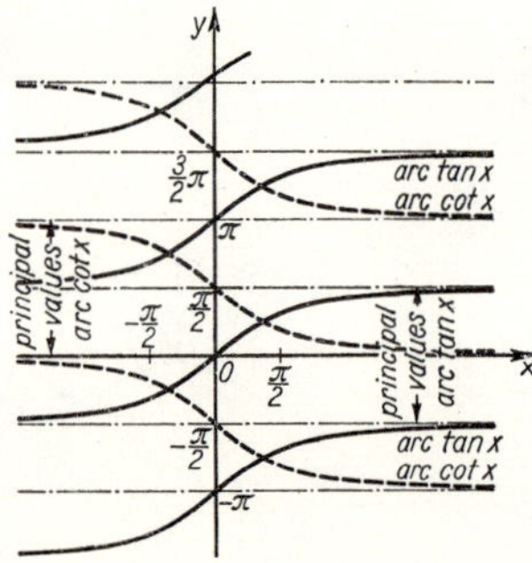

Calculation of one inverse trigonometric function in terms of another (applies to the principal values)

$$\text{arc sin}\, x = \frac{\pi}{2} - \text{arc cos}\, x = \text{arc tan}\, \frac{x}{\sqrt{1 - x^2}}$$

$$\text{arc cos}\, x = \frac{\pi}{2} - \text{arc sin}\, x = \text{arc cot}\, \frac{x}{\sqrt{1 - x^2}}$$

$$\text{arc tan}\, x = \frac{\pi}{2} - \text{arc cot}\, x = \text{arc sin}\, \frac{x}{\sqrt{1 + x^2}}$$

$$\text{arc cot}\, x = \frac{\pi}{2} - \text{arc tan}\, x = \text{arc cos}\, \frac{x}{\sqrt{1 + x^2}}$$

$$\left.\begin{aligned} \text{arc cot}\, x &= \text{arc tan}\, \frac{1}{x} \qquad \text{for} \quad x > 0 \\ &= \text{arc tan}\, \frac{1}{x} + \pi \ \text{for} \quad x < 0 \end{aligned}\right\} \text{(generally accepted)}$$

Inverse trigonometric functions of negative arguments

$$\text{arc sin}\, (-x) = -\text{arc sin}\, x$$

$$\text{arc cos}\, (-x) = \pi - \text{arc cos}\, x$$

$$\text{arc tan}\, (-x) = -\text{arc tan}\, x$$

$$\text{arc cot}\, (-x) = \pi - \text{arc cot}\, x$$

Sums and differences of inverse trigonometric functions

$$\operatorname{arc\,sin} x_1 + \operatorname{arc\,sin} x_2 = \operatorname{arc\,sin}\left(x_1\sqrt{1-x_2^2} + x_2\sqrt{1-x_1^2}\right) \quad \text{for} \quad x_1^2 + x_2^2 \leqq 1$$

$$\operatorname{arc\,sin} x_1 - \operatorname{arc\,sin} x_2 = \operatorname{arc\,sin}\left(x_1\sqrt{1-x_2^2} - x_2\sqrt{1-x_1^2}\right) \quad \text{for} \quad x_1^2 + x_2^2 \leqq 1$$

$$\operatorname{arc\,cos} x_1 + \operatorname{arc\,cos} x_2 = \operatorname{arc\,cos}\left(x_1x_2 - \sqrt{1-x_1^2}\sqrt{1-x_2^2}\right) \quad \text{for} \quad x_1 + x_2 \geqq 0$$

$$\operatorname{arc\,cos} x_1 - \operatorname{arc\,cos} x_2 = -\operatorname{arc\,cos}\left(x_1x_2 + \sqrt{1-x_1^2}\sqrt{1-x_2^2}\right) \quad \text{for} \quad x_1 \geqq x_2$$

$$= \operatorname{arc\,cos}\left(x_1x_2 + \sqrt{1-x_1^2}\sqrt{1-x_2^2}\right) \quad \text{for} \quad x_1 < x_2$$

$$\operatorname{arc\,tan} x_1 + \operatorname{arc\,tan} x_2 = \operatorname{arc\,tan}\frac{x_1 + x_2}{1 - x_1x_2} \quad \text{for} \quad x_1x_2 < 1$$

$$\operatorname{arc\,tan} x_1 - \operatorname{arc\,tan} x_2 = \operatorname{arc\,tan}\frac{x_1 - x_2}{1 + x_1x_2} \quad \text{for} \quad x_1x_2 > -1$$

$$\operatorname{arc\,cot} x_1 + \operatorname{arc\,cot} x_2 = \operatorname{arc\,cot}\frac{x_1x_2 - 1}{x_1 + x_2} \quad \text{for} \quad x_1 \neq -x_2$$

$$\operatorname{arc\,cot} x_1 - \operatorname{arc\,cot} x_2 = \operatorname{arc\,cot}\frac{x_1x_2 + 1}{x_2 - x_1} \quad \text{for} \quad x_1 \neq x_2$$

Relations between the inverse trigonometric functions and the logarithmic function

$$\left.\begin{aligned}
\operatorname{arc\,sin} x &= -j\ln\left(xj + \sqrt{1-x^2}\right)\\
\operatorname{arc\,cos} x &= -j\ln\left(x + \sqrt{x^2-1}\right)\\
\operatorname{arc\,tan} x &= \frac{1}{2j}\ln\frac{1+xj}{1-xj}\\
\operatorname{arc\,cot} x &= -\frac{1}{2j}\ln\frac{xj+1}{xj-1}
\end{aligned}\right\} \text{(hold for principal values)}$$

3.4.5. Hyperbolic functions

Definition

$$\sinh x = \frac{e^x - e^{-x}}{2} \qquad \tanh x = \frac{e^x - e^{-x}}{e^x + e^{-x}}$$

$$\cosh x = \frac{e^x + e^{-x}}{2} \qquad \coth x = \frac{e^x + e^{-x}}{e^x - e^{-x}}$$

Behavior of the hyperbolic functions

$$\left.\begin{array}{l} \sinh x \in (-\infty, \infty) \\ \cosh x \in [1, \infty) \\ \tanh x \in (-1, +1) \\ \coth x \in (1, \infty) \quad \text{for} \quad x \in (0, \infty) \\ \coth x \in (-\infty, -1) \quad \text{for} \quad x \in (-\infty, 0) \\ \lim\limits_{x\to\infty} \coth x = 1; \qquad \lim\limits_{x\to-\infty} \coth x = -1 \end{array}\right\} x \in R$$

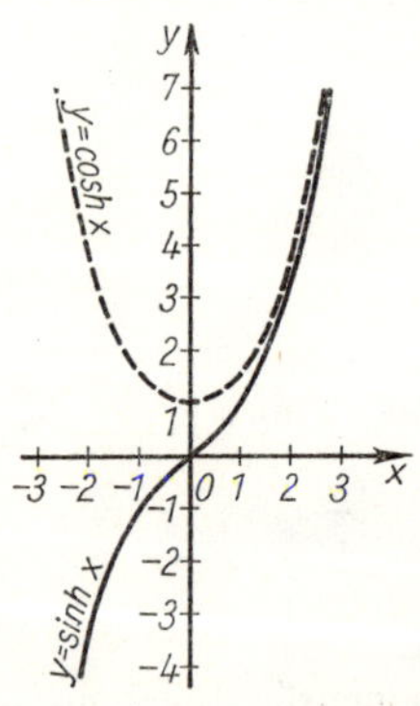

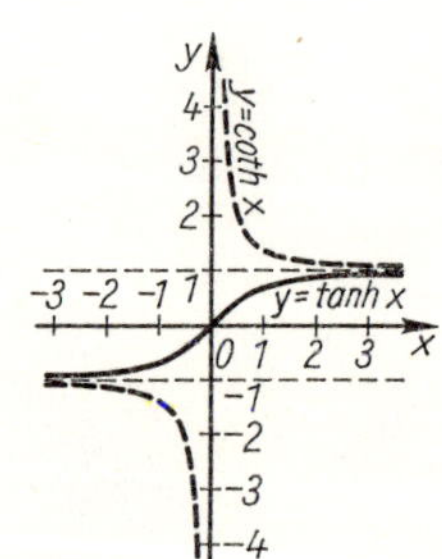

Period of the hyperbolic functions

$$\sinh (x + 2k\pi j) = \sinh x \qquad \tanh (x + k\pi j) = \tanh x$$

$$\cosh (x + 2k\pi j) = \cosh x \qquad \coth (x + k\pi j) = \coth x$$

Special values

$\sinh 0 = 0$

$\cosh 0 = 1$

$\tanh 0 = 0$

$\coth 0 = \pm\infty$

Negative arguments

$$\sinh(-x) = -\sinh x$$

$$\cosh(-x) = \cosh x$$

$$\tanh(-x) = -\tanh x$$

$$\coth(-x) = -\coth x$$

Relations between the functions of the same argument

$$\sinh x + \cosh x = e^x \qquad \sinh x - \cosh x = -e^{-x}$$

$$\cosh^2 x - \sinh^2 x = 1$$

$$\tanh x = \frac{\sinh x}{\cosh x} \qquad \coth x = \frac{\cosh x}{\sinh x}$$

$$\coth x = \frac{1}{\tanh x}$$

$$1 - \tanh^2 x = \frac{1}{\cosh^2 x} \qquad \coth^2 x - 1 = \frac{1}{\sinh^2 x}$$

Calculation of one function from another with the same argument

	sinh	cosh	tanh	coth
$\sinh x =$	—	$\sqrt{\cosh^2 x - 1}$	$\frac{\tanh x}{\sqrt{1 - \tanh^2 x}}$	$\frac{1}{\sqrt{\coth^2 x - 1}}$
$\cosh x =$	$\sqrt{\sinh^2 x + 1}$	—	$\frac{1}{\sqrt{1 - \tanh^2 x}}$	$\frac{\coth x}{\sqrt{\coth^2 x - 1}}$
$\tanh x =$	$\frac{\sinh x}{\sqrt{\sinh^2 x + 1}}$	$\frac{\sqrt{\cosh^2 x - 1}}{\cosh x}$	—	$\frac{1}{\coth x}$
$\coth x =$	$\frac{\sqrt{\sinh^2 x + 1}}{\sinh x}$	$\frac{\cosh x}{\sqrt{\cosh^2 x - 1}}$	$\frac{1}{\tanh x}$	—

ADDITION THEOREMS

Functions of the sum and difference of two arguments

$$\sinh (x \pm y) = \sinh x \cosh y \pm \cosh x \sinh y$$

$$\cosh (x \pm y) = \cosh x \cosh y \pm \sinh x \sinh y$$

$$\tanh (x \pm y) = \frac{\tanh x \pm \tanh y}{1 \pm \tanh x \tanh y}$$

$$\coth (x \pm y) = \frac{1 \pm \coth x \coth y}{\coth x \pm \coth y}$$

Functions of double argument

$$\sinh 2x = 2 \sinh x \cosh x$$

$$\cosh 2x = \sinh^2 x + \cosh^2 x$$

$$\tanh 2x = \frac{2 \tanh x}{1 + \tanh^2 x}$$

$$\coth 2x = \frac{1 + \coth^2 x}{2 \coth x}$$

Functions of further multiples of the argument

$$\sinh 3x = \sinh x(4 \cosh^2 x - 1)$$

$$\sinh 4x = \sinh x \cosh x(8 \cosh^2 x - 4)$$

$$\sinh 5x = \sinh x(1 - 12 \cosh^2 x + 16 \cosh^4 x)$$

$$\cosh 3x = \cosh x(4 \cosh^2 x - 3)$$

$$\cosh 4x = 1 - 8 \cosh^2 x + 8 \cosh^4 x$$

$$\cosh 5x = \cosh x(5 - 20 \cosh^2 x + 16 \cosh^4 x)$$

$$\sinh nx = \binom{n}{1}\cosh^{n-1} x \sinh x + \binom{n}{3}\cosh^{n-3} x \sinh^3 x + \binom{n}{5}\cosh^{n-5} x \sinh^5 x + \cdots$$

$$\cosh nx = \cosh^n x + \binom{n}{2} \cosh^{n-2} x \sinh^2 x + \binom{n}{4} \cosh^{n-4} x \sinh^4 x + \cdots$$

Functions of half argument

$$\sinh\frac{x}{2} = \sqrt{\frac{\cosh x - 1}{2}} = \frac{\sinh x}{\sqrt{2(\cosh x + 1)}}$$

$$\cosh\frac{x}{2} = \sqrt{\frac{\cosh x + 1}{2}} = \frac{\sinh x}{\sqrt{2(\cosh x - 1)}}$$

$$\tanh\frac{x}{2} = \frac{\sinh x}{\cosh x + 1} = \frac{\cosh x - 1}{\sinh x} = \sqrt{\frac{\cosh x - 1}{\cosh x + 1}}$$

$$\coth\frac{x}{2} = \frac{\sinh x}{\cosh x - 1} = \frac{\cosh x + 1}{\sinh x} = \sqrt{\frac{\cosh x + 1}{\cosh x - 1}}$$

Powers of hyperbolic functions

$$\sinh^2 x = \frac{1}{2}(\cosh 2x - 1)$$

$$\cosh^2 x = \frac{1}{2}(\cosh 2x + 1)$$

$$\sinh^3 x = \frac{1}{4}(-3\sinh x + \sinh 3x)$$

$$\cosh^3 x = \frac{1}{4}(3\cosh x + \cosh 3x)$$

$$\sinh^4 x = \frac{1}{8}(3 - 4\cosh 2x + \cosh 4x)$$

$$\cosh^4 x = \frac{1}{8}(3 + 4\cosh 2x + \cosh 4x)$$

$$\sinh^5 x = \frac{1}{16}(10\sinh x - 5\sinh 3x + \sinh 5x)$$

$$\cosh^5 x = \frac{1}{16}(10\cosh x + 5\cosh 3x + \cosh 5x)$$

$$\sinh^6 x = \frac{1}{32}(-10 + 15\cosh 2x - 6\cosh 4x + \cosh 6x)$$

$$\cosh^6 x = \frac{1}{32}(10 + 15\cosh 2x + 6\cosh 4x + \cosh 6x)$$

Sums and differences of hyperbolic functions

$$\sinh x + \sinh y = 2 \sinh \frac{x+y}{2} \cosh \frac{x-y}{2}$$

$$\sinh x - \sinh y = 2 \sinh \frac{x-y}{2} \cosh \frac{x+y}{2}$$

$$\cosh x + \cosh y = 2 \cosh \frac{x+y}{2} \cosh \frac{x-y}{2}$$

$$\cosh x - \cosh y = 2 \sinh \frac{x+y}{2} \sinh \frac{x-y}{2}$$

$$\tanh x + \tanh y = \frac{\sinh (x+y)}{\cosh x \cosh y}$$

$$\tanh x - \tanh y = \frac{\sinh (x-y)}{\cosh x \cosh y}$$

$$\coth x + \coth y = \frac{\sinh (x+y)}{\sinh x \sinh y}$$

$$\coth x - \coth y = \frac{\sinh (y-x)}{\sinh x \sinh y}$$

Theorem of De Moivre

$$(\sinh x + \cosh x)^n = \sinh nx + \cosh nx$$

$$(\cosh x - \sinh x)^n = \cosh nx - \sinh nx$$

Products of hyperbolic functions

$$\sinh x \sinh y = \frac{1}{2} [\cosh (x+y) - \cosh (x-y)]$$

$$\cosh x \cosh y = \frac{1}{2} [\cosh (x+y) + \cosh (x-y)]$$

$$\sinh x \cosh y = \frac{1}{2} [\sinh (x+y) + \sinh (x-y)]$$

$$\tanh x \tanh y = \frac{\tanh x + \tanh y}{\coth x + \coth y}$$

Relations between hyperbolic and exponential functions

$$\sinh x + \cosh x = e^x \qquad \sinh x - \cosh x = -e^{-x}$$

$$e^x = \frac{1 + \tanh \frac{x}{2}}{1 - \tanh \frac{x}{2}}$$

(for further relations see definition on page 196)

Relations between hyperbolic and trigonometric functions

$$\cosh xj = \cos x \qquad \cosh x = \cos xj$$

$$\sinh xj = j \sin x \qquad \sinh x = -j \sin xj$$

$$\tanh xj = j \tan x \qquad \tanh x = -j \tan xj$$

$$\coth xj = -j \cot x \qquad \coth x = j \cot xj$$

$$\sin (x \pm yj) = \sin x \cosh y \pm j \cos x \sinh y$$

$$\cos (x \pm yj) = \cos x \cosh y \mp j \sin x \sinh y$$

$$\tan (x \pm yj) = \frac{\sin 2x \pm j \sinh 2y}{\cos 2x + \cosh 2y} = \frac{\sin 2x \pm j \sinh 2y}{2 (\cos^2 x + \sinh^2 y)}$$

$$\cot (x \pm yj) = \frac{\sin 2x \mp j \sinh 2y}{2(\sin^2 x + \sinh^2 y)} = -\frac{\sin 2x \mp j \sinh 2y}{\cos 2x - \cosh 2y}$$

$$\sinh (x \pm yj) = \sinh x \cos y \pm j \cosh x \sin y$$

$$\cosh (x \pm yj) = \cosh x \cos y \pm j \sinh x \sin y$$

$$\tanh (x \pm yj) = \frac{\sinh 2x \pm j \sin 2y}{\cosh 2x + \cos 2y}$$

$$\coth (x \pm yj) = \frac{\sinh 2x \mp j \sin 2y}{\cosh 2x - \cos 2y}$$

3.4.6. Inverse hyperbolic functions

The inverse function to $y = \sinh x$ is $y = \sinh^{-1} x$

The inverse function to $y = \cosh x$ is $y = \cosh^{-1} x$

For each $x > 1$, $\cosh^{-1} x$ has two opposite equal values.

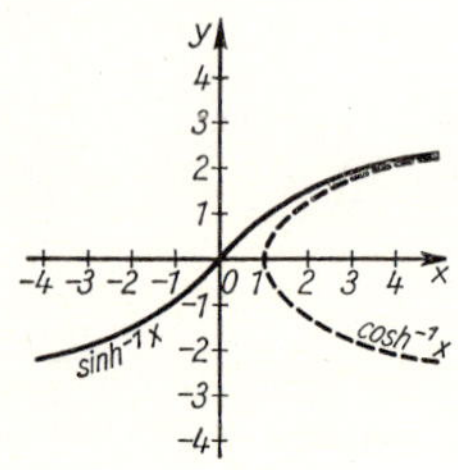

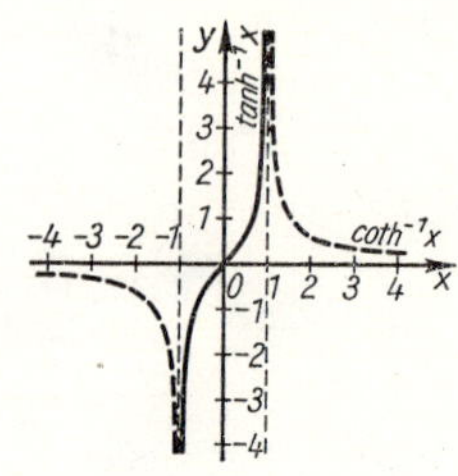

The inverse function to $y = \tanh x$ is $\tanh^{-1} x$

Domain of definition $|x| < 1$

The inverse function to $y = \coth x$ is $y = \coth^{-1} x$

Domain of definition $|x| > 1$

Representation of one inverse hyperbolic function by another one

$$\sinh^{-1} x = \cosh^{-1} \sqrt{x^2 + 1} \quad \text{for} \quad x > 0$$

$$= -\cosh^{-1} \sqrt{x^2 + 1} \quad \text{for} \quad x < 0$$

$$= \tanh^{-1} \frac{x}{\sqrt{x^2 + 1}} = \coth^{-1} \frac{\sqrt{x^2 + 1}}{x}$$

$$\cosh^{-1} x = \pm \sinh^{-1} \sqrt{x^2 - 1}$$

$$= \pm \tanh^{-1} \frac{\sqrt{x^2 - 1}}{x} = \pm \coth^{-1} \frac{x}{\sqrt{x^2 - 1}}$$

$$\tanh^{-1} x = \sinh^{-1} \frac{x}{\sqrt{1 - x^2}}$$

$$= \cosh^{-1} \frac{1}{\sqrt{1 - x^2}} \qquad \text{for} \quad x > 0$$

$$= -\cosh^{-1} \frac{1}{\sqrt{1 - x^2}} \qquad \text{for} \quad x < 0$$

$$= \coth^{-1} \frac{1}{x}$$

$$\coth^{-1} x = \sinh^{-1} \frac{1}{\sqrt{x^2 - 1}}$$

$$= \cosh^{-1} \frac{x}{\sqrt{x^2 - 1}} \qquad \text{for} \quad x > 0$$

$$= -\cosh^{-1} \frac{x}{\sqrt{x^2 - 1}} \qquad \text{for} \quad x < 0$$

$$= \tanh^{-1} \frac{1}{x}$$

Sums and differences of inverse hyperbolic functions

$$\sinh^{-1} x \pm \sinh^{-1} y = \sinh^{-1} \left(x \sqrt{1 + y^2} \pm y \sqrt{1 + x^2}\right)$$

$$\cosh^{-1} x \pm \cosh^{-1} y = \cosh^{-1} \left[xy \pm \sqrt{(x^2 - 1)(y^2 - 1)}\right]$$

$$\tanh^{-1} x \pm \tanh^{-1} y = \tanh^{-1} \frac{x \pm y}{1 \pm xy}$$

Relations between inverse hyperbolic functions and inverse trigonometric functions

$$\sinh^{-1} xj = j \operatorname{arc} \sin x$$

$$\cosh^{-1} xj = j \operatorname{arc} \cos x$$

$$\tanh^{-1} xj = j \operatorname{arc} \tan x$$

$$\coth^{-1} xj = -j \operatorname{arc} \cot x$$

Relations between inverse hyperbolic functions and the logarithmic function

$$\sinh^{-1} x = \ln \left(x + \sqrt{x^2 + 1}\right) \qquad \text{for} \quad x \in (-\infty, +\infty)$$

$$\cosh^{-1} x = \ln \left(x \pm \sqrt{x^2 - 1}\right) = \pm \ln \left(x + \sqrt{x^2 - 1}\right) \qquad \text{for} \quad x \in [1, \infty)$$

$$\tanh^{-1} x = \frac{1}{2} \ln \frac{1 + x}{1 - x} \quad \text{for} \quad |x| < 1$$

$$\coth^{-1} x = \frac{1}{2} \ln \frac{x + 1}{x - 1} \quad \text{for} \quad |x| > 1$$

3.5. Spherical trigonometry

3.5.1. General

Explanation

Great circles are the lines of intersection of planes through the center of a sphere with the surface of the sphere.
Small circles are the lines of intersection of a plane which does not pass through the center of a sphere with the surface of the sphere.
Spherical lunes are bounded by two great circles.
Spherical triangles are bounded by three great circles.
Since great circles on a sphere form several spherical triangles, here we shall consider the spherical triangle with sides and angles smaller than 180°.

Notations in the spherical triangle

$$\left.\begin{array}{l}\text{sides } a, b, c\\ \text{angles } \alpha, \beta, \gamma\end{array}\right\}\ \text{(according to notations in a plane triangle)}$$

Conditions for the sides and angles of the spherical triangle

$$0 < a + b + c < 360°$$

$$180° < \alpha + \beta + \gamma < 540°$$

$$a \gtreqless b, \text{ according as } \alpha \gtreqless \beta$$

$$a + b \gtreqless 180°, \text{ according as } \alpha + \beta \gtreqless 180°$$ (for other sides and angles accordingly).

$$\alpha + \beta + \gamma - 180° = \varepsilon \qquad \textit{(spherical excess)}$$

$$360° - a - b - c = d \qquad \textit{(spherical defect)}$$

3.5.2. Right spherical triangle

Notations are as in the plane (i.e. if $\gamma = 90°$, then a and b are the legs and c is the hypotenuse).

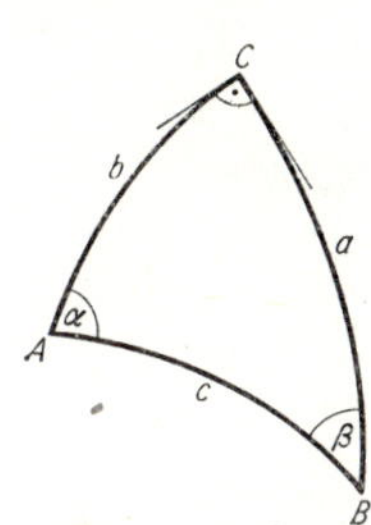

$$\sin a = \sin\alpha \sin c$$
$$\sin a = \tan b \cot\beta$$
$$\sin b = \sin c \sin\beta$$
$$\sin b = \tan a \cot\alpha$$
$$\cos c = \cos a \cos b$$
$$\cos c = \cot\alpha \cot\beta$$
$$\cos\alpha = \cos a \sin\beta$$
$$\cos\alpha = \cot c \tan b$$
$$\cos\beta = \sin\alpha \cos b$$
$$\cos\beta = \cot c \tan a$$

The above formulas are condensed into **Napier's rule:**
If we exclude the right angle and instead of the legs use their complements, then the cosine of any element is equal to the product of the sines of the element separated from it or
equal to the product of the cotangents of the two adjacent elements.
Any right-angled spherical triangle can be calculated from two given elements with the help of Napier's rule.

3.5.3. Oblique spherical triangle

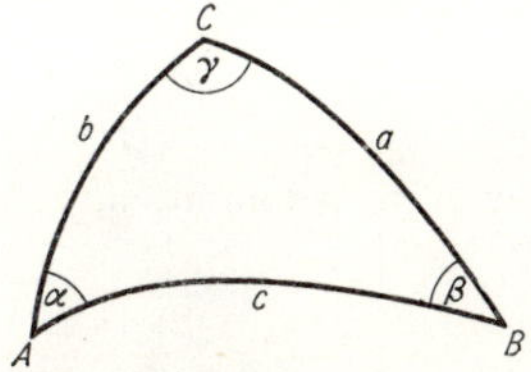

Sine theorem

$$\sin a : \sin b : \sin c = \sin\alpha : \sin\beta : \sin\gamma$$

Side-cosine theorem

$$\cos a = \cos b \cos c + \sin b \sin c \cos\alpha$$
$$\cos b = \cos c \cos a + \sin c \sin a \cos\beta$$
$$\cos c = \cos a \cos b + \sin a \sin b \cos\gamma$$

Angle-cosine theorem

$$\cos\alpha = -\cos\beta \cos\gamma + \sin\beta \sin\gamma \cos a$$
$$\cos\beta = -\cos\gamma \cos\alpha + \sin\gamma \sin\alpha \cos b$$
$$\cos\gamma = -\cos\alpha \cos\beta + \sin\alpha \sin\beta \cos c$$

Half-side theorem on a spherical triangle

$$\sin\frac{a}{2} = \sqrt{-\frac{\cos\sigma\cos(\sigma-\alpha)}{\sin\beta\sin\gamma}} \qquad \sigma = \frac{\alpha+\beta+\gamma}{2}$$

$$\cos\frac{a}{2} = \sqrt{\frac{\cos(\sigma-\beta)\cos(\sigma-\gamma)}{\sin\beta\sin\gamma}}$$

$$\tan\frac{a}{2} = \sqrt{-\frac{\cos\sigma\cos(\sigma-\alpha)}{\cos(\sigma-\beta)\cos(\sigma-\gamma)}}$$

$$\cot\frac{a}{2} = \sqrt{-\frac{\cos(\sigma-\beta)\cos(\sigma-\gamma)}{\cos\sigma\cos(\sigma-\alpha)}}$$

Formulas for $\frac{b}{2}$ and $\frac{c}{2}$ are obtained by cyclic permutation.

Half-angle theorem on a spherical triangle

$$\sin\frac{\alpha}{2} = \sqrt{\frac{\sin(s-b)\sin(s-c)}{\sin b\sin c}} \qquad s = \frac{a+b+c}{2}$$

$$\cos\frac{\alpha}{2} = \sqrt{\frac{\sin s\sin(s-a)}{\sin b\sin c}}$$

$$\tan\frac{\alpha}{2} = \sqrt{\frac{\sin(s-b)\sin(s-c)}{\sin s\sin(s-a)}}$$

$$\cot\frac{\alpha}{2} = \sqrt{\frac{\sin s\sin(s-a)}{\sin(s-b)\sin(s-c)}}$$

The formulas for $\frac{\beta}{2}$ and $\frac{\gamma}{2}$ are obtained by cyclic permutation.

Gauss's formulas

$$\frac{\sin\frac{\alpha+\beta}{2}}{\cos\frac{\gamma}{2}} = \frac{\cos\frac{a-b}{2}}{\cos\frac{c}{2}} \qquad \frac{\cos\frac{\alpha+\beta}{2}}{\sin\frac{\gamma}{2}} = \frac{\cos\frac{a+b}{2}}{\cos\frac{c}{2}}$$

$$\frac{\sin\frac{\alpha-\beta}{2}}{\cos\frac{\gamma}{2}} = \frac{\sin\frac{a-b}{2}}{\sin\frac{c}{2}} \qquad \frac{\cos\frac{\alpha-\beta}{2}}{\sin\frac{\gamma}{2}} = \frac{\sin\frac{a+b}{2}}{\sin\frac{c}{2}}$$

Eight similar formulas are obtained by cyclic permutation.

Napier's analogies

$$\frac{\tan\dfrac{a+b}{2}}{\tan\dfrac{c}{2}} = \frac{\cos\dfrac{\alpha-\beta}{2}}{\cos\dfrac{\alpha+\beta}{2}} \qquad \frac{\tan\dfrac{\alpha+\beta}{2}}{\cot\dfrac{\gamma}{2}} = \frac{\cos\dfrac{a-b}{2}}{\cos\dfrac{a+b}{2}}$$

$$\frac{\tan\dfrac{a-b}{2}}{\tan\dfrac{c}{2}} = \frac{\sin\dfrac{\alpha-\beta}{2}}{\sin\dfrac{\alpha+\beta}{2}} \qquad \frac{\tan\dfrac{\alpha-\beta}{2}}{\cot\dfrac{\gamma}{2}} = \frac{\sin\dfrac{a-b}{2}}{\sin\dfrac{a+b}{2}}$$

Further formulas are obtained by cyclic permutation.

Circumcircle radius r and incircle radius ϱ of the spherical triangle

$$\cot r = \sqrt{-\frac{\cos(\sigma-\alpha)\cos(\sigma-\beta)\cos(\sigma-\gamma)}{\cos\sigma}}$$

σ see page 206

$$\tan\varrho = \sqrt{\frac{\sin(s-a)\sin(s-b)\sin(s-c)}{\sin s}}$$ s see page 206

$$\left.\begin{aligned} \cot r &= \cot\frac{a}{2}\cos(\sigma-\alpha) \\ \tan\varrho &= \tan\frac{\alpha}{2}\sin(s-a) \end{aligned}\right\}$$ further formulas are obtained by cyclic permutation

L'Huilier's formulas

$$\tan\frac{\varepsilon}{4} = \sqrt{\tan\frac{s}{2}\tan\frac{s-a}{2}\tan\frac{s-b}{2}\tan\frac{s-c}{2}}$$

s see page 206

$$\tan\left(\frac{\alpha}{2}-\frac{\varepsilon}{2}\right) = \sqrt{\frac{\tan\dfrac{s-b}{2}\tan\dfrac{s-c}{2}}{\tan\dfrac{s}{2}\tan\dfrac{s-a}{2}}}$$ ε see page 204

Spherical defect

$$\tan\frac{d}{4}$$

$$=\sqrt{-\tan(45^\circ-\sigma)\tan\left(45^\circ-\frac{\sigma-\alpha}{2}\right)\tan\left(45^\circ-\frac{\sigma-\beta}{2}\right)\tan\left(45^\circ-\frac{\sigma-\gamma}{2}\right)}$$

$$\text{for}\quad \sigma=\frac{\alpha+\beta+\gamma}{2}$$

(For areas of spherical triangles and lunes see Section "Stereometry", page 167)

CALCULATION OF THE OBLIQUE SPHERICAL TRIANGLE

Basic problem 1:

Given the three sides

Solution 1:

One angle according to the side-cosine theorem (e.g. α)

$$\cos\alpha=\frac{\cos a-\cos b\cos c}{\sin b\sin c}$$

$$\sin\beta=\frac{\sin b\sin\alpha}{\sin a}\qquad\text{(sine theorem)}$$

$$\sin\gamma=\frac{\sin c\sin\alpha}{\sin a}\qquad\text{(sine theorem)}$$

Solution 2:

All angles according to the side-cosine theorem

Solution 3:

All angles according to the half-angle theorem or one angle according to the half-angle theorem and the other ones according to the sine-theorem as in Solution 1

Basic problem 2:

Given two sides and the enclosed angle (e.g. b, c, α)

Solution 1:

$$\cos a = \cos b \cos c + \sin b \sin c \cos \alpha \quad \text{(side cosine theorem)}$$

$$\sin \beta = \frac{\sin b \sin \alpha}{\sin a} \quad \text{(sine theorem)}$$

$$\sin \gamma = \frac{\sin c \sin \alpha}{\sin a} \quad \text{(sine theorem)}$$

Solution 2:

$$\tan \frac{\beta + \gamma}{2} = \frac{\cos \frac{b - c}{2} \cot \frac{\alpha}{2}}{\cos \frac{b + c}{2}} \quad \text{(Napier's analogy)}$$

$$\tan \frac{\beta - \gamma}{2} = \frac{\sin \frac{b - c}{2} \cot \frac{\alpha}{2}}{\sin \frac{b + c}{2}} \quad \text{(Napier's analogy)}$$

From $\frac{\beta + \gamma}{2}$ and $\frac{\beta - \gamma}{2}$ the two angles β and γ are obtained.

$$\tan \frac{a}{2} = \frac{\tan \frac{b + c}{2} \cos \frac{\beta + \gamma}{2}}{\cos \frac{\beta - \gamma}{2}} \quad \text{(Napier's analogy)}$$

Basic problem 3:

Given two sides and the angle opposite to one side (e.g. b, c, β)

Solution:

$$\sin \gamma = \frac{\sin c \sin \beta}{\sin b} \quad \text{(sine theorem)}$$

$$\tan \frac{a}{2} = \frac{\tan \frac{b + c}{2} \cos \frac{\beta + \gamma}{2}}{\cos \frac{\beta - \gamma}{2}} \quad \text{(Napier's analogy)}$$

$$\cot \frac{\alpha}{2} = \frac{\tan \frac{\beta + \gamma}{2} \cos \frac{b + c}{2}}{\cos \frac{b - c}{2}} \quad \text{(Napier's analogy)}$$

or angle α with the sine theorem

Basic problem 4:

Given one side and the two adjacent angles (e.g. a, β, γ)

Solution 1:

$$\cos\alpha = -\cos\beta\cos\gamma + \sin\beta\sin\gamma\cos a \qquad \text{(angle cosine theorem)}$$

$$\sin b = \frac{\sin\beta\sin a}{\sin\alpha} \qquad \text{(sine theorem)}$$

$$\sin c = \frac{\sin\gamma\sin a}{\sin\alpha} \qquad \text{(sine theorem)}$$

Solution 2:

$$\tan\frac{b+c}{2} = \frac{\cos\frac{\beta-\gamma}{2}\tan\frac{a}{2}}{\cos\frac{\beta+\gamma}{2}} \qquad \text{(Napier's analogy)}$$

$$\tan\frac{b-c}{2} = \frac{\sin\frac{\beta-\gamma}{2}\tan\frac{a}{2}}{\sin\frac{\beta+\gamma}{2}} \qquad \text{(Napier's analogy)}$$

From $\frac{b+c}{2}$ and $\frac{b-c}{2}$ the two sides b and c are obtained.

$$\sin\alpha = \frac{\sin a\sin\beta}{\sin b} \qquad \text{(sine theorem)}$$

Basic problem 5:

Given one side, one adjacent and the opposite angle (e.g. b, β, γ)

Solution:

$$\sin c = \frac{\sin\gamma\sin b}{\sin\beta} \qquad \text{(sine theorem)}$$

$$\tan\frac{a}{2} = \frac{\tan\frac{b+c}{2}\cos\frac{\beta+\gamma}{2}}{\cos\frac{\beta-\gamma}{2}} \qquad \text{(Napier's analogy)}$$

$$\sin\alpha = \frac{\sin a\sin\beta}{\sin b} \qquad \text{(sine theorem)}$$

Basic problem 6:

Given the three angles

Solution 1:

$$\cos a = \frac{\cos \alpha + \cos \beta \cos \gamma}{\sin \beta \sin \gamma} \qquad \text{(angle cosine theorem)}$$

$$\sin b = \frac{\sin \beta \sin a}{\sin \alpha} \qquad \text{(sine theorem)}$$

$$\sin c = \frac{\sin \gamma \sin a}{\sin \alpha} \qquad \text{(sine theorem)}$$

Solution 2:

One side according to the half-angle theorem; then proceed as above.

3.5.4. Mathematical geography

Measures of length

The earth is considered a sphere.

Mean radius of the earth $r \approx 6370$ km

Circumference of the earth ≈ 40000 km

Length of one *meridian degree* ≈ 111.3 km (applies to great circle)
Length of one *minute of arc* ≈ 1.852 km $= 1$ *nautical mile* (applies to great circle)

1 *geographical mile* = 4 nautical miles ≈ 7.420 km

1 *point* of a compass card $= 11\frac{1}{4}^\circ$

Coordinate system of the earth

The abscissa axis is the *equator*. The ordinate axis is the *zero meridian* (meridian through Greenwich).

Geographical coordinates

The *geographical longitude* λ is measured either
at the equator as the arc between the zero meridian and the meridian of the point in question
or
as the angle between the plane of the meridian of the point in question and the plane of the zero meridian (measured from 0° to 180°, positive to the East, negative to the West).

Geographical latitude φ is the spherical distance of the point in question from the equator (measured from 0° to 90°, positive to the North, negative to the South).

The shortest distance between two points

The arc of the great circle between the points $P_1(\varphi_1, \lambda_1)$ and $P_2(\varphi_2, \lambda_2)$ (*orthodrome*) determines the shortest distance (orthodromic distance).

In the triangle P_1P_2N are known

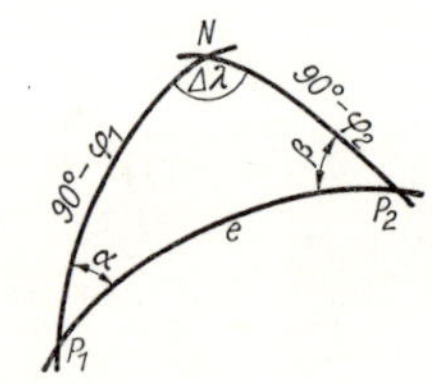

$$NP_1 = 90° - \varphi_1$$

$$NP_2 = 90° - \varphi_2$$

$$\text{angles } P_1NP_2 = \lambda_2 - \lambda_1 = \Delta\lambda$$

(see *basic problem 2* of the spherical triangle)

$$\cos e = \cos(90° - \varphi_1)\cos(90° - \varphi_2) + \sin(90° - \varphi_1)\sin(90° - \varphi_2)\cos\Delta\lambda;$$

$$\cos e = \sin\varphi_1 \sin\varphi_2 + \cos\varphi_1 \cos\varphi_2 \cos\Delta\lambda$$

(side-cosine theorem)

Calculation of the course angles

The angles α and β of the spherical triangle P_1P_2N (figure see above) can be calculated if the distance e has been determined.

$$\sin\alpha = \frac{\sin\Delta\lambda\sin(90° - \varphi_2)}{\sin e} = \frac{\sin\Delta\lambda\cos\varphi_2}{\sin e} \quad \text{(sine theorem)}$$

$$\sin\beta = \frac{\sin\Delta\lambda\cos\varphi_1}{\sin e}$$

Distance between two points of the same geographical latitude φ

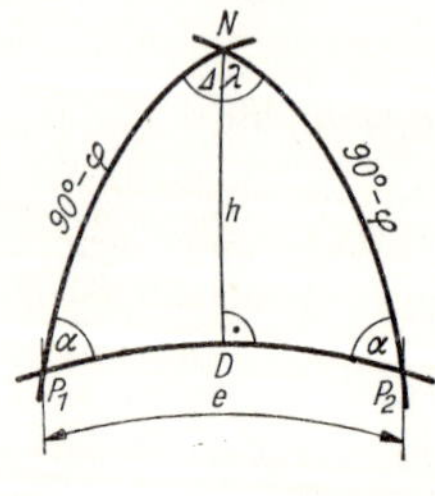

Orthodromic distance $e = \widehat{P_1P_2}$ can be easily calculated from the right-angled subtriangle P_1DN which is formed by the altitude h within the isosceles triangle.

$$\sin\frac{e}{2} = \cos\varphi\sin\frac{\Delta\lambda}{2}$$

(NAPIER's rule)

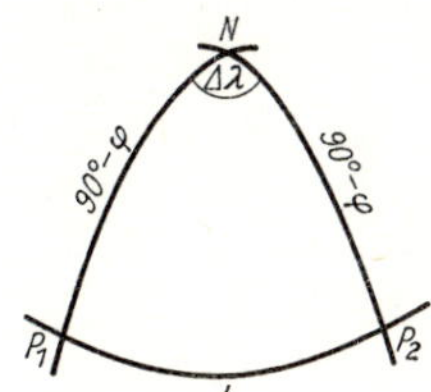

Loxodromic distance (arc on the latitude) l, it is calculated from the formula

$$l = \Delta\lambda \cos \varphi \qquad \text{(in degrees)}$$

or

$$l = \frac{\pi r \, \Delta\lambda \cos \varphi}{180°} \text{ (in km)}$$

$$r \approx 6370 \text{ km}$$

Note: The *loxodrome* is a line on the surface of the sphere that cuts all meridians at the same angle. If the angle differs from 90°, the loxodrome approaches the pole spirally. Each circle of latitude is a loxodrome that intersects the meridians at right angles.

4. Analytical geometry

4.1. Analytical geometry of the plane

4.1.1. The various systems of coordinates

O origin of coordinates, zero

Oblique coordinate system

The position of point P is defined by two lines parallel to the oblique-angled coordinates axes.

x abscissa, y ordinate

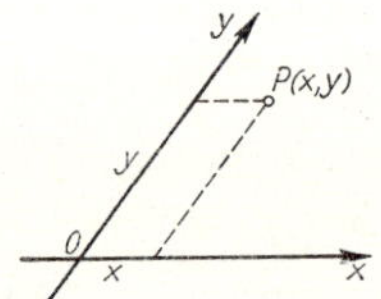

Rectangular (Cartesian) system of coordinates

The coordinates of point P are the abscissa x and the ordinate y.
We write: $P(x, y)$
The points in the plane are put into one-to-one correspondence with pairs of numbers $(x, y) \in R \times R$.

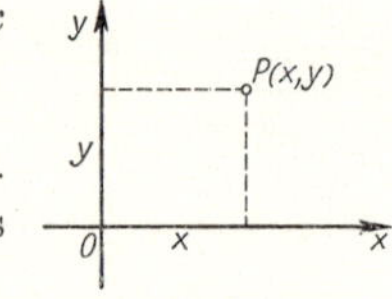

System of polar coordinates

The position of point P is defined by its distance from a fixed point O (pole) and by the angle φ that the line segment OP makes with a given straight line passing through O (*polar axis*) (φ positive if measured in the mathematical sence of positive rotation).
r *radius vector*,
φ *amplitude, polar angle*

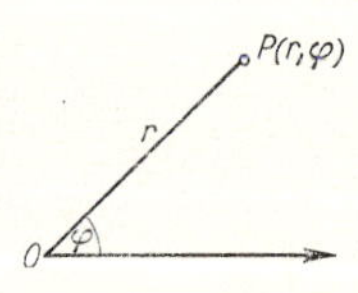

Parallel translation of the rectangular system of coordinates

Coordinates of the point P in the original system x, y

Coordinates in the new system ξ, η

$$x = \xi + c \qquad \xi = x - c$$

$$y = \eta + d \qquad \eta = y - d$$

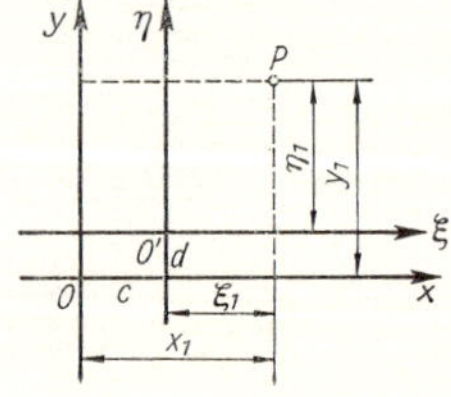

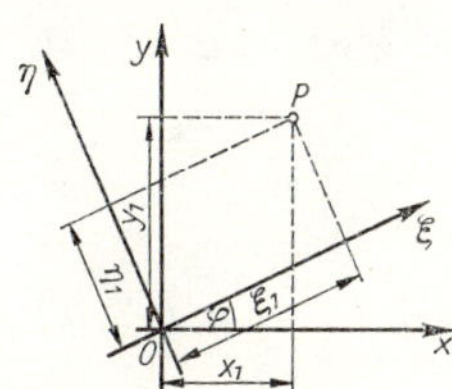

Rotation of the rectangular coordinate system about the angle φ

$$x = \xi \cos \varphi - \eta \sin \varphi$$
$$y = \xi \sin \varphi + \eta \cos \varphi$$
$$\xi = x \cos \varphi + y \sin \varphi$$
$$\eta = -x \sin \varphi + y \cos \varphi$$

Parallel translation and rotation of the rectangular coordinate system through the angle φ

$$x = \xi \cos \varphi - \eta \sin \varphi + c \qquad \xi = x \cos \varphi + y \sin \varphi - c$$
$$y = \xi \sin \varphi + \eta \cos \varphi + d \qquad \eta = -x \sin \varphi + y \cos \varphi - d$$

Transition from a rectangular to an oblique coordinate system

x, y coordinates in the rectangular system
ξ, η coordinates in the oblique system
φ_1 angle between x-axis and ξ-axis
φ_2 angle between x-axis and η-axis

$$x = \xi \cos \varphi_1 + \eta \cos \varphi_2 \qquad \xi = \frac{-x \sin \varphi_2 + y \cos \varphi_2}{\sin (\varphi_1 - \varphi_2)}$$

$$y = \xi \sin \varphi_1 + \eta \sin \varphi_2 \qquad \eta = \frac{x \sin \varphi_1 - y \cos \varphi_1}{\sin (\varphi_1 - \varphi_2)}$$

Transition from rectangular coordinates to polar coordinates

$$\left.\begin{array}{l} x = r \cos \varphi \\ y = r \sin \varphi \end{array}\right\} \Rightarrow \frac{y}{x} = \tan \varphi$$

$$r = \sqrt{x^2 + y^2}; \qquad \varphi = \arctan \frac{y}{x}$$

4.1.2. Points and line segments

Distance e between two points $P_1(x_1, y_1)$ and $P_2(x_2, y_2)$

$$e = \sqrt{(x_2 - x_1)^2 + (y_2 - y_1)^2}$$

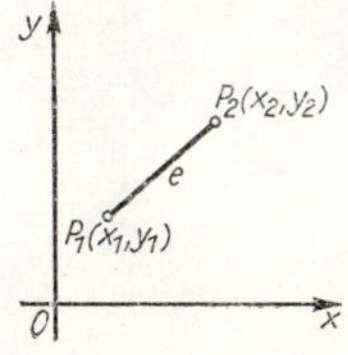

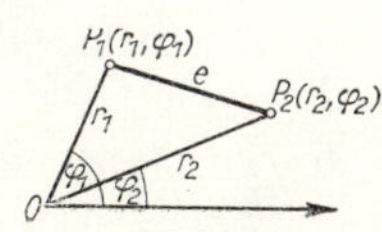

Distance e between two points $P_1(r_1, \varphi_1)$ and $P_2(r_2, \varphi_2)$

$$e = \sqrt{r_1^2 + r_2^2 - 2r_1r_2\cos(\varphi_2 - \varphi_1)}$$

Division of a line segment P_1P_2 in the ratio λ

$P_1(x_1, y_1)$, $P_2(x_2, y_2)$, point of division $T(x_t, y_t)$, $\lambda = \overline{P_1T} : \overline{TP_2}$

$$x_t = \frac{x_1 + \lambda x_2}{1 + \lambda} \qquad \lambda > 0 \quad \text{for inner point of division}$$

$$y_t = \frac{y_1 + \lambda y_2}{1 + \lambda} \qquad \lambda < 0 \quad \text{for outer point of division}$$

Midpoint $P_0(x_0, y_0)$ of the line segment P_1P_2

$$x_0 = \frac{x_1 + x_2}{2}; \quad y_0 = \frac{y_1 + y_2}{2}$$

Centroid $S(x_c, y_c)$ of the triangle $P_1P_2P_3$

$$x_c = \frac{x_1 + x_2 + x_3}{3}; \quad y_c = \frac{y_1 + y_2 + y_3}{3}$$

If the points are material with the masses m_1, m_2, m_3, then we have

$$x_c = \frac{m_1x_1 + m_2x_2 + m_3x_3}{m_1 + m_2 + m_3}; \quad y_c = \frac{m_1y_1 + m_2y_2 + m_3y_3}{m_1 + m_2 + m_3}$$

and in general, with n mass points

$$x_c = \frac{\sum_1^n m_k x_k}{\sum_1^n m_k}; \quad y_c = \frac{\sum_1^n m_k y_k}{\sum_1^n m_k}$$

Direction of the line segment P_1P_2

$$\tan\alpha = m = \frac{y_2 - y_1}{x_2 - x_1}$$

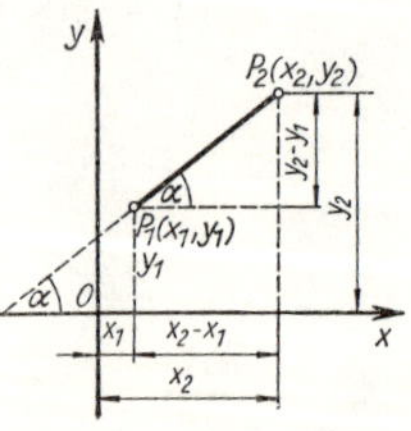

α angle of the line-segment with the positive direction of the abscissa axis

$\tan\alpha = m$ *direction factor* or *gradient* or *slope* of the line segment

Condition for three points P_1, P_2, P_3 on a straight line

$$\begin{vmatrix} x_1 & y_1 & 1 \\ x_2 & y_2 & 1 \\ x_3 & y_3 & 1 \end{vmatrix} = 0$$

Area of a triangle

$$A = \frac{1}{2}\left[x_1(y_2 - y_3) + x_2(y_3 - y_1) + x_3(y_1 - y_2)\right] = \frac{1}{2}\begin{vmatrix} x_1 & y_1 & 1 \\ x_2 & y_2 & 1 \\ x_3 & y_3 & 1 \end{vmatrix}$$

Area of a n-gon

$$A = \frac{1}{2}\left[x_1(y_2 - y_n) + x_2(y_3 - y_1) + x_3(y_4 - y_2) + \cdots + x_n(y_1 - y_{n-1})\right]$$

$$= \frac{1}{2}\left[(x_1 - x_2)(y_1 + y_2) + (x_2 - x_3)(y_2 + y_3) + \cdots + (x_n - x_1)(y_n + y_1)\right]$$

4.1.3. Straight line

Cartesian normal form

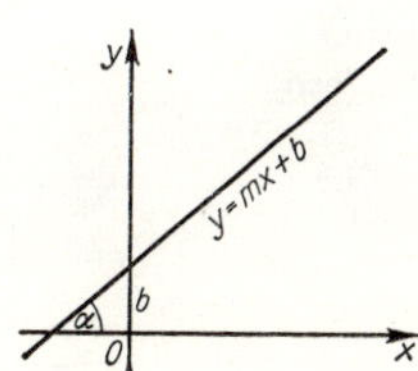

$$y = mx + b$$

b intercept of the y-axis
$m = \tan\alpha$ direction factor, direction coefficient, gradient, slope

$m > 0$: increasing straight line
$m < 0$: decreasing straight line

Intercept form

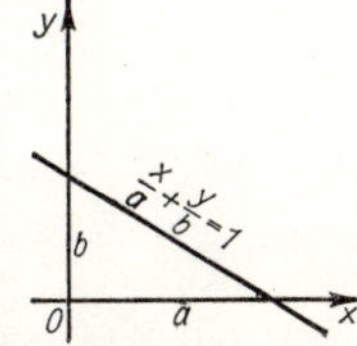

$$\frac{x}{a} + \frac{y}{b} = 1$$

a intercept of the x-axis
b intercept of the y-axis

Point slope form

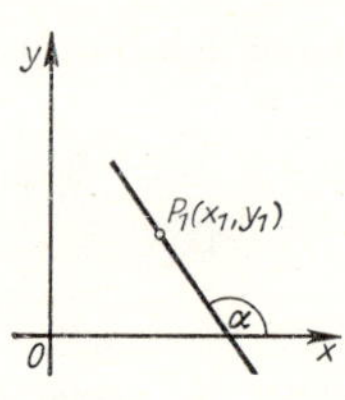

$$y - y_1 = m(x - x_1)$$

Two-point form

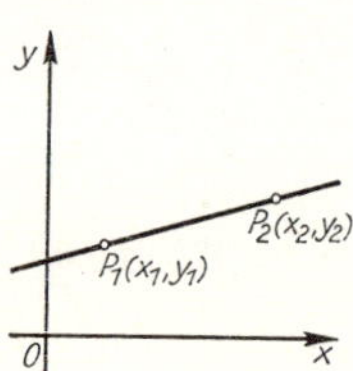

$$\frac{y - y_1}{x - x_1} = \frac{y_2 - y_1}{x_2 - x_1}$$

in the form of a determinant:

$$\begin{vmatrix} x & y & 1 \\ x_1 & y_1 & 1 \\ x_2 & y_2 & 1 \end{vmatrix} = 0$$

Hessian normal form

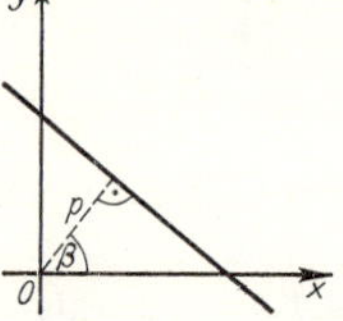

$$x \cos \beta + y \sin \beta - p = 0$$

p distance of the origin from the straight line
β angle which p makes with the positive direction of the x-axis

General form

$Ax + By + C = 0$ A, B, C constants (A, B not both zero)

Polar form

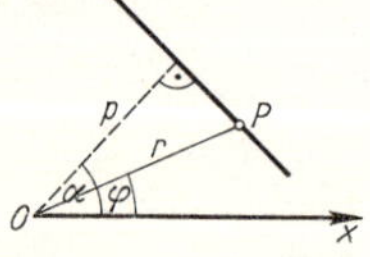

$$r = \frac{p}{\cos(\alpha - \varphi)}$$

Transformation of the general form into other forms

into the Cartesian form:

$$y = -\frac{A}{B}x - \frac{C}{B} \qquad B \neq 0$$

$$m = -\frac{A}{B}; \quad b = -\frac{C}{B}$$

into the intercept form:

$$\frac{x}{-\frac{C}{A}} + \frac{y}{-\frac{C}{B}} = 1 \qquad A, B, C \neq 0$$

$$a = -\frac{C}{A}; \quad b = -\frac{C}{B}$$

into the HESSIAN normal form:

$$\frac{Ax + By + C}{\pm\sqrt{A^2 + B^2}} = 0$$

The sign of the root must be opposite to the sign of C.

Special straight lines

Straight line through the origin

$$y = mx \qquad Ax + By = 0$$

Line parallel to the x-axis

$$y = b \qquad By + C = 0$$

Line parallel to the y-axis

$$x = a \qquad Ax + C = 0$$

Equation of the y-axis

$$x = 0$$

Equation of the x-axis

$$y = 0$$

Distance d of point $P_1(x_1, y_1)$ from the straight line in the Hessian normal form

$$d = x_1 \cos\beta + y_1 \sin\beta - p = \frac{Ax_1 + By_1 + C}{\pm\sqrt{A^2 + B^2}}$$

Sign of the root see above.
d is positive if the point P_1 and the origin are on different sides of the line, otherwise the distance has a negative sign. In the latter case, d must be taken as absolute.

The intersection $S(x_i, y_i)$ of two straight lines

$$\left.\begin{aligned} y &= m_1x + b_1 \\ y &= m_2x + b_2 \end{aligned}\right\} x_i = \frac{b_1 - b_2}{m_2 - m_1}; \qquad y_i = \frac{b_1m_2 - b_2m_1}{m_2 - m_1}$$

$$\left.\begin{aligned} A_1x + B_1y + C_1 &= 0 \\ A_2x + B_2y + C_2 &= 0 \end{aligned}\right\}$$

$$x_i = \frac{B_1C_2 - B_2C_1}{A_1B_2 - A_2B_1} = \begin{vmatrix} B_1 & C_1 \\ B_2 & C_2 \end{vmatrix} : \begin{vmatrix} A_1 & B_1 \\ A_2 & B_2 \end{vmatrix}$$

$$y_i = \frac{C_1A_2 - C_2A_1}{A_1B_2 - A_2B_1} = \begin{vmatrix} C_1 & A_1 \\ C_2 & A_2 \end{vmatrix} : \begin{vmatrix} A_1 & B_1 \\ A_2 & B_2 \end{vmatrix}$$

Condition for three straight lines intersecting in a point

$$\left.\begin{aligned} A_1x + B_1y + C_1 &= 0 \\ A_2x + B_2y + C_2 &= 0 \\ A_3x + B_3y + C_3 &= 0 \end{aligned}\right\} \begin{vmatrix} A_1 & B_1 & C_1 \\ A_2 & B_2 & C_2 \\ A_3 & B_3 & C_3 \end{vmatrix} = 0$$

Angle φ between two straight lines g_1 and g_2

$$\left.\begin{aligned} y &= m_1x + b_1 \\ y &= m_2x + b_2 \end{aligned}\right\} \tan\varphi = \left|\frac{m_2 - m_1}{1 + m_1m_2}\right|$$

φ acute angle

$$\left.\begin{aligned} A_1x + B_1y + C_1 &= 0 \\ A_2x + B_2y + C_2 &= 0 \end{aligned}\right\} \tan\varphi = \left|\frac{A_1B_2 - A_2B_1}{A_1A_2 + B_1B_2}\right|$$

$\varphi < 90°$

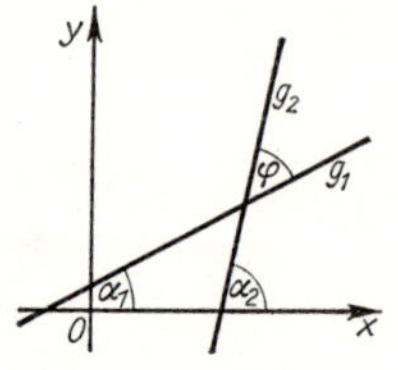

Condition for parallel straight lines

$$\left.\begin{aligned} y &= m_1x + b_1 \\ y &= m_2x + b_2 \end{aligned}\right\} m_1 = m_2$$

$$\left.\begin{aligned} A_1x + B_1y + C_1 &= 0 \\ A_2x + B_2y + C_2 &= 0 \end{aligned}\right\} A_1 : A_2 = B_1 : B_2$$

Condition for perpendicular straight lines

$$\left.\begin{aligned} y &= m_1x + b_1 \\ y &= m_2x + b_2 \end{aligned}\right\} m_2 = -\frac{1}{m_1} \quad \text{or} \quad m_1m_2 = -1$$

$$\left.\begin{aligned} A_1x + B_1y + C_1 &= 0 \\ A_2x + B_2y + C_2 &= 0 \end{aligned}\right\} A_1A_2 + B_1B_2 = 0$$

Pencil of straight lines

$$\left.\begin{array}{l} g_1 \equiv A_1x + B_1y + C_1 = 0 \\ g_2 \equiv A_2x + B_2y + C_2 = 0 \end{array}\right\} \text{given straight lines}$$

Equation of the pencil of straight lines through the point of intersection of the straight lines g_1 and g_2:

$$g_1 + \lambda g_2 = 0, \qquad \lambda \in R$$

Angle-bisectors w_1 and w_2 between two straight lines g_1 and g_2

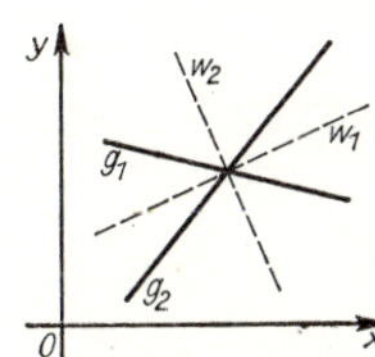

$$\left.\begin{array}{l} g_1 \equiv A_1x + B_1y + C_1 = 0 \\ g_2 \equiv A_2x + B_2y + C_2 = 0 \end{array}\right\} \begin{array}{l}\text{given straight lines} \\ \text{(general form)}\end{array}$$

Equations of the angle-bisectors:

$$\frac{A_1x + B_1y + C}{\pm\sqrt{A_1^2 + B_1^2}} \pm \frac{A_2x + B_2y + C_2}{\pm\sqrt{A_2^2 + B_2^2}} = 0$$

Sign of root (see page 219)

$$\left.\begin{array}{l} g_1 \equiv x\cos\beta_1 + y\sin\beta_1 - p_1 = 0 \\ g_2 \equiv x\cos\beta_2 + y\sin\beta_2 - p_2 = 0 \end{array}\right\} \begin{array}{l}\text{given straight lines} \\ \text{(HESSIAN form)}\end{array}$$

Equations of the angle-bisectors:

$$x(\cos\beta_1 \pm \cos\beta_2) + y(\sin\beta_1 \pm \sin\beta_2) - (p_1 \pm p_2) = 0$$

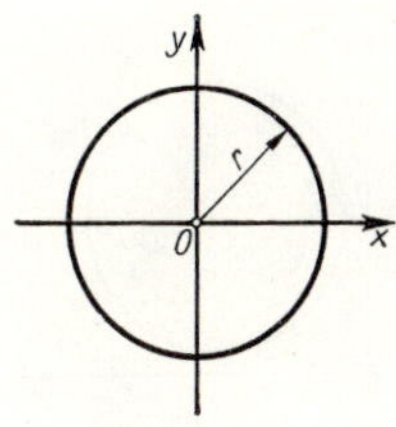

4.1.4. Circle

Equations of the circle

Equation for center at origin

$$x^2 + y^2 = r^2$$

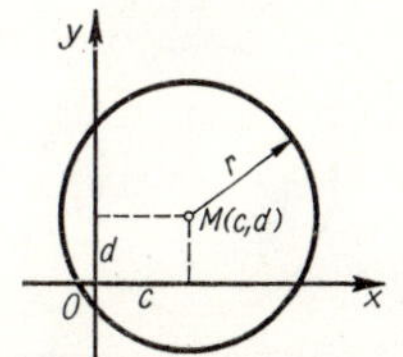

General equation of the circle

$$(x - c)^2 + (y - d)^2 = r^2$$

Equation of a circle with center lying on x-axis, origin on circle (vertex equation)

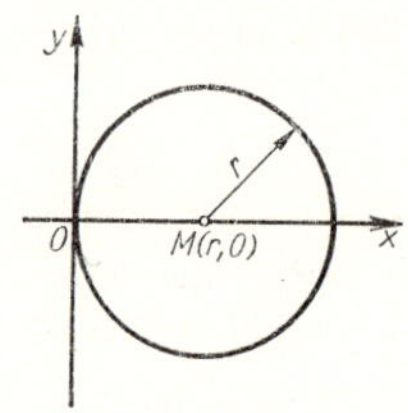

$$y^2 = 2rx - x^2$$

General equation of the second degree as a circle

$$Ax^2 + 2Bxy + Cy^2 + 2Dx + 2Ey + F = 0$$

(general equation of the second degree in x and y)

Condition for the circle:

$$A = C; \qquad B = 0; \qquad D^2 + E^2 - AF > 0$$

thus, the equation of the circle has the form

$$Ax^2 + Ay^2 + 2Dx + 2Ey + F = 0$$

center $M\left(-\frac{D}{A}, -\frac{E}{A}\right)$

radius $r = \frac{1}{A}\sqrt{D^2 + E^2 - AF}$

Parametric equation

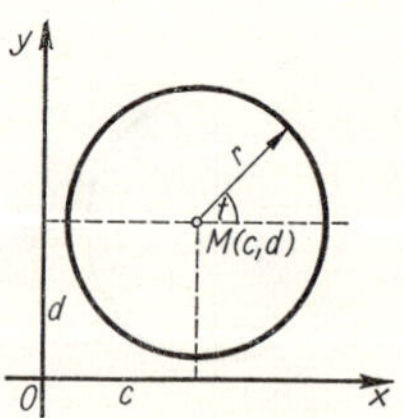

$$x = r\cos t + c$$

$$y = r\sin t + d$$

Equation of the circle in polar coordinates

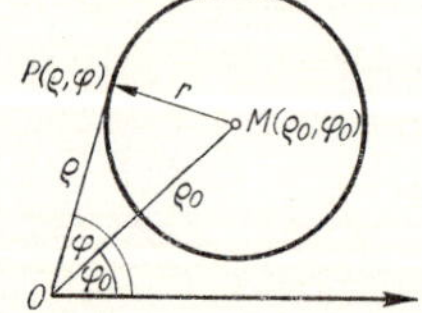

Center $M(\varrho_0, \varphi_0)$ in arbitrary position:

$$\varrho^2 - 2\varrho\varrho_0 \cos(\varphi - \varphi_0) + \varrho_0{}^2 = r^2$$

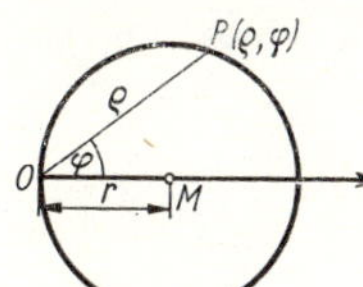

The pole O lies on the circle; the diameter passing through O is the polar axis:

$$\varrho = 2r \cos \varphi$$

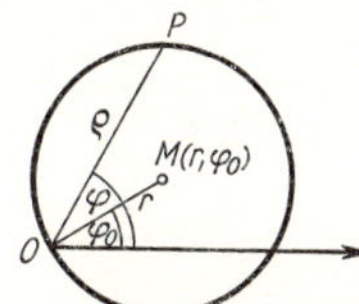

The pole O lies on the circle; the angle φ_0 is included between the polar axis and the diameter passing through O:

$$\varrho = 2r \cos (\varphi - \varphi_0)$$

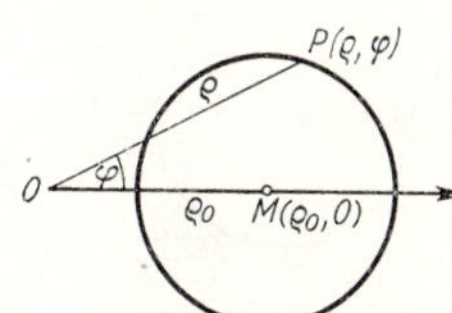

The pole does not lie on the circle; the diameter passing through the pole is the polar axis:

$$r^2 = \varrho^2 - 2\varrho\varrho_0 \cos \varphi + {\varrho_0}^2$$

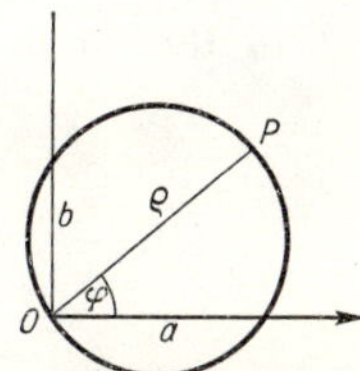

The circle passes through O; intercepts on the rectangular coordinate axes a and b:

$$\varrho = a \cos \varphi + b \sin \varphi$$

Points of intersection of the straight line $y = mx + b$ with the circle $x^2 + y^2 = r^2$

$$x_{1;2} = -\frac{bm}{1+m^2} \pm \frac{1}{1+m^2}\sqrt{r^2(1+m^2) - b^2}$$

$$y_{1;2} = \frac{b}{1+m^2} \pm \frac{m}{1+m^2}\sqrt{r^2(1+m^2) - b^2}$$

Radicand $r^2(1 + m^2) - b^2 = D$ (discriminant)

$D > 0$ The straight line intersects the circle in two points.
$D = 0$ The straight line touches the circle in one point (double point).
$D < 0$ The straight line has no points in common with the circle.

Tangent and normal to the circle $x^2 + y^2 = r^2$ in point $P_1(x_1, y_1)$

Equation of the tangent: $xx_1 + yy_1 = r^2$

Direction factor $m_t = -\dfrac{x_1}{y_1}$

Equation of the normal: $yx_1 - xy_1 = 0$

Direction factor $m_n = \dfrac{y_1}{x_1}$

Length of tangent $t = \left|\dfrac{ry_1}{x_1}\right|$

Length of normal $n = r$

Subtangent $s_t = \left|\dfrac{y_1^2}{x_1}\right|$

Subnormal $s_n = x_1$

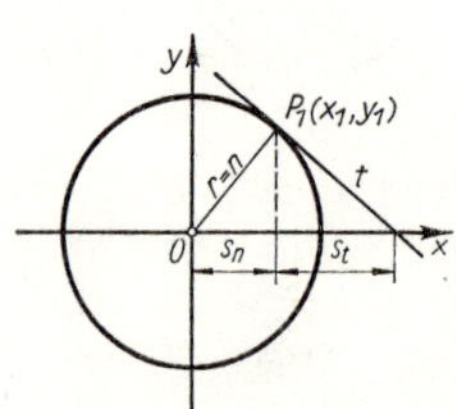

Polar line of the point $P_0(x_0, y_0)$ with respect to the circle $x^2 + y^2 = r^2$

$xx_0 + yy_0 = r_2$ $\quad P_0$ is called pole

To the straight line $Ax + By + C = 0$ as polar line belongs pole P_0 with the coordinates $x_0 = -\dfrac{Ar^2}{C}$, $y_0 = -\dfrac{Br^2}{C}$.

Power p of the point $P_0(x_0, y_0)$ with respect to the circle $x^2+y^2=r^2$

$p = x_0^2 + y_0^2 - r^2$

Tangent and normal to the circle $(x - c)^2 + (y - d)^2 = r^2$ in point $P_1(x_1, y_1)$

Equation of the tangent:

$$(x - c)(x_1 - c) + (y - d)(y_1 - d) = r^2$$

Gradient $m_t = -\dfrac{x_1 - c}{y_1 - d}$

Equation of the normal:

$$(y - y_1)(x_1 - c) = (x - x_1)(y_1 - d)$$

$$\text{Gradient } m_n = \frac{y_1 - d}{x_1 - c}$$

Equation of the circle through three points $P_1(x_1, y_1)$, $P_2(x_2, y_2)$, $P_3(x_3, y_3)$

$$\begin{vmatrix} x^2 + y^2 & x & y & 1 \\ x_1^2 + y_1^2 & x_1 & y_1 & 1 \\ x_2^2 + y_2^2 & x_2 & y_2 & 1 \\ x_3^2 + y_3^2 & x_3 & y_3 & 1 \end{vmatrix} = 0$$

Polar line of the point $P_0(x_0, y_0)$ with respect to the circle $(x - c)^2 + (y - d)^2 = r^2$

$$(x - c)(x_0 - c) + (y - d)(y_0 - d) = r^2$$

Power p of the point $P_0(x_0, y_0)$ with respect to the circle $(x - c)^2 + (y - d)^2 = r^2$

$$p = (x_0 - c)^2 + (y_0 - d)^2 - r^2$$

Common chord (radical axis) of two circles

$$\left.\begin{aligned} K_1 &\equiv (x - c_1)^2 + (y - d_1)^2 - r_1^2 = 0 \\ K_2 &\equiv (x - c_2)^2 + (y - d_2)^2 - r_2^2 = 0 \end{aligned}\right\} \text{given circles}$$

$K_1 - K_2 = 0$ equation of the common chord

The radical axis is perpendicular to the line of centers.

Family of circles

$$\left.\begin{aligned} K_1 &\equiv (x - c_1)^2 + (y - d_1)^2 - r_1^2 = 0 \\ K_2 &\equiv (x - c_2)^2 + (y - d_2)^2 - r_2^2 = 0 \end{aligned}\right\} \text{given circles}$$

$K_1 + \lambda K_2 = 0$ equation of the family of circles for $\lambda \neq -1$

4.1.5. Parabola

Definition

A parabola is the set of all points equidistant from a fixed point (the *focus*) and a straight line (the *directrix*).

Notations

O vertex of the parabola

x-axis = axis of the parabola

l directrix $\left(\text{equation } x = -\dfrac{p}{2}\right)$

$p = \overline{DF}$ semifocal chord

$2p$ parameter

$\overline{OF} = \overline{OD} = \dfrac{p}{2}$

$F\left(\dfrac{p}{2}, 0\right)$ focus

$\overline{PF}$ focal line (focal distance), $\overline{PL}$ distance to directrix $\left(\overline{PF} = \overline{PL}\right)$

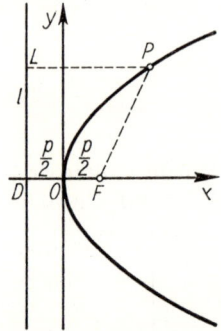

Equations of the parabola

Vertex equation

$$y^2 = 2px$$

$p > 0$ parabola opening to the right

$p < 0$ parabola opening to the left

General equation with axially parallel position

$$(y - d)^2 = 2p(x - c)$$

Vertex $V(c, d)$

Axis of parabola $\parallel$ x-axis

$p > 0$ parabola opening to the right

$p < 0$ parabola opening to the left

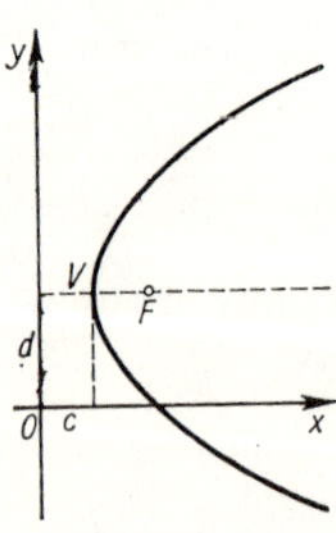

General equation of the second degree as a parabola with axially parallel position

$$Ax^2 + 2Bxy + Cy^2 + 2Dx + 2Ey + F = 0$$

(general equation of the second degree in x and y)

Condition for parabola: $A = 0$; $B = 0$; hence, parabola equation

$$Cy^2 + 2Dx + 2Ey + F = 0$$

Parameter $2p = -\frac{2D}{C}$; Vertex $V\left(\frac{E^2 - CF}{2CD}, -\frac{E}{C}\right)$

Parabola parallel to the x-axis

Inverse equations

$x^2 = 2py$ — Parabola with y-axis as axis of parabola
Vertex $V(0, 0)$
$p > 0$ (opening above)
$p < 0$ (opening below)

$(x - c)^2 = 2p(y - d)$ — Axis of parabola parallel to the y-axis
Vertex $V(c, d)$

$Cx^2 + 2Dy + 2Ex + F = 0$ — Axis of parabola parallel to the y-axis
Vertex $V\left(-\frac{E}{C}, \frac{E^2 - CF}{2CD}\right)$
and
Parameter $2p = -\frac{2D}{C}$

Polar equation

$$r = \frac{p}{1 + \varepsilon \cos \varphi},$$

where $\varepsilon = 1$.
(pole = focus; polar axis through vertex; figure on the left)

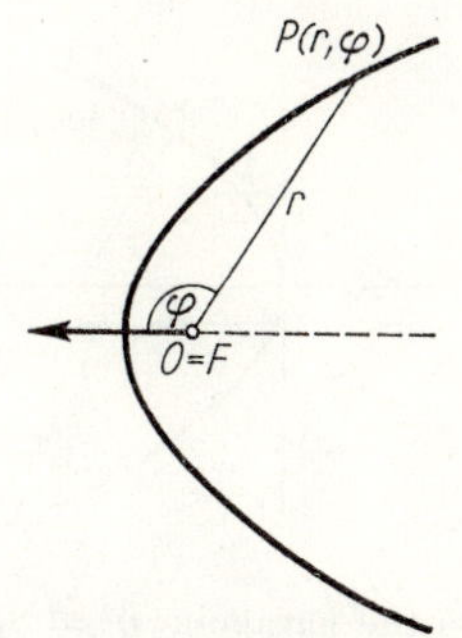

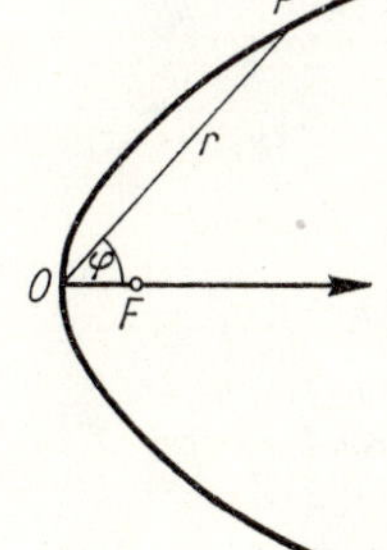

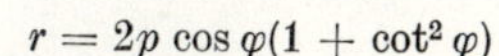

$$r = 2p \cos \varphi (1 + \cot^2 \varphi)$$

(pole = vertex; polar axis = axis of parabola; figure on the right)

Intersections of the line $y = mx + b$ with the parabola $y^2 = 2px$ ($p > 0$)

$$x_{1;2} = \frac{p - bm}{m^2} \pm \frac{1}{m^2}\sqrt{p(p - 2bm)}$$

$$y_{1;2} = \frac{p}{m} \pm \frac{1}{m}\sqrt{p(p - 2bm)}$$

Radicand $p(p - 2bm) = D$ (discriminant)

$D > 0$ The straight line intersects the parabola

$D = 0$ The straight line touches the parabola

$D < 0$ The straight line has no point in common with the parabola.

Tangent and normal to the parabola $y^2 = 2px$ in the point $P_1(x_1, y_1)$

Equation of the tangent: $yy_1 = p(x + x_1)$

Gradient $m_t = \dfrac{p}{y_1}$

Equation of the normal: $p(y - y_1) + y_1(x - x_1) = 0$

Gradient $m_n = -\dfrac{y_1}{p}$

Length of tangent

$$t = \sqrt{y_1{}^2 + 4x_1{}^2}$$

Length of normal

$$n = \sqrt{y_1{}^2 + p^2}$$

Subtangent $s_t = 2x_1$

Subnormal $s_n = p$

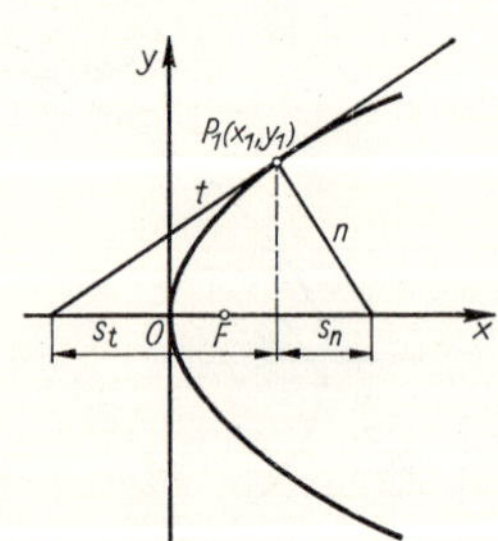

Polar of point $P_0(x_0, y_0)$ with respect to the parabola $y^2 = 2px$

$yy_0 = p(x + x_0)$ $\qquad P_0$ pole

Diameter of the parabola $y^2 = 2px$

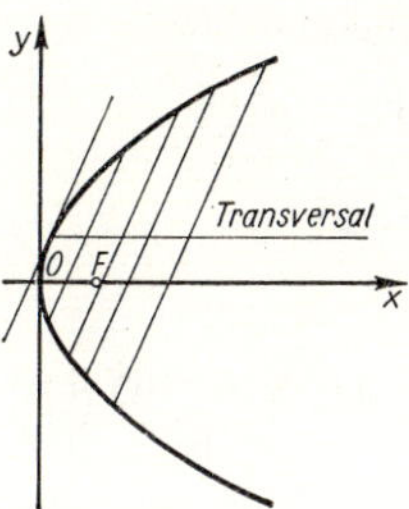

$$y = \frac{p}{m}$$

m gradient of the associated parallel chords which are bisected by the diameter.

Tangent and normal to the parabola $(y - d)^2 = 2p(x - c)$ in point $P_1(x_1, y_1)$

Equation of the tangent:

$$(y - d)(y_1 - d) = p(x + x_1 - 2c)$$

$$\text{Gradient } m_t = \frac{p}{y_1 - d}$$

Equation of the normal:

$$p(y - y_1) + (y_1 - d)(x - x_1) = 0$$

$$\text{Gradient } m_n = -\frac{y_1 - d}{p}$$

Segment of a parabola

Chord P_1P_2 is of arbitrary direction.

$$A = \left|\frac{(y_1 - y_2)^3}{12p}\right| = \left|\frac{(x_1 - x_2)(y_1 - y_2)^2}{6(y_1 + y_2)}\right|$$

Chord perpendicular to the axis of parabola

$$A = \frac{4}{3}x_1y_1$$

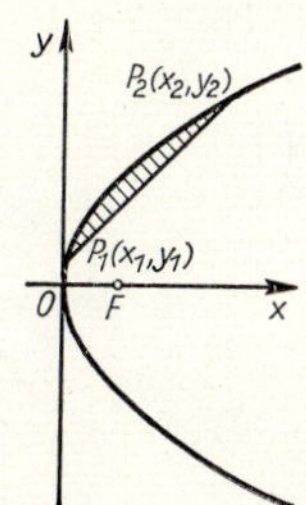

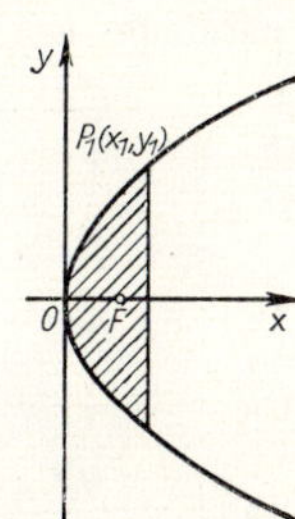

Paraboloid of revolution

The parabolic arc rotates about the x-axis.

$$V = \frac{1}{2}\pi x_1 y_1^2$$

Radius of curvature ϱ and center of curvature $M_c(\xi, \eta)$ of the parabola $y^2 = 2px$ in point $P_1(x_1, y_1)$ and in vertex V

For P_1 holds:

$$\varrho = \frac{\sqrt{(y_1^2 + p^2)^3}}{p^2} = \frac{n^3}{p^2}$$ n length of normal (see page 228)

$$\xi = 3x_1 + p$$

$$\eta = -\frac{y_1^3}{p^2}$$

For vertex $V(0, 0)$ holds:

$$\varrho = |p|; \qquad \xi = p; \qquad \eta = 0$$

Evolute of the parabola $y^2 = 2px$

$$\eta^2 = \frac{8(\xi - p)^3}{27p} \quad \text{for} \quad \xi \geqq p$$

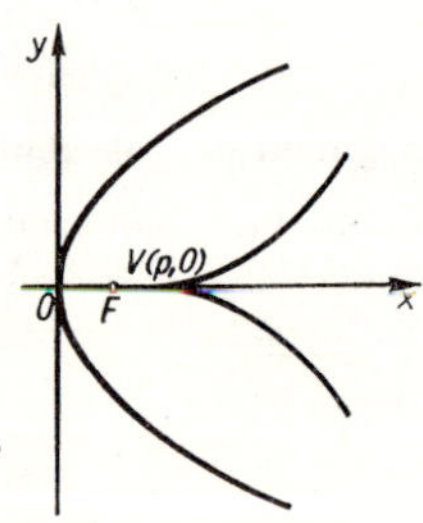

NEIL'S or semicubic parabola

Length of the parabolic arc OP_1 of the parabola $y^2 = 2px$

$$\widehat{OP_1} = \frac{p}{2}\left[\sqrt{\frac{2x_1}{p}\left(1 + \frac{2x_1}{p}\right)} + \ln\left(\sqrt{\frac{2x_1}{p}} + \sqrt{1 + \frac{2x_1}{p}}\right)\right]$$

$$= \frac{y_1}{2p}\sqrt{p^2 + y_1^2} + \frac{p}{2}\ln\frac{y_1 + \sqrt{p^2 + y_1^2}}{p}$$

$$= \frac{y_1}{2p}\sqrt{p^2 + y_1^2} + \frac{p}{2}\sinh^{-1}\frac{y_1}{p}$$

Approximate value for small $\frac{x_1}{y_1}$:

$$\widehat{OP_1} \approx y_1 \left[1 + \frac{2}{3}\left(\frac{x_1}{y_1}\right)^2 - \frac{2}{5}\left(\frac{x_1}{y_1}\right)^4\right]$$

Length l of the focal line to point $P_1(x_1, y_1)$

$$l = x_1 + \frac{p}{2}$$

CONSTRUCTIONS OF PARABOLA

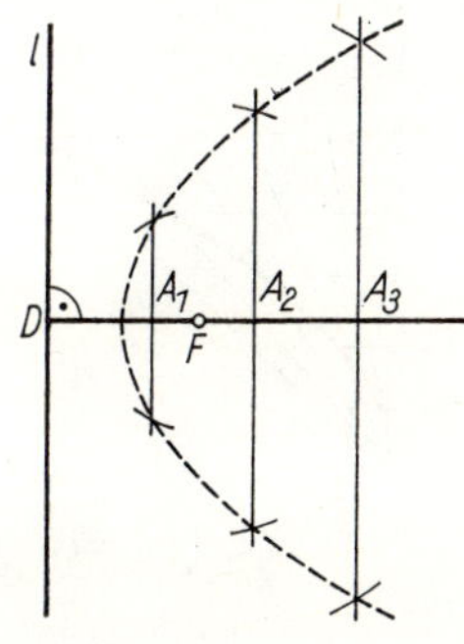

Given: *Focus and directrix*

We drop the perpendicular FD (axis of parabola) from point F (focus) onto the directrix l and erect perpendiculars to the axis at arbitrary points A_1, A_2, A_3, ... of this axis. Then we describe circles with radii DA_1, DA_2, DA_3, ... about F; the points of intersection of the circles with the perpendiculars are points of the parabola.

The midpoint of $\overline{FD}$ is the vertex of the parabola.

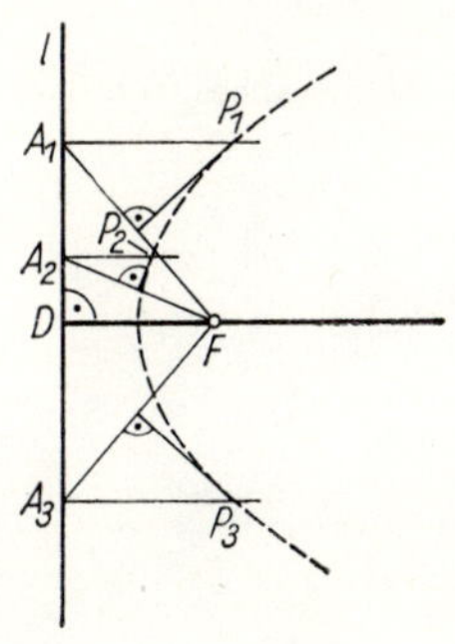

Given: *Focus and directrix*

We draw a line joining an arbitrary point A_1 of the directrix with the focus and erect the midperpendicular on $\overline{A_1F}$; the point of intersection of the midperpendicular with the perpendicular through A_1 on the directrix is a point of the parabola.

Given: *Focus and tangent to the vertex*

We join different points of the tangent to the vertex with the focus F and erect perpendiculars on the various points of these joins; the perpendiculars envelope the parabola.

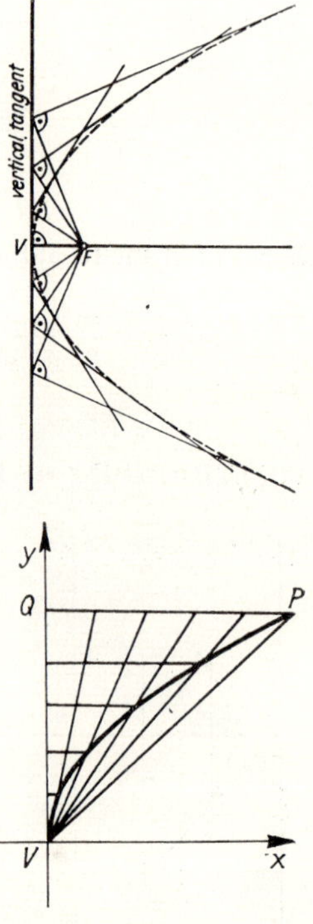

Given: *Coordinate axes, vertex at the origin and one point P of the parabola*

We drop the perpendicular PQ from P on tothe y-axis and divide $\overline{PQ}$ and $\overline{VQ}$ into the same number of parts equal to one another. We join the points of division on $\overline{PQ}$ with lines to V and draw lines parallel to the x-axis through the points of division on $\overline{VQ}$. The points of intersection of the parallel lines with the joins from V are points of the parabola.
The points of the parabola below the x-axis will be readily found since they are symmetrical with respect to the axis.

4.1.6. Ellipse

Definition

An ellipse is the set of all points the sum of whose distances from two fixed points (foci) is constant.

Notations

F_1, F_2 foci

A, B vertices on the major axis

C, D vertices on the minor axis

M center

$\overline{PF_1} + \overline{PF_2} = 2a$ (constant sum)

$\overline{PF_1}, \overline{PF_2}$ focal distances

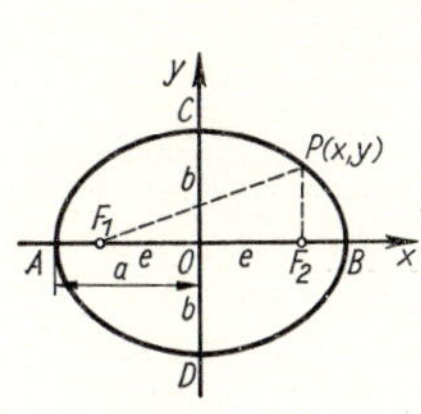

$\overline{AB} = 2a$ major axis; a major semiaxis

$\overline{CD} = 2b$ minor axis; b minor semiaxis

$\overline{F_1F_2} = 2e$

$e = \sqrt{a^2 - b^2}$ *linear eccentricity*

$\dfrac{e}{a} = \varepsilon$ *numerical eccentricity* $(\varepsilon < 1)$

$2p = \dfrac{2b^2}{a}$ *parameter* = perpendicular to the major axis in the focus

Equations of the ellipse

Central equation

$$\frac{x^2}{a^2} + \frac{y^2}{b^2} = 1 \quad \text{or} \quad b^2x^2 + a^2y^2 - a^2b^2 = 0$$ see figure above

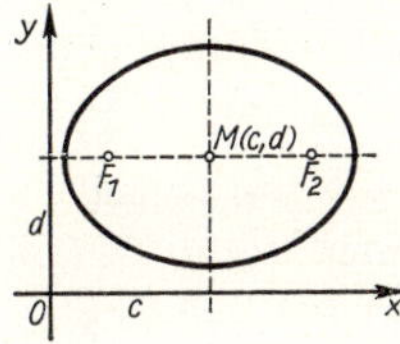

General equation with axially parallel position

$$\frac{(x - c)^2}{a^2} + \frac{(y - d)^2}{b^2} = 1$$

Center $M\ (c, d)$
Axes of the ellipse || to the coordinate axes

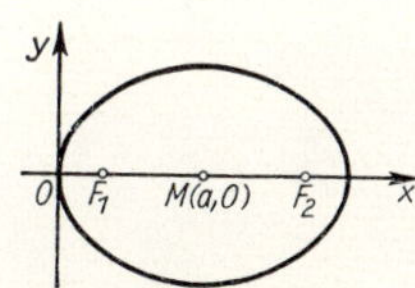

Vertex equation

$$y^2 = 2px - \frac{p}{a}x^2$$

General equation of the second degree as an ellipse with axially parallel position

$$Ax^2 + 2Bxy + Cy^2 + 2Dx + 2Ey + F = 0$$

(general equation of the second degree in x and y)

Condition for an ellipse in axially parallel position:

$$\operatorname{sgn} A = \operatorname{sgn} C \quad \text{and} \quad B = 0, \quad A \neq C$$

Thus, we have the equation of the ellipse

$$Ax^2 + Cy^2 + 2Dx + 2Ey + F = 0$$

major semiaxis $a = \sqrt{\dfrac{CD^2 + AE^2 - ACF}{A^2C}}$

minor semiaxis $b = \sqrt{\dfrac{CD^2 + AE^2 - ACF}{AC^2}}$

center $M\left(-\dfrac{D}{A}, -\dfrac{E}{C}\right)$

axes of the ellipse || coordinate axes

Inverse equations

$\dfrac{y^2}{a^2} + \dfrac{x^2}{b^2} = 1$ Ellipse whose major axis is the y-axis
a major semiaxis, b minor semiaxis, center in the origin

$x^2 = 2py - \dfrac{p}{a} y^2$ Ellipse whose major axis is the y-axis
a major semiaxis, center $M(0, a)$

Polar equations

$$r = \frac{p}{1 + \varepsilon \cos \varphi}, \quad \varepsilon < 1$$

(focus F_2 pole; polar axis in the direction of B)

$$r^2 = \frac{b^2}{1 - \varepsilon^2 \cos^2 \varphi}, \quad \varepsilon < 1$$

(pole at the center; major axis is polar axis)

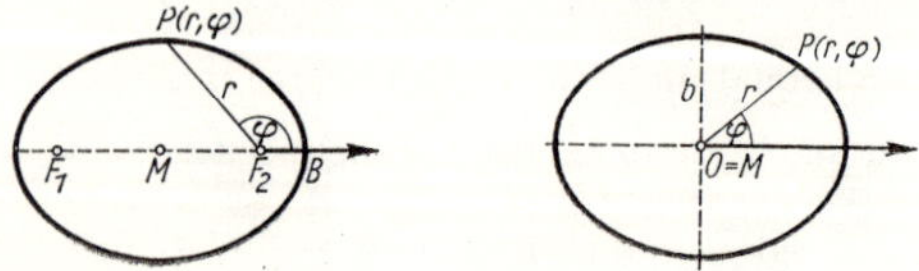

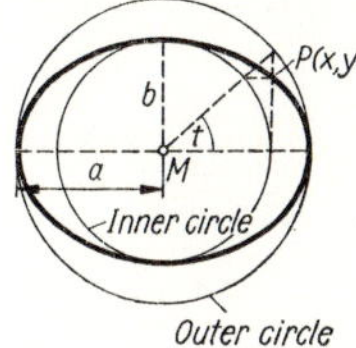

Parametric equation

$$x = a \cos t$$

$$y = b \sin t$$

Points of intersection of the straight line $y = mx + b_1$ with the ellipse $\frac{x^2}{a^2} + \frac{y^2}{b^2} = 1$

$$x_{1;2} = -\frac{a^2 m b_1}{b^2 + a^2 m^2} \pm \frac{ab}{b^2 + a^2 m^2} \sqrt{a^2 m^2 + b^2 - b_1^2}$$

$$y_{1;2} = \frac{b^2 b_1}{b^2 + a^2 m^2} \pm \frac{abm}{b^2 + a^2 m^2} \sqrt{a^2 m^2 + b^2 - b_1^2}$$

Radicand $a^2 m^2 + b^2 - b_1^2 = D$ (discriminant)

$D > 0$ The straight line intersects the ellipse.

$D = 0$ The straight line touches the ellipse.

$D < 0$ The straight line has no point in common with the ellipse.

The length of the focal distances PF_1 and PF_2

$$\left.\begin{array}{l} \overline{PF_1} = a + \varepsilon x \\ \overline{PF_2} = a - \varepsilon x \end{array}\right\} \overline{PF_1} + \overline{PF_2} = 2a$$

Tangent and normal to the ellipse $\frac{x^2}{a^2} + \frac{y^2}{b^2} = 1$ in point $P_1(x_1, y_1)$

Equation of the tangent: $\frac{x x_1}{a^2} + \frac{y y_1}{b^2} = 1$

Gradient $m_t = -\frac{b^2 x_1}{a^2 y_1}$

Equation of the normal: $y - y_1 = \frac{a^2 y_1}{b^2 x_1}(x - x_1)$

Gradient $m_n = \frac{a^2 y_1}{b^2 x_1}$

Length of tangent $t = \sqrt{{y_1}^2 + \left(\frac{a^2}{x_1} - x_1\right)^2}$

Length of normal $n = \frac{b\sqrt{a^4 - e^2{x_1}^2}}{a^2}$

Subtangent $s_t = \left|\frac{a^2}{x_1} - x_1\right|$

Subnormal $s_n = \left|\frac{b^2 x_1}{a^2}\right|$

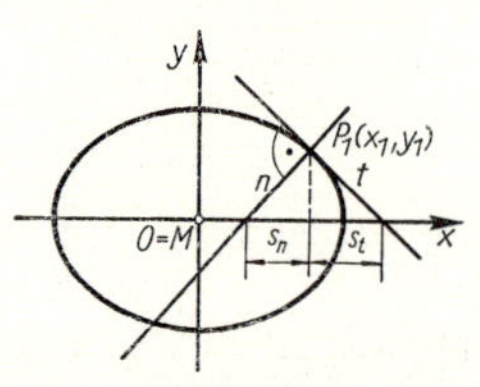

Tangent and normal to the ellipse $\frac{(x-c)^2}{a^2} + \frac{(y-d)^2}{b^2} = 1$ in point $P_1(x_1, y_1)$

Equation of the tangent:

$$\frac{(x - c)(x_1 - c)}{a^2} + \frac{(y - d)(y_1 - d)}{b^2} = 1$$

Gradient $m_t = -\frac{b^2(x_1 - c)}{a^2(y_1 - d)}$

Equation of the normal:

$$y - y_1 = \frac{a^2(y_1 - d)}{b^2(x_1 - c)}(x - x_1)$$

Gradient $m_n = \frac{a^2(y_1 - d)}{b^2(x_1 - c)}$

Polar of point $P_0(x_0, y_0)$ with respect to the ellipse $\frac{x^2}{a^2} + \frac{y^2}{b^2} = 1$

$$\frac{xx_0}{a^2} + \frac{yy_0}{b^2} = 1 \qquad P_0 \text{ pole}$$

Diameter of the ellipse

$$y = -\frac{b^2}{a^2 m}x$$

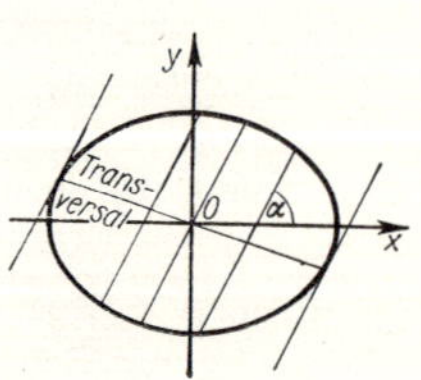

m slope of the associated parallel chords which are bisected by the diameter

The diameter through a family of parallel chords passes through the points of contact of the tangents parallel to these chords. The diameter of a tangential chord passes through the point of intersection of its tangents.

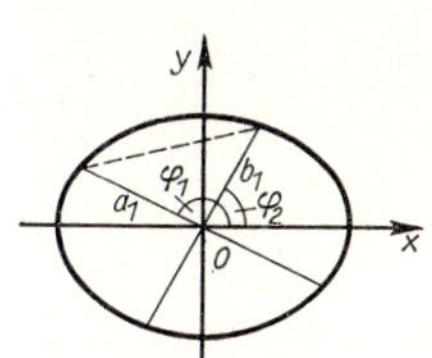

Conjugate diameters are diameters each of which bisects the chords parallel to the other one. The two axes are conjugate diameters.

$y = m_1 x$ and $y = m_2 x$ are conjugate diameters if

$$m_1 m_2 = -\frac{b^2}{a^2}$$

For two conjugate diameters $2a_1$ and $2b_1$ we have:

$$a_1^2 + b_1^2 = a^2 + b^2$$

$$a_1 b_1 \sin(\varphi_1 - \varphi_2) = ab$$

In words:

The area of a triangle formed by two conjugate semidiameters of an ellipse and the line joining their endpoints is constant.
The equation of the ellipse related to the two conjugate diameters $2a_1$ and $2b_1$ is as follows:

$$\frac{x^2}{a_1^2} + \frac{y^2}{b_1^2} = 1$$

Radius of curvature ϱ and center of curvature $M_c(\xi, \eta)$ for the ellipse $\frac{x^2}{a^2} + \frac{y^2}{b^2} = 1$

In the point $P_1(x_1, y_1)$

$$\varrho = \frac{1}{a^4 b^4}\sqrt{(a^4 y_1^2 + b^4 x_1^2)^3} = \frac{\sqrt{(a^4 - e^2 x_1^2)^3}}{a^4 b} = \frac{n^3}{p^2}$$

n length of normal (see page 236)

$$\xi = \frac{e^2 x_1^3}{a^4}; \quad \eta = -\frac{e^2 y_1^3}{b^4} = -\frac{\varepsilon^2 a^2 y_1^3}{b^4}$$

In the vertex on the major axis $A(-a, 0)$

$$\varrho = \frac{b^2}{a} = p; \quad \xi = -\frac{e^2}{a}; \quad \eta = 0$$

In the vertex on the major axis $B(a, 0)$

$$\varrho = \frac{b^2}{a} = p; \quad \xi = \frac{e^2}{a}; \quad \eta = 0$$

In the vertex on the minor axis $D(0, -b)$

$$\varrho = \frac{a^2}{b}; \quad \xi = 0; \quad \eta = \frac{e^2}{b}$$

In the vertex on the minor axis $C(0, b)$

$$\varrho = \frac{a^2}{b}; \quad \xi = 0; \quad \eta = -\frac{e^2}{b}$$

Circle on the major axis and circle on the minor axis of the ellipse $\frac{x^2}{a^2} + \frac{y^2}{b^2} = 1$

$$x^2 + y^2 = a^2 \quad \text{and} \quad x^2 + y^2 = b^2$$

Evolute of the ellipse $\frac{x^2}{a^2} + \frac{y^2}{b^2} = 1$

$$\left(\frac{a\xi}{e^2}\right)^{\frac{2}{3}} + \left(\frac{b\eta}{e^2}\right)^{\frac{2}{3}} = 1 \qquad \text{(astroid)}$$

for $\quad |\xi| \leqq \frac{e^2}{a}$

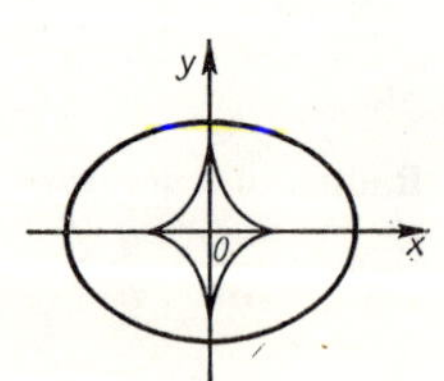

Area of the whole ellipse, of the elliptical segment and elliptical sector

Ellipse: $A = \pi ab$

Elliptical segment P_1P_2C:

$$A = \frac{1}{2}(x_1y_2 - x_2y_1) + \frac{ab}{2}\left(\text{arc sin}\frac{x_2}{a} - \text{arc sin}\frac{x_1}{a}\right)$$

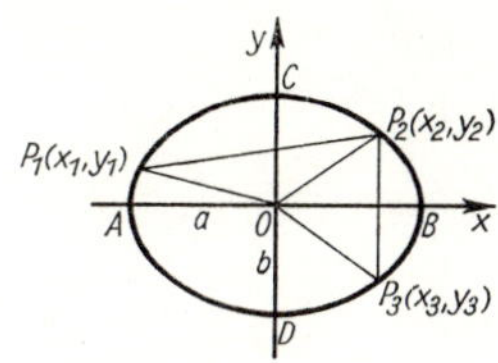

Elliptical segment P_2P_3B:

$$A = ab \arccos \frac{x_2}{a} - x_2 y_2$$

Elliptical sector P_2OP_3B:

$$A = ab \arccos \frac{x_2}{a}$$

Elliptical sector P_1OP_2C:

$$A = \frac{ab}{2} \left(\arcsin \frac{x_2}{a} - \arcsin \frac{x_1}{a} \right)$$

Ellipsoid of revolution

Revolution about the x-axis: $V = \frac{4}{3} \pi a b^2$ (prolate ellipsoid)

Revolution about the y-axis: $V = \frac{4}{3} \pi a^2 b$ (spheroid)

Circumference of the ellipse

$$u = 2\pi a \left[1 - \left(\frac{1}{2}\right)^2 \varepsilon^2 - \left(\frac{1 \cdot 3}{2 \cdot 4}\right)^2 \frac{\varepsilon^4}{3} - \left(\frac{1 \cdot 3 \cdot 5}{2 \cdot 4 \cdot 6}\right)^2 \frac{\varepsilon^6}{5} - \cdots \right]$$

Formulas of approximation:

$$u \approx \pi \left[\frac{3}{2} (a + b) - \sqrt{ab} \right]; \quad u \approx \frac{\pi}{2} \left[a + b + \sqrt{2(a^2 + b^2)} \right]$$

CONSTRUCTIONS OF THE ELLIPSE

Given: *Foci and major axis 2a*

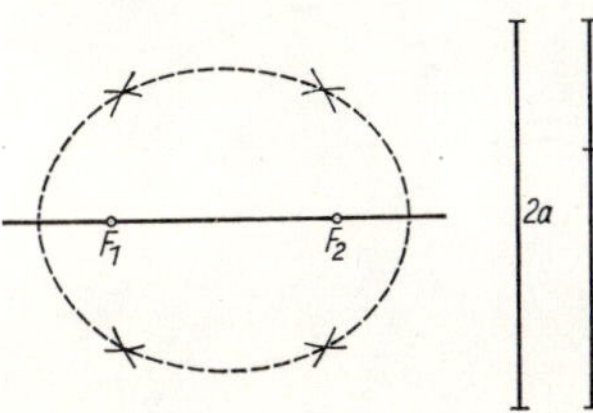

We describe circles with radii $p < 2a$ and $(2a - p)$ about the foci F_1 and F_2 and obtain as their points of intersection four symmetrical points of the ellipse. By varying $p < 2a$, further points of the ellipse are obtained.

Given: *Foci and major axis 2a*

We draw a circle with $2a$ as the radius and F_1 as the center, then draw a line joining an arbitrary point P of this circle with the second focus F_2 and erect the midperpendicular of this join. The point of intersection of this midperpendicular with $\overline{PF_1}$ is a point of the ellipse. The same procedure can be repeated with a circle about F_2.

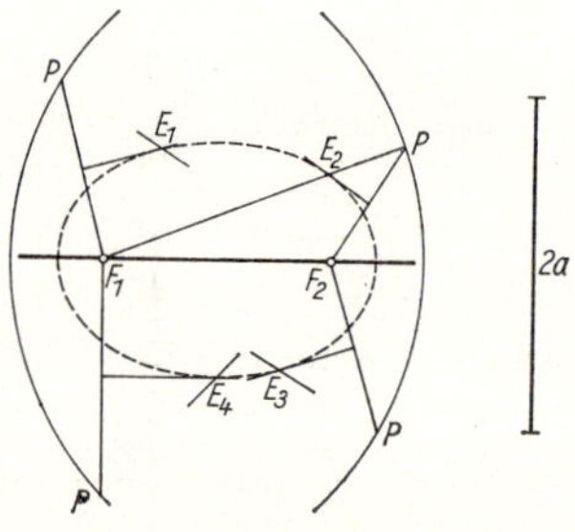

Given: *Semiaxes a and b*

Solution 1:

We describe the circles on the minor and major axes about O. Then we erect an arbitrary perpendicular g to the horizontal axis which intersects the circle on the major axis in A_1 and A_2. The joins OA_1 and OA_2 intersect the circle on the minor axis in B_1 and B_2. The lines drawn through B_1 and B_2 and parallel to the horizontal axis intersect the perpendicular g in two points of the ellipse E_1 and E_2.

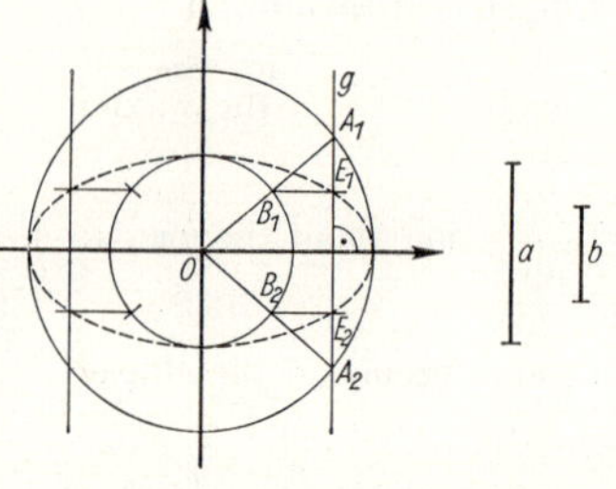

Solution 2:

We draw lines through the vertices A, B, C, D and parallel to the coordinate axes which intersect one another in E_1, E_2, E_3, E_4. Then we divide the line-segments OC and E_3C into the same number of equal parts. We draw rays from A through the points of division of $\overline{OC}$ and from B through the points of division of $\overline{E_3C}$. The points of intersection of the respective rays are points of the ellipse. The whole ellipse is obtained by constructing the points symmetrical with respect to the x-axis and y-axis.

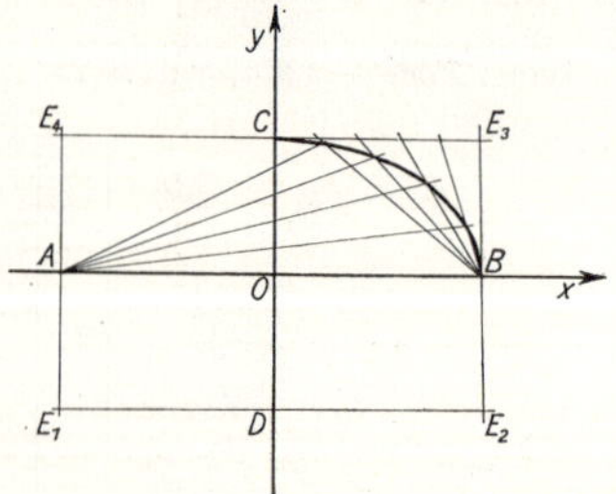

Solution 3:

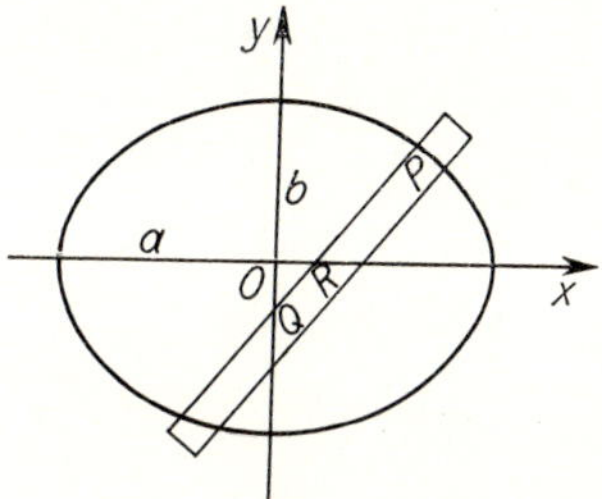

On a strip of paper with straight edges, we measure off from point P the two semiaxes $PQ = a$ and $PR = b$, one on top of the other $\left(\overline{QR} = a - b\right)$. When moving the paper-strip in such a way that point Q moves on the y-axis and point R on the x-axis, point P describes an ellipse (principle of the elliptical trammel).

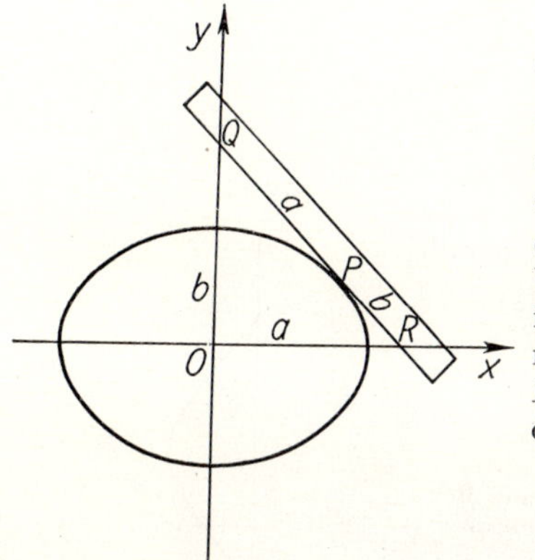

Solution 4:

On a paper-strip with straight edges, we measure off the two semiaxes $PQ = a$ and $PR = b$, one after another $\left(\overline{QR} = a + b\right)$. When moving the paper-strip in such a way that point Q moves on the y-axis and point R on the x-axis, point P describes an ellipse.

4.1.7. Hyperbola

Definition

The hyperbola is the set of all points for which the difference of their distances from two fixed points (the foci) is constant.

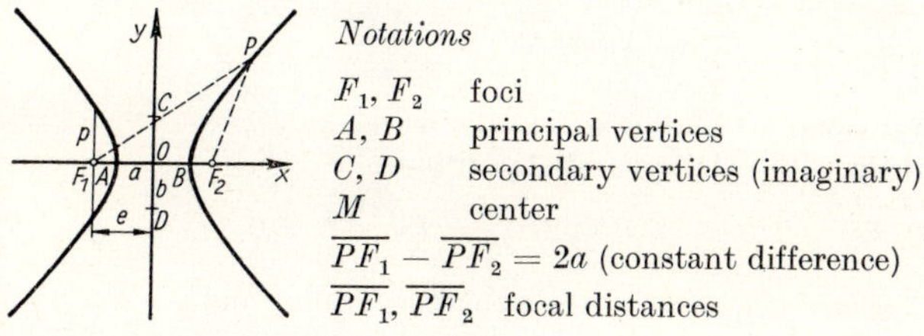

Notations

F_1, F_2	foci
A, B	principal vertices
C, D	secondary vertices (imaginary)
M	center

$\overline{PF_1} - \overline{PF_2} = 2a$ (constant difference)

$\overline{PF_1}, \overline{PF_2}$ focal distances

$\overline{AB} = 2a$ transverse axis

$\overline{CD} = 2b$ conjugate axis

$\overline{F_1F_2} = 2e$

$e = \sqrt{a^2 + b^2}$ *linear eccentricity*

$\frac{e}{a} = \varepsilon$ *numerical eccentricity*; $\varepsilon > 1$

$2p = \frac{2b^2}{a}$ *parameter* = chord through the focus and perpendicular to the major axis

Equations of the hyperbola

Central equation

$$\frac{x^2}{a^2} - \frac{y^2}{b^2} = 1 \quad \text{or} \quad b^2x^2 - a^2y^2 - a^2b^2 = 0$$

see figure page 241, bottom

General equation with axially parallel position

$$\frac{(x-c)^2}{a^2} - \frac{(y-d)^2}{b^2} = 1$$

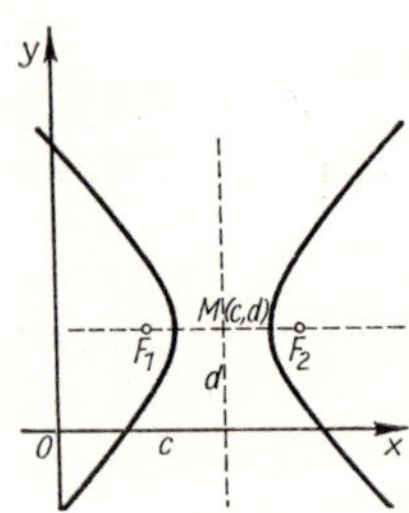

Center $M(c, d)$
Axes of the hyperbola || coordinate axes

Vertex equation

$$y^2 = 2px + \frac{p}{a}x^2$$

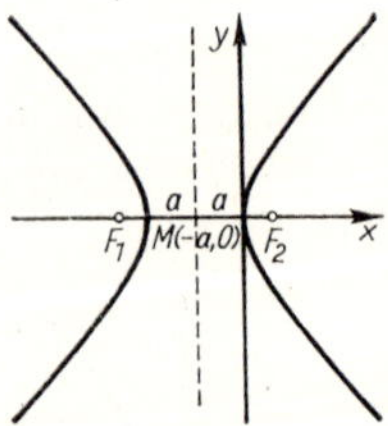

Center $M(-a, 0)$

General equation of the second degree as a hyperbola with axially parallel position

$$Ax^2 + 2Bxy + Cy^2 + 2Dx + 2Ey + F = 0$$

(general equation of the second degree in x and y)

Condition for hyperbola: $\operatorname{sgn} A \neq \operatorname{sgn} C$ and $B = 0$
Thus, we have as equation of the hyperbola

$$Ax^2 - Cy^2 + 2Dx + 2Ey + F = 0$$

transverse semiaxis $a = \sqrt{\dfrac{CD^2 - AE^2 - ACF}{A^2C}}$

conjugate semiaxis $b = \sqrt{\dfrac{CD^2 - AE^2 - ACF}{AC^2}}$

center $M\left(-\dfrac{D}{A}, \dfrac{E}{C}\right)$

Inverse equations

$\dfrac{y^2}{a^2} - \dfrac{x^2}{b^2} = 1$ Hyperbola whose transverse axis is the y-axis

a transverse semiaxis, b conjugate semiaxis, center in origin

$x^2 = 2py + \dfrac{p}{a} y^2$ Hyperbola whose transverse axis is the y-axis

a transverse axis, p semiparameter, center $M(0, -a)$

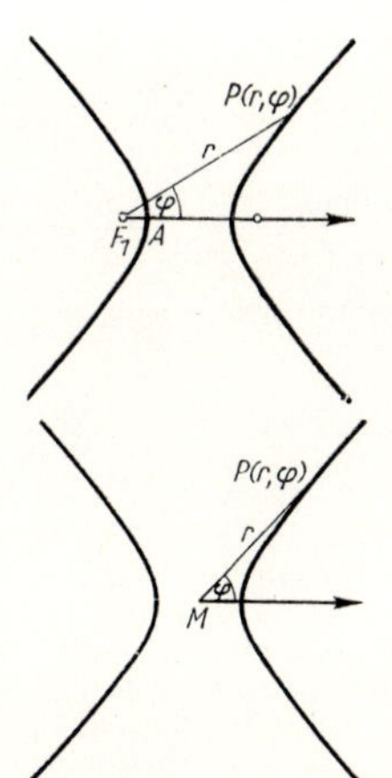

Polar equations

$$r = \frac{p}{1 + \varepsilon \cos \varphi}, \quad \varepsilon > 1$$

(focus F_1 as pole; $\overline{F_1A}$ polar axis)

$$r^2 = \frac{b^2}{\varepsilon^2 \cos^2 \varphi - 1}, \quad \varepsilon > 1$$

(center as pole; x-axis as polar axis)

Parametric equations

$$x = \frac{a}{\cos t}; \quad y = \pm b \tan t$$

$$x = \pm a \cosh t; \quad y = b \sinh t$$

Equation of the rectangular hyperbola

$$x^2 - y^2 = a^2 \qquad (b = a)$$

Focal lengths of the hyperbola $\frac{x^2}{a^2} - \frac{y^2}{b^2} = 1$

$$\left.\begin{aligned} \overline{PF_1} &= \varepsilon x + a \\ \overline{PF_2} &= \varepsilon x - a \end{aligned}\right\} \overline{PF_1} - \overline{PF_2} = 2a$$

Intersections of the straight line $y = mx + b_1$ **with the hyperbola** $\frac{x^2}{a^2} - \frac{y^2}{b^2} = 1$

$$\left.\begin{aligned} x_{1;2} &= \frac{a^2 m b_1}{b^2 - a^2 m^2} \pm \frac{ab}{b^2 - a^2 m^2}\sqrt{b^2 + b_1^2 - a^2 m^2} \\ y_{1;2} &= \frac{b^2 b_1}{b^2 - a^2 m^2} \pm \frac{abm}{b^2 - a^2 m^2}\sqrt{b^2 + b_1^2 - a^2 m^2} \end{aligned}\right\} b^2 - a^2 m^2 \neq 0$$

Radicand $b^2 + b_1^2 - a^2m^2 = D$ (discriminant)

$D > 0$ The straight line intersects the hyperbola.

$D = 0$ The straight line touches the hyperbola.

$D < 0$ The straight line has no point in common with the hyperbola.

Special cases

1. $b^2 - a^2m^2 = 0; \quad m \neq 0; \quad b_1 \neq 0$

The straight line intersects the hyperbola only in one point (x_s, y_s) and is parallel to one of the two asymptotes:

$$x_s = -\frac{b_1^2 + b^2}{2mb_1}; \quad y_s = \frac{b_1^2 - b^2}{2b_1}$$

2. $\quad b^2 - a^2m^2 = 0; \quad m \neq 0; \quad b_1 = 0$

The straight line is an asymptote and has the form of one of the two equations of asymptotes: $y = \pm \frac{b}{a} x$

Tangent and normal to the hyperbola $\frac{x^2}{a^2} - \frac{y^2}{b^2} = 1$ in point $P_1(x_1, y_1)$

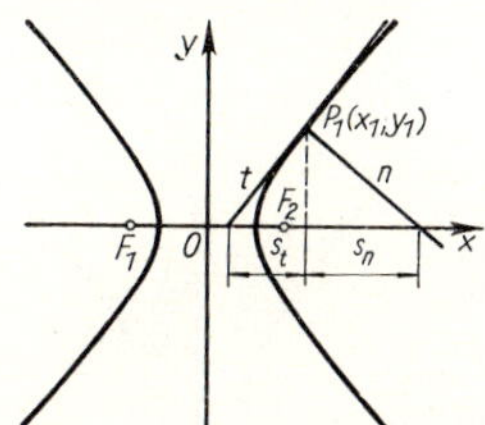

Equation of the tangent:

$$\frac{xx_1}{a^2} - \frac{yy_1}{b^2} = 1$$

Gradient $m_t = \frac{b^2x_1}{a^2y_1}$

Equation of the normal:

$$y - y_1 = -\frac{a^2y_1}{b^2x_1}(x - x_1)$$

Gradient $m_n = -\frac{a^2y_1}{b^2x_1}$

Length of tangent $\quad t = \sqrt{{y_1}^2 + \left(x_1 - \frac{a^2}{x_1}\right)^2}$

Length of normal $\quad n = \frac{b}{a^2}\sqrt{e^2{x_1}^2 - a^4}$

Subtangent $\quad s_t = \left| x_1 - \frac{a^2}{x_1} \right|$

Subnormal $\quad s_n = \left| \frac{b^2x_1}{a^2} \right|$

Tangent and normal to the hyperbola $\frac{(x - c)^2}{a^2} - \frac{(y - d)^2}{b^2} = 1$ in point $P_1(x_1, y_1)$

Equation of the tangent:

$$\frac{(x - c)(x_1 - c)}{a^2} - \frac{(y - d)(y_1 - d)}{b^2} = 1$$

Gradient $m_t = \frac{b^2(x_1 - c)}{a^2(y_1 - d)}$

Equation of the normal:

$$y - y_1 = -\frac{a^2(y_1 - d)}{b^2(x_1 - c)}(x - x_1)$$

$$\text{Gradient } m_n = -\frac{a^2(y_1 - d)}{b^2(x_1 - c)}$$

Asymptotes

The tangents of a hyperbola at infinitely remote points are called asymptotes. They pass through the center.

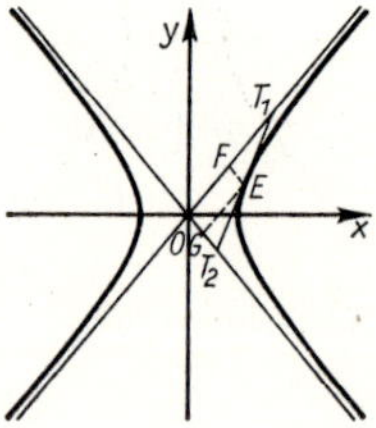

$$\text{Equations: } y = \pm\frac{b}{a}x$$

The length of the tangent between the asymptotes is bisected by the point of contact:

$$\overline{T_1E} = \overline{T_2E}$$

Theorem on the constant triangle

The area of the triangle T_1OT_2 is constant ($A = ab$).

Theorem on the constant parallelogram

If $\overline{EF}$ and $\overline{EG}$ are lines parallel to the asymptotes, then the area of the parallelogram thus formed $OGEF$ is constant $\left(A = \frac{ab}{2}\right)$.

Asymptotic equation of the hyperbola

When choosing the asymptotes as coordinate axes, we obtain an oblique-angled system for the hyperbola ($a \neq b$) in which the equation of the hyperbola is

$$x'y' = \frac{e^2}{4}$$

The asymptotes of the rectangular hyperbola ($a = b$) are perpendicular to one another so that a rectangular system of coordinates is given if we choose the asymptotes as axes. The equation of the rectangular hyperbola then is $x'y' = \frac{a^2}{2}$.

An equation in the form $y = \dfrac{Ax + B}{Cx + D}$ with $AD - BC \neq 0$ and $C \neq 0$ represents a hyperbola whose asymptotes are parallel to the coordinate axes.

Polar of point $P_0(x_0, y_0)$ with respect to the hyperbola $\dfrac{x^2}{a^2} - \dfrac{y^2}{b^2} = 1$

$$\frac{xx_0}{a^2} - \frac{yy_0}{b^2} = 1 \qquad P_0 \text{ pole}$$

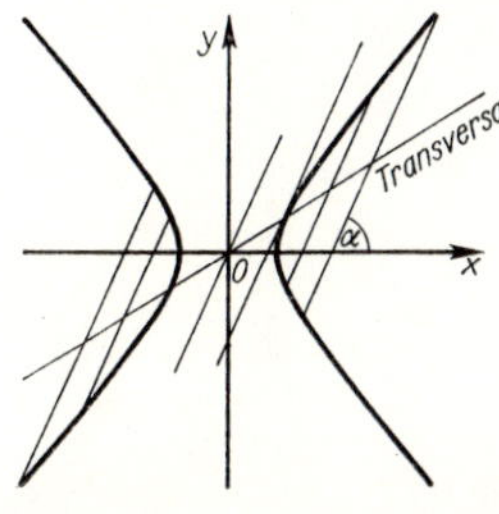

Diameter of the hyperbola

$$y = \frac{b^2}{a^2 m} x$$

m slope of the associated parallel chords which are bisected by the diameter

Conjugate diameters are diameters each of which bisects the chords parallel to the other diameter.

$y = m_1 x$ and $y = m_2 x$ are two conjugate diameters if

$$m_1 m_2 = \frac{b^2}{a^2}$$

For two conjugate diameters $2a_1$ and $2b_1$ we have

$$a_1^2 - b_1^2 = a^2 - b^2$$

Equation of the hyperbola with respect to the two conjugate diameters $2a_1$ and $2b_1$:

$$\frac{x^2}{a_1^2} - \frac{y^2}{b_1^2} = 1$$

Radius of curvature ϱ and center of curvature $M_c(\xi, \eta)$ of the hyperbola $\dfrac{x^2}{a^2} - \dfrac{y^2}{b^2} = 1$

In point $P_1(x_1, y_1)$

$$\varrho = \frac{1}{a^4b^4}\sqrt{(b^4{x_1}^2 + a^4{y_1}^2)^3} = \frac{\sqrt{(e^2{x_1}^2 - a^4)^3}}{a^4b} = \frac{n^3}{p^2}$$

n length of the normal (see page 245)

$$\xi = \frac{e^2{x_1}^3}{a^4}; \quad \eta = -\frac{e^2{y_1}^3}{b^4} = -\frac{\varepsilon^2a^2{y_1}^3}{b^4}$$

In vertex $A(-a, 0)$

$$\varrho = \frac{b^2}{a} = p; \quad \xi = -\frac{e^2}{a}; \quad \eta = 0$$

In vertex $B(a, 0)$

$$\varrho = \frac{b^2}{a} = p; \quad \xi = \frac{e^2}{a}; \quad \eta = 0$$

Principal circle of the hyperbola

$$x^2 + y^2 = a^2$$

Evolute of the hyperbola $\dfrac{x^2}{a^2} - \dfrac{y^2}{b^2} = 1$

$$\left(\frac{a\xi}{e^2}\right)^{\frac{2}{3}} - \left(\frac{b\eta}{e^2}\right)^{\frac{2}{3}} = 1$$

for $\quad |\xi| \geqq \dfrac{e^2}{a}$

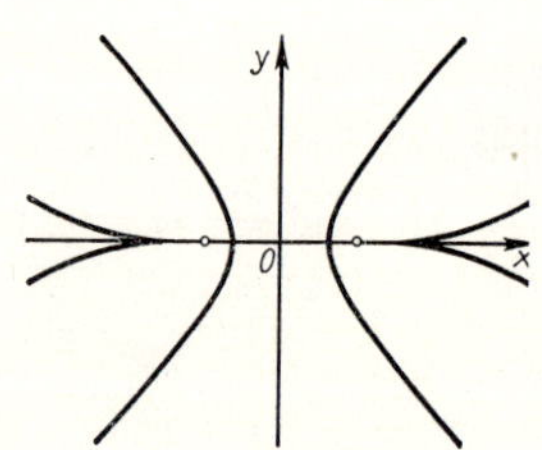

Area of the segment P_1BP_2 and of the sector OP_2BP_1 of the hyperbola

$$A = x_1y_1 - ab\ln\left(\frac{x_1}{a} + \frac{y_1}{b}\right)$$

$$= x_1y_1 - \cosh^{-1}\frac{x_1}{a} \text{ (segment)}$$

$$A = ab\ln\left(\frac{x_1}{a} + \frac{y_1}{b}\right) = \cosh^{-1}\frac{x_1}{a}$$

(sector)

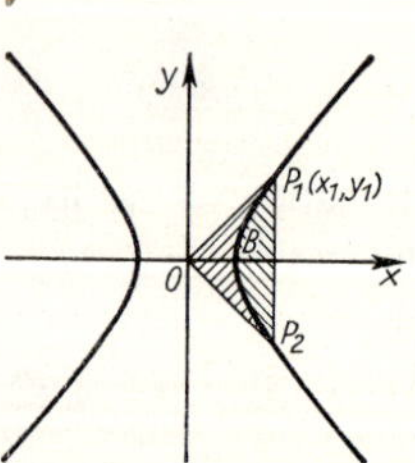

Hyperboloids of revolution $\left(\text{rotating curve } \frac{x^2}{a^2} - \frac{y^2}{b^2} = 1\right)$

Rotation about the x-axis within the limits a to x_1 and $-a$ to $-x_1$:

$$V = \frac{2\pi b^2(x_1 - a)^2 (x_1 + 2a)}{3a^2} \quad \text{(hyperboloid of } \textit{two sheets}\text{)}$$

Rotation about the y-axis within the limits from y_1 to $-y_1$:

$$V = \frac{2\pi a^2 y_1(y_1^2 + 3b^2)}{3b^2} \quad \text{(hyperboloid of } \textit{one sheet}\text{)}$$

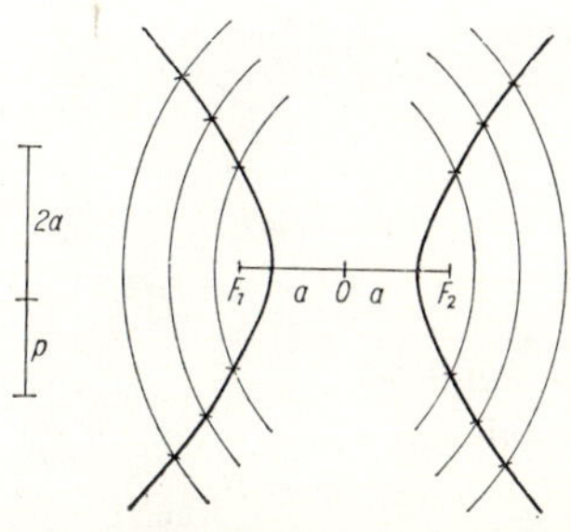

CONSTRUCTIONS OF HYPERBOLAS

Given: *Foci and principal axis* $2a$

We describe circles with an arbitrary radius p and the radius $2a + p$ about the foci F_1 and F_2; in this way we obtain four symmetrical points of the hyperbola where the circles intersect each other. Further points of the hyperbola are obtained by varying p.

Given: *Semiaxes a and b*

We describe circles with a and b as radii and O as the center. From these "directing circles" we draw the tangents s and t perpendicular to the x-axis. An arbitrary straight line g is drawn through O to intersect the tangents in A and B. Then we describe an arc with radius OB about O and erect the perpendicular to the x-axis from the point of intersection of the arc with the x-axis. The point of intersection of this perpendicular with the line drawn through A and parallel to the x-axis is one point of the hyperbola. By varying the straight line g, further points of the hyperbola are obtained.

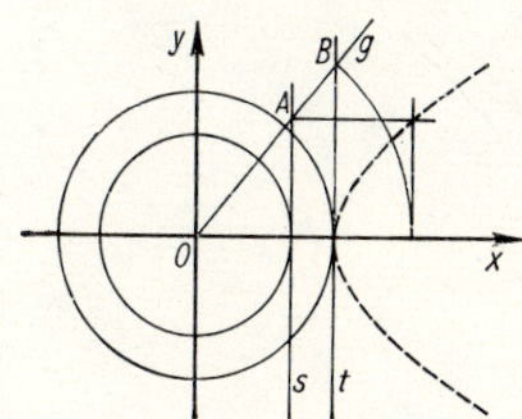

Given: *Coordinate axes as asymptotes and a point P of the rectangular hyperbola*

We drop perpendiculars PQ and PR from P onto the coordinate axes, extend $\overline{PQ}$ through P and draw lines joining arbitrary points

on the extension with the origin O. We draw lines parallel to the x-axis through the points of intersection of these joins with $\overline{PQ}$. Then we draw lines parallel to the y-axis through the points of division on the extension of $\overline{PR}$. Their points of intersection with the respective lines parallel to the x-axis are points of the hyperbola.
By similar constructions in the third quadrant, we obtain the second branch of the hyperbola.

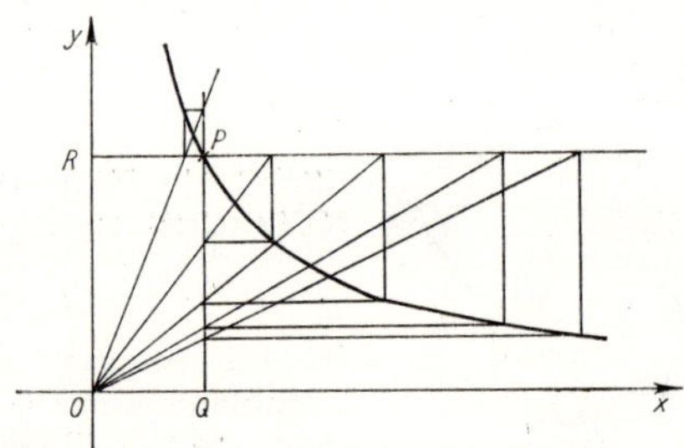

Given: *Asymptotes and a point P_1 of the hyperbola*
We draw an arbitrary straight line through P_1 to intersect the asymptotes in Q_1 and Q_2, and then we measure off $\overline{Q_2P_2} = \overline{Q_1P_1}$ on this line. Then point P_2 is another point of the hyperbola.
Further points are obtained by varying the straight line.

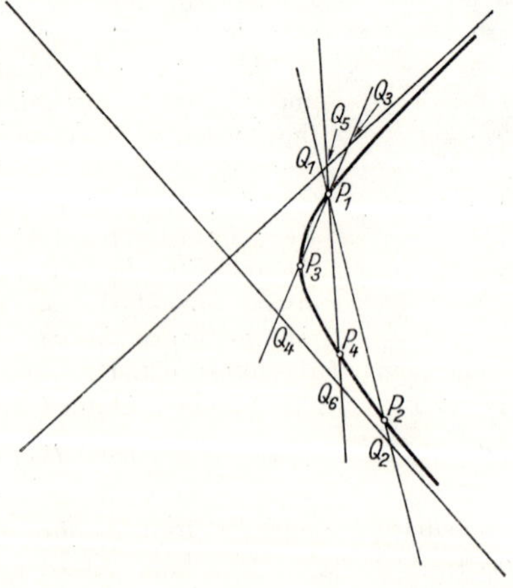

4.1.8. The general equation of the second degree in x and y

$$F(x, y) = a_{11}x^2 + 2a_{12}xy + a_{22}y^2 + 2a_{13}x + 2a_{23}y + a_{33} = 0$$

The *invariants* of a curve of the second order:

$$\varDelta = \begin{vmatrix} a_{11} & a_{12} & a_{13} \\ a_{21} & a_{22} & a_{23} \\ a_{31} & a_{32} & a_{33} \end{vmatrix}; \qquad \delta = \begin{vmatrix} a_{11} & a_{12} \\ a_{21} & a_{22} \end{vmatrix}$$

Note: In the determinants, the relation $a_{ik} = a_{ki}$ holds.

Invariants are quantities which remain unchanged in coordinate transformations.

The general equation of the second degree represents a proper conic section if $\varDelta \neq 0$, whereas it represents a degenerate conic section if $\varDelta = 0$.

For $\delta \neq 0$, the general equation of the second degree defines an ellipse or hyperbola, whereas a parabola is defined for $\delta = 0$.

Case 1: $\delta \neq 0$

Parallel translation of the coordinate system

$x = x' + u \qquad x, y \quad$ original coordinates

$y = y' + v \qquad x', y' \quad$ new coordinates

(u, v) new origin referred to the old system.

$$u = \frac{a_{12}a_{23} - a_{22}a_{13}}{\delta}; \qquad v = \frac{a_{12}a_{13} - a_{11}a_{23}}{\delta}$$

Conditional equations for u and v are

$$\frac{\partial F(x, y)}{\partial x} = 0 \quad \text{and} \quad \frac{\partial F(x, y)}{\partial y} = 0$$

(u, v) becomes the center of the conic section.
The transformed equation has the form

$$a_{11}x'^2 + 2a_{12}x'y' + a_{22}y'^2 + a_{44} = 0$$

where

$$a_{44} = a_{13}u + a_{23}v + a_{33} = \frac{\Delta}{\delta}$$

Rotation of the x', y'-coordinate system about the angle α:

$$\tan 2\alpha = \frac{2a_{12}}{a_{11} - a_{22}} \qquad \alpha \leqq 90^\circ$$

$$x' = \xi \cos \alpha - \eta \sin \alpha$$

$$y' = \xi \text{ in } \alpha + \eta \cos \alpha \qquad \xi, \eta \text{ new coordinates}$$

The transformed equation is of the form

(I) $\quad b_{11}\xi^2 + b_{22}\eta^2 + a_{44} = 0, \quad$ where

$$b_{11} = \frac{1}{2}\left[a_{11} + a_{22} \pm \sqrt{(a_{11} - a_{22})^2 + 4a_{12}^2}\right]$$

$$b_{22} = \frac{1}{2}\left[a_{11} + a_{22} \mp \sqrt{(a_{11} - a_{22})^2 + 4a_{12}^2}\right]$$

The upper sign before the root must be used if a_{12} is positive, the lower sign before the root if a_{12} is negative.

$$b_{11} + b_{22} = a_{11} + a_{22} = S \quad \text{(invariant)}$$

Equation (I) represents an ellipse for $\delta > 0; \Delta \neq 0$, namely,

$$\text{real ellipse} \begin{cases} b_{11} > 0 \\ b_{22} > 0 \\ a_{44} < 0 \end{cases} \qquad \text{imaginary ellipse} \begin{cases} b_{11} > 0 \\ b_{22} > 0 \\ a_{44} > 0 \end{cases}$$

If $\delta > 0$, but $\Delta = 0$, the ellipse degenerates into an imaginary pair of straight lines with real finite points of intersection. If $\Delta = 0$, then $a_{44} = 0$.

Equation (I) represents a hyperbola for $\delta < 0$ and $\Delta \neq 0$ which with $\Delta = 0$ degenerates into a real intersecting pair of straight lines.

Case 2: $\delta = 0$

Rotation of the coordinate system about the angle α:

$$x = x' \cos\alpha - y' \sin\alpha \qquad \tan 2\alpha = \frac{2a_{12}}{a_{11} - a_{22}}; \quad \alpha \leqq 90°$$
$$y = x' \sin\alpha + y' \cos\alpha$$

x, y original coordinates
x', y' new coordinates

The transformed equation is

$$b_{11}x'^2 + 2b_{13}x' + 2b_{23}y' + a_{33} = 0, \text{ where}$$

$$b_{11} = a_{11} + a_{22}$$

$$b_{13} = \frac{a_{13}\sqrt{a_{11}} + a_{23}\sqrt{a_{22}}}{\sqrt{a_{11} + a_{22}}}$$

$$b_{23} = \frac{a_{23}\sqrt{a_{11}} - a_{13}\sqrt{a_{22}}}{\sqrt{a_{11} + a_{22}}}$$

Parallel translation of the x', y'-system:

$$x' = \xi - \frac{b_{13}}{b_{11}}; \quad y' = \eta - \frac{b_{11}a_{33} - b_{13}^2}{2b_{11}b_{23}}$$

The transformed equation is

$$\text{(II)} \quad \xi^2 = -\frac{2b_{23}}{b_{11}}\eta$$

For $\Delta \neq 0$, equation (II) represents a parabola. If $\Delta = 0$, the curve degenerates into a pair of straight lines, namely,
into two real and different parallels for $a_{13}^2 - a_{11}a_{33} > 0$,
into a coinciding pair of parallels (double straight line) for $a_{13}^2 - a_{11}a_{33} = 0$,
into two imaginary parallels for $a_{13}^2 - a_{11}a_{33} < 0$.

SURVEY

<table>
<tr><td rowspan="2"></td><td colspan="2">$\delta \neq 0$
conic sections with center</td><td colspan="3" rowspan="2">$\delta = 0$
conic section without center</td></tr>
<tr><td>$\delta > 0$</td><td>$\delta < 0$</td></tr>
<tr><td rowspan="2">$\Delta \neq 0$
proper conic sections</td><td>$\Delta \cdot S < 0$
real ellipse</td><td rowspan="2">hyperbola</td><td colspan="3" rowspan="2">parabola</td></tr>
<tr><td>$\Delta \cdot S > 0$
imaginary ellipse</td></tr>
<tr><td rowspan="4">$\Delta = 0$
degenerate conic sections</td><td colspan="2">nonparallel (intersecting) pair of straight lines</td><td colspan="3">pair of parallel lines</td></tr>
<tr><td rowspan="3">imaginary with real finite intersection</td><td rowspan="3">real</td><td colspan="3">$a_{13}^2 - a_{11}a_{33}$</td></tr>
<tr><td>> 0
two different real</td><td>$= 0$
two coinciding</td><td>< 0
two imaginary</td></tr>
<tr><td colspan="3">parallel lines</td></tr>
</table>

Example 1:

$$5x^2 + 4xy + 2y^2 - 18x - 12y + 15 = 0$$

$$a_{11} = 5; \quad a_{12} = 2; \quad a_{22} = 2; \quad a_{13} = -9;$$

$$a_{23} = -6; \quad a_{33} = 15$$

$$\Delta = \begin{vmatrix} 5 & 2 & -9 \\ 2 & 2 & -6 \\ -9 & -6 & 15 \end{vmatrix} \neq 0; \qquad \delta = \begin{vmatrix} 5 & 2 \\ 2 & 2 \end{vmatrix} = 10 - 4 > 0$$

The equation defines an **ellipse.**

Determination of u and v:

$$\frac{\partial F(x, y)}{\partial x} = 10x + 4y - 18 = 0$$

$$\frac{\partial F(x, y)}{\partial y} = 4x + 4y - 12 = 0$$

$$x = u = 1; \quad y = v = 2$$

Center of the ellipse M (1, 2) in the original system

Determination of a_{44}:

$$a_{44} = a_{13}u + a_{23}v + a_{33} = (-9) \cdot 1 + (-6) \cdot 2 + 15 = -6$$

Transformed equation after parallel translation

$$5x'^2 + 4x'y' + 2y'^2 - 6 = 0$$

Determination of the angle of rotation α:

$$\tan 2\alpha = \frac{2a_{12}}{a_{11} - a_{22}} = \frac{4}{3}; \quad \alpha = 26°34'$$

$$b_{11} = \frac{1}{2}\left[a_{11} + a_{22} + \sqrt{(a_{11} - a_{22})^2 + 4a_{12}^2}\right]$$

(upper sign, since a_{12} is positive)

$$b_{11} = \frac{1}{2}\left[5 + 2 + \sqrt{(5 - 2)^2 + 4 \cdot 2^2}\right] = 6$$

$$b_{22} = \frac{1}{2}\left[a_{11} + a_{22} - \sqrt{(a_{11} - a_{22})^2 + 4a_{12}^2}\right] = \frac{1}{2}(7 - 5) = 1$$

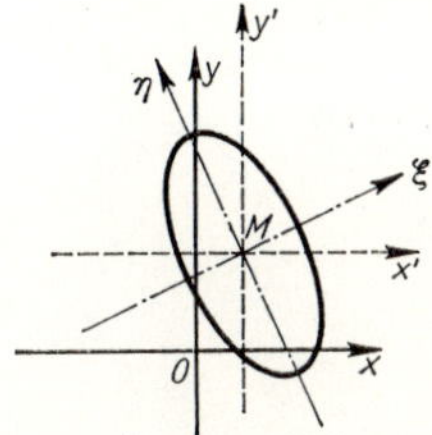

Transformed equation after rotation:

$$6\xi^2 + 1\eta^2 - 6 = 0;$$

$$\frac{\xi^2}{1} + \frac{\eta^2}{6} = 1$$

i.e. ellipse with $a = \sqrt{6}$ and $b = 1$ and with center at the origin of the ξ,η-system. The major axis is the η-axis.

Question: What must be the absolute term a_{33} in the equation in order that the ellipse degenerates?

The condition is: $\Delta = 0$ or $a_{44} = 0$

$$a_{44} = a_{13}u + a_{23}v + a_{33} = 0; \qquad -9 - 12 + a_{33} = 0;$$

$$a_{33} = 21$$

The equation $5x^2 + 4xy + 2y^2 - 18x - 12y + 21 = 0$ represents an ellipse which degenerates into its center.

Example 2:

$$9x^2 - 24xy + 16y^2 + 220x - 40y - 100 = 0$$

$$a_{11} = 9; \quad a_{12} = -12; \quad a_{22} = 16; \quad a_{13} = 110;$$

$$a_{23} = -20; \quad a_{33} = -100$$

$$\Delta = \begin{vmatrix} 9 & -12 & 110 \\ -12 & 16 & -20 \\ 110 & -20 & -100 \end{vmatrix} \neq 0$$

$$\delta = \begin{vmatrix} 9 & -12 \\ -12 & 16 \end{vmatrix} = 144 - 144 = 0$$

The equation defines a **parabola**.

Rotation of the coordinate system:

$$\tan 2\alpha = \frac{2a_{12}}{a_{11} - a_{22}} = \frac{-24}{-7} = \frac{24}{7}; \quad \alpha = 36^\circ 52'$$

$$b_{11} = a_{11} + a_{22} = 25$$

$$b_{13} = \frac{a_{13}\sqrt{a_{11}} + a_{23}\sqrt{a_{22}}}{\sqrt{a_{11} + a_{22}}} = \frac{110\sqrt{9} + (-20)\sqrt{16}}{\sqrt{25}} = 50$$

$$b_{23} = \frac{a_{23}\sqrt{a_{11}} - a_{13}\sqrt{a_{22}}}{\sqrt{a_{11} + a_{22}}} = \frac{-20\sqrt{9} - 110\sqrt{16}}{\sqrt{25}} = -100$$

Transformed equation after rotation:

$$25x'^2 + 100x' - 200y' - 100 = 0$$

$$x'^2 + 4x' - 8y' - 4 = 0$$

Transformed equation after parallel translation:

$$\xi^2 = -\frac{-200}{25}\eta; \quad \xi^2 = 8\eta$$

i.e. parabola with vertex at the origin of the ξ,η-system and with parameter $2p = 8$. The axis of the parabola is the η-axis.

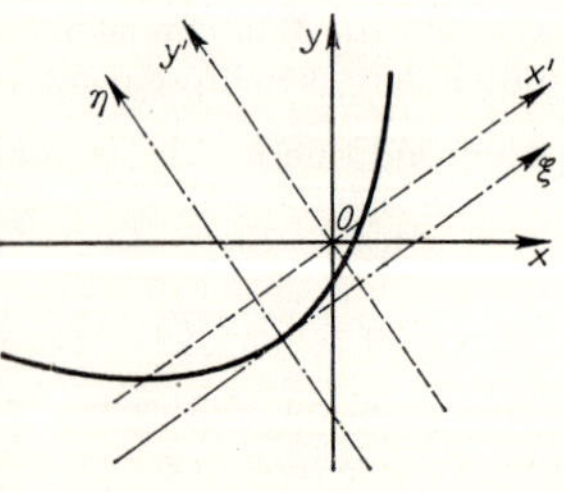

4.2. Analytical geometry of space

4.2.1. The various systems of coordinate

Rectangular (Cartesian) coordinate system

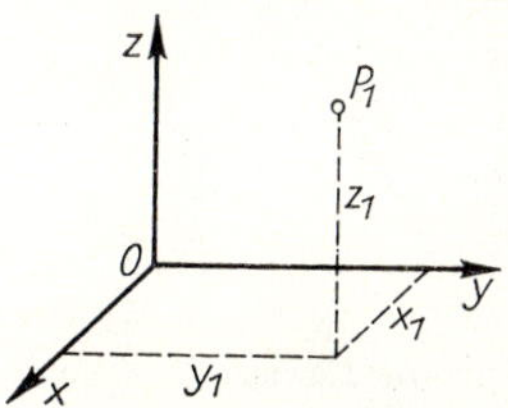

Right-handed system:

Rotation of the positive x-axis toward the positive y-axis with simultaneous displacement toward the positive z-direction represents the motion of a *right-handed screw.*
The coordinates of the point P are x, y, z.
We write: $P(x, y, z)$

The x-, y-, z-lines, i.e. the lines along which only one coordinate changes are *axially parallel straight lines.* The areas for which one coordinate is constant, that is, the planes $x = \text{const.}$, $y = \text{const.}$, $z = \text{const.}$, are *axially parallel planes.*

Cylindrical coordinate system

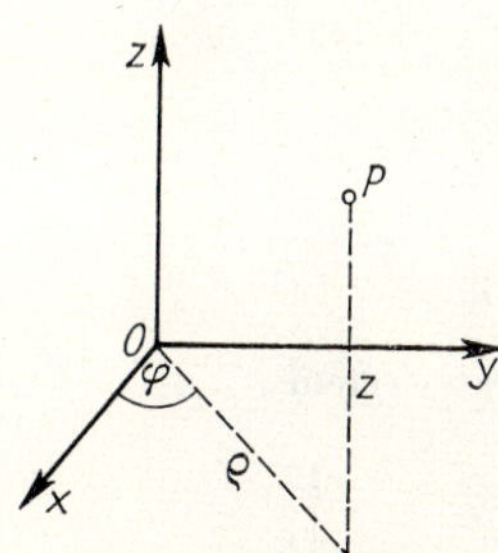

The coordinates of the point P are ϱ, φ, z.
We write: $P(\varrho, \varphi, z)$
ϱ, φ are the polar coordinates of the projection of the point onto the x,y-plane. z is the distance of the point from the x, y-plane. The areas $\varrho = \text{const.}$ are cylindrical surfaces with common z-axis, the areas $\varphi = \text{const.}$ are the planes through the z-axis, the areas $z = \text{const.}$ are planes perpendicular to the z-axis.

Relations between cylindrical coordinates and rectangular coordinates

$$\varrho = \sqrt{x^2 + y^2}$$

$$\varphi = \operatorname{arc\,tan} \frac{y}{x} = \operatorname{arc\,sin} \frac{y}{\varrho} = \operatorname{arc\,cos} \frac{x}{\varrho} \quad (0 \leqq \varphi \leqq 2\pi)$$

$$x = \varrho \cos \varphi; \quad y = \varrho \sin \varphi; \quad z = z$$

pherical (polar) coordinate system

The coordinates of the point P are r, φ, ϑ.

We write: $P(r, \varphi, \vartheta)$

r value of the radius vector OP

φ angle which the projection of $\overline{OP}$ onto the x, y-plane makes with the x-axis $(0 \leqq \varphi < 2\pi)$.

ϑ angle between $\overline{OP}$ and positive z-axis $(0 \leqq \vartheta < \pi)$.

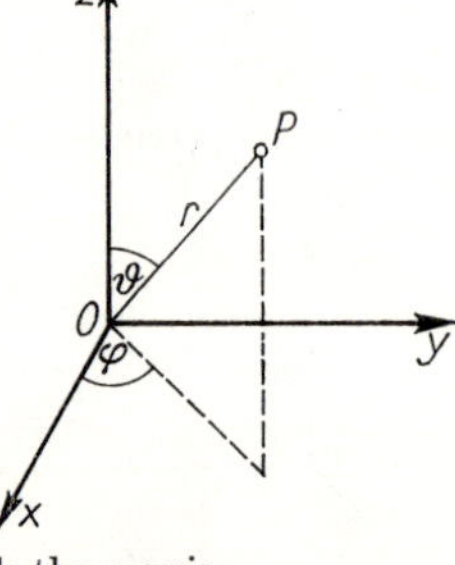

The areas $r =$ const. are concentric spheres with the pole O as center.
The areas $\varphi =$ const. are half-planes through the z-axis.
The areas $\vartheta =$ const. are cones with the vertex at O and the z-axis as axis.

Relationship between spherical and rectangular coordinates

$$r = \sqrt{x^2 + y^2 + z^2}$$

$$\varphi = \arctan \frac{y}{x} = \arcsin \frac{y}{\sqrt{x^2 + y^2}} = \arccos \frac{x}{\sqrt{x^2 + y^2}}$$

$$\vartheta = \arccos \frac{z}{r} = \arctan \frac{\sqrt{x^2 + y^2}}{z}$$

$$x = r \sin \vartheta \cos \varphi; \qquad y = r \sin \vartheta \sin \varphi; \qquad z = r \cos \vartheta$$

Transformation of the rectangular coordinate system

1. Parallel translation

$$x = x' + a \qquad x' = x - a$$

$$y = y' + b \qquad y' = y - b$$

$$z = z' + c \qquad z' = z - c$$

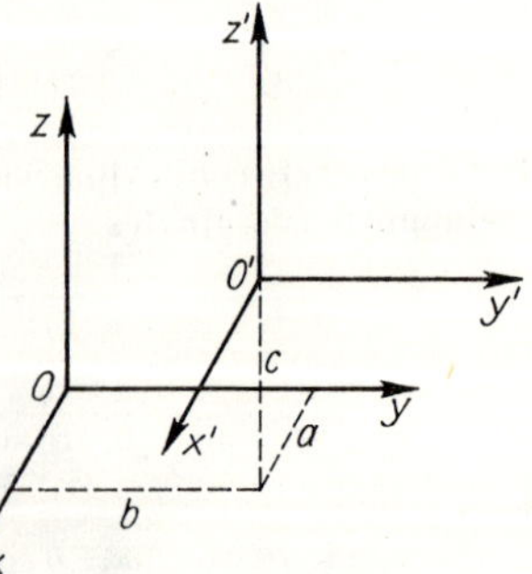

x, y, z original coordinates
x', y', z' new coordinates
a, b, c coordinates of the new origin in the first coordinate system

2. Rotation

$$x = x' \cos \alpha_1 + y' \cos \alpha_2 + z' \cos \alpha_3$$

$$y = x' \cos \beta_1 + y' \cos \beta_2 + z' \cos \beta_3$$

$$z = x' \cos \gamma_1 + y' \cos \gamma_2 + z' \cos \gamma_3$$

$$x' = x \cos \alpha_1 + y \cos \beta_1 + z \cos \gamma_1$$

$$y' = x \cos \alpha_2 + y \cos \beta_2 + z \cos \gamma_2$$

$$z' = x \cos \alpha_3 + y \cos \beta_3 + z \cos \gamma_3$$

$\alpha_1, \beta_1, \gamma_1$ angles which the x'-axis makes with the original axes.

$\alpha_2, \beta_2, \gamma_2$ angles which the y'-axis makes with the original axes.

$\alpha_3, \beta_3, \gamma_3$ angles which the z'-axis makes with the original axes.

Relationship between the direction cosine of the new axes:

$$\cos^2 \alpha_1 + \cos^2 \beta_1 + \cos^2 \gamma_1 = 1$$

$$\cos^2 \alpha_1 + \cos^2 \alpha_2 + \cos^2 \alpha_3 = 1$$

$$\cos \alpha_1 \cos \alpha_2 + \cos \beta_1 \cos \beta_2 + \cos \gamma_1 \cos \gamma_2 = 0$$

$$\cos \alpha_1 \cos \beta_1 + \cos \alpha_2 \cos \beta_2 + \cos \alpha_3 \cos \beta_3 = 0$$

Further formulas are obtained by cyclic permutation.

$$\Delta = \begin{vmatrix} \cos \alpha_1 & \cos \beta_1 & \cos \gamma_1 \\ \cos \alpha_2 & \cos \beta_2 & \cos \gamma_2 \\ \cos \alpha_3 & \cos \beta_3 & \cos \gamma_3 \end{vmatrix} = 1$$

4.2.2. Points and line segments in space

The distance between two points $P_1(x_1, y_1, z_1)$ and $P_2(x_2, y_2, z_2)$

$$\overline{P_1P_2} = \sqrt{(x_1 - x_2)^2 + (y_1 - y_2)^2 + (z_1 - z_2)^2} = e$$

Projection of the line segment e onto the coordinate axes

$$e_x = e \cos \alpha; \qquad e_y = e \cos \beta; \qquad e_z = e \cos \gamma$$

α, β, γ angles which the line through the origin and parallel to the straight line which contains the line-segment e makes with the coordinate axes.

$$e^2 = e_x^2 + e_y^2 + e_z^2$$

$$e = e_x \cos\alpha + e_y \cos\beta + e_z \cos\gamma$$

$$\cos^2\alpha + \cos^2\beta + \cos^2\gamma = 1$$

$\cos\alpha$, $\cos\beta$, $\cos\gamma$ direction cosine of e.

Division of a line segment P_1P_2 in the ratio λ

$$x_t = \frac{x_1 + \lambda x_2}{1 + \lambda} \qquad \begin{array}{l} \lambda > 0 \text{ for inner point of division} \\ \lambda < 0 \text{ for outer point of division} \end{array}$$

$$y_t = \frac{y_1 + \lambda y_2}{1 + \lambda}$$

$$z_t = \frac{z_1 + \lambda z_2}{1 + \lambda}$$

Center $P_0(x_0, y_0, z_0)$ of the line segment P_1P_2

$$x_0 = \frac{x_1 + x_2}{2}; \quad y_0 = \frac{y_1 + y_2}{2}; \quad z_0 = \frac{z_1 + z_2}{2}$$

Centroid $C(x_C, y_C, z_C)$ of the triangle $P_1P_2P_3$

$$x_C = \frac{x_1 + x_2 + x_3}{3}; \quad y_C = \frac{y_1 + y_2 + y_3}{3}; \quad z_C = \frac{z_1 + z_2 + z_3}{3}$$

For material points in the vertices of the triangle holds:

$$x_C = \frac{m_1x_1 + m_2x_2 + m_3x_3}{m_1 + m_2 + m_3}; \quad y_C = \frac{m_1y_1 + m_2y_2 + m_3y_3}{m_1 + m_2 + m_3};$$

$$z_C = \frac{m_1z_1 + m_2z_2 + m_3z_3}{m_1 + m_2 + m_3}$$

and generally with n mass points

$$x_C = \frac{\sum_1^n m_k x_k}{\sum_1^n m_k}; \quad y_C = \frac{\sum_1^n m_k y_k}{\sum_1^n m_k}; \quad z_C = \frac{\sum_1^n m_k z_k}{\sum_1^n m_k}$$

Area of the triangle $P_1P_2P_3$

$$A = \sqrt{A_1^2 + A_2^2 + A_3^2}, \text{ where}$$

$$A_1 = \frac{1}{2}\begin{vmatrix} y_1 & z_1 & 1 \\ y_2 & z_2 & 1 \\ y_3 & z_3 & 1 \end{vmatrix} \qquad A_2 = \frac{1}{2}\begin{vmatrix} z_1 & x_1 & 1 \\ z_2 & x_2 & 1 \\ z_3 & x_3 & 1 \end{vmatrix}$$

$$A_3 = \frac{1}{2}\begin{vmatrix} x_1 & y_1 & 1 \\ x_2 & y_2 & 1 \\ x_3 & y_3 & 1 \end{vmatrix}$$

A becomes positive if the succession of the vectors $\overrightarrow{OP_1}$, $\overrightarrow{OP_2}$, $\overrightarrow{OP_3}$ corresponds to that of a right-handed system.

Volume of the tetrahedron $P_1P_2P_3P_4$ (P_1 vertex)

$$V = \frac{1}{6}\begin{vmatrix} x_1 & y_1 & z_1 & 1 \\ x_2 & y_2 & z_2 & 1 \\ x_3 & y_3 & z_3 & 1 \\ x_4 & y_4 & z_4 & 1 \end{vmatrix} = \frac{1}{6}\begin{vmatrix} x_1 - x_2 & y_1 - y_2 & z_1 - z_2 \\ x_1 - x_3 & y_1 - y_3 & z_1 - z_3 \\ x_1 - x_4 & y_1 - y_4 & z_1 - z_4 \end{vmatrix}$$

V becomes positive if the succession of the vectors $\overrightarrow{P_1P_2}$, $\overrightarrow{P_1P_3}$, $\overrightarrow{P_1P_4}$ corresponds to that of a right-handed system.
Four points in a plane: $V = 0$.

Angle τ between two radius vectors r_1 and r_2

$$\cos\tau = \cos\alpha_1\cos\alpha_2 + \cos\beta_1\cos\beta_2 + \cos\gamma_1\cos\gamma_2$$
$$= \frac{x_1x_2 + y_1y_2 + z_1z_2}{r_1r_2}$$

$\left.\begin{matrix}\alpha_1, \beta_1, \gamma_1 \\ \alpha_2, \beta_2, \gamma_2\end{matrix}\right\}$ slope angles of the radius vector $\mathbf{r}_1$ and $\mathbf{r}_2$

$$\mathbf{r}_1 \perp \mathbf{r}_2 \Rightarrow \cos\tau = 0$$

4.2.3. Planes in space

General equation of the plane

$$E \equiv Ax + By + Cz + D = 0$$

(A, B, C not zero coincidentally)

Intercepts on the coordinate axes

$$a = -\frac{D}{A}; \quad b = -\frac{D}{B}; \quad c = -\frac{D}{C}$$

Perpendicular from the origin onto the plane

$$p = -\frac{D}{\pm\sqrt{A^2 + B^2 + C^2}}$$

The root must be given the sign opposite to that of D.

Direction cosine of the plane

$$\cos\alpha = \frac{A}{\pm\sqrt{A^2 + B^2 + C^2}}$$

α, β, γ are angles which the perpendicular p makes with the positive direction of the axes.

$$\cos\beta = \frac{B}{\pm\sqrt{A^2 + B^2 + C^2}}$$

$$\cos\gamma = \frac{C}{\pm\sqrt{A^2 + B^2 + C^2}}$$

Sign of the root the same as above.

Special cases

$D = 0$	The plane passes through the origin: $Ax + By + Cz = 0$
$A = 0$	The plane is parallel to the x-axis: $By + Cz + D = 0$
$B = 0$	The plane is parallel to the y-axis: $Ax + Cz + D = 0$
$C = 0$	The plane is parallel to the z-axis: $Ax + By + D = 0$
$A = B = 0$	The plane is parallel to the x,y-plane: $Cz + D = 0$
$A = C = 0$	The plane is parallel to the x,z-plane: $By + D = 0$
$B = C = 0$	The plane is parallel to the y,z-plane: $Ax + D = 0$
$A = D = 0$	The plane passes through the x-axis: $By + Cz = 0$
$B = D = 0$	The plane passes through the y-axis: $Ax + Cz = 0$
$C = D = 0$	The plane passes through the z-axis: $Ax + By = 0$

Specific planes

$z = 0$ equation of the x,y-plane

$y = 0$ equation of the x,z-plane

$x = 0$ equation of the y,z-plane

Intercept form of the equation of the plane

$$\frac{x}{a} + \frac{y}{b} + \frac{z}{c} = 1$$ a, b, c intercepts of the axes (see page 262)

Hessian normal form of the equation of the plane

$$x \cos \alpha + y \cos \beta + z \cos \gamma - p = 0$$

p perpendicular from the origin onto the plane
$\cos \alpha$, $\cos \beta$, $\cos \gamma$ direction cosine of the plane (see page 262)
Transformation of the general form into the HESSIAN normal form:

$$\frac{Ax + By + Cz + D}{\pm\sqrt{A^2 + B^2 + C^2}} = 0,$$ where the root must be given the sign opposite to that of D.

Distance of point $P_0(x_0, y_0, z_0)$ from the plane $Ax + By + Cz + D = 0$

$$d = \frac{Ax_0 + By_0 + Cz_0 + D}{\pm\sqrt{A^2 + B^2 + C^2}}$$ $d < 0$, if P_0 lies on the same side of the plane as the origin, otherwise positive.

$$d = x_0 \cos \alpha + y_0 \cos \beta + z_0 \cos \gamma - p$$

Plane through three points $P_1(x_1, y_1, z_1)$, $P_2(x_2, y_2, z_2)$, $P_3(x_3, y_3, z_3)$

$$\begin{vmatrix} x & y & z & 1 \\ x_1 & y_1 & z_1 & 1 \\ x_2 & y_2 & z_2 & 1 \\ x_3 & y_3 & z_3 & 1 \end{vmatrix} = 0 \text{ or } \begin{vmatrix} x - x_1 & y - y_1 & z - z_1 \\ x_2 - x_1 & y_2 - y_1 & z_2 - z_1 \\ x_3 - x_1 & y_3 - y_1 & z_3 - z_1 \end{vmatrix} = 0$$

Plane through point $P_0(x_0, y_0, z_0)$

$$A(x - x_0) + B(y - y_0) + C(z - z_0) = 0$$

Family of planes

$$\left.\begin{aligned} E_1 &\equiv A_1x + B_1y + C_1z + D_1 = 0 \\ E_2 &\equiv A_2x + B_2y + C_2z + D_2 = 0 \end{aligned}\right\} \text{given planes}$$

Family of planes through the straight line of intersection between the two planes $E_1 = 0$ and $E_2 = 0$:

$$E_1 + \lambda E_2 = 0, \quad \lambda \in R \setminus \{0\}$$

Angle-bisecting planes with respect to two planes

$$E_1 = 0, \quad E_2 = 0 \text{ given planes}$$

$$\frac{A_1x + B_1y + C_1z + D_1}{\pm\sqrt{A_1^2 + B_1^2 + C_1^2}} \pm \frac{A_2x + B_2y + C_2z + D_2}{\pm\sqrt{A_2^2 + B_2^2 + C_2^2}} = 0$$

The signs of the roots are opposite to those of D_1 and D_2, respectively.

Point of intersection $S(x_i, y_i, z_i)$ of the three planes

$$E_1 = 0, \quad E_2 = 0, \quad E_3 = 0 \quad \text{given planes}$$

$$x_i = -\frac{\Delta_x}{\Delta}; \quad y_i = -\frac{\Delta_y}{\Delta}; \quad z_i = -\frac{\Delta_z}{\Delta},$$

where

$$\Delta = \begin{vmatrix} A_1 & B_1 & C_1 \\ A_2 & B_2 & C_2 \\ A_3 & B_3 & C_3 \end{vmatrix},$$

$$\Delta_x = \begin{vmatrix} D_1 & B_1 & C_1 \\ D_2 & B_2 & C_2 \\ D_3 & B_3 & C_3 \end{vmatrix}, \quad \Delta_y = \begin{vmatrix} A_1 & D_1 & C_1 \\ A_2 & D_2 & C_2 \\ A_3 & D_3 & C_3 \end{vmatrix},$$

$$\Delta_z = \begin{vmatrix} A_1 & B_1 & D_1 \\ A_2 & B_2 & D_2 \\ A_3 & B_3 & D_3 \end{vmatrix}$$

Condition: $\Delta \neq 0$

Four planes through a point

The four planes are given in the general form.

$$\begin{vmatrix} A_1 & B_1 & C_1 & D_1 \\ A_2 & B_2 & C_2 & D_2 \\ A_3 & B_3 & C_3 & D_3 \\ A_4 & B_4 & C_4 & D_4 \end{vmatrix} = 0$$

Angle τ between two planes

$E_1 = 0, \quad E_2 = 0 \qquad$ given planes

$$\cos\tau = \left| \frac{A_1A_2 + B_1B_2 + C_1C_2}{\sqrt{(A_1^2 + B_1^2 + C_1^2)(A_2^2 + B_2^2 + C_2^2)}} \right| \qquad \tau \leqq 90°$$

The planes are *parallel* if

$$A_1:B_1:C_1 = A_2:B_2:C_2$$

The planes are *perpendicular to one another* if

$$A_1A_2 + B_1B_2 + C_1C_2 = 0$$

Equation of two parallel planes

$$A_1x + B_1y + C_1z + D_1 = 0$$
$$A_1x + B_1y + C_1z + D_2 = 0$$

Projection of the plane surface A onto the x,y-, y,z-, x,z-plane

$$A_{xy} = A\cos\gamma, \qquad A_{yz} = A\cos\alpha, \qquad A_{xz} = A\cos\beta$$

α, β, γ are the angles which the perpendicular from the origin onto plane which contains the surface A makes with the coordinate axes.

$$A^2 = A_{xy}^2 + A_{yz}^2 + A_{xz}^2$$
$$A = A_{xy}\cos\gamma + A_{yz}\cos\alpha + A_{xz}\cos\beta$$

4.2.4. Straight lines in space

General form of the equation of the straight line

$$\begin{cases} A_1x + B_1y + C_1z + D_1 = 0, & \text{in short } E_1 = 0 \\ A_2x + B_2y + C_2z + D_2 = 0, & \text{in short } E_2 = 0 \end{cases}$$

The straight line is the intersection of the two arbitrary planes.

Equation of straight line in two projected planes

$$\begin{cases} y = mx + b & \text{(plane perpendicular to the } x,y\text{-plane)} \\ z = nx + c & \text{(plane perpendicular to the } x,z\text{-plane)} \end{cases}$$

Conversion of the general form into the latter one:

$$A_1 = -m; \quad B_1 = 1; \quad C_1 = 0; \quad D_1 = -b$$

$$A_2 = -n; \quad B_2 = 0; \quad C_2 = 1; \quad D_2 = -c$$

Special cases

Straight line parallel to the x,y-plane $\begin{cases} y = mx + b \\ z = c \end{cases}$

Straight line parallel to the x,z-plane $\begin{cases} z = nx + c \\ y = b \end{cases}$

Straight line parallel to the y,z-plane $\begin{cases} z = py + q \\ x = a \end{cases}$

Straight line parallel to the x-axis $\begin{cases} y = b \\ z = c \end{cases}$

Straight line parallel to the y-axis $\begin{cases} x = a \\ z = c \end{cases}$

Straight line parallel to the z-axis $\begin{cases} x = a \\ y = b \end{cases}$

Straight line through the origin $\begin{cases} y = mx \\ z = nx \end{cases}$

Equation of the axes:

x-axis $\begin{cases} y = 0 \\ z = 0 \end{cases}$

y-axis $\begin{cases} x = 0 \\ z = 0 \end{cases}$

z-axis $\begin{cases} x = 0 \\ y = 0 \end{cases}$

Angles α, β, γ between the straight lines $\begin{cases} E_1 = 0 \\ E_2 = 0 \end{cases}$ and the axes

$$\cos\alpha = \frac{1}{N}\begin{vmatrix} B_1 & C_1 \\ B_2 & C_2 \end{vmatrix}; \quad \cos\beta = \frac{1}{N}\begin{vmatrix} C_1 & A_1 \\ C_2 & A_2 \end{vmatrix};$$

$$\cos\gamma = \frac{1}{N}\begin{vmatrix} A_1 & B_1 \\ A_2 & B_2 \end{vmatrix}$$

$$N^2 = \begin{vmatrix} B_1 & C_1 \\ B_2 & C_2 \end{vmatrix}^2 + \begin{vmatrix} C_1 & A_1 \\ C_2 & A_2 \end{vmatrix}^2 + \begin{vmatrix} A_1 & B_1 \\ A_2 & B_2 \end{vmatrix}^2$$

Angles α, β, γ between the straight lines $\begin{cases} y = mx + b \\ z = nx + c \end{cases}$ and the axes

$$\cos\alpha = \frac{1}{\sqrt{1 + m^2 + n^2}}; \quad \cos\beta = \frac{m}{\sqrt{1 + m^2 + n^2}};$$

$$\cos\gamma = \frac{n}{\sqrt{1 + m^2 + n^2}}$$

$$\cos^2\alpha + \cos^2\beta + \cos^2\gamma = 1$$

Equation of the straight line through $P_1(x_1, y_1, z_1)$ with direction α, β, γ

$$\frac{x - x_1}{\cos\alpha} = \frac{y - y_1}{\cos\beta} = \frac{z - z_1}{\cos\gamma}$$

Equation of straight line through $P_1(x_1, y_1, z_1)$ in parametric form

$$x = x_1 + t\cos\alpha$$

$$y = y_1 + t\cos\beta$$

$$z = z_1 + t\cos\gamma$$

General parametric representation of a straight line

$$x = a_1 t + a_2; \quad y = b_1 t + b_2; \quad z = c_1 t + c_2$$

where a_1, b_1, c_1 need not satisfy the direction cosine relation but are arbitrary given numbers as are a_2, b_2, c_2.

Straight line through two points $P_1(x_1, y_1, z_1)$ and $P_2(x_2, y_2, z_2)$

$$\frac{x - x_1}{x_2 - x_1} = \frac{y - y_1}{y_2 - y_1} = \frac{z - z_1}{z_2 - z_1}$$

Special cases

1. Straight line through $P_1(x_1, y_1, z_1)$ and the origin

$$\frac{x}{x_1} = \frac{y}{y_1} = \frac{z}{z_1}$$

2. Straight line through $P_1(x_1, y_1, z_1)$ and perpendicular to plane $E = 0$

$$\frac{x - x_1}{A} = \frac{y - y_1}{B} = \frac{z - z_1}{C}$$

Point of intersection of the straight line $\frac{x - x_1}{\cos\alpha} = \frac{y - y_1}{\cos\beta} = \frac{z - z_1}{\cos\gamma}$ with the plane $E = 0$

$$x_i = x_1 - t\cos\alpha$$

$$y_i = y_1 - t\cos\beta \qquad t = \frac{Ax_1 + By_1 + Cz_1 + D}{A\cos\alpha + B\cos\beta + C\cos\gamma}$$

$$z_i = z_1 - t\cos\gamma$$

Straight line parallel to the plane:

$$A\cos\alpha + B\cos\beta + C\cos\gamma = 0$$

Point of intersection of the straight line $\begin{cases} y = mx + b \\ z = nx + c \end{cases}$ with the plane $E = 0$

$$x_i = -\frac{bB + cC + D}{A + mB + nC}$$

y_i and z_i by substitution in the original equation

Straight line parallel to the plane:

$$A + mB + nC = 0$$

Condition for the intersection of two straight lines

$$\left.\begin{array}{l} E_1 \equiv A_1x + B_1y + C_1z + D_1 = 0 \\ E_2 \equiv A_2x + B_2y + C_2z + D_2 = 0 \end{array}\right\} \quad \left.\begin{array}{l} E_3 \equiv A_3x + B_3y + C_3z + D_3 = 0 \\ E_4 \equiv A_4x + B_4y + C_4z + D_4 = 0 \end{array}\right\} \quad \text{given straight lines}$$

$$\begin{vmatrix} A_1 & B_1 & C_1 & D_1 \\ A_2 & B_2 & C_2 & D_2 \\ A_3 & B_3 & C_3 & D_3 \\ A_4 & B_4 & C_4 & D_4 \end{vmatrix} = 0$$

Point of intersection of two straight lines

$$\begin{cases} \boldsymbol{y = m_1x + b_1} \\ \boldsymbol{z = n_1x + c_1} \end{cases} \quad \textbf{and} \quad \begin{cases} \boldsymbol{y = m_2x + b_2} \\ \boldsymbol{z = n_2x + c_2} \end{cases}$$

$$x_i = \frac{b_2 - b_1}{m_1 - m_2} = \frac{c_2 - c_1}{n_1 - n_2}; \qquad y_i = \frac{m_1b_2 - m_2b_1}{m_1 - m_2};$$

$$z_i = \frac{n_1c_2 - n_2c_1}{n_1 - n_2}$$

Condition for point of intersection: $\dfrac{b_1 - b_2}{c_1 - c_2} = \dfrac{m_1 - m_2}{n_1 - n_2}$

Angle τ between the straight line

$$\frac{x - x_1}{\cos\alpha} = \frac{y - y_1}{\cos\beta} = \frac{z - z_1}{\cos\gamma} \quad \textbf{and the plane } E = 0$$

$$\sin\tau = \left| \frac{A\cos\alpha + B\cos\beta + C\cos\gamma}{\sqrt{A^2 + B^2 + C^2}} \right| \qquad \tau \leqq 90°$$

Straight line *parallel* to the plane: $A\cos\alpha + B\cos\beta + C\cos\gamma = 0$

Straight line *perpendicular* to the plane: $\dfrac{A}{\cos\alpha} = \dfrac{B}{\cos\beta} = \dfrac{C}{\cos\gamma}$

Intersecting angle τ of the straight lines

$$\frac{x - x_1}{\cos\alpha_1} = \frac{y - y_1}{\cos\beta_1} = \frac{z - z_1}{\cos\gamma_1} \quad \textbf{and} \quad \frac{x - x_2}{\cos\alpha_2} = \frac{y - y_2}{\cos\beta_2} = \frac{z - z_2}{\cos\gamma_2}$$

$$\cos\tau = \cos\alpha_1\cos\alpha_2 + \cos\beta_1\cos\beta_2 + \cos\gamma_1\cos\gamma_2$$

Parallel straight lines:

$$\cos\alpha_1 = \cos\alpha_2, \qquad \cos\beta_1 = \cos\beta_2, \qquad \cos\gamma_1 = \cos\gamma_2$$

Perpendicular straight lines:

$$\cos\alpha_1\cos\alpha_2+\cos\beta_1\cos\beta_2+\cos\gamma_1\cos\gamma_2=0$$

Intersecting angle τ of the straight lines
$\begin{cases} y=m_1x+b_1 \\ z=n_1x+c_1 \end{cases}$ **and** $\begin{cases} y=m_2x+b_2 \\ z=n_2x+c_2 \end{cases}$

$$\cos\tau=\frac{1+m_1m_2+n_1n_2}{\sqrt{(1+m_1^2+n_1^2)(1+m_2^2+n_2^2)}}$$

Parallel straight lines: $m_1=m_2;\ n_1=n_2$

Perpendicular straight lines: $1+m_1m_2+n_1n_2=0$

Distance between two skew straight lines in space

$$\left.\begin{aligned}\frac{x-x_1}{\cos\alpha_1}=\frac{y-y_1}{\cos\beta_1}=\frac{z-z_1}{\cos\gamma_1}\\ \frac{x-x_2}{\cos\alpha_2}=\frac{y-y_2}{\cos\beta_2}=\frac{z-z_2}{\cos\gamma_2}\end{aligned}\right\}\text{given straight lines}$$

$$e=\left|\frac{\begin{vmatrix}x_1-x_2 & y_1-y_2 & z_1-z_2\\ \cos\alpha_1 & \cos\beta_1 & \cos\gamma_1\\ \cos\alpha_2 & \cos\beta_2 & \cos\gamma_2\end{vmatrix}}{\sqrt{\begin{vmatrix}\cos\beta_1 & \cos\gamma_1\\ \cos\beta_2 & \cos\gamma_2\end{vmatrix}^2+\begin{vmatrix}\cos\gamma_1 & \cos\alpha_1\\ \cos\gamma_2 & \cos\alpha_2\end{vmatrix}^2+\begin{vmatrix}\cos\alpha_1 & \cos\beta_1\\ \cos\alpha_2 & \cos\beta_2\end{vmatrix}^2}}\right|$$

4.2.5. Surfaces of the second order

Sphere

General equation:

$$(x-a)^2+(y-b)^2+(z-c)^2-r^2=0$$

Center $M(a, b, c)$
r radius

Central equation

$$x^2+y^2+z^2-r^2=0 \qquad \text{Center in the origin}$$

Quadratic equation as a sphere:

$$A(x^2 + y^2 + z^2) + Bx + Cy + Dz + E = 0 \quad \text{for} \quad A \neq 0$$

Center $M\left(-\frac{B}{2A};\ -\frac{C}{2A};\ -\frac{D}{2A}\right)$

Radius $r = \frac{1}{2A}\sqrt{B^2 + C^2 + D^2 - 4AE}$

$$\text{for} \quad B^2 + C^2 + D^2 > 4AE$$

Tangential plane in point $P_0(x_0, y_0, z_0)$ to the sphere $(x - a)^2 + (y - b)^2 + (z - c)^2 - r^2 = 0$:

$$(x - a)(x_0 - a) + (y - b)(y_0 - b) + (z - c)(z_0 - c) - r^2 = 0$$

If P_0 does not lie on the sphere, the equation represents the *polar plane* of P_0 with respect to the sphere.
Power p of the point $P_0(x_0, y_0, z_0)$ with respect to the sphere $(x - a)^2 + (y - b)^2 + (z - c)^2 - r^2 = 0$

$$p = (x_0 - a)^2 + (y_0 - b)^2 + (z_0 - c)^2 - r^2$$

Power plane

$$\left.\begin{aligned} K_1 &\equiv (x - a_1)^2 + (y - b_1)^2 + (z - c_1)^2 - r_1^2 = 0 \\ K_2 &\equiv (x - a_2)^2 + (y - b_2)^2 + (z - c_2)^2 - r_2^2 = 0 \end{aligned}\right\} \text{given spheres}$$

$$K_1 - K_2 = 0$$

The power plane is perpendicular to the line of centers of the two spheres.

Radical axis with respect to three spheres

$$K_1 = 0; \quad K_2 = 0; \quad K_3 = 0 \quad \text{three given spheres}$$

$$\left.\begin{aligned} K_1 - K_2 &= 0 \\ K_1 - K_3 &= 0 \end{aligned}\right\} \text{radical axis}$$

Ellipsoid

$$\frac{x^2}{a^2} + \frac{y^2}{b^2} + \frac{z^2}{c^2} = 1$$

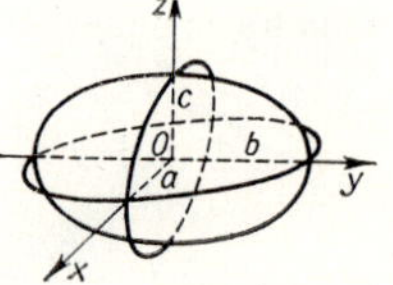

Center at the origin; a, b, c semiaxes of the principal sections

If two axes are equal, we have an ellipsoid of revolution.
If three axes are equal, we have a sphere.

Polar plane to pole $P_0(x_0, y_0, z_0)$:

$$\frac{xx_0}{a^2} + \frac{yy_0}{b^2} + \frac{zz_0}{c^2} = 1$$

If P_0 lies on the surface of the ellipsoid, the equation represents the tangential plane.

Diametral plane:

$$\frac{x \cos\alpha}{a^2} + \frac{y \cos\beta}{b^2} + \frac{z \cos\gamma}{c^2} = 0$$

α, β, γ angles of direction of the associated diameters

Three *conjugate diameters:*

$$\frac{\cos\alpha_1 \cos\alpha_2}{a^2} + \frac{\cos\beta_1 \cos\beta_2}{b^2} + \frac{\cos\gamma_1 \cos\gamma_2}{c^2} = 0$$

$$\frac{\cos\alpha_2 \cos\alpha_3}{a^2} + \frac{\cos\beta_2 \cos\beta_3}{b^2} + \frac{\cos\gamma_2 \cos\gamma_3}{c^2} = 0$$

$$\frac{\cos\alpha_3 \cos\alpha_1}{a^2} + \frac{\cos\beta_3 \cos\beta_1}{b^2} + \frac{\cos\gamma_3 \cos\gamma_1}{c^2} = 0$$

$\alpha_1, \beta_1, \gamma_1; \alpha_2, \beta_2, \gamma_2; \alpha_3, \beta_3, \gamma_3$ angles of direction of the three conjugate diameters

Each plane cuts the ellipsoid in a real or imaginary ellipse.

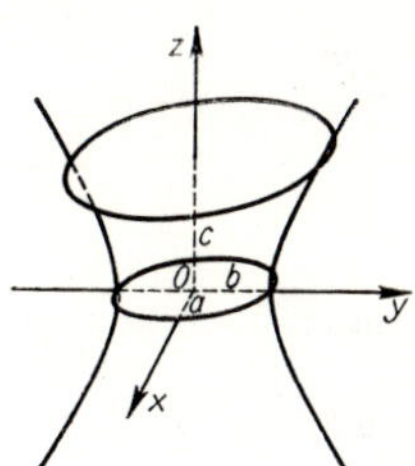

Hyperboloid

$$\frac{x^2}{a^2} + \frac{y^2}{b^2} - \frac{z^2}{c^2} - 1 = 0$$

Hyperboloid of *one sheet*
(a, b transverse semiaxes, c conjugate semiaxis)

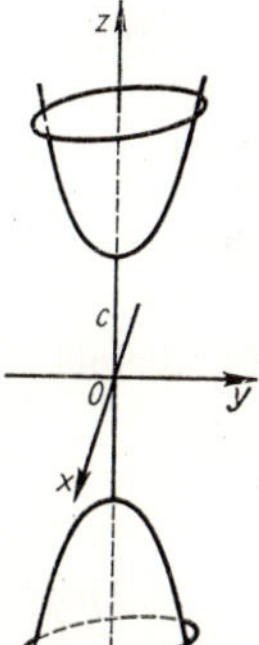

$$\frac{x^2}{a^2} + \frac{y^2}{b^2} - \frac{z^2}{c^2} + 1 = 0$$

Hyperboloid of *two sheets*

(c transverse semiaxis, a and b conjugate semiaxes)

$$\frac{x^2}{a^2} + \frac{y^2}{b^2} - \frac{z^2}{c^2} = 0$$ *Asymptotic cone* for both hyperboloids

$a = b$ hyperboloid of revolution (z-axis is axis of revolution)

a, b, c semiaxes of the principal sections

Diametral plane:

$$\frac{x \cos \alpha}{a^2} + \frac{y \cos \beta}{b^2} - \frac{z \cos \gamma}{c^2} = 0$$ applies to both hyperboloids

α, β, γ angles of direction of the associated diameter

Polar plane to pole $P_0(x_0, y_0, z_0)$:

$$\frac{xx_0}{a^2} + \frac{yy_0}{b^2} - \frac{zz_0}{c^2} \pm 1 = 0$$ positive sign for hyperboloid of two sheets, negative sign for hyperboloid of one sheet

If P_0 lies in the surface, the equation represents the *tangential plane.*
The straight rulings of the hyperboloid of one sheet:

$$\text{First family} \quad \begin{cases} \dfrac{x}{a} + \dfrac{z}{c} = \varkappa\left(1 + \dfrac{y}{b}\right) \\ \dfrac{x}{a} - \dfrac{z}{c} = \dfrac{1}{\varkappa}\left(1 - \dfrac{y}{b}\right) \end{cases} \qquad \varkappa \in R \setminus \{0\}$$

$$\text{Second family} \begin{cases} \dfrac{x}{a} + \dfrac{z}{c} = \lambda\left(1 - \dfrac{y}{b}\right) \\ \dfrac{x}{a} - \dfrac{z}{c} = \dfrac{1}{\lambda}\left(1 + \dfrac{y}{b}\right) \end{cases} \qquad \lambda \in R \setminus \{0\}$$

Each ruling of the first family intersects each ruling of the second family.
A plane cuts the hyperboloid in a hyperbola, a parabola or ellipse, depending on whether the plane is parallel to two elements, to one element, or to no element of the asymptotic cone.

Paraboloid

$$\frac{x^2}{a^2} + \frac{y^2}{b^2} - 2z = 0 \qquad \frac{x^2}{a^2} - \frac{y^2}{b^2} - 2z = 0$$

Elliptical paraboloid — *Hyperbolic* paraboloid

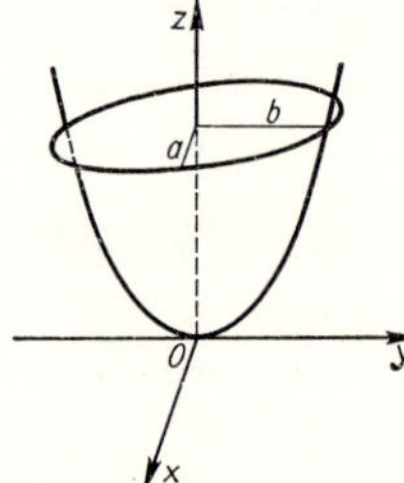

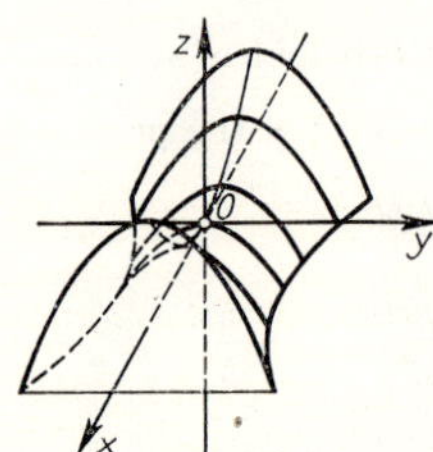

The axis of the paraboloid is the z-axis.
Vertex at the origin
$a = b$ in an elliptical paraboloid yields a *paraboloid of revolution* with the z-axis as axis of revolution.
The x,y-plane is the tangential plane to the two paraboloids at the origin.
Each plane parallel to the z-axis cuts the elliptical paraboloid in a parabola, each of the other planes cuts it in a real or imaginary ellipse.
Each plane parallel to the z-axis cuts the hyperbolic paraboloid in a parabola, each of the other planes cuts it in a hyperbola. The x,y-plane and the y,z-plane are planes of symmetry for both paraboloids.

Tangential plane in $P_0(x_0, y_0, z_0)$:

$$\frac{xx_0}{a^2} \pm \frac{yy_0}{b^2} - (z + z_0) = 0$$

If P_0 does not lie in the surface, then the equation represents the *polar plane*.

The straight *rulings* of the hyperbolic paraboloid:

First family $\begin{cases} \dfrac{x}{a} + \dfrac{y}{b} = \varkappa \\ \dfrac{x}{a} - \dfrac{y}{b} = \dfrac{1}{\varkappa} 2z \end{cases}$ parallel to the plane $\dfrac{x}{a} + \dfrac{y}{b} = 0$

Second family $\begin{cases} \dfrac{x}{a} - \dfrac{y}{b} = \lambda \\ \dfrac{x}{a} + \dfrac{y}{b} = \dfrac{1}{\lambda} 2z \end{cases}$ parallel to the plane $\dfrac{x}{a} - \dfrac{y}{b} = 0$

$\varkappa, \lambda \in R \setminus \{0\}$

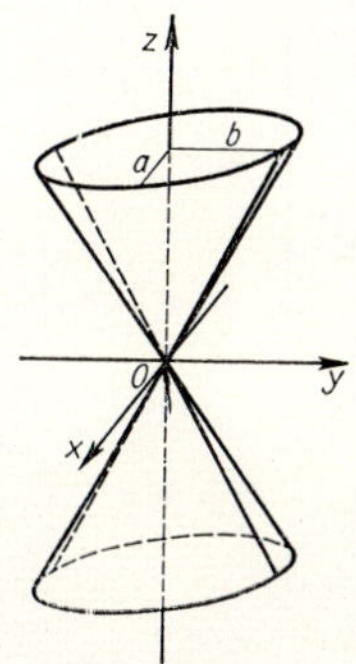

Cone

$$\frac{x^2}{a^2} + \frac{y^2}{b^2} - \frac{z^2}{c^2} = 0 \quad \text{(I)}$$

a, b semiaxes of the ellipse which is the directing curve of the cone and whose plane is perpendicular to the z-axis

c distance of the plane of the ellipse from the x,y-plane;

vertex at origin

Tangential plane to the cone (I) in $P_0(x_0, y_0, z_0)$-

$$\frac{xx_0}{a^2} + \frac{yy_0}{b^2} - \frac{zz_0}{c^2} = 0$$

Right circular cone

For $a = b$, the directing curve becomes the circle; the surface area of the circular cone then is represented by the equation

$$\frac{x^2 + y^2}{a^2} - \frac{z^2}{c^2} = 0$$

and its *tangential plane*

$$\frac{xx_0 + yy_0}{a^2} - \frac{zz_0}{c^2} = 0$$

Equations of the *rulings* of the cone:

$$\begin{cases} \dfrac{x}{a} + \dfrac{z}{c} = \dfrac{1}{\lambda}\dfrac{y}{b} \\ \dfrac{x}{a} - \dfrac{z}{c} = -\lambda\dfrac{y}{b} \end{cases}$$

The regulus passes through the vertex of the cone
$\lambda \in R \setminus \{0\}$

Cylinder

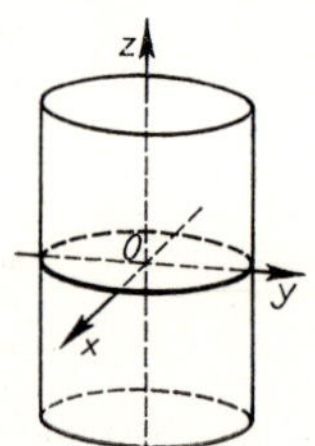

$$\frac{x^2}{a^2} + \frac{y^2}{b^2} = 1$$

Elliptical cylinder perpendicular to the x,y-plane.

x,z- and y,z-plane are planes of symmetry.

The equation is tantamount to the ellipse as a section in the x,y-plane.

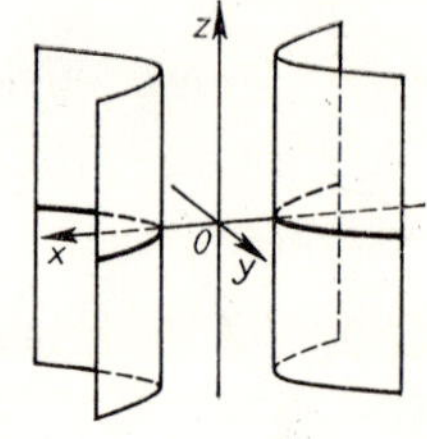

$$\frac{x^2}{a^2} - \frac{y^2}{b^2} = 1$$

Hyperbolic cylinder perpendicular to the x,y-plane.

x,z- and y,z-planes are planes of symmetry.

The section in the x,y-plane is a hyperbola.

$y^2 = 2px$ *parabolic cylinder* perpendicular to the x,y-plane

The x,z-plane is a plane of symmetry.

The y,z-plane is tangential plane which touches the surface in the z-axis.

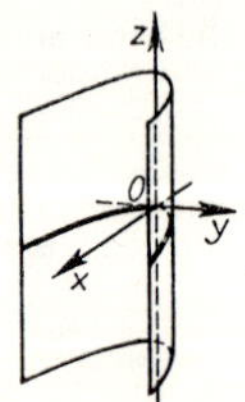

Sections perpendicular to the x,y-plane result in a real or imaginary pair of straight lines. Any other plane cuts the elliptical, hyperbolic or parabolic cylinder so that an ellipse, a hyperbola or parabola is obtained.

Equation of the *tangential plane* in $P_0(x_0, y_0, z_0)$

for an elliptical cylinder $\dfrac{xx_0}{a^2} + \dfrac{yy_0}{b^2} = 1$

for a hyperbolic cylinder $\frac{xx_0}{a^2} - \frac{yy_0}{b^2} = 1$

for a parabolic cylinder $yy_0 = p(x + x_0)$

4.2.6. The general equation of the second degree in x, y, and z

$$f(x, y, z) = a_{11}x^2 + 2a_{12}xy + a_{22}y^2 + 2a_{13}xz + 2a_{23}yz + a_{33}z^2 + 2a_{14}x + 2a_{24}y + 2a_{34}z + a_{44} = 0$$

The *invariants* of the surface of the second order:

$$\Delta = \begin{vmatrix} a_{11} & a_{12} & a_{13} & a_{14} \\ a_{21} & a_{22} & a_{23} & a_{24} \\ a_{31} & a_{32} & a_{33} & a_{34} \\ a_{41} & a_{42} & a_{43} & a_{44} \end{vmatrix}; \qquad \delta = \begin{vmatrix} a_{11} & a_{12} & a_{13} \\ a_{21} & a_{22} & a_{23} \\ a_{31} & a_{32} & a_{33} \end{vmatrix}$$

$$s = a_{11} + a_{22} + a_{33}$$

$$t = \begin{vmatrix} a_{11} & a_{12} \\ a_{12} & a_{22} \end{vmatrix} + \begin{vmatrix} a_{22} & a_{23} \\ a_{23} & a_{33} \end{vmatrix} + \begin{vmatrix} a_{33} & a_{31} \\ a_{31} & a_{11} \end{vmatrix}$$

Note: In the determinants, the relation $a_{ik} = a_{ki}$ holds.

The general functional equation of the second degree represents a **surface of the second order.** It has a finite center if $\delta \neq 0$ (so-called *central surface*). Chords through the center are called *diameters.* The locus of the midpoints of parallel chords is a *diametral plane.* The diameter associated with the chords is conjugate with the diametral plane.

The surface of the second order *degenerates* into a pair of planes if

$$\begin{vmatrix} a_{11} & a_{12} & a_{14} \\ a_{21} & a_{22} & a_{24} \\ a_{41} & a_{42} & a_{44} \end{vmatrix} + \begin{vmatrix} a_{11} & a_{13} & a_{14} \\ a_{31} & a_{33} & a_{34} \\ a_{41} & a_{43} & a_{44} \end{vmatrix} + \begin{vmatrix} a_{22} & a_{23} & a_{24} \\ a_{32} & a_{33} & a_{34} \\ a_{42} & a_{43} & a_{44} \end{vmatrix} = 0$$

The following conditions hold for nondegenerate surfaces of the second order:

Case 1: $\delta \neq 0$ (central planes)

	$\delta s > 0, \quad t > 0$	δs and t not both > 0
$\Delta < 0$	ellipsoid	hyperboloid of two sheets
$\Delta > 0$	imaginary ellipsoid	hyperboloid of one sheet
$\Delta = 0$	imaginary cone	cone

Case 2: $\delta = 0$

<table>
<tr><td></td><td colspan="2">$\Delta < 0, \quad t > 0$</td><td colspan="2">$\Delta > 0, \quad t < 0$</td></tr>
<tr><td>$\Delta \neq 0$</td><td colspan="2">elliptical paraboloid</td><td colspan="2">hyperbolic paraboloid</td></tr>
<tr><td></td><td>$t > 0$</td><td colspan="2">$t < 0$</td><td>$t = 0$</td></tr>
<tr><td>$\Delta = 0$</td><td>elliptical cylinder</td><td colspan="2">hyperbolic cylinder</td><td>parabolic cylinder</td></tr>
</table>

5. Differential calculus

5.1. Limits

A sequence of numbers $\{a_k\}$ has limit C for sufficiently large k if, with respect to an arbitrary number $\varepsilon > 0$, a $K = K(\varepsilon)$ can be stated in such a way that for all $k > K$ the following holds:

$$|a_k - C| < \varepsilon$$

We write: $\lim\limits_{k\to\infty} a_k = C$; $\{a_k\} \underset{k\to\infty}{\to} C = \{a_k - C\}$ null sequence

A sequence is said to be convergent if the limit exists, otherwise it is said to be divergent.

A function $y = f(x)$ has limit C at the point $x = c$ if, with unlimited approach of x to the point c, the function $f(x)$ approaches the value C uniquely.

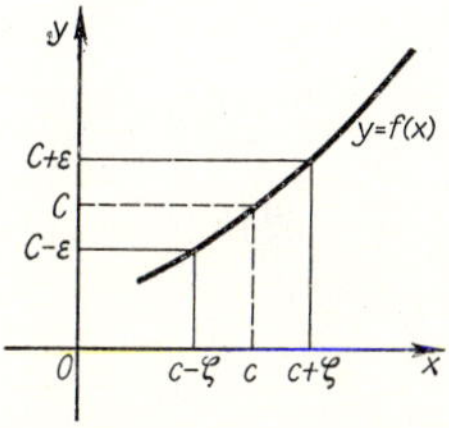

We write: $\lim\limits_{x\to c} f(x) = C$

Exact definition

A function $f(x)$ has the limit C at the point $x = c$ if, for some quantity $\varepsilon > 0$, however small, a number $\zeta > 0$ can always be found so that function values in the interval $f(x) \in (C - \varepsilon;\ C + \varepsilon)$ correspond to x-values in the interval $x \in (c - \zeta;\ c + \zeta)$.

$\lim\limits_{x\to c} f(x)$ must not be mistaken for $f(c)$.

Left-hand and right-hand limits

C is a *left-hand* limit if, with increasing x-values, the function $f(x)$ approaches the value C unlimitedly.

We write: $C = \lim\limits_{x\to c-0} f(x)$ or $C = \lim\limits_{\substack{x\to c\\ x<c}} f(x)$.

C is a *right-hand* limit if, with decreasing x-values, the function $f(x)$ approaches the value C unlimitedly.

We write: $C = \lim\limits_{x\to c+0} f(x)$ or $C = \lim\limits_{\substack{x\to c\\ x>c}} f(x)$.

Calculating with limits

On condition that the limiting values occurring in the rules exist, the following holds:

$$\lim_{x\to a} [f(x) \pm g(x)] = \lim_{x\to a} f(x) \pm \lim_{x\to a} g(x)$$

$$\lim_{x\to a} [(fx)\, g(x)] = \lim_{x\to a} f(x) \lim_{x\to a} g(x)$$

$$\lim_{x\to a} [cf(x)] = c \lim_{x\to a} f(x)$$

$$\lim_{x\to a} \frac{f(x)}{g(x)} = \frac{\lim\limits_{x\to a} f(x)}{\lim\limits_{x\to a} g(x)} \quad \text{for} \quad \lim_{x\to a} g(x) \neq 0$$

$$\lim_{x\to a} \sqrt[n]{f(x)} = \sqrt[n]{\lim_{x\to a} f(x)}$$

$$\lim_{x\to a} [f(x)]^n = \left[\lim_{x\to a} f(x)\right]^n$$

$$\lim_{x\to a} c^{f(x)} = c^{\lim\limits_{x\to a} f(x)} \qquad c \in R$$

$$\lim_{x\to a} [\log_c f(x)] = \log_c \left[\lim_{x\to a} f(x)\right]$$

If $g(x) < f(x) < h(x)$ and $\lim\limits_{x\to a} g(x) = C$, $\lim\limits_{x\to a} h(x) = C$, then $\lim\limits_{x\to a} f(x) = C$ holds.

Examples of limits:

$$\lim_{n\to\infty} \frac{1}{1+a^n} = \begin{cases} 1 & \text{for } |a| < 1 \\ \frac{1}{2} & \text{for } a = 1, \qquad \text{divergent for } a = -1 \\ 0 & \text{for } |a| > 1 \end{cases}$$

$$\lim_{x\to +0} \arctan \frac{1}{x} = \frac{\pi}{2}$$

$$\lim_{x\to -0} \arctan \frac{1}{x} = -\frac{\pi}{2}$$

$$\lim_{x\to 0} \frac{\sin x}{x} = 1; \qquad \lim_{n\to 0} \frac{\sin nx}{n} = x$$

$$\lim_{x\to\infty} \frac{\sin x}{x} = 0$$

$$\lim_{x\to 0} \frac{\tan x}{n} = 1; \qquad \lim_{n\to 0} \frac{\tan nx}{n} = x$$

$$\lim_{h\to 0} (1 + h)^{\frac{1}{h}} = \lim_{n\to\infty} \left(1 + \frac{1}{n}\right)^n = e = 2.71828\ldots$$

$$\lim_{n\to\infty} \left(1 + \frac{x}{n}\right)^n = e^x$$

$$\lim_{x\to\infty} \frac{x^n}{a^x} = 0 \quad \text{for} \quad a > 1; \quad n \in N; \quad \lim_{x\to\infty} \frac{x^n}{e^x} = 0$$

$$\lim_{n\to\infty} \frac{a^n}{n!} = 0; \qquad \lim_{x\to a} \frac{x^n - a^n}{x - a} = na^{n-1}$$

$$\lim_{x\to\infty} \frac{a^x - 1}{x} = \ln a \quad \text{for} \quad a > 0$$

$$\lim_{x\to\infty} \left(1 + \frac{1}{2} + \frac{1}{3} + \cdots + \frac{1}{n} - \ln n\right) = C = 0.5772\ldots$$

(*Euler's constant*)

$$\lim_{x\to\infty} \frac{n!}{n^n e^{-n} \sqrt{n}} = \sqrt{2\pi} \qquad (\textit{Stirling's formula})$$

$$\lim_{x\to\infty} \left[\frac{2\cdot 4\cdot 6\cdots(2n)}{1\cdot 3\cdot 5\cdots(2n-1)}\right]^2 \cdot \frac{1}{2n} = \frac{\pi}{2} \qquad (\textit{Wallis's product})$$

$$\lim_{x\to 0} \frac{\sin x}{x\sqrt[3]{\cos x}} = 1 \qquad (\textit{Maskelyne's rule})$$

5.2. Difference quotient, differential quotient, differential

We write: *Difference quotient* $\dfrac{\Delta y}{\Delta x}$

Differential quotient $\dfrac{dy}{dx}$, $\quad y', f'(x), \dfrac{df(x)}{dx}$

$$\frac{\Delta y}{\Delta x} = \frac{y - y_0}{x - x_0} = \frac{f(x) - f(x_0)}{x - x_0}$$

$$= \frac{f(x_0 + \Delta x) - f(x_0)}{\Delta x} = \tan \sigma_0$$

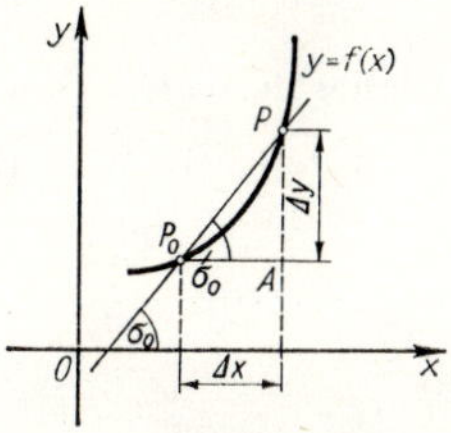

The difference quotient indicates the *gradient of the secant* P_0P. The triangle P_0AP is called *secant triangle.*

$$\frac{dy}{dx} = \lim_{\Delta x \to 0} \frac{\Delta y}{\Delta x} = \lim_{x \to x_0} \frac{f(x) - f(x_0)}{x - x_0} = \lim_{\Delta x \to 0} \frac{f(x_0 + \Delta x) - f(x_0)}{\Delta x}$$

$$= \tan \tau_0 \qquad x_0 \in X; \qquad y_0 = f(x_0) \in Y$$

The differential quotient (the *derivative*) of the function at the point x_0 indicates the *gradient of the tangent* in point P_0. The triangle P_0BC is called *tangent triangle.*

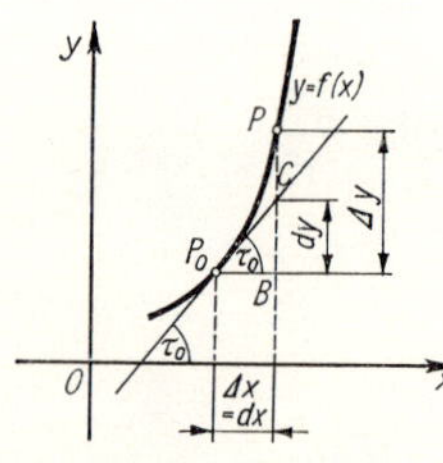

The function $y = f(x)$ is differentiable at the point $x_0 \in X$ if it is defined there and if the limit $\dfrac{dy}{dx}$ exists at this point.

The function $y = f(x)$ is differentiable in an interval which belongs to the domain of definition ($x \in X$) if it is differentiable at every point of the interval.

Every function which is differentiable at the point x_0 is continuous there (this does not hold vice versa!).

Left-hand derivative $\displaystyle\lim_{x \to x_0 - 0} \frac{f(x) - f(x_0)}{x - x_0}$

Right-hand derivative $\displaystyle\lim_{x \to x_0 + 0} \frac{f(x) - f(x_0)}{x - x_0}$

If, at the point x_0, right-hand and left-hand derivatives differ, the function is not differentiable at x_0.

Concept of the differential

$$\frac{dy}{dx} = f'(x) \Rightarrow dy = f'(x)\,dx$$

dy is called the differential of y that belongs to the differential dx.

Higher derivatives and differentials

Each derivative that represents itself a differentiable function of x can again be differentiated.

Second derivative:

$$\frac{d^2y}{dx^2} = y'' = f''(x) = \frac{d^2f(x)}{dx^2} = \frac{df'(x)}{dx}$$

Third derivative:

$$\frac{d^3y}{dx^3} = y''' = f'''(x) = \frac{d^3f(x)}{dx^3} = \frac{df''(x)}{dx}$$

nth derivatives:

$$\frac{d^ny}{dx^n} = y^{(n)} = f^{(n)}(x) = \frac{d^nf(x)}{dx^n} = \frac{df^{(n-1)}(x)}{dx}$$

Accordingly,

second differential: $d^2y = d(dy) \quad = f''(x)\,dx^2$

third differential: $d^3y = d(d^2y) \quad = f'''(x)\,dx^3$

$\vdots \qquad\qquad \vdots$

nth differential: $d^n\,y = d(d^{n-1}y) = f^n(x)\,dx^n$

5.3. Rules for differentiation

1. $y = C; \quad y' = 0$

2. $y = u_1 + u_2 + \cdots + u_n; \quad y' = u_1' + u_2' + \cdots + u_n'$

   $u_1, u_2, \ldots, u_n$ functions of x

3. $y = uv; \quad y' = u'v + uv'$ (*product rule*)

   $y = uvw; \quad y' = u'vw + uv'w + uvw'$

4. $y = cu; \quad y' = cu'$

5. $y = \frac{u}{v}; \quad y' = \frac{u'v - uv'}{v^2}$ (*quotient rule*) for $v \neq 0$

6. $y = f(u); \quad u = g(v); \quad v = h(x)$

   $\frac{dy}{dx} = f'(u)\,g'(v)\,h'(x)$ (*chain rule*)

7. Simplified case: $y = f(u); \quad u = g(x)$

   $y = f(u)$ *outer function*, $u = g(x)$ *inner function*

   $\frac{dy}{dx} = f'(u)\,g'(x)$

8. If $x = g(y)$ is the inverse of $y = f(x)$, the following holds

$$g'(y) = \frac{1}{f'(x)} \quad \text{or} \quad \frac{dx}{dy} = \frac{1}{\frac{dy}{dx}} \quad \text{for} \quad \begin{array}{l} f'(x) \neq 0 \\ g'(y) \neq 0 \end{array}$$

Examples:

1. $y = -9; \quad \underline{\underline{y' = 0}}$

2. $y = x^5 + x^2 - x^7; \quad \underline{\underline{y' = 5x^4 + 2x - 7x^6}}$

3. $y = (x^3 + a)(x^2 + 3b)$ $\quad \left| \begin{array}{ll} x^3 + a = u; & u' = 3x^2 \\ x^2 + 3b = v; & v' = 2x \end{array} \right|$

$$y' = 3x^2(x^2 + 3b) + (x^3 + a)\,2x$$

$$\underline{\underline{y' = 5x^4 + 9bx^2 + 2ax}}$$

4. $y = 10x^6; \quad y' = 10 \cdot 6x^5$

$$\underline{\underline{y' = 60x^5}}$$

5. $y = \dfrac{x^3 + 2x}{4x^2 - 7}; \quad \begin{array}{ll} x^3 + 2x = u; & u' = 3x^2 + 2 \\ 4x^2 - 7 = v; & v' = 8x \end{array}$

$$y' = \frac{(3x^2 + 2)(4x^2 - 7) - (x^3 + 2x)\,8x}{(4x^2 - 7)^2}$$

$$\underline{\underline{y' = \frac{4x^4 - 29x^2 - 14}{(4x^2 - 7)^2}}}$$

6. $y = (1 - \cos^4 x)^2 = u^2 = f(u), \quad \text{where} \quad u = 1 - \cos^4 x$

$$f'(u) = 2u = 2(1 - \cos^4 x)$$

$$u = 1 - \cos^4 x = 1 - v^4 = g(v), \quad \text{where} \quad v = \cos x$$

$$g'(v) = -4v^3 = -4 \cos^3 x$$

$$v = \cos x = h(x); \quad h'(x) = -\sin x$$

$$\frac{dy}{dx} = f'(u)\,g'(v)\,h'(x) = 2(1 - \cos^4 x)(-4 \cos^3 x)(-\sin x)$$

$$\underline{\underline{\frac{dy}{dx} = 8 \sin x \cos^3 x(1 - \cos^4 x)}}$$

7. $y = \text{arc tan}\, x$ inversion: $x = \tan y = g(y)$

$$g'(y) = \frac{1}{\cos^2 y} = 1 + \tan^2 y$$

$$y' = f'(x) = \frac{1}{g'(y)} = \frac{1}{1 + \tan^2 y} = \underline{\underline{\frac{1}{1 + x^2}}}$$

Logarithmic differentiation

Occasionally it is of advantage to determine the derivative of the function $y = f(x)$ by finding the natural logarithm of the function first and then to differentiate with respect to x (logarithmic derivative).

This procedure is particularly convenient in the case of functions in the form $y = u(x)^{v(x)}$.

$$\ln y = v(x) \ln u(x)$$

$$\frac{1}{y} y' = v'(x) \ln u(x) + v(x) \frac{u'(x)}{u(x)}$$

$$y' = y \left[v'(x) \ln u(x) + v(x) \frac{u'(x)}{u(x)} \right]$$

$$y' = u(x)^{v(x)} \left[v'(x) \ln u(x) + v(x) \frac{u'(x)}{u(x)} \right]$$

Example:

$$y = (\text{arc tan}\, x)^x$$

$$\ln y = x \ln (\text{arc tan}\, x)$$

$$\frac{1}{y} y' = \ln (\text{arc tan}\, x) + \frac{x}{\text{arc tan}\, x} \cdot \frac{1}{1 + x^2}$$

$$y' = (\text{arc tan}\, x)^x \left[\ln (\text{arc tan}\, x) + \frac{x}{(1 + x^2)\, \text{arc tan}\, x} \right]$$

Differentiation of functions of several variables $z = f(x, y)$

Partial derivatives

$$\frac{\partial z}{\partial x} = z_x = f_x = \lim_{\Delta x \to 0} \frac{f(x + \Delta x, y) - f(x, y)}{\Delta x}$$

$$\frac{\partial z}{\partial y} = z_y = f_y = \lim_{\Delta y \to 0} \frac{f(x, y + \Delta y) - f(x, y)}{\Delta y}$$

$$\frac{\partial^2 z}{\partial x^2} = \frac{\partial \left(\dfrac{\partial z}{\partial x}\right)}{\partial x} = f_{xx} = z_{xx}$$

$$\frac{\partial^2 z}{\partial y^2} = \frac{\partial \left(\dfrac{\partial z}{\partial y}\right)}{\partial y} = f_{yy} = z_{yy}$$

$$\frac{\partial^2 z}{\partial x\, \partial y} = \frac{\partial \left(\dfrac{\partial z}{\partial x}\right)}{\partial y} = f_{xy} = z_{xy}$$

$$\frac{\partial^2 z}{\partial y\, \partial x} = \frac{\partial \left(\dfrac{\partial z}{\partial y}\right)}{\partial x} = f_{yx} = z_{yx}$$

On condition that the latter two derivatives are continuous at the point x, y, the **theorem of Schwarz** holds:

$$\frac{\partial^2 z}{\partial x\, \partial y} = \frac{\partial^2 z}{\partial y\, \partial x}$$

Total derivatives:

$$\frac{dz}{dx} = \frac{\partial z}{\partial x} + \frac{\partial z}{\partial y} \frac{dy}{dx}$$

$$\frac{d^2 z}{dx^2} = \frac{\partial^2 z}{\partial x^2} + 2 \frac{\partial^2 z}{\partial x\, \partial y} \frac{dy}{dx} + \frac{\partial^2 z}{\partial y^2} \left(\frac{dy}{dx}\right)^2$$

Total differentials

$$dz = \frac{\partial z}{\partial x}\, dx + \frac{\partial z}{\partial y}\, dy$$

$$d^2 z = \frac{\partial^2 z}{\partial x^2}\, \partial x^2 + 2 \frac{\partial^2 z}{\partial x\, \partial y}\, dx\, dy + \frac{\partial^2 z}{\partial y^2}\, dy^2$$

Example:

$$z = y^2 e^x$$

$$\frac{\partial z}{\partial x} = y^2 e^x; \qquad \frac{\partial z}{\partial y} = 2ye^x$$

$$\frac{\partial^2 z}{\partial x^2} = y^2 e^x; \qquad \frac{\partial^2 z}{\partial y^2} = 2e^x; \qquad \frac{\partial^2 z}{\partial x\, \partial y} = 2ye^x$$

Total differentials

$$dz = y^2 e^x\, dx + 2ye^x\, dy = ye^x(y\, dx + 2dy)$$

$$d^2 z = y^2 e^x\, dx^2 + 2 \cdot 2ye^x\, dx\, dy + 2e^x\, dy^2$$

$$= e^x(y^2\, dx^2 + 4y\, dx\, dy + 2dy^2)$$

Differentiation of implicit functions $\boldsymbol{f(x, y) = 0}$

$$\frac{dy}{dx} = y' = -\frac{\dfrac{\partial f}{\partial x}}{\dfrac{\partial f}{\partial y}} = -\frac{f_x}{f_y}$$

$$\frac{d^2 y}{dx^2} = y'' = -\frac{\dfrac{\partial^2 f}{\partial x^2}\left(\dfrac{\partial f}{\partial y}\right)^2 - 2\dfrac{\partial^2 f}{\partial x\, \partial y} \cdot \dfrac{\partial f}{\partial x} \cdot \dfrac{\partial f}{\partial y} + \dfrac{\partial^2 f}{\partial y^2}\left(\dfrac{\partial f}{\partial x}\right)^2}{\left(\dfrac{\partial f}{\partial y}\right)^3}$$

$$= -\frac{f_{xx} f_y^{\,2} - 2 f_{xy} f_x f_y + f_{yy} f_x^{\,2}}{f_y^{\,3}}$$

Example:

$$f(x, y) = x^3 - x^2 y + y^5 = 0$$

$$\frac{\partial f}{\partial x} = 3x^2 - 2xy = f_x \qquad \frac{\partial f}{\partial y} = 5y^4 - x^2 = f_y$$

$$\frac{\partial^2 f}{\partial x^2} = 6x - 2y = f_{xx} \qquad \frac{\partial^2 f}{\partial y^2} = 20y^3 = f_{yy}$$

$$\frac{\partial^2 f}{\partial x\,\partial y} = -2x = f_{xy} = f_{yx}$$

$$\frac{dy}{dx} = -\frac{3x^2 - 2xy}{5y^4 - x^2}$$

$$\frac{d^2y}{dx^2} = -\frac{(6x - 2y)(5y^4 - x^2)^2 - 2(-2x)(3x^2 - 2xy)(5y^4 - x^2) + 20y^3(3x^2 - 2xy)^2}{(5y^4 - x^2)^3}$$

Differentiation of functions in parametric form

$x = (t)$; $y = \psi(t)$

$$\frac{dy}{dx} = \frac{\frac{dy}{dt}}{\frac{dx}{dt}} = \frac{\dot\psi(t)}{\dot\varphi(t)} \qquad \dot\varphi(t) \neq 0$$

$$\frac{d^2y}{dx^2} = \frac{\frac{d^2y}{dt^2}\cdot\frac{dx}{dt} - \frac{d^2x}{dt^2}\cdot\frac{dy}{dt}}{\left(\frac{dx}{dt}\right)^3} = \frac{\ddot\psi(t)\,\dot\varphi(t) - \ddot\varphi(t)\,\dot\psi(t)}{[\dot\varphi(t)]^3}$$

or

$$\frac{d^2y}{dx^2} = \frac{d\left(\frac{dy}{dx}\right)}{dt}\cdot\frac{dt}{dx}$$

Example:

$$x = \ln t$$

$$y = \frac{1}{1-t}$$

$$\frac{dx}{dt} = \frac{1}{t}; \qquad \frac{d^2x}{dt^2} = -\frac{1}{t^2}; \qquad \frac{dt}{dx} = t$$

$$\frac{dy}{dt} = \frac{1}{(1-t)^2}; \qquad \frac{d^2y}{dt^2} = \frac{2}{(1-t)^3}$$

$$\frac{dy}{dx} = \frac{1}{(1-t)^2} : \frac{1}{t} = \frac{t}{(1-t)^2}$$

$$\frac{d^2y}{dx^2} = \frac{\frac{2}{(1-t)^3} \cdot \frac{1}{t} + \frac{1}{t^2} \cdot \frac{1}{(1-t)^2}}{\left(\frac{1}{t}\right)^3} = \frac{2t^2 + t - t^2}{(1-t)^3} = \frac{t^2 + t}{(1-t)^3}$$

or

$$\frac{d^2y}{dx^2} = \frac{d}{dt}\left[\frac{t}{(1-t)^2}\right] t = \frac{1+t}{(1-t)^3} t = \frac{t^2+t}{(1-t)^3}$$

Differentiation of functions in polar coordinates $r = f(\varphi)$

Differential quotient $\frac{\Delta r}{\Delta \varphi} = \frac{r}{\tan \sigma}$ $= r \cot \sigma$, where σ represents the angle which the secant PP_1 makes with the position vector OP_1.

Differential quotient (derivative)

$$\frac{dr}{d\varphi} = \lim_{\Delta\varphi \to 0} \frac{\Delta r}{\Delta \varphi} = \frac{r}{\tan \tau} = r \cot \tau$$

τ angle which the tangent in P makes with the position vector OP.

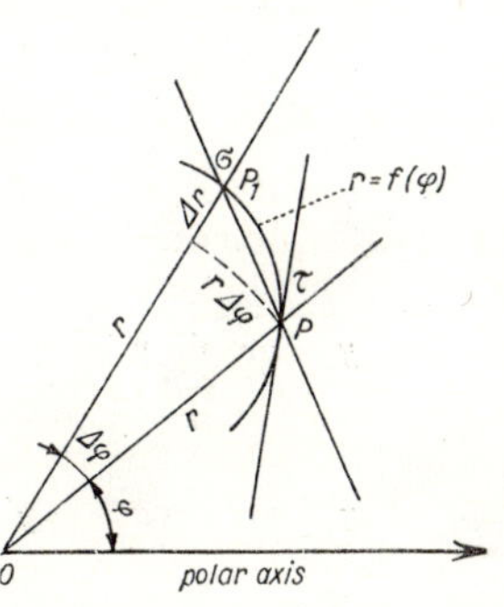

Connection

$$\left.\begin{array}{l} x = r\cos\varphi \\ y = r\sin\varphi \end{array}\right\} \quad y' = \frac{\frac{dy}{d\varphi}}{\frac{dx}{d\varphi}} = \frac{\frac{dr}{d\varphi}\sin\varphi + r\cos\varphi}{\frac{dr}{d\varphi}\cos\varphi - r\sin\varphi}$$

5.4. Derivatives of the elementary functions

$y = x^n$ $\quad y' = nx^{n-1} \quad n \in R \setminus \{0\}$

$y = x$ $\quad y' = 1$

$y = e^x$ $\quad y' = e^x$

$y = a^x$ $\quad y' = a^x \ln a$

$$y = \ln x \qquad y' = \frac{1}{x}$$

$$y = \log_a x \qquad y' = \frac{1}{x \ln a} = \frac{1}{x} \log_a e$$

$$y = \lg x \qquad y' = \frac{1}{x} \lg e \approx \frac{0.4343}{x}$$

$$y = \sin x \qquad y' = \cos x$$

$$y = \cos x \qquad y' = -\sin x$$

$$y = \tan x \qquad y' = \frac{1}{\cos^2 x} = 1 + \tan^2 x$$

$$y = \cot x \qquad y' = -\frac{1}{\sin^2 x} = -(1 + \cot^2 x)$$

$$y = \arcsin x \qquad y' = \frac{1}{\sqrt{1 - x^2}} \quad \text{for} \quad |x| < 1$$

$$y = \arccos x \qquad y' = -\frac{1}{\sqrt{1 - x^2}} \quad \text{for} \quad |x| < 1$$

$$y = \arctan x \qquad y' = \frac{1}{1 + x^2}$$

$$y = \operatorname{arc\,cot} x \qquad y' = -\frac{1}{1 + x^2}$$

$$y = \sinh x \qquad y' = \cosh x$$

$$y = \cosh x \qquad y' = \sinh x$$

$$y = \tanh x \qquad y' = \frac{1}{\cosh^2 x} = 1 - \tanh^2 x$$

$$y = \coth x \qquad y' = -\frac{1}{\sinh^2 x} = 1 - \coth^2 x$$

$$y = \sinh^{-1} x \qquad y' = \frac{1}{\sqrt{1 + x^2}}$$

$$y = \cosh^{-1} x \qquad y' = \pm \frac{1}{\sqrt{x^2 - 1}} \quad \text{for} \quad x > 1$$

$$y = \tanh^{-1} x \qquad y' = \frac{1}{1 - x^2} \quad \text{for} \quad |x| < 1$$

$$y = \coth^{-1} x \qquad y' = \frac{1}{1 - x^2} \quad \text{for} \quad |x| > 1$$

$$y = \ln f(x) \qquad y' = \frac{f'(x)}{f(x)}$$

A few derivatives of higher order

$$y = x^m \qquad y^{(n)} = m(m-1)(m-2)\cdots(m-n+1)x^{m-n}$$

$$y = x^n \qquad y^{(n)} = n! \quad \text{für} \quad n \in N$$

$$y = a_n x^n + a_{n-1}x^{n-1} + a_{n-2}x^{n-2} + \cdots + a_1 x + a_0;\ y^{(n)} = a_n n!$$

$$y = \ln x \qquad y^{(n)} = (-1)^{n+1} \frac{(n-1)!}{x^n}$$

$$y = \log_a x \qquad y^{(n)} = (-1)^{n+1} \cdot \frac{(n-1)!}{x^n \ln a}$$

$$y = e^x \qquad y^{(n)} = e^x$$

$$y = e^{mx} \qquad y^{(n)} = m^n e^{mx}$$

$$y = a^x \qquad y^{(n)} = a^x (\ln a)^n$$

$$y = \sin x \qquad y^{(n)} = \sin\left(x + \frac{n\pi}{2}\right)$$

$$y = \cos x \qquad y^{(n)} = \cos\left(x + \frac{n\pi}{2}\right)$$

$$y = \sin mx \qquad y^{(n)} = m^n \sin\left(mx + \frac{n\pi}{2}\right)$$

$$y = \cos mx \qquad y^{(n)} = m^n \cos\left(mx + \frac{n\pi}{2}\right)$$

$$y = \sinh x \qquad y^{(n)} = \begin{cases} \sinh x & \text{for even } n \\ \cosh x & \text{for odd } n \end{cases}$$

$$y = \cosh x \qquad y^{(n)} = \begin{cases} \cosh x & \text{for even } n \\ \sinh x & \text{for odd } n \end{cases}$$

$$y = uv \qquad y^{(n)} = u^{(n)}v + \binom{n}{1} u^{(n-1)}v' + \binom{n}{2} u^{(n-2)}v'' + \cdots + \binom{n}{n-1} u'v^{(n-1)} + uv^{(n)}$$

(LEIBNIZ's formula)

(The structure of the formula corresponds to the binomial theorem.)

5.5. Differentiation of a vector function

Explanation

$\mathbf{v} = \mathbf{v}(t)$ is called a vector function of the scalar variable t if a certain value of $\mathbf{v}$ corresponds to each value of t.
Vector function represented in terms of components:

$$\mathbf{v} = \mathbf{v}_x + \mathbf{v}_y + \mathbf{v}_z = v_x\mathbf{i} + v_y\mathbf{j} + v_z\mathbf{k}$$

v_x, v_y, v_z scalar function of t

Derivative of the vector function:

$$\frac{d\mathbf{v}}{dt} = \dot{\mathbf{v}} = \lim_{\Delta t \to 0} \frac{f(t + \Delta t) - f(t)}{\Delta t} = \frac{dv_x}{dt}\mathbf{i} + \frac{dv_y}{dt}\mathbf{j} + \frac{dv_z}{dt}\mathbf{k}$$

Differential of the vector function:

$$d\mathbf{v} = \dot{\mathbf{v}}\, dt$$

Rules for the differentiation of vectors:

$$\frac{d(\mathbf{v}_1 + \mathbf{v}_2 + \mathbf{v}_3)}{dt} = \frac{d\mathbf{v}_1}{dt} + \frac{d\mathbf{v}_2}{dt} + \frac{d\mathbf{v}_3}{dt}$$

$$\frac{d(g\mathbf{v})}{dt} = \frac{dg}{dt}\mathbf{v} + g\frac{d\mathbf{v}}{dt}, \qquad \text{where } g \text{ is a scalar function of } t$$

$$\frac{d(\mathbf{v}_1\mathbf{v}_2)}{dt} = \frac{d\mathbf{v}_1}{dt}\mathbf{v}_2 + \mathbf{v}_1\frac{d\mathbf{v}_2}{dt} = \frac{d\mathbf{v}_1}{dt}\mathbf{v}_2 + \frac{d\mathbf{v}_2}{dt}\mathbf{v}_1$$

$$\frac{d(\mathbf{v}_1 \times \mathbf{v}_2)}{dt} = \frac{d\mathbf{v}_1}{dt} \times \mathbf{v}_2 + \mathbf{v}_1 \times \frac{d\mathbf{v}_2}{dt} = \frac{d\mathbf{v}_1}{dt} \times \mathbf{v}_2 - \frac{d\mathbf{v}_2}{dt} \times \mathbf{v}_1$$

$$\frac{d}{dt}\mathbf{v}[\varphi(t)] = \frac{d\mathbf{v}}{d\varphi} \cdot \frac{d\varphi}{dt}$$

5.6. Graphical differentiation

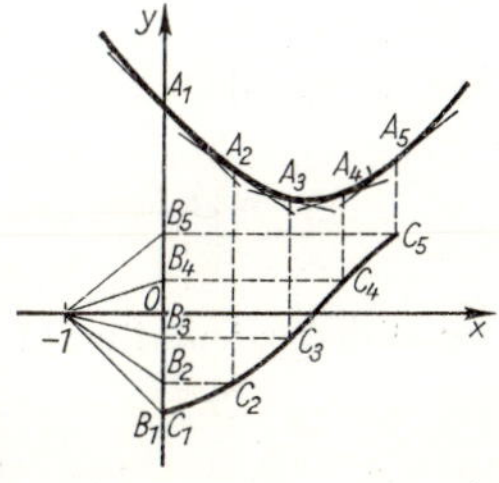

If a function $y = f(x)$ is given by its curve, the derived curve can approximately be constructed in the following way: We draw lines tangent to the original curve at a conveniently large number of points $A_1, A_2, A_3, \ldots$ of this curve and draw through an arbitrary point [point $(-1; 0)$ in the accompanying illustration], the so-called pole, lines parallel to them which intersect the y-axis in the respective points $B_1, B_2, B_3, \ldots$ Through points $B_1, B_2, B_3, \ldots$ we draw lines parallel to the x-axis which intersect the perpendiculars dropped from $A_1, A_2, A_3, \ldots$ on the x-axis in $C_1, C_2, C_3, \ldots$ The points $C_1, C_2, C_3, \ldots$ lie on the derived curve.

5.7. Extrema of functions (maxima and minima)

Extrema of functions in explicit form

At point $x = x_0$, the function $y = f(x)$ has

a **maximum** if $f'(x_0) = 0;\quad f''(x_0) < 0$

a **minimum** if $f'(x_0) = 0;\quad f''(x_0) > 0$

General: If the first $n - 1$ derivatives of the function $y = f(x)$ vanish for $x = x_0$, the function has, at point $x = x_0$, a maximum for $f^{(n)}(x_0) < 0$, a minimum (n even) for $f^{(n)}(x_0) > 0$.

Example:

Find the locations of extrema of the function $y = x^3 - 15x^2 + 48x - 3$

$$y' = 3x^2 - 30x + 48$$

$$y'' = 6x - 30$$

$$y' = 0 \Rightarrow 3x^2 - 30x + 48 = 0$$

$$x_1 = 8; \quad x_2 = 2 \quad \text{or} \quad E = \{8; 2\}$$

$$f''(x_1) = 6 \cdot 8 - 30 = 18 \Rightarrow \text{minimum}$$

$$f''(x_2) = 6 \cdot 2 - 30 = -18 \Rightarrow \text{maximum}$$

Associated extreme values: $y_1 = -67; \quad y_2 = 41$

The extrema are: $E_1(8, -67)$ minimum

$E_2(2, 41)$ maximum

Simplified calculation of extrema of fractional functions

If $f(x) = \dfrac{g(x)}{h(x)}$, then the first derivative $f'(x) = \dfrac{p(x)}{q(x)}$ also is a fractional function so that the condition for the occurrence of an extremum is $p(x) = 0;\ q(x) \neq 0$.
The kind of extremum is defined by the sign of the second derivative which takes the simple form $f''(x) = \dfrac{p'(x)}{q(x)}$ for all zeros of $p(x)$.

Example:

Find the locations of extrema of the function $y = \dfrac{2 - 3x + x^2}{2 + 3x + x^2}$,

$$y' = \frac{(-3 + 2x)(2 + 3x + x^2) - (2 - 3x + x^2)(3 + 2x)}{(2 + 3x + x^2)^2}$$

$$= \frac{6x^2 - 12}{(2 + 3x + x^2)^2} = \frac{p(x)}{q(x)}$$

$$y'' = \frac{12x}{(2 + 3x + x^2)^2} \quad \text{[in simplified form for the zeros of } p(x)]$$

$$y' = 0 \Rightarrow 6x^2 - 12 = 0 \quad \text{with} \quad E = \left\{\sqrt{2}, -\sqrt{2}\right\}$$

$$f''\left(\sqrt{2}\right) = \frac{12\sqrt{2}}{(2 + 3x + x^2)^2} > 0 \quad \Rightarrow \text{minimum}$$

$$f''\left(-\sqrt{2}\right) = \frac{-12\sqrt{2}}{(2 + 3x + x^2)^2} < 0 \quad \Rightarrow \text{maximum}$$

Associated extreme values:

$$y_1 = \frac{4 - 3\sqrt{2}}{4 + 3\sqrt{2}}; \qquad y_2 = \frac{4 + 3\sqrt{2}}{4 - 3\sqrt{2}}$$

The extrema are

$$E_1\left(\sqrt{2}, \frac{4 - 3\sqrt{2}}{4 + 3\sqrt{2}}\right) \qquad \text{minimum}$$

$$E_2\left(-\sqrt{2}, \frac{4 + 3\sqrt{2}}{4 - 3\sqrt{2}}\right) \qquad \text{maximum}$$

Extrema of implicit functions

The function $f(x, y) = 0$ has extrema if the following conditional equations apply to it:

$$\boldsymbol{f(x, y) = 0}$$

$$\boldsymbol{f_x = 0, \; f_y \neq 0}$$

A **maximum** is given if for the point in question $\boldsymbol{f_{xx} : f_y > 0}$, whereas a **minimum** is given if for the point in question $\boldsymbol{f_{xx} : f_y < 0}$.

Example:

Find the extrema of the function $f(x, y) = x^3 - 3a^2x + y^3 = 0$.

$$f_x = 3x^2 - 3a^2; \qquad f_y = 3y^2; \qquad f_{xx} = 6x$$

$$f(x, y) = 0 \quad \text{and} \quad f_x = 0 \Rightarrow x^3 - 3a^2x + y^3 = 0$$

$$3x^2 - 3a^2 = 0$$

with $E = \left\{\left(a, a\sqrt[3]{2}\right); \left(-a, -a\sqrt[3]{2}\right)\right\}$

For $\left(a, a\sqrt[3]{2}\right)$ we have:

$$\left.\begin{aligned} f_y &= 3a^2\sqrt[3]{4} \neq 0 \\ f_{xx} &= 6a \end{aligned}\right\} f_{xx}:f_y = \frac{6a}{3a^2\sqrt[3]{4}} > 0 \Rightarrow \text{maximum}$$

for $\left(-a, -a\sqrt[3]{2}\right)$ we have

$$\left.\begin{aligned} f_y &= 3a^2\sqrt[3]{4} \neq 0 \\ f_{xx} &= -6a \end{aligned}\right\} f_{xx}:f_y = \frac{-6a}{3a^2\sqrt[3]{4}} < 0 \Rightarrow \text{minimum}$$

Extrema of functions in parametric representation

The function $x = \varphi(t)$; $y = \psi(t)$ has extrema if the following conditional equations apply to it:

$$\dot{\psi}(t) = 0; \quad \dot{\varphi}(t) \neq 0$$

A maximum is given if for the point in question $\ddot{\varphi}(t) < 0$;
a minimum is given if $\ddot{\varphi}(t) > 0$.

Example:

Find the extrema of the function $x = a\cos t = \varphi(t)$

$$y = b\sin t = \psi(t)$$

$$\dot{\varphi}(t) = -a\sin t; \quad \dot{\psi}(t) = b\cos t; \quad \ddot{\psi}(t) = -b\sin t$$

$$\dot{\psi}(t) = 0 \Rightarrow b\cos t = 0 \quad \text{with} \quad t_1 = \frac{\pi}{2}; \; t_2 = \frac{3\pi}{2}$$

$$\dot{\varphi}(t_1) \neq 0; \quad \dot{\varphi}(t_2) \neq 0$$

$$\ddot{\psi}(t_1) = -b\sin\frac{\pi}{2} = -b < 0 \Rightarrow \text{maximum}$$

$$\ddot{\psi}(t_2) = +b > 0 \Rightarrow \text{minimum}$$

Associated extrema:

$$x_1 = a\cos\frac{\pi}{2} = 0; \qquad y_1 = b\sin\frac{\pi}{2} = b$$

$$x_2 = a\cos\frac{3\pi}{2} = 0; \qquad y_2 = b\sin\frac{3\pi}{2} = -b$$

We obtain $E_1 = \{0, b\}$ as a maximum and $E_2 = \{0, -b\}$ as a minimum.

Extrema of the function $z = f(x, y)$

(*Minimum* and *maximum points* of a surface)
Conditional equations

$$f_x = 0; \quad f_y = 0; \quad f_{xx}f_{yy} - (f_{xy})^2 > 0$$

Maximum for $f_{xx} < 0$

Minimum for $f_{xx} > 0$

For $f_{xx}f_{yy} - (f_{xy})^2 < 0$ the point in question is a *saddle point.*
For $f_{xx}f_{yy} - (f_{xy})^2 = 0$ it cannot be decided whether a maximum or minimum or neither of them is existing.

Maxima and minima with supplementary conditions (Method of multipliers)

If the extrema are to be found for the function $z = f(x, y)$ and the extrema are linked by the equation (supplementary condition) $\varphi(x, y) = 0$, then the following *conditional equations* hold:

$$\varphi(x, y) = 0; \quad \frac{\partial}{\partial x}[f(x, y) + \lambda\varphi(x, y)] = 0$$

$$\frac{\partial}{\partial y}[f(x, y) + \lambda\varphi(x, y)] = 0 \quad \lambda \in R$$

The three unknowns x, y, λ are found with the help of the three equations.
Decision on the kind of extremum:

$$\Delta = \frac{\partial^2(f + \lambda\varphi)}{\partial x^2}\left[\frac{\partial\varphi}{\partial y}\right]^2 - 2\,\frac{\partial^2(f + \lambda\varphi)}{\partial x\,\partial y}\cdot\frac{\partial\varphi}{\partial x}\cdot\frac{\partial\varphi}{\partial y}$$

$$+\frac{\partial^2(f + \lambda\varphi)}{\partial y^2}\left[\frac{\partial\varphi}{\partial x}\right]^2 \begin{array}{l} < 0 \Rightarrow \text{maximum} \\ > 0 \Rightarrow \text{minimum} \end{array}$$

[where f means $f(x, y)$ and φ means $\varphi(x, y)$.]

Example:

Find the extrema of the function $z = f(x, y) = x^2 + xy + y^2$. Supplementary condition:

$$\varphi(x, y) = xy - 9 = 0$$

Conditional equations from which x, y, λ can be calculated:

$$\begin{array}{l|l} xy - 9 = 0 & [\varphi(x, y) = 0] \\ 2x + y + \lambda y = 0 & \left(\frac{\partial}{\partial x}[f(x, y) + \lambda\varphi(x, y)] = 0\right) \\ x + 2y + \lambda x = 0 & \left(\frac{\partial}{\partial y}[f(x, y) + \lambda\varphi(x, y)] = 0\right) \end{array}$$

$$x_{1;2} = \pm 3$$

$$y_{1;2} = \pm 3; \qquad \lambda = -3$$

Extreme values at $P_1(3, 3, 27)$ and at $P_2(-3, -3, -27)$
Decision on the kind of extremum:

$$\frac{\partial^2(f + \lambda\varphi)}{\partial x^2} = 2; \quad \left(\frac{\partial\varphi}{\partial y}\right)^2 = x^2; \quad \frac{\partial^2(f + \lambda\varphi)}{\partial x\,\partial y} = 1 + \lambda;$$

$$\frac{\partial\varphi}{\partial x} = y; \quad \frac{\partial\varphi}{\partial y} = x; \quad \frac{\partial^2(f + \lambda\varphi)}{\partial y^2} = 2$$

Hence,

$$\Delta = 2x^2 - 2(1 + \lambda)xy + 2y^2$$

For P_1 holds:

$$\Delta = 2 \cdot 9 - 2(1 - 3) \cdot 3 \cdot 3 + 2 \cdot 9 = 72 > 0 \Rightarrow \text{minimum}$$

For P_2 holds:

$$\Delta = 2 \cdot 9 - 2(1 - 3) \cdot 9 + 2 \cdot 9 = 72 > 0 \Rightarrow \text{minimum}$$

5.8. Mean-value theorems

Rolle's theorem

If $y = f(x)$ is singlevalued, continuous in the interval $[a, b]$ and differentiable in the interval (a, b), and if $f(a) = f(b)$ in addition, then in the interval (a, b) there is at least one point $x = \xi$ for which holds

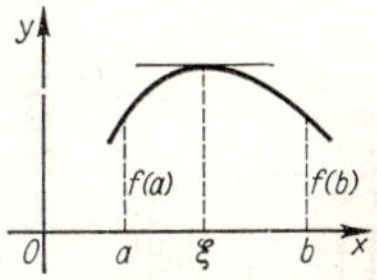

$$f'(\xi) = 0$$

Interval

(a, b) means $a < x < b$ (*open interval*)

$[a, b]$ means $a \leqq x \leqq b$ (*closed interval*)

$[a, b)$ means $a \leqq x < b$ (*half-open interval*)

$(a, b]$ means $a < x \leqq b$ (*half-open interval*)

Geometrically the theorem of ROLLE states that, in the interval under consideration, there is at least one point with a tangent parallel to the x-axis.

Mean-value theorem of the differential calculus

If $y = f(x)$ is singlevalued, continuous in the interval $[a, b]$, and differentiable in the interval (a, b), then there is at least one value $x = \xi$ in the interval for which holds

$$\frac{f(b) - f(a)}{b - a} = f'(\xi) \quad \text{for} \quad \xi \in (a, b)$$

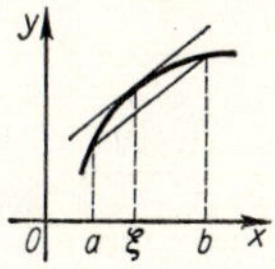

Other form

$$\frac{f(x + h) - f(x)}{h} = f'(x + \vartheta h) \quad \text{for} \quad \vartheta \in (0, 1)$$

Geometrically the mean-value theorem states that, under the given conditions in the interval, there is a point where the tangent to the curve of the chord between the endpoints of the interval is parallel to the chord.

Generalized mean-value theorem of the differential calculus

If the two functions $y = f(x)$ and $y = g(x)$ are continuous in the interval $[a, b]$ and differentiable in the interval (a, b), where $g'(x) \neq 0$, then there is in the interval under consideration at least one point $x = \xi$ for which holds

$$\frac{f(b) - f(a)}{g(b) - g(a)} = \frac{f'(\xi)}{g'(\xi)} \quad \text{for} \quad \xi \in (a, b)$$

5.9. Indeterminate expressions

Expressions of the form $\frac{\text{“}0\text{”}}{0}$ or $\frac{\text{“}\infty\text{”}}{\infty}$

If $\varphi(x) = \frac{f(x)}{g(x)}$ for $x = a$ assumes the indeterminate form $\frac{\text{“}0\text{”}}{0}$ or $\frac{\text{“}\infty\text{”}}{\infty}$, **l'Hospital's rule** applies:

$$\lim_{x \to a} \frac{f(x)}{g(x)} = \lim_{x \to a} \frac{f'(x)}{g'(x)}$$

if this limit exists.
If the new limit again appears in an indeterminate form, the procedure must be repeated.

Example:

$$\varphi(x) = \frac{\sin 2x - 2 \sin x}{2e^x - x^2 - 2x - 2}; \qquad \varphi(0) \text{ has the form } \frac{\text{“}0\text{”}}{0}.$$

$$\lim_{x \to 0} \varphi(x) = \lim_{x \to 0} \frac{2 \cos 2x - 2 \cos x}{2e^x - 2x - 2}$$

$$= \lim_{x \to 0} \frac{-4 \sin 2x + 2 \sin x}{2e^x - 2}$$

$$= \lim_{x \to 0} \frac{-8 \cos 2x + 2 \cos x}{2e^x} = -3$$

Expressions of the form "$0 \cdot \infty$"

If $\varphi(x) = f(x)g(x)$ for $x = a$ takes the indeterminate form "$0 \cdot \infty$", then

$$\lim_{x\to a} \varphi(x) = \lim_{x\to a} \frac{f(x)}{\dfrac{1}{g(x)}} \quad \text{or} \quad \lim_{x\to a} \frac{g(x)}{\dfrac{1}{f(x)}}$$

holds whereby the case is reduced to the form $\dfrac{\text{"}0\text{"}}{0}$ or $\dfrac{\text{"}\infty\text{"}}{\infty}$.

Example:

$$\varphi(x) = (1 - \sin x) \tan x; \varphi\left(\frac{\pi}{2}\right) \text{ then has the form "}0 \cdot \infty\text{".}$$

$$\lim_{x\to\frac{\pi}{2}} \frac{1 - \sin x}{\dfrac{1}{\tan x}} = \lim_{x\to\frac{\pi}{2}} \frac{-\cos x}{-\dfrac{1}{1 \tan^2 x} \cdot \dfrac{1}{\cos^2 x}}$$

$$= \lim_{x\to\frac{\pi}{2}} \frac{\cos^3 x}{\dfrac{\cos^2 x}{\sin^2 x}} = \lim_{x\to\frac{\pi}{2}} (\cos x \sin^2 x) = 0$$

Expressions of the form "$\infty - \infty$"

If $\varphi(x) = f(x) - g(x)$ for $x = a$ takes the indeterminate form "$\infty - \infty$", then

$$\lim_{x\to a} [f(x) - g(x)] = \lim_{x\to a} \frac{\dfrac{1}{g(x)} - \dfrac{1}{f(x)}}{\dfrac{1}{f(x)} \cdot \dfrac{1}{g(x)}}$$

holds whereby the case is reduced to the form $\dfrac{\text{"}0\text{"}}{0}$.

Example:

$$\varphi(x) = \frac{1}{x - 1} - \frac{1}{\ln x}; \quad \varphi(1) \text{ has the form "}\infty - \infty\text{".}$$

$$\lim_{x\to 1} \frac{\ln x - (x - 1)}{(x - 1) \ln x} = \lim_{x\to 1} \frac{\dfrac{1}{x} - 1}{1 + \ln x - \dfrac{1}{x}}$$

$$= \lim_{x\to 1} \frac{1 - x}{x + x \ln x - 1} = \lim_{x\to 1} \frac{-1}{1 + 1 + \ln x} = -\frac{1}{2}$$

Expressions of the form "0^0", "∞^0", "1^∞"

If $\varphi(x) = f(x)^{g(x)}$ for $x = a$ takes the indeterminate form "0^0" or "∞^0" or "1^∞", then

$$\lim_{x\to a} f(x)^{g(x)} = \lim_{x\to a} e^{g(x)\ln f(x)}$$

holds whereby the exponent is reduced to the form "$0 \cdot \infty$".

Example:

$$\varphi(x) = (\sin x)^{\tan x}; \quad \varphi\left(\frac{\pi}{2}\right) \text{ has the form "}1^\infty\text{".}$$

$$\lim_{x\to\frac{\pi}{2}} e^{\tan x \ln \sin x} \quad \text{has the form "}e^{\infty\cdot 0}\text{".}$$

For the exponent, the following holds:

$$\lim_{x\to\frac{\pi}{2}} \frac{\ln(\sin x)}{\dfrac{1}{\tan x}} = \lim_{x\to\frac{\pi}{2}} \frac{\dfrac{1}{\sin x}\cos x}{-\dfrac{1}{\tan^2 x}\cdot\dfrac{1}{\cos^2 x}}$$

$$= \lim_{x\to\frac{\pi}{2}} [-\cot x \tan^2 x \cos^2 x] = \lim_{x\to\frac{\pi}{2}} (-\sin x \cos x) = 0;$$

hence

$$\underline{\underline{\lim_{x\to\frac{\pi}{2}} (\sin x)^{\tan x} = e^0 = 1}}$$

Note: Sometimes an expansion in increasing powers of x (series expansion) may more readily lead to a result.

Example:

$$\varphi(x) = \frac{1-\cos x}{\sin^2 x}; \quad \varphi(0) \quad \text{has the form } \frac{\text{"}0\text{"}}{0}$$

$$1 - \cos x = \frac{x^2}{2!} - \frac{x^4}{4!} + - \cdots = x^2\left(\frac{1}{2!} - \frac{x^2}{4!} + - \cdots\right)$$

$$\sin^2 x = \left(\frac{x}{1!} - \frac{x^3}{3!} + - \cdots\right)^2 = x^2\left(\frac{1}{1!} - \frac{x^2}{3!} + - \cdots\right)^2$$

$$\underline{\underline{\lim_{x\to 0} \frac{1-\cos x}{\sin^2 x}}} = \lim_{x\to 0} \frac{\dfrac{1}{2!} - \dfrac{x^2}{4!} + - \cdots}{\left(\dfrac{1}{1!} - \dfrac{x^2}{3!} + - \cdots\right)^2} = \underline{\underline{\frac{1}{2}}}$$

6. Differential geometry

Differential geometry is the application of differential calculus to the study of curves and surfaces. A continuous curve corresponds to each continuous function. A *smooth* curve, i.e. a curve without discontinuities and breaks, corresponds to each differentiable function.

6.1. Plane curves

6.1.1. Main elements of plane curves

Arc element of a curve

$$ds = \sqrt{1 + y'^2}\, dx \qquad \text{for curve} \quad y = f(x)$$

$$ds = \sqrt{\dot{\varphi}(t)^2 + \dot{\psi}(t)^2}\, dt \qquad \text{for curve} \quad x = \varphi(t); \quad y = \psi(t)$$

$$ds = \sqrt{r^2 + \left(\frac{dr}{d\varphi}\right)^2}\, d\varphi \qquad \text{for curve} \quad r = f(\varphi)$$

Arc length and curve have a positive direction (corresponds to increasing values of x, t, φ).

Tangent and normal

A positive direction of the tangent corresponds to the positive direction of the curve. A positive direction of the normal is obtained by rotating the positive tangent through 90° in the positive sense (anticlockwise rotation).

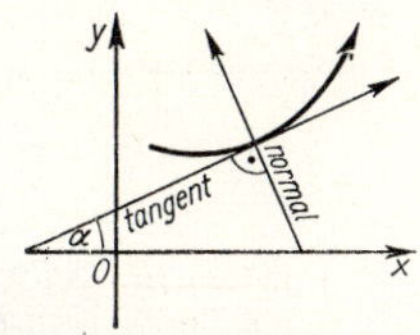

For the angle α which the positive tangent makes with the positive direction of the x-axis.

$$\sin\alpha = \frac{dy}{ds}; \quad \cos\alpha = \frac{dx}{ds}; \quad \tan\alpha = \frac{dy}{dx}$$

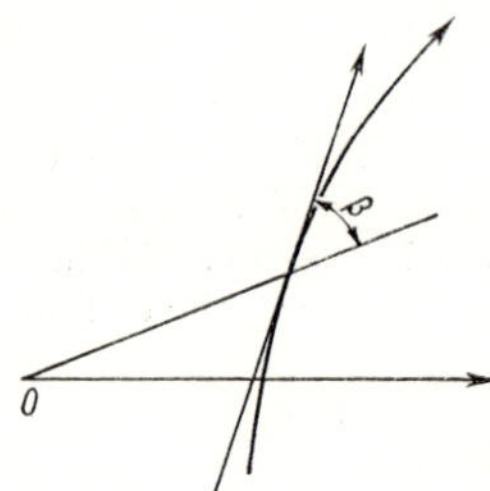

Angle β which the positive tangent makes with the positive direction of the radius vector, is calculated from

$$\sin\beta = r\frac{d\varphi}{ds}; \qquad \cos\beta = \frac{dr}{ds};$$

$$\tan\beta = \frac{r}{\left(\dfrac{dr}{d\varphi}\right)}$$

Equation of the *tangent*; point of contact $P(x, y)$; consecutive coordinates ξ, η

$$\eta - y = y'(\xi - x) \qquad \text{for curve } y = f(x)$$

$$(\xi - x)\frac{\partial f}{\partial x} + (\eta - y)\frac{\partial f}{\partial y} = 0 \qquad \text{for curve } f(x, y) = 0$$

$$(\xi - x)\dot{\psi} - (\eta - y)\dot{\varphi} = 0 \qquad \text{for curve } x = \varphi(t); \quad y = \psi(t)$$

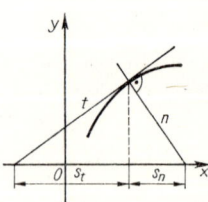

Equation of the *normal*; point of curve $P(x, y)$; consecutive coordinates ξ, η

$$\eta - y = -\frac{1}{y'}(\xi - x) \qquad \text{for curve } y = f(x)$$

$$(\xi - x)\frac{\partial f}{\partial y} - (\eta - y)\frac{\partial f}{\partial x} = 0 \qquad \text{for curve } f(x, y) = 0$$

$$(\xi - x)\dot{\varphi} + (\eta - y)\dot{\psi} = 0 \qquad \text{for curve } x = \varphi(t); \quad y = \psi(t)$$

for curve $y = f(x)$:

Tangent length $t = \left|\dfrac{y}{y'}\sqrt{1 + y'^2}\right|$

Length of normal $n = \left| y\sqrt{1 + y'^2} \right|$

Subtangent $s_t = \left|\dfrac{y}{y'}\right|$

Subnormal $s_n = | yy' |$

Tangent length (*polar tangent length*)

$$t = \left|\frac{r}{\dfrac{dr}{d\varphi}}\sqrt{r^2 + \left(\frac{dr}{d\varphi}\right)^2}\right| \qquad \text{for curve } r = f(\varphi)$$

Length of normal (*length of polar normal*)	$n = \left\| \sqrt{r^2 + \left(\frac{dr}{d\varphi}\right)^2} \right\|$	for curve $r = f(\varphi)$
Subtangent (*polar subtangent*)	$s_t = \left\| \frac{r^2}{\frac{dr}{d\varphi}} \right\|$	
Subnormal (*polar subnormal*)	$s_n = \left\| \frac{dr}{d\varphi} \right\|$	

Tangency of two curves

The two curves $y = f(x)$ and $y = g(x)$ have a tangency of the nth order in point $P_0(x_0, y_0)$ if

$$f(x_0) = g(x_0), \quad f'(x_0) = g'(x_0), \quad f''(x_0) = g''(x_0), \ldots$$

$$f^{(n)}(x_0) = g^{(n)}(x_0), \quad \text{but} \quad f^{(n+1)}(x_0) \neq g^{(n+1)}(x_0)$$

If n is even, the curves intersect each other in the common point of contact; if n is odd, they touch each other without intersection.

Circle of curvature, radius of curvature, curvature, center of curvature

The *circle of curvature* of a curve at a point is defined as the circle that has a contact of the second or higher order with the curve at the point concerned. The radius of this circle is called the *radius of curvature* ϱ.

Its center (*center of curvature*) $M_C(\xi, \eta)$ lies on the normal to the curve at the point concerned.

The reciprocal of ϱ is called *curvature* $k = \dfrac{1}{\varrho}$.

For curve $y = f(x)$ [point of contact $P(x, y)$] we have

$$\left| \varrho = \frac{(1 + y'^2)^{\frac{3}{2}}}{y''} \right|; \qquad k = \frac{y''}{(1 + y'^2)^{\frac{3}{2}}}$$

$$\xi = x - \frac{y'(1 + y'^2)}{y''}; \qquad \eta = y + \frac{1 + y'^2}{y''}$$

For curve $f(x, y) = 0$ we have

$$\varrho = m\sqrt{f_x^2 + f_y^2}; \qquad k = \frac{1}{m\sqrt{f_x^2 + f_y^2}}$$

$$\xi = x - mf_x; \qquad \eta = y - mf_y$$

where

$$m = \frac{f_x^2 + f_y^2}{f_{xx}f_y^2 - 2f_{xy}f_xf_y + f_{yy}f_x^2}$$

For curve $x = \varphi(t)$; $y = \psi(t)$ we have

$$\varrho = \left| \frac{(\dot\varphi^2 + \dot\psi^2)^{\frac{3}{2}}}{\begin{vmatrix} \dot\varphi & \dot\psi \\ \ddot\varphi & \ddot\psi \end{vmatrix}} \right|; \qquad k = \frac{\begin{vmatrix} \dot\varphi & \dot\psi \\ \ddot\varphi & \ddot\psi \end{vmatrix}}{(\dot\varphi^2 + \dot\psi^2)^{\frac{3}{2}}}$$

$$\xi = x - \frac{\dot\psi(\dot\varphi^2 + \dot\psi^2)}{\begin{vmatrix} \dot\varphi & \dot\psi \\ \ddot\varphi & \ddot\psi \end{vmatrix}}; \qquad \eta = y + \frac{\dot\varphi(\dot\varphi^2 + \dot\psi^2)}{\begin{vmatrix} \dot\varphi & \dot\psi \\ \ddot\varphi & \ddot\psi \end{vmatrix}}$$

For curve $r = f(\varphi)$ we have

$$\varrho = \left| \frac{\left[r^2 + \left(\frac{dr}{d\varphi}\right)^2\right]^{\frac{3}{2}}}{r^2 + 2\left(\frac{dr}{d\varphi}\right)^2 - r\frac{d^2r}{d\varphi^2}} \right|; \qquad k = \frac{r^2 + 2\left(\frac{dr}{d\varphi}\right)^2 - r\frac{d^2r}{d\varphi^2}}{\left[r^2 + \left(\frac{dr}{d\varphi}\right)^2\right]^{\frac{3}{2}}}$$

$$\xi = r\cos\varphi - \frac{\left[r^2 + \left(\frac{dr}{d\varphi}\right)^2\right]\left[r\cos\varphi + \frac{dr}{d\varphi}\sin\varphi\right]}{r^2 + 2\left(\frac{dr}{d\varphi}\right)^2 - r\frac{d^2r}{d\varphi^2}}$$

$$\eta = r\sin\varphi - \frac{\left[r^2 + \left(\frac{dr}{d\varphi}\right)^2\right]\left[r\sin\varphi - \frac{dr}{d\varphi}\cos\varphi\right]}{r^2 + 2\left(\frac{dr}{d\varphi}\right)^2 - r\frac{d^2r}{d\varphi^2}}$$

For ϱ or k we also have

$$\varrho = \frac{1}{|k|} = \lim_{\Delta\tau \to 0} \frac{\Delta s}{\Delta\tau} = \frac{ds}{d\tau}$$

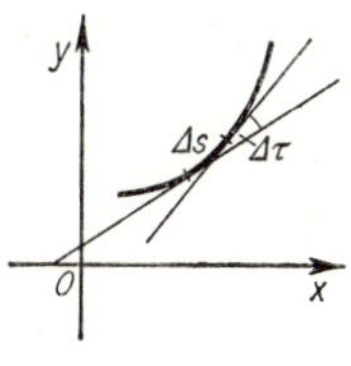

$\Delta\tau$ represents the angle through which one of the two positive tangents must be rotated to cause it to coincide with the second one. Δs is the arc of the curve between the two points of contact.

$d\tau$ is called *angle of contingence.*

The radius of curvature and the curvature are taken as positive or negative, depending on whether the center of curvature is to the left or to the right of the curve when tracing the curve in the positive sense.

The points of a curve in which the curvature exhibits a maximum or minimum are called *vertices* (principal vertex and secondary vertex, respectively).

Convex and concave behavior of a curve

At point P a curve is convex from the outside (clockwise curvature) if $k < 0$; it is concave (anticlockwise curvature) if $k > 0$.

Points of inflection

Points of inflection are points in a curve where the convex behavior of the curve changes to a concave behavior, or conversely.

A point of inflection is given if $f''(x_1) = 0$; $f'''(x_1) = 0; \ldots;$ $f^{(n-1)}(x_1) = 0$; $f^{(n)}(x_1) \neq 0$ for odd $n \geqq 3$.

If, at the same time, $f'(x_1) = 0$ at the inflection point, the tangent to the point of inflection is horizontal.

The condition $f''(x) = 0$ for inflection points in the curve $x = \varphi(t)$; $y = \psi(t)$ is represented by

$$\begin{vmatrix} \dot{\varphi} & \dot{\psi} \\ \ddot{\varphi} & \ddot{\psi} \end{vmatrix} = 0$$

in the curve $r = f(\varphi)$ by

$$r^2 + 2\left(\frac{dr}{d\varphi}\right)^2 - r\frac{d^2r}{d\varphi^2} = 0$$

in the curve $f(x, y) = 0$ by

$$\begin{vmatrix} f_{xx} & f_{xy} & f_x \\ f_{yx} & f_{yy} & f_y \\ f_x & f_y & 0 \end{vmatrix} = 0$$

Singular points

Conditional equations:

$$f(x, y) = 0; \quad f_x = 0; \quad f_y = 0$$

Crunode if additionally $f_{xy}^2 > f_{xx}f_{yy}$,

Cusp if additionally $f_{xy}^2 = f_{xx}f_{yy}$,

Acnode if additionally $f_{xy}^2 < f_{xx}f_{yy}$.

Crunodes have two real distinct tangents.

Cusps have one common tangent.

Acnodes have no real tangents.

Example 1:
Check the curve $f(x, y) = x^3 + y^3 - 3axy = 0$ for singular points.

$$f_x = 3x^2 - 3ay$$

$$f_y = 3y^2 - 3ax$$

$$f_{xy} = -3a; \quad f_{xx} = 6x; \quad f_{yy} = 6y$$

$$f_x = 0 \quad \text{and} \quad f_y = 0 \Rightarrow E = \{0, 0\} \Rightarrow P(0, 0)$$

$$\Rightarrow f_{xy}^2 = 9a^2 \quad \text{and} \quad f_{xx} = 0 \quad \text{and} \quad f_{yy} = 0$$

$$\Rightarrow f_{xy}^2 > f_{xx}f_{yy} \Rightarrow \text{crunode}$$

See Cartesian oval, page 319.

Example 2:
Check the curve $f(x, y) = x^3 - y^2(a - x) = 0$ for singular points.

$$f_x = 3x^2 + y^2$$

$$f_y = 2xy - 2ay$$

$$f_{xy} = 2y; \quad f_{xx} = 6x; \quad f_{yy} = 2x - 2a$$

$$f_x = 0 \quad \text{and} \quad f_y = 0 \Rightarrow E = \{0, 0\} \Rightarrow P(0, 0)$$

$$\Rightarrow f_{xy}^2 = 0 \quad \text{and} \quad f_{xx} = 0 \quad \text{and} \quad f_{yy} = -2a$$

$$\Rightarrow f_{xy}^2 = f_{xx} f_{yy} \Rightarrow \text{cusp}$$

See cissoid, page 319.

Example 3:

Check the function $f(x, y) = y^2 - x^3 + 7x^2 - 16x + 12 = 0$ for singular points.

$$f_y = -3x^2 + 14x - 16$$

$$f_y = 2y$$

$$f_{xy} = 0; \quad f_{xx} = -6x + 14; \quad f_{yy} = 2$$

$$f_x = 0 \text{ and } f_y = 0 \Rightarrow E = \{2, 0\} \Rightarrow P(2, 0)$$

$$\Rightarrow f_{xy}^2 = 0 \quad \text{and} \quad f_{xx} = 2 \quad \text{and} \quad f_{yy} = 2$$

$$\Rightarrow f_{xy}^2 < f_{xx} f_{yy} \Rightarrow \text{acnode } I(2, 0)$$

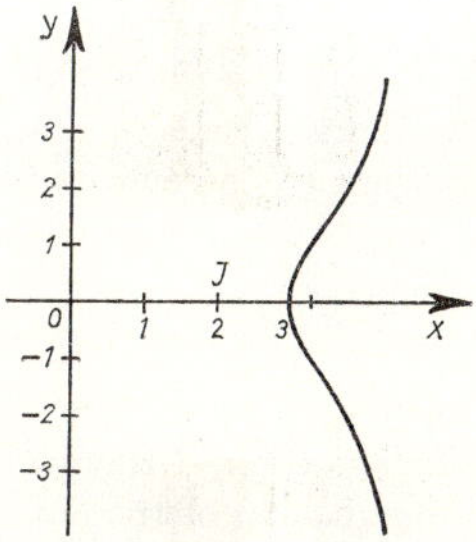

Asymptotes

A straight line is called an asymptote of a curve extending to infinity if the distance between a point P of the curve and the straight line converges to zero if P approaches infinity along the curve.

Asymptotes of the curve $y = f(x)$ that are parallel to the axes have equations

$$\eta = \lim_{x \to \infty} y \quad \text{or} \quad \xi = \lim_{y \to \infty} x \quad (\xi, \eta \text{ consecutive coordinates})$$

Asymptotes of arbitrary direction have equation

$$\eta = m\xi + b, \quad \text{where} \quad m = \lim_{x \to \infty} \frac{y}{x}; \quad b = \lim_{x \to \infty} (y - mx)$$

Asymptotes in the case of polar coordinates:

If $\lim_{\varphi \to \alpha} r = \infty$, α determines the direction of the asymptote. The distance of the asymptote from the pole is given by

$$p = \lim_{\varphi \to \alpha} [r \sin (\alpha - \varphi)].$$

Envelope

This is a curve that is tangent to each member of a given family of curves. The envelope of the family of curves $f(x, y, p) = 0$, where p is the variable parameter independent of x and y, is given by the pair of equations:

$$f(x, y, p) = 0; \quad \frac{\partial f(x, y, p)}{\partial p} = 0$$

These parametric equations of the envelope may be used to eliminate the parameter. The tangent to one point of the envelope is also tangent to one curve of the family of curves.

Evolute

The *evolute* of a curve is the set of all centers of curvature.
The equation of the evolute is obtained by eliminating x and y from the equation of the curve and from the equations of the coordinates ξ, η of the center of curvature, where ξ, η then are the consecutive coordinates.
The tangents of the evolute are also normals to the given curve.
The difference between two radii of curvature is equal to the length of the evolute arc between the associated centers of curvature. (For equations of the evolute with respect to the parabola, ellipse, and hyperbola see pages 230, 238, 240.)

Involute

When unwinding the tangent to an evolute from the evolute, any point of the tangent describes a curve parallel to the original curve. This family of parallel curves including the original curve are called *involutes* of the given curve.
Each radius of curvature is normal to the involute and tangent to the evolute.
The radii of curvature of the evolute and involute are to each other as the associated arc elements.

Involute of a circle

When unwinding the tangent to a given circle, any point of the tangent describes an involute of the circle.

$$x = a(\cos t + t \sin t)$$

$$y = a(\sin t - t \cos t)$$

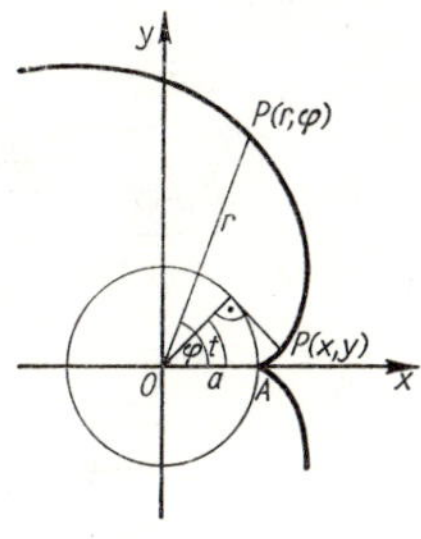

a radius of the given circle;
t angle rolled through

In polar coordinates

$$\varphi = \sqrt{\frac{r^2}{a^2} - 1} - \operatorname{arc\,tan} \sqrt{\frac{r^2}{a^2} - 1}$$

(Unwinding starts at A)

6.1.2. A few important plane curves

Cycloids

Common cycloid

If a circle is rolled along a straight line, the curve traced out by a point on its circumference is called a (common) cycloid.

$$x = a(t - \sin t); \qquad y = a(1 - \cos t)$$

$$x = a \operatorname{arc\,cos} \frac{a - y}{a} - \sqrt{y(2a - y)}$$

a radius of the circle
t angle rolled through

Length of arc OP:

$$l_1 = 8a \sin^2 \frac{t}{4}$$

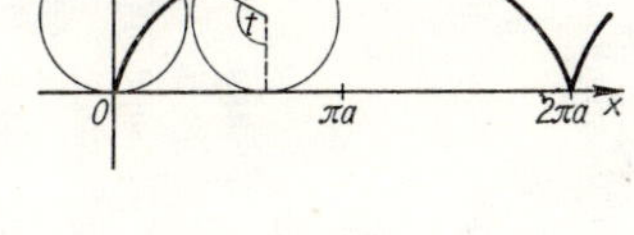

Length of a full arc: $l = 8a$
Area under a full cycloidal arc:
$A = 3\pi a^2$

Period $= 2\pi a$

The evolute of a cycloid is a congruent cycloid, displaced by πa in the direction of the positive x-axis, displaced by $2a$ in the direction of the negative y-axis.

Prolate cycloid (trochoid)

The path described by a point on a diameter of a circle as the circle rolls along a straight line. The generating point is a distance c from

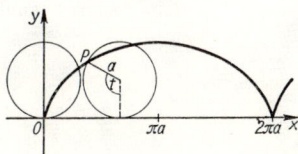

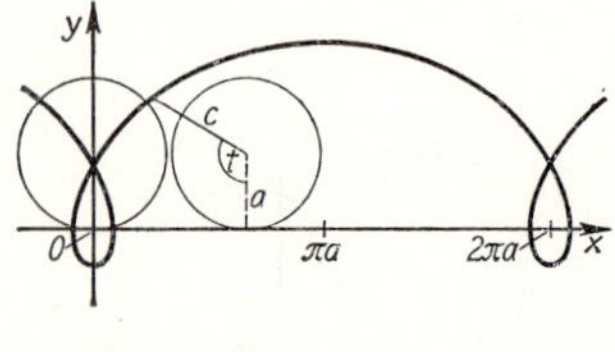

the center ($c > a$).

$$x = at - c \sin t$$

$$y = a - c \cos t$$

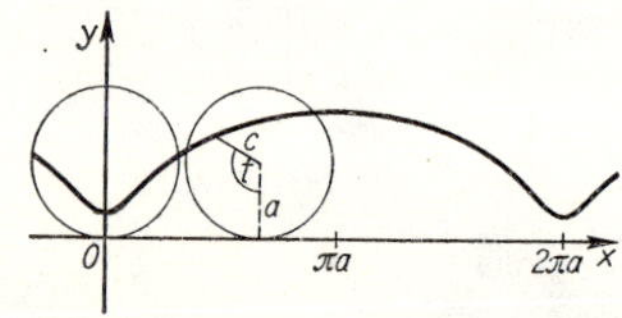

Curtate cycloid (trochoid)

The generating point on the diameter of the circle is within the circle a distance c from the center ($c < a$).

$$x = at - c \sin t$$

$$y = a - c \cos t$$

Epicycloids

Common epicycloid

The path described by a point on the circumference of a circle rolled on the exterior of another circle is called an epicycloid.

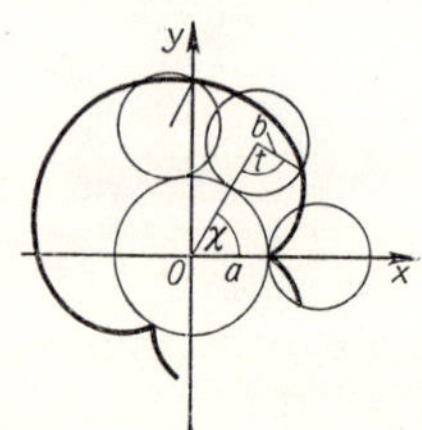

$$x = (a + b) \cos \frac{b}{a} t - b \cos \frac{a + b}{a} t$$

$$y = (a + b) \sin \frac{b}{a} t - b \sin \frac{a + b}{a} t$$

a radius of the fixed circle
b radius of the rolled circle
t angle through which the circle has rolled

or

$$x = (a + b) \cos \chi - b \cos \frac{a + b}{b} \chi$$

$$y = (a + b) \sin \chi - b \sin \frac{a + b}{b} \chi \quad \chi \text{ angle of rotation}$$

If the ratio $\dfrac{a}{b} = m$ is an integer, the curve consists of m arcs linked together, otherwise the arcs overlap.

Length of an arc: $l_1 = \dfrac{8(a + b)}{m}$

Length of the whole curve (if m is an integer): $l = 8(a + b)$
Area under one complete arc (between epicycloid and fixed circle):

$$A = \frac{\pi b^2(3a + 2b)}{a}$$

Curtate and prolate epicycloids (*epitrochoids*)

The path described by a point on the diameter of a circle rolled on the exterior of another circle is either a curtate or a prolate epicycloid, depending on the distance c of the generating point from the center of the rolling circle.

$c < b$: curtate epicycloid
$c > b$: prolate epicycloid

$$x = (a + b)\cos\frac{b}{a}t - c\cos\frac{a + b}{a}t$$

$$y = (a + b)\sin\frac{b}{a}t - c\sin\frac{a + b}{a}t$$

or

$$x = (a + b)\cos\chi - c\cos\frac{a + b}{b}\chi$$

$$y = (a + b)\sin\chi - c\sin\frac{a + b}{b}\chi$$

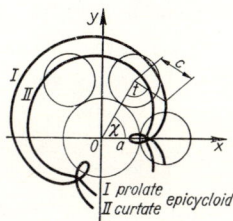

Special case:

If $m = 1$, hence $a = b$, the common epicycloid changes into a *cardioid*.

$$x = a(2\cos t - \cos 2t)$$

$$y = a(2\sin t - \sin 2t)$$

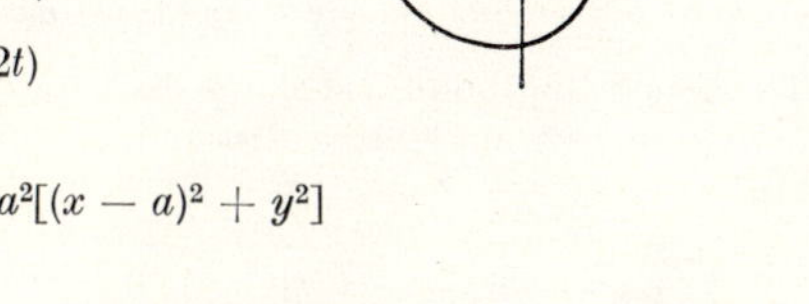

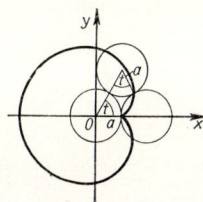

or

$$(x^2 + y^2 - a^2)^2 = 4a^2[(x - a)^2 + y^2]$$

or

$$r = 2a(1 - \cos\varphi)$$

(pole on the circumference of the fixed circle, polar axis as extension of the diameter belonging to the pole)

Hypocycloids

The path described by a point on the circumference of a circle that is rolled on the inner side of the circumference of another circle is called a hypocycloid.

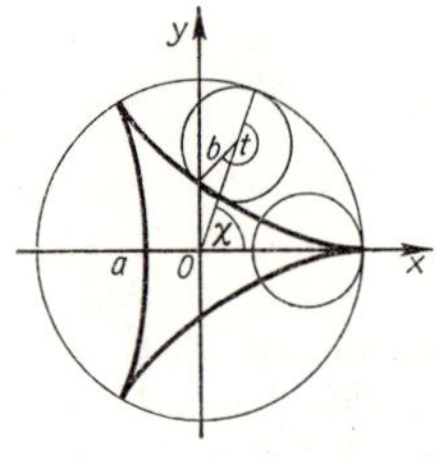

$$x = (a - b) \cos \frac{b}{a} t + b \cos \frac{a - b}{a} t$$

$$y = (a - b) \sin \frac{b}{a} t - b \sin \frac{a - b}{a} t$$

or

$$x = (a - b) \cos \chi + b \cos \frac{a - b}{b} \chi$$

$$y = (a - b) \sin \chi - b \sin \frac{a - b}{b} \chi$$

a radius of the fixed circle

b radius of the rolling circle

t angle through which the circle has rolled

χ angle of rotation

If the ratio $\frac{a}{b} = m$ is an integer (in the figure, $m = 3$), the curve consists of m arcs linked together, otherwise the arcs overlap each other.

Length of an arc: $l_1 = \frac{8(a - b)}{m}$

Length of the whole curve (if m is an integer): $l = 8(a - b)$

Area under a complete arc (between hypocycloid and fixed circle):
$A = \frac{\pi b^2(3a - 2b)}{a}$

Curtate and prolate hypocycloids (hypotrochoids)

The generating point lies on the diameter of the rolling circle outside or within this circle at a distance c from the center of the rolling circle.

$c < b$: curtate hypocycloid

$c > b$: prolate hypocycloid

$$x = (a - b) \cos \frac{b}{a} t + c \cos \frac{a - b}{a} t$$

$$y = (a - b) \sin \frac{b}{a} t - c \sin \frac{a - b}{a} t$$

or

$$x = (a - b) \cos \chi + c \cos \frac{a - b}{b} \chi$$

$$y = (a - b) \sin \chi - c \sin \frac{a - b}{b} \chi$$

Special cases:

If $m = 4$, hence $b = \frac{1}{4} a$, the common hypocycloid changes into an *astroid*.

$$x = a \cos^3 \frac{1}{4} t$$

$$y = a \sin^3 \frac{1}{4} t$$

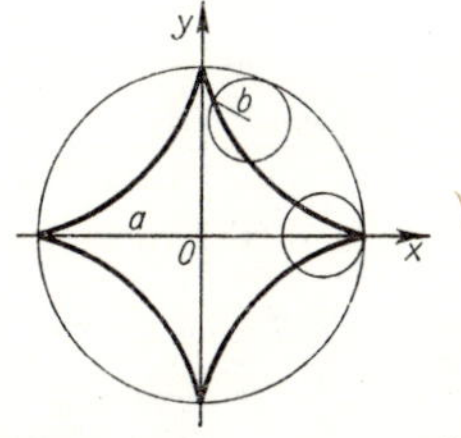

or

$$x^{\frac{2}{3}} + y^{\frac{2}{3}} = a^{\frac{2}{3}}$$

or

$$(x^2 + y^2 - a^2)^3 + 27a^2x^2y^2 = 0$$

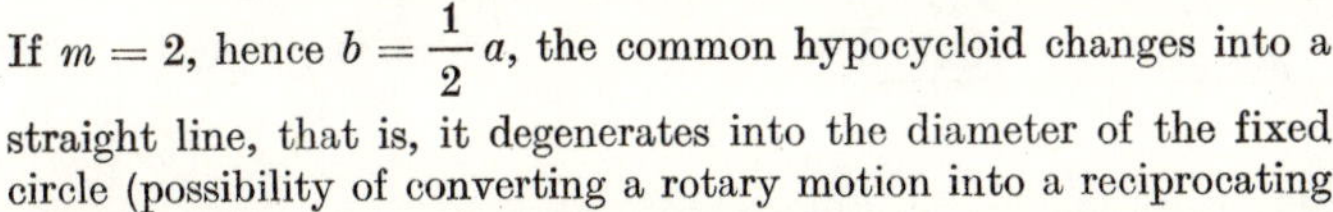

If $m = 2$, hence $b = \frac{1}{2} a$, the common hypocycloid changes into a straight line, that is, it degenerates into the diameter of the fixed circle (possibility of converting a rotary motion into a reciprocating motion).

If $m = 2$, hence $b = \frac{1}{2} a$, the curtate and prolate hypocycloids become ellipses with the equations

$$x = \left(\frac{a}{2} + c\right) \cos \frac{t}{2}; \quad y = \left(\frac{a}{2} - c\right) \sin \frac{t}{2}$$

(possibility of converting a rotary motion into an elliptical motion)

Cassini ovals

A Cassini oval is the set of all points whose distances from two fixed points have a constant product a^2.

$$(x^2 + y^2)^2 - 2e^2(x^2 - y^2) = a^4 - e^4$$

or

$$r^2 = e^2 \cos 2\varphi \pm \sqrt{e^4 \cos^2 2\varphi + a^4 - e^4} \qquad (\overline{F_1F_2} = 2e)$$

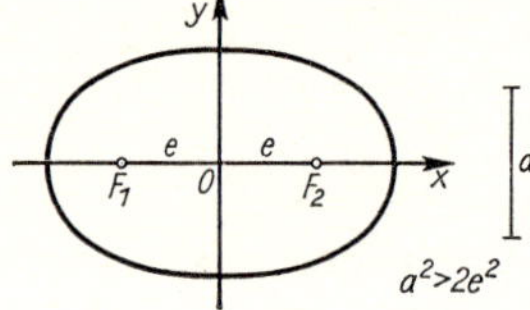

For $a^2 > 2e^2$ and for $a^2 = 2e^2$ the oval has an ellipselike shape.

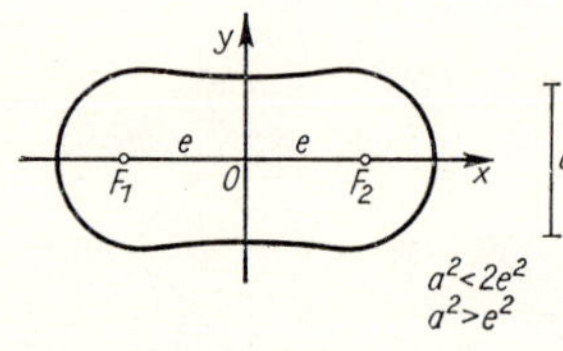

For $a^2 < 2e^2$ but $a^2 > e^2$ the curve has two flattened parts.

For $a^2 = e^2$ we obtaine a *lemniscate* (see below).

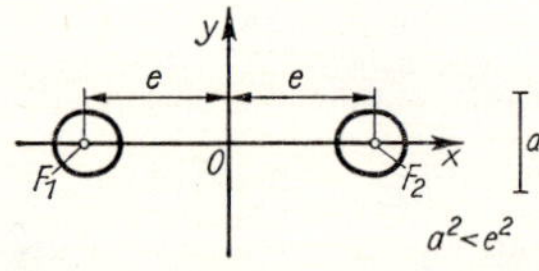

For $a^2 < e^2$ the curve consists of two separate egg-shaped parts about one of the two fixed points each.

Lemniscate

The lemniscate is a *special case of the Cassini oval*, namely, for $a^2 = e^2$.

$$(x^2 + y^2)^2 = 2a^2(x^2 - y^2)$$

or

$$r = a\sqrt{2 \cos 2\varphi}$$

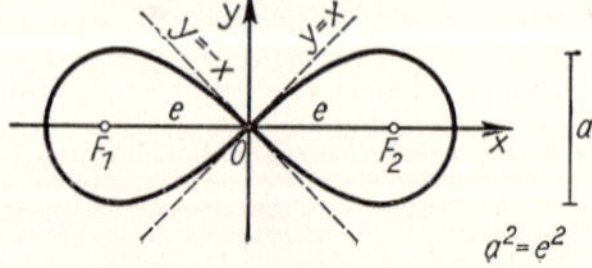

Origin is a crunode which is a point of inflection at the same time.

Radius of curvature: $\varrho = \dfrac{2a^2}{3r}$

Area of one loop: $A = a^2$

Helical lines

Logarithmic spiral

$$r = e^{k\varphi} \quad (k > 0)$$

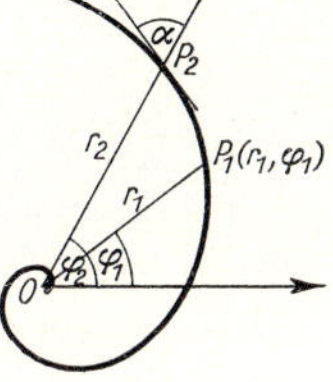

A logarithmic spiral cuts all rays emitted from the origin at the same angle α.

$$\cot \alpha = k$$

The pole is an asymptotic point.
Length of the arc P_1P_2:

$$l = \frac{1}{k}\sqrt{1 + k^2}\,(r_2 - r_1)$$

Limiting case: P_1 approaches the origin.

$$\widehat{OP}_2 = \frac{1}{k}\sqrt{1 + k^2} \cdot r^2$$

Area of the sector in this limiting case: $A = \dfrac{r^2}{4k}$

Radius of curvature: $\varrho = r\sqrt{1 + k^2}$

Archimedean spiral

A point moving on a radius vector (from the origin) at constant speed while the radius vector rotates about a pole at constant speed describes an Archimedean spiral.

$$r = a\varphi$$

Length of arc OP:

$$l = \frac{a}{2}\left(\varphi\sqrt{\varphi^2 + 1} + \sinh^{-1}\varphi\right) \approx \frac{a\varphi^2}{2} \quad \text{(for large } \varphi\text{)}$$

Area of sector P_1OP_2:

$$A = \frac{a^2}{6}\,(\varphi_2{}^3 - \varphi_1{}^3)$$

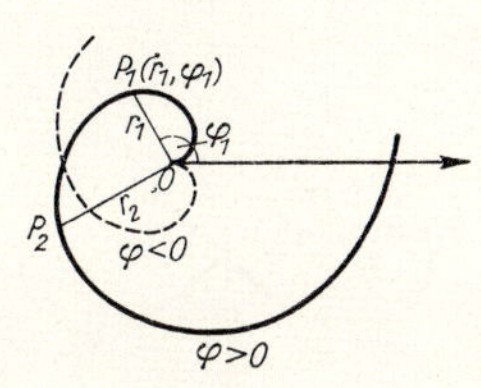

Radius of curvature:

$$\varrho = \frac{(a^2 + r^2)^{\frac{3}{2}}}{2a^2 + r^2} = \frac{a(\varphi^2 + 1)^{\frac{3}{2}}}{\varphi^2 + 2}$$

Hyperbolic spiral

$$r = \frac{a}{\varphi}$$

Asymptote: $y = a$
The pole is an asymptotic point.
Area of sector P_1OP_2:

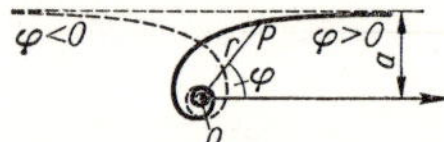

$$A = \frac{a^2}{2}\left(\frac{1}{\varphi_1} - \frac{1}{\varphi_2}\right)$$

Radius of curvature:

$$\varrho = \frac{a}{\varphi}\left(\frac{\sqrt{1+\varphi^2}}{\varphi}\right)^3 = r\left(\frac{r^2}{a^2} + 1\right)^{\frac{3}{2}}$$

Catenary

A cable, perfectly flexible and heavy, suspended at two points in equilibrium takes the form of a catenary.

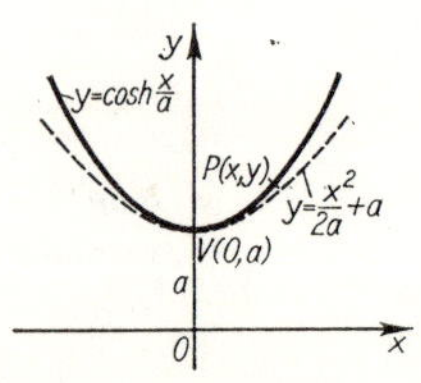

$$y = \frac{a}{2}\left(e^{\frac{x}{a}} + e^{-\frac{x}{a}}\right) = a\cosh\frac{x}{a}$$

Vertex $V(0; a)$
Near the lowest point (V), the parabola $y = \frac{x^2}{2a} + a$ closely follows the catenary (contact of the third order).

Area between the catenary, the x-axis and the straight lines $x = 0$ and $x = x$:

$$A = a^2 \sinh\frac{x}{2}$$

Length of arc VP:

$$l = a\sinh\frac{x}{a} = a\,\frac{e^{\frac{x}{a}} - e^{-\frac{x}{a}}}{2}$$

Radius of curvature:

$$\varrho = \frac{y^2}{a} = a\cosh^2\frac{x}{a}$$

Tractrix

A material point at the end of a nonextensible thread of length a describes a tractrix if the starting point of the thread is moved along the straight line $y = 0$; initial position of the point $(0, a)$.

$$x = a \cosh^{-1} \frac{a}{y} \mp \sqrt{a^2 - y^2} = a \ln \left| \frac{a \pm \sqrt{a^2 - y^2}}{y} \right| \mp \sqrt{a^2 - y^2}$$

Asymptote: $y = 0$

The points $S(0, a)$ and $S(0, -a)$ are *cusps*.

Length of arc SP: $l = a \ln \left| \frac{a}{y} \right|$

Radius of curvature: $\varrho = a \cot \frac{x}{y}$

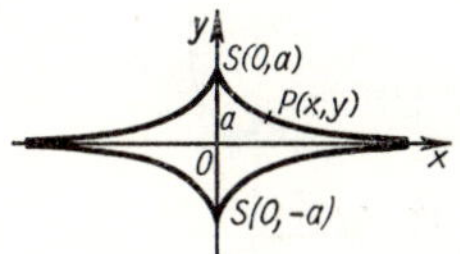

Cartesian oval

$$x^3 + y^3 - 3axy = 0$$

or

$$r = \frac{3a \sin \varphi \cos \varphi}{\sin^3 \varphi + \cos^3 \varphi}$$

Asymptote of the curve: $y = -x - a$

Vertex $V \left(\frac{3}{2} a, \frac{3}{2} a \right)$

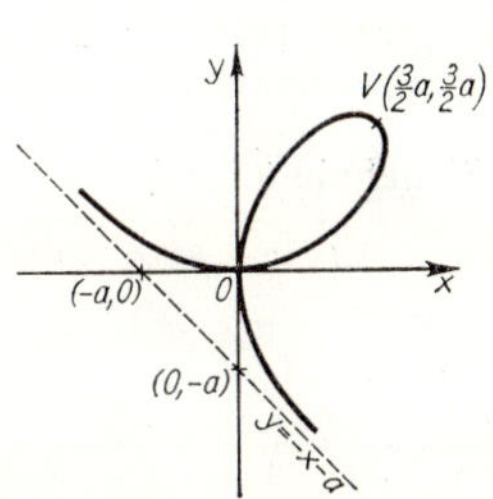

Area of the loop = area between the curve and its asymptote:

$$A = \frac{3}{2} a^2$$

Cissoid

$$y^2(a - x) = x^3$$

or

$$r = \frac{a \sin^2 \varphi}{\cos \varphi} = a \sin \varphi \tan \varphi$$

Asymptote: $x = a$

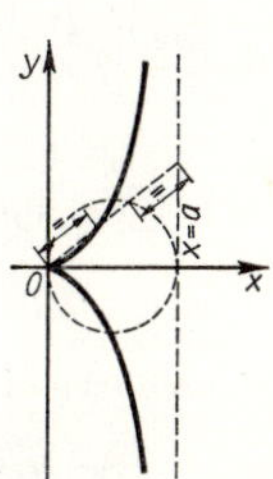

Area between the curve and the asymptote:

$$A = \frac{3}{4} \pi a^2$$

Strophoid

$$(a - x) y^2 = (a + x) x^2$$

or

$$r = \frac{-a \cos 2\varphi}{\cos \varphi}$$

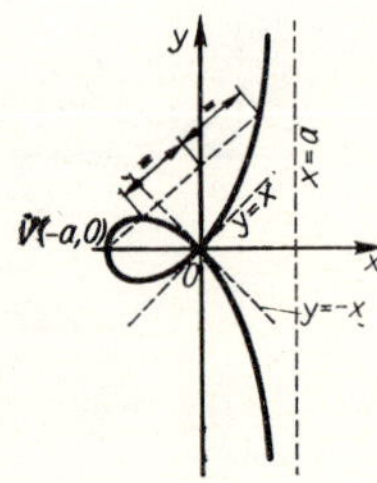

Vertex $V(-a, 0)$
Asymptote: $x = a$
Area of the loop:

$$A_1 = 2a^2 - \frac{\pi a^2}{2}$$

Area between the curve and the asymptote:

$$A_2 = 2a^2 + \frac{\pi a^2}{2}$$

Conchoid (of Nikomedes)

$$(x - a)^2 (x^2 + y^2) = b^2 x^2$$

or

$$r = \frac{a}{\cos \varphi} \pm b$$

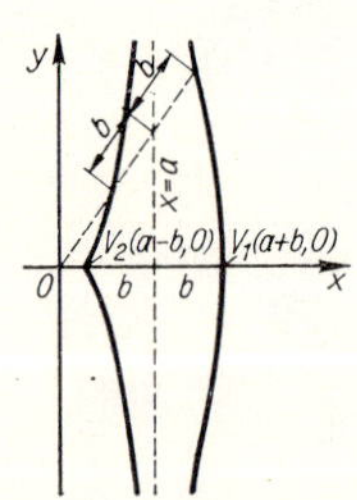

Vertices $V_1(a + b, 0)$ and $V_2(a - b, 0)$
Asymptote: $x = a$
Origin for $b < a$ an acnode (cf. illustration)
for $b > a$ a crunode
for $b = a$ a cusp
Area between the outer branch and the asymptote: $A = \infty$

6.2. Space curves

Representation in rectangular coordinates

As a curve of intersection of two surfaces

$$f(x, y, z) = 0; \quad g(x, y, z) = 0$$

By projection of the curve onto two planes, e.g. the x,y-plane and the x,z-plane

$$y = \varphi_1(x); \quad z = \varphi_2(x)$$

In parametric representation

$$x = \varphi(t);\ y = \psi(t);\ z = \chi(t) \qquad (t \text{ arbitrary parameter})$$

$$x = x(s);\ y = y(s);\ z = z(s)$$

$$s = \int_{t_0}^{t} \sqrt{\dot\varphi^2 + \dot\psi^2 + \dot\chi^2}\, dt$$

(parameter s is the arc length from a fixed starting point P_0 to the point P concerned)

Representation in vector form

$$\mathbf{r} = \mathbf{r}(t) = x(t)\mathbf{i} + y(t)\mathbf{j} + z(t)\mathbf{k} \qquad (t \text{ see above})$$

$$\mathbf{r} = \mathbf{r}(s) = x(s)\mathbf{i} + y(s)\mathbf{j} + z(s)\mathbf{k} \qquad (s \text{ see above})$$

Arc element of a space curve

$$ds = \sqrt{dx^2 + dy^2 + dz^2}$$

$$ds = \sqrt{\dot\varphi^2 + \dot\psi^2 + \dot\chi^2}\, dt$$

$$ds = |d\mathbf{r}| = |\dot{\mathbf{r}}(t)\, dt| = \left| \frac{d\mathbf{r}(s)}{ds}\, ds \right|$$

In cylindrical coordinates

$$ds = \sqrt{d\varrho^2 + \varrho^2\, d\varphi^2 + dz^2}$$

In spherical coordinates

$$ds = \sqrt{dr^2 + r^2\, d\vartheta^2 + r^2 \sin^2\vartheta\, d\varphi^2}$$

Definitions

The *tangent* at a point P_0 is the limiting position of a secant P_0P_1 for $P_1 \to P_0$.
The *positive direction* of a tangent corresponds to the positive direction of the curve (in the sense of increasing values of the variable or the parameter t).
The *osculating plane* in point P_0 is the limiting position of a plane through the tangent at P_0 and a point of curve P_1 for $P_1 \to P_0$.

The *normal plane* is the plane perpendicular to the tangent at its point of contact. Any straight line passing through the point of tangency and lying in the normal plane is called *normal*. The normal that at the same time belongs to the osculating plane is called the *principal normal*.
The normal that is perpendicular to the osculating plane is called the *binormal*.
A plane formed by the tangent and the binormal is called *rectification plane*.

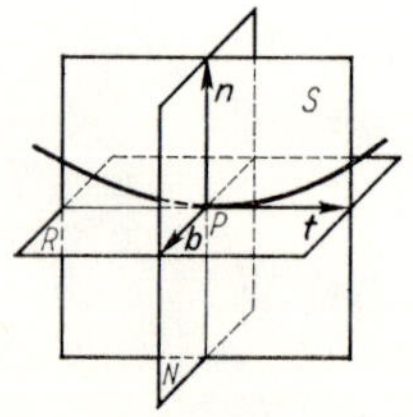

t *tangential vector* (unit vector in the direction of the tangent)

n *principal normal vector* (unit vector in the direction of the principal normal)

b *binormal vector* (unit vector in the direction of the binormal)

N normal plane, S osculating plane, R rectification plane

Slope angles of the tangent, the principal normal, the binormal

For the slope angles α, β, γ of the tangent, the following relations hold:

$$\cos\alpha = \frac{dx}{ds}; \qquad \cos\beta = \frac{dy}{ds}; \qquad \cos\gamma = \frac{dz}{ds}$$

For the slope angles l, m, n of the principal normal, the following relations hold:

$$\cos l = \varrho\frac{d^2x}{ds^2}; \qquad \cos m = \varrho\frac{d^2y}{ds^2}; \qquad \cos n = \varrho\frac{d^2z}{ds^2}$$

For the slope angles λ, μ, ν of the binormal, the following relations hold:

$$\cos\lambda = \varrho\left(\frac{dy}{ds}\cdot\frac{d^2z}{ds^2} - \frac{dz}{ds}\cdot\frac{d^2y}{ds^2}\right)$$

$$\cos\mu = \varrho\left(\frac{dz}{ds}\cdot\frac{d^2x}{ds^2} - \frac{dx}{ds}\cdot\frac{d^2z}{ds^2}\right)$$

$$\cos\nu = \varrho\left(\frac{dx}{ds}\cdot\frac{d^2y}{ds^2} - \frac{dy}{ds}\cdot\frac{d^2x}{ds^2}\right)$$

ϱ radius of curvature

Note: In the formulas given on pages 322 ... 326 all occurring derivatives must be calculated at point P_0.

Equation of the tangent to the space curve at $P_0(x_0, y_0, z_0)$

$$\frac{x - x_0}{\begin{vmatrix} f_y & f_z \\ g_y & g_z \end{vmatrix}} = \frac{y - y_0}{\begin{vmatrix} f_z & f_x \\ g_z & g_x \end{vmatrix}} = \frac{z - z_0}{\begin{vmatrix} f_x & f_y \\ g_x & g_y \end{vmatrix}}$$

for curve $f(x, y, z) = 0; \quad g(x, y, z) = 0$

$$\frac{x - x_0}{\varphi'} = \frac{y - y_0}{\psi'} = \frac{z - z_0}{\chi'}$$

for curve $x = \varphi(t); \quad y = \psi(t); \quad z = \chi(t)$

$$\mathbf{r} = \mathbf{r}_0 + \lambda \frac{d\mathbf{r}}{dt} \quad \text{for curve } \mathbf{r} = \mathbf{r}(t); \quad \lambda \in R$$

($\mathbf{r}$ variable radius vector; $\mathbf{r}_0$ radius vector to P_0)

Equation of the normal plane in $P_0(x_0, y_0, z_0)$

$$\begin{vmatrix} x - x_0 & y - y_0 & z - z_0 \\ f_x & f_y & f_z \\ g_x & g_y & g_z \end{vmatrix} = 0 \qquad \text{to curve } \begin{aligned} f(x, y, z) &= 0 \\ g(x, y, z) &= 0 \end{aligned}$$

$$\dot\varphi(x - x_0) + \dot\psi(y - y_0) + \dot\chi(z - z_0) = 0$$

to curve $x = \varphi(t); \quad y = \psi(t); \quad z = \chi(t)$

$$(\mathbf{r} - \mathbf{r}_0) \frac{d\mathbf{r}}{dt} = 0 \quad \text{to curve} \quad \mathbf{r} = \mathbf{r}(t)$$

($\mathbf{r}$ variable radius vector; $\mathbf{r}_0$ radius vector to P_0)

Equation of the osculating plane in $P_0(x_0, y_0, z_0)$

$$\begin{vmatrix} x - x_0 & y - y_0 & z - z_0 \\ \dot\varphi & \dot\psi & \dot\chi \\ \ddot\varphi & \ddot\psi & \ddot\chi \end{vmatrix} = 0 \qquad \text{to curve } \begin{aligned} x &= \varphi(t) \\ y &= \psi(t) \\ z &= \chi(t) \end{aligned}$$

$$(\mathbf{r} - \mathbf{r}_0) \frac{d\mathbf{r}}{dt} \cdot \frac{d^2\mathbf{r}}{dt^2} = 0 \qquad \text{to curve } \mathbf{r} = \mathbf{r}(t)$$

Equation of the binormal in $P_0(x_0, y_0, z_0)$

$$\frac{x - x_0}{\begin{vmatrix} \dot{\psi} & \dot{\chi} \\ \ddot{\psi} & \ddot{\chi} \end{vmatrix}} = \frac{y - y_0}{\begin{vmatrix} \dot{\chi} & \dot{\varphi} \\ \ddot{\chi} & \ddot{\varphi} \end{vmatrix}} = \frac{z - z_0}{\begin{vmatrix} \dot{\varphi} & \dot{\psi} \\ \ddot{\varphi} & \ddot{\psi} \end{vmatrix}} \quad \text{for curve} \quad \begin{aligned} x &= \varphi(t) \\ y &= \psi(t) \\ z &= \chi(t) \end{aligned}$$

$$\mathbf{r} = \mathbf{r}_0 + \lambda \left(\frac{d\mathbf{r}}{dt} \times \frac{d^2\mathbf{r}}{dt^2} \right) \qquad \text{for curve } \mathbf{r} = \mathbf{r}(t)$$

Equation of the principal normal in $P_0(x_0, y_0, z_0)$

$$\frac{x - x_0}{\begin{vmatrix} \dot{\psi} & \dot{\chi} \\ \cos\mu & \cos\nu \end{vmatrix}} = \frac{y - y_0}{\begin{vmatrix} \dot{\chi} & \dot{\varphi} \\ \cos\nu & \cos\chi \end{vmatrix}} = \frac{z - z_0}{\begin{vmatrix} \dot{\varphi} & \dot{\psi} \\ \cos\lambda & \cos\mu \end{vmatrix}} \quad \text{for curve} \quad \begin{aligned} x &= \varphi(t) \\ y &= \psi(t) \\ z &= \chi(t) \end{aligned}$$

λ, μ, ν slope angles of the binormal (see page 322)

$$\mathbf{r} = \mathbf{r}_0 + \lambda \frac{d\mathbf{r}}{dt} \times \left(\frac{d\mathbf{r}}{dt} \times \frac{d^2\mathbf{r}}{dt^2} \right) \qquad \text{for curve } \mathbf{r} = \mathbf{r}(t);\ \lambda \in R$$

Equation of the tangential plane to $P_0(x_0, y_0, z_0)$

$$f_x(x - x_0) + f_y(y - y_0) + f_z(z - z_0) = 0$$

to surface $f(x, y, z) = 0$

$$z - z_0 = \frac{\partial z}{\partial x}(x - x_0) + \frac{\partial z}{\partial y}(y - y_0) \qquad \text{to surface } z = f(x, y)$$

$$\begin{vmatrix} x - x_0 & y - y_0 & z - z_0 \\ \dfrac{\partial x}{\partial u} & \dfrac{\partial y}{\partial u} & \dfrac{\partial z}{\partial u} \\ \dfrac{\partial x}{\partial v} & \dfrac{\partial y}{\partial v} & \dfrac{\partial z}{\partial v} \end{vmatrix} = 0 \qquad \text{to surface} \quad \begin{aligned} x &= x(u, v) \\ y &= y(u, v) \\ z &= z(u, v) \end{aligned}$$

$$(\mathbf{r} - \mathbf{r}_0)\mathbf{n} = 0 \qquad \text{to surface } \mathbf{r} = \mathbf{r}(u, v)$$

($\mathbf{r}$ variable radius vector; $\mathbf{r}_0$ radius vector to P_0; $\mathbf{n}$ vector in the direction of the normal to the surface)

Equation of the normal to a surface at $P_0(x_0, y_0, z_0)$

$$\frac{x - x_0}{f_x} = \frac{y - y_0}{f_y} = \frac{z - z_0}{f_z} \qquad \text{for surface } f(x, y, z) = 0$$

$$\frac{x - x_0}{\dfrac{\partial z}{\partial x}} = \frac{y - y_0}{\dfrac{\partial z}{\partial y}} = z_0 - z \qquad \text{for surface } z = f(x, y)$$

$$\frac{x - x_0}{\begin{vmatrix} \dfrac{\partial y}{\partial u} & \dfrac{\partial z}{\partial u} \\ \dfrac{\partial y}{\partial v} & \dfrac{\partial z}{\partial v} \end{vmatrix}} = \frac{y - y_0}{\begin{vmatrix} \dfrac{\partial z}{\partial u} & \dfrac{\partial x}{\partial u} \\ \dfrac{\partial z}{\partial v} & \dfrac{\partial x}{\partial v} \end{vmatrix}} = \frac{z - z_0}{\begin{vmatrix} \dfrac{\partial x}{\partial u} & \dfrac{\partial y}{\partial u} \\ \dfrac{\partial x}{\partial v} & \dfrac{\partial y}{\partial v} \end{vmatrix}} \qquad \text{for surface } \begin{aligned} x &= x(u, v) \\ y &= y(u, v) \\ z &= z(u, v) \end{aligned}$$

$$\mathbf{r} = \mathbf{r}_0 + \lambda \mathbf{n} \qquad \text{for curve } \mathbf{r} = \mathbf{r}(u, v)$$

($\mathbf{r}$ variable vector; $\mathbf{r}_0$ radius vector to P_0; $\mathbf{n}$ vector in the direction of the normal to the surface)

Equation of the rectification plane with $P_0(x_0, y_0, z_0)$ as point of contact of the tangent

$$\begin{vmatrix} x - x_0 & y - y_0 & z - z_0 \\ \dot{\varphi} & \dot{\psi} & \dot{\chi} \\ \cos\lambda & \cos\mu & \cos\nu \end{vmatrix} = 0 \qquad \text{for curve } \begin{aligned} x &= \varphi(t) \\ y &= \psi(t) \\ z &= \chi(t) \end{aligned}$$

λ, μ, ν angles of slope of the binormal

$$(\mathbf{r} - \mathbf{r}_0) \frac{d\mathbf{r}}{dt} \left(\frac{d\mathbf{r}}{dt} \times \frac{d^2\mathbf{r}}{dt^2} \right) = 0 \qquad \text{for curve } \mathbf{r} = \mathbf{r}(t)$$

Circle of curvature, radius of curvature, curvature, center of curvature

The *circle of curvature* of a space curve in point P_0 is the limiting position of a circle through the points of curve P_1, P_0, P_2 for $P_1 \to P_0$ and $P_2 \to P_0$. Its center (*center of curvature*) lies on the principal normal. Its radius is the radius of curvature ϱ.

The reciprocal value of ϱ is called *curvature*: $k = \frac{1}{\varrho}$.
k and ϱ are always positive.

$$\frac{1}{\varrho} = k = \lim_{\Delta s \to 0} \frac{\Delta \tau}{\Delta s} = \frac{d\tau}{ds}$$

where $\Delta\tau$ represents the angle through which the tangent rotates if the points of contact are separated by a distance Δs. $d\tau$ is called *angle of contingence.*

$$k = \sqrt{\left(\frac{dx}{ds}\right)^2 + \left(\frac{dy}{ds}\right)^2 + \left(\frac{dz}{ds}\right)^2} \qquad \text{for curve } x = x(s),\ y = y(s),\ z = z(s)$$

$$k^2 = \frac{\left(\frac{d\mathbf{r}}{dt}\right)^2 \left(\frac{d^2\mathbf{r}}{dt^2}\right)^2 - \left(\frac{d\mathbf{r}}{dt} \cdot \frac{d^2\mathbf{r}}{dt^2}\right)^2}{\left[\left(\frac{d\mathbf{r}}{dt}\right)^2\right]^3}$$

$$= \frac{(\dot{x}^2 + \dot{y}^2 + \dot{z}^2)(\ddot{x}^2 + \ddot{y}^2 + \ddot{z}^2) - (\dot{x}\ddot{x} + \dot{y}\ddot{y} + \dot{z}\ddot{z})^2}{(\dot{x}^2 + \dot{y}^2 + \dot{z}^2)^3}$$

for curve $\mathbf{r} = \mathbf{r}(t) = x(t)\mathbf{i} + y(t)\mathbf{j} + z(t)\mathbf{k}$

$$k = \left|\frac{d^2\mathbf{r}}{ds^2}\right| = \sqrt{\left(\frac{d^2x}{ds^2}\right)^2 + \left(\frac{d^2y}{ds^2}\right)^2 + \left(\frac{d^2z}{ds^2}\right)^2}$$

for curve $\mathbf{r} = \mathbf{r}(s) = x(s)\mathbf{i} + y(s)\mathbf{j} + z(s)\mathbf{k}$

Coordinates of the center of curvature

$$\xi = x + \varrho^2 \frac{d^2x}{ds^2}; \qquad \eta = y + \varrho^2 \frac{d^2y}{ds^2}; \qquad \zeta = z + \varrho^2 \frac{d^2z}{ds^2}$$

Torsion

If the arc element between two adjacent points of a curve P_1 and P_2 are denoted by Δs and the angle that the binormals make at P_1 and P_2 by $\Delta\varepsilon$, then we have

$$\lim_{\Delta s \to 0} \frac{\Delta \varepsilon}{\Delta s} = \frac{d\varepsilon}{ds} = \frac{1}{\tau} = T$$

T torsion; τ *radius of torsion*; $d\varepsilon$ *angle of torsion*

A torsion is *positive* or *negative* according as the curve is *right-handed* or *left-handed* (sense of twist anticlockwise or clockwise)

$$T = 0 \Rightarrow \text{plane curve}$$

$$T \neq 0 \Rightarrow \text{skew curve}$$

$$T = \varrho^2 \left(\frac{d\mathbf{r}}{ds} \cdot \frac{d^2\mathbf{r}}{ds^2} \cdot \frac{d^3\mathbf{r}}{ds^3}\right) = \frac{\begin{vmatrix} \dot{x} & \dot{y} & \dot{z} \\ \ddot{x} & \ddot{y} & \ddot{z} \\ \dddot{x} & \dddot{y} & \dddot{z} \end{vmatrix}}{\ddot{x}^2 + \ddot{y}^2 + \ddot{z}^2}$$

with

$$\dot{x} = \frac{dx}{ds}; \quad \ddot{x} = \frac{d^2x}{ds^2}; \quad \dddot{x} = \frac{d^3x}{ds^3}; \quad \varrho \text{ radius of curvature}$$

for curve $\mathbf{r} = \mathbf{r}(s) = x(s)\mathbf{i} + y(s)\mathbf{j} + z(s)\mathbf{k}$

$$T = \varrho^2 \frac{\dfrac{d\mathbf{r}}{dt} \cdot \dfrac{d^2\mathbf{r}}{dt^2} \cdot \dfrac{d^3\mathbf{r}}{dt^3}}{\left[\left(\dfrac{d\mathbf{r}}{dt}\right)^2\right]^3} = \varrho^2 \frac{\begin{vmatrix} \dot{x} & \dot{y} & \dot{z} \\ \ddot{x} & \ddot{y} & \ddot{z} \\ \dddot{x} & \dddot{y} & \dddot{z} \end{vmatrix}}{(\dot{x}^2 + \dot{y}^2 + \dot{z}^2)^3}$$

for curve $\mathbf{r} = \mathbf{r}(t) = x(t)\mathbf{i} + y(t)\mathbf{j} + z(t)\mathbf{k}$

Example:

Common *helix*

$$\left\{(x, y, z) \,\middle|\, x = r\cos t,\ y = r\sin t,\ z = \frac{h}{2\pi} t\right\} \quad \begin{matrix} (h \text{ pitch}) \\ h, t \in R \end{matrix}$$

Calculation of the curvature and torsion

$$k^2 = \frac{(\dot{x}^2 + \dot{y}^2 + \dot{z}^2)(\ddot{x}^2 + \ddot{y}^2 + \ddot{z}^2) - (\dot{x}\ddot{x} + \dot{y}\ddot{y} + \dot{z}\ddot{z})^2}{(\dot{x}^2 + \dot{y}^2 + \dot{z}^2)^3}$$

$$= \frac{\left(r^2\sin^2 t + r^2\cos^2 t + \dfrac{h^2}{4\pi^2}\right)^3 (r^2\cos^2 t + r^2\sin^2 t)}{\left(r^2\sin^2 t + r^2\cos^2 t + \dfrac{h^2}{4\pi^2}\right)^3}$$

$$- \frac{(r^2\sin t\cos t - r^2\sin t\cos t)^2}{\left(r^2\sin^2 t + r^2\cos^2 t + \dfrac{h^2}{4\pi^2}\right)^3} = \frac{r^2}{\left(r^2 + \dfrac{h^2}{4\pi^2}\right)^2}$$

$$k = \frac{r}{r^2 + \dfrac{h^2}{4\pi^2}}$$

$$T = \varrho^2 \frac{\begin{vmatrix} \dot{x} & \dot{y} & \dot{z} \\ \ddot{x} & \ddot{y} & \ddot{z} \\ \dddot{x} & \dddot{y} & \dddot{z} \end{vmatrix}}{(\dot{x}^2 + \dot{y}^2 + \dot{z}^2)^3}$$

$$= \frac{\left(r^2 + \dfrac{h^2}{4\pi^2}\right)^2}{r^2 \left(r^2 + \dfrac{h^2}{4\pi^2}\right)^3} \cdot \begin{vmatrix} -r \sin t & r \cos t & \dfrac{h}{2\pi} \\ -r \cos t & -r \sin t & 0 \\ r \sin t & -r \cos t & 0 \end{vmatrix}$$

$$= \frac{1}{r^2 \left(r^2 + \dfrac{h^2}{4\pi^2}\right)} \cdot \frac{hr^2}{2\pi} = \underline{\underline{\frac{\dfrac{h}{2\pi}}{r^2 + \dfrac{h^2}{4\pi^2}}}}$$

6.3. Curved surfaces

Representation

$f(x, y, z) = 0$ (implicit form)

$z = f(x, y)$ (explicit form)

$x = x(u, v); \quad y = y(u, v); \quad z = z(u, v)$ (parametric form)

$u; v \in R$

$\mathbf{r} = \mathbf{r}(u, v) = x(u, v)\mathbf{i} + y(u, v)\,\mathbf{j} + z(u, v)\mathbf{k}$ (vector form)

The pair of parameter values u, v are referred to as curvilinear coordinates of the point $P(x, y, z)$ of a surface.
The so-called v-lines or u-lines are obtained for constant u and variable v or for constant v and variable u.
For extreme values of the function $z = f(x, y)$ see page 297.

Singular points of surface

If point $P_0(x_0, y_0, z_0)$ is a singular point of the surface $f(x, y, z) = 0$, its coordinates satisfy the equations

$$f_x = 0; \quad f_y = 0; \quad f_z = 0$$

Tangents through an ordinary point of a surface lie in the tangential plane, whereas the tangents through a singular point form a *cone of the second order.*

7. Integral calculus

7.1. Definition of the indefinite integral

$\int f(x)\,dx = F(x) + C$ where $F'(x) = f(x)$. $f(x)$ is called the integrand.

$F(x)$ is called an *antiderivative* of the given function $f(x)$. Due to the occurrence of the arbitrary constant C (*constant of integration*), an infinite number of integrals may be obtained from a given integrand.
Geometrically this means that an infinite number of integral curves exist which follow from one another by parallel displacement in the direction of the ordinate axis.

7.2. Basic integrals

1. $\displaystyle\int x^n\,dx = \frac{x^{n+1}}{n+1} + C \quad \text{for} \quad n \in R \setminus \{-1\}$

2. $\displaystyle\int \frac{dx}{x} = \ln|x| + C \quad \text{for} \quad x \neq 0$

3. $\int e^x\,dx = e^x + C$

4. $\displaystyle\int a^x\,dx = \frac{a^x}{\ln a} + C = a^x \log_a e + C$

5. $\int \sin x\,dx = -\cos x + C$

6. $\int \cos x\,dx = \sin x + C$

7. $\displaystyle\int \frac{dx}{\cos^2 x} = \tan x + C \quad \text{for} \quad x \neq \frac{(2n+1)\pi}{2} \quad n \in I$

8. $\displaystyle\int \frac{dx}{\sin^2 x} = -\cot x + C \quad \text{for} \quad x \neq n\pi \quad n \in I$

9. $\displaystyle\int \frac{dx}{\sqrt{1-x^2}} = \arcsin x + C = -\arccos x + C_1 \quad \text{for} \quad |x| < 1$

10. $\int \frac{dx}{1+x^2} = \arc\tan x + C = -\operatorname{arc}\cot x + C_1$

11. $\int \sinh x\, dx = \cosh x + C$

12. $\int \cosh x\, dx = \sinh x + C$

13. $\int \frac{dx}{\cosh^2 x} = \tanh x + C$

14. $\int \frac{dx}{\sinh^2 x} = -\coth x + C \quad \text{for} \quad x \neq 0$

15. $\int \frac{dx}{\sqrt{x^2+1}} = \sinh^{-1} x + C = \ln\left(x + \sqrt{x^2+1}\right) + C$

16. $\int \frac{dx}{\sqrt{x^2-1}} = \cosh^{-1} x + C = \ln\left(x \pm \sqrt{x^2-1}\right) + C \text{ for } |x| > 1$

17. $\int \frac{dx}{1-x^2} = \tanh^{-1} x + C = \frac{1}{2} \ln \frac{1+x}{1-x} + C \quad \text{for} \quad |x| < 1$

18. $\int \frac{dx}{1-x^2} = \coth^{-1} x + C = \frac{1}{2} \ln \frac{x+1}{x-1} + C \quad \text{for} \quad |x| > 1$

19. $\int \frac{dx}{x^2-1} = -\coth^{-1} x + C = \frac{1}{2} \ln \frac{x-1}{x+1} + C \quad \text{for} \quad |x| > 1$

7.3. Rules of integration

Integration of a sum or difference:

$$\int [f(x) + g(x) - h(x)]\, dx = \int f(x)\, dx + \int g(x)\, dx - \int h(x)\, dx$$

Integration of a function with constant factor:

$$\int cf(x)\, dx = c \int f(x)\, dx$$

Integration by substitution:

$$\int f(x)\, dx = \int f[\varphi(t)]\, \dot{\varphi}(t)\, dt$$

where $x = \varphi(t)$ and $dx = \dot{\varphi}(t)\, dt$

Frequently occurring *substitutions:*

$$ax = t \qquad dx = \frac{1}{a}\,dt$$

$$\frac{x}{a} = t \qquad dx = a\,dt$$

$$\frac{a}{x} = t \qquad dx = -\frac{a}{t^2}\,dt$$

$$a^x = t \qquad dx = \frac{dt}{t \ln a}$$

$$\sqrt{x} = t \qquad dx = 2t\,dt$$

$$e^x = t \qquad dx = \frac{1}{t}\,dt$$

$$\ln x = t \qquad dx = e^t\,dt$$

$$a + bx = t \qquad dx = \frac{1}{b}\,dt$$

$$a^2 + x^2 = t \qquad dx = \frac{dt}{2\sqrt{t - a^2}}$$

$$\sqrt{a + bx} = t \qquad dx = \frac{2t\,dt}{b}$$

$$a + bx^2 = t \qquad dx = \frac{dt}{2\sqrt{bt - ab}}$$

$$\sqrt{a^2 + x^2} = t \qquad dx = \frac{t\,dt}{\sqrt{t^2 - a^2}}$$

$$\sqrt{a^2 - x^2} = t \qquad dx = -\frac{t\,dt}{\sqrt{a^2 - t^2}}$$

$$\sqrt[n]{a + bx} = t \qquad dx = \frac{nt^{n-1}\,dt}{b}$$

$$\sqrt{x^2 - a^2} = t \qquad dx = \frac{t\,dt}{\sqrt{t^2 + a^2}}$$

Substitutions with respect to special integrals

1. $\int f\left(x, \sqrt{a^2 - x^2}\right) dx$

Substitution: $x = a \sin t; \quad dx = a \cos t \, dt$

yields $\int f(a \sin t, a \cos t) \, a \cos t \, dt$

where:

$$\sin t = \frac{x}{a}; \qquad \tan t = \frac{x}{\sqrt{a^2 - x^2}}$$

$$\cos t = \frac{\sqrt{a^2 - x^2}}{a}; \quad \cot t = \frac{\sqrt{a^2 - x^2}}{x}$$

2. $\int f\left(x, \sqrt{a^2 - x^2}\right) dx$

Substitution: $x = a \tanh t; \quad dx = \dfrac{a \, dt}{\cosh^2 t}$

yields $\displaystyle\int f\left(a \tanh t, \frac{a}{\cosh t}\right) \frac{a \, dt}{\cosh^2 t}$

where:

$$\sinh t = \frac{x}{\sqrt{a^2 - x^2}}; \quad \tanh t = \frac{x}{a}$$

$$\cosh t = \frac{a}{\sqrt{a^2 - x^2}}; \quad \coth t = \frac{a}{x}$$

3. $\int f\left(x, \sqrt{a^2 + x^2}\right) dx$

Substitution: $x = a \tan t; \quad dx = \dfrac{a \, dt}{\cos^2 t}$

yields $\displaystyle\int f\left(a \tan t, \frac{a}{\cos^2 t}\right) \frac{a \, dt}{\cos^2 t}$

where:

$$\sin t = \frac{x}{\sqrt{a^2 + x^2}}; \quad \tan t = \frac{x}{a}$$

$$\cos t = \frac{a}{\sqrt{a^2 + x^2}}; \quad \cot t = \frac{a}{x}$$

4. $\int f\left(x, \sqrt{a^2 + x^2}\right) dx$

Substitution: $x = a \sinh t; \quad dx = a \cosh t \, dt$

yields $\int f(a \sinh t, a \cosh t)\, a \cosh t \, dt$

where:

$$\sinh t = \frac{x}{a}; \quad \tanh t = \frac{x}{\sqrt{a^2 + x^2}}$$

$$\cosh t = \frac{\sqrt{a^2 + x^2}}{a}; \quad \coth t = \frac{\sqrt{a^2 + x^2}}{x}$$

5. $\int f\left(x, \sqrt{x^2 - a^2}\right) dx$

Substitution: $x = \dfrac{a}{\cos t}; \quad dx = \dfrac{a \sin t \, dt}{\cos^2 t}$

yields $\displaystyle\int f\left(\frac{a}{\cos t}, a \tan t\right) \frac{a \sin t \, dt}{\cos^2 t}$

where:

$$\sin t = \frac{\sqrt{x^2 - a^2}}{x}; \quad \tan t = \frac{\sqrt{x^2 - a^2}}{a}$$

$$\cos t = \frac{a}{x}; \quad \cot t = \frac{a}{\sqrt{x^2 - a^2}}$$

6. $\int f\left(x, \sqrt{x^2 - a^2}\right) dx$

Substitution: $x = a \cosh t; \quad dx = a \sinh t \, dt$

yields $\int f(a \cosh t, a \sinh t)\, a \sinh t \, dt$

where:

$$\sinh t = \frac{\sqrt{x^2 - a^2}}{a}; \quad \tanh t = \frac{\sqrt{x^2 - a^2}}{x}$$

$$\cosh t = \frac{x}{a}; \quad \coth t = \frac{x}{\sqrt{x^2 - a^2}}$$

7. $\int f(\sin x, \cos x, \tan x, \cot x)\, dx$

Substitution: $\tan \frac{x}{2} = t; \quad dx = \frac{2}{1 + t^2}\, dt$

yields $\int f\left(\frac{2t}{1 + t^2}, \frac{1 - t^2}{1 + t^2}, \frac{2t}{1 - t^2}, \frac{1 - t^2}{2t}\right) \frac{2dt}{1 + t^2}$

In this case, the integrand of the initial integral must be a rational function of the four trigonometric functions.

8. $\int f(\sinh x, \cosh x, \tanh x, \coth x)\, dx$

Substitution: $\tanh \frac{x}{2} = t; \quad dx = \frac{2dt}{1 - t^2}$

yields $\int f\left(\frac{2t}{1 - t^2}, \frac{1 + t^2}{1 - t^2}, \frac{2t}{1 + t^2}, \frac{1 + t^2}{2t}\right) \frac{2dt}{1 - t^2}$

where the integrand of the initial integral must be a rational function of the hyperbolic functions.

9. $\int f(e^x)\, dx$

Substitution: $e^x = t; \quad dx = \frac{dt}{t}$

yields $\int f(t)\, dt$

10. $\int f\left(x, \sqrt[k]{ax + b}\right) dx$

Substitution: $ax + b = t; \quad dx = \frac{kt^{k-1}\, dt}{a}$

yields $\int f(t)\, dt$

11. $\int f\left[x, \left(\frac{ax + b}{cx + d}\right)^p, \left(\frac{ax + b}{cx + d}\right)^q, \ldots\right] dx$

Substitution: $\frac{ax + b}{cx + d} = t^n$

yields $\int f(t)\, dt$

(n is the smallest common multiple of the exponents $p, q, \ldots$; $ad - bc \neq 0$)

12. $\int f\left(x, \sqrt{ax^2 + bx + c}\right) dx$

Substitutions according to EULER result in $\int f(t)\, dt$.

Case 1

$a > 0$, substitution: $\sqrt{ax^2 + bx + c} = x\sqrt{a} + t$

$$x = \frac{t^2 - c}{b - 2t\sqrt{a}}; \quad dx = 2\,\frac{-t^2\sqrt{a} + bt - c\sqrt{a}}{\left(b - 2t\sqrt{a}\right)^2}\, dt$$

Case 2

$c > 0$, $x \neq 0$, substitution: $\sqrt{ax^2 + bx + c} = xt + \sqrt{c}$

$$x = \frac{2t\sqrt{c} - b}{a - t^2}; \quad dx = \frac{2a\sqrt{c} - 2bt + 2t^2\sqrt{c}}{(a - t^2)^2}\, dt$$

Case 3

The radicand has the real roots x_1 and x_2.

Substitution: $\sqrt{ax^2 + bx + c} = t(x - x_1)$

$$x = \frac{t^2 x_1 - a x_2}{t^2 - a}; \quad dx = \frac{2at(x_2 - x_1)}{(t^2 - a)^2}\, dt$$

Integration by parts

$$\int uv'\, dx = uv - \int vu'\, dx \quad \text{where} \quad u = u(x); \quad v = v(x)$$

This may also be written as $\int u\, dv = uv - \int v\, du$

Example:

$$\int x^3 \ln x\, dx \quad u = \ln x \quad \Big| \quad dv = x^3\, dx$$

$$du = \frac{1}{x}\, dx \quad \Big| \quad v = \frac{x^4}{4}$$

$$\int x^3 \ln x\, dx = \frac{x^4}{4} \ln x - \frac{1}{4} \int \frac{x^4}{4}\, dx = \frac{x^4}{4} \ln x - \frac{1}{4} \int x^3\, dx$$

$$= \frac{x^4}{4} \ln x - \frac{1}{4} \cdot \frac{x^4}{4} + C$$

$$= \frac{x^4}{4}\left(\ln x - \frac{1}{4}\right) + C$$

Integration by conversion into partial fractions

Decomposition of a proper fraction $\frac{f(x)}{g(x)}$ into partial fractions

Case 1

The equation $g(x) = 0$ has only simple real roots $x_1, x_2, \ldots$ Then $\frac{f(x)}{g(x)}$ can be expanded into partial fractions in the following way:

$$\frac{f(x)}{g(x)} = \frac{A}{x - x_1} + \frac{B}{x - x_2} + \frac{C}{x - x_3} + \cdots$$

where

$$A = \frac{f(x_1)}{g'(x_1)}, \quad B = \frac{f(x_2)}{g'(x_2)}, \quad C = \frac{f(x_3)}{g'(x_3)} \quad \text{etc.}$$

$$\int \frac{f(x)}{g(x)}\, dx = A \int \frac{dx}{x - x_1} + B \int \frac{dx}{x - x_2} + C \int \frac{dx}{x - x_3} + \cdots$$

The denominators $A, B, C, \ldots$ of the partial fractions can also be found by the *method of the indefinite coefficients* (*comparison of coefficients*); frequently, this procedure may lead to a solution more quickly.

Example:

$$\int \frac{15x^2 - 70x - 95}{x^3 - 6x^2 - 13x + 42}\, dx$$

$$x^3 - 6x^2 - 13x + 42 = 0 \Rightarrow x_1 = 2; \quad x_2 = -3; \quad x_3 = 7$$

$$\frac{15x^2 - 70x - 95}{x^3 - 6x^2 - 13x + 42} = \frac{A}{x - 2} + \frac{B}{x + 3} + \frac{C}{x - 7}$$

$$= \frac{A(x + 3)(x - 7) + B(x - 2)(x - 7) + C(x - 2)(x + 3)}{(x - 2)(x + 3)(x - 7)}$$

$$= \frac{(A + B + C)x^2 - (4A + 9B - C)x - (21A - 14B + 6C)}{(x - 2)(x + 3)(x - 7)}$$

By equating the coefficients of equal powers of x in the numerator of the first and last fractions, we obtain the following system of equations:

$$\left.\begin{array}{l} A + B + C = 15 \\ 4A + 9B - C = 70 \\ 21A - 14B + 6C = 95 \end{array}\right\} \Rightarrow A = 7;\ B = 5;\ C = 3$$

or

$$A = \frac{f(x_1)}{g'(x_1)} = \frac{-175}{-25} = 7 \quad \text{etc.}$$

Hence, the following holds:

$$\int \frac{15x^2 - 70x - 95}{x^3 - 6x^2 - 13x + 42}\,dx = 7\int \frac{dx}{x-2} + 5\int \frac{dx}{x+3}$$

$$+ 3\int \frac{dx}{x-7} = \underline{\underline{7\ln|x-2| + 5\ln|x+3| + 3\ln|x-7| + C}}$$

Case 2

The roots of the equation $g(x) = 0$ are real, but they occur several times, e.g. root x_1 α times, root x_2 β times, etc.

$$\frac{f(x)}{g(x)} = \frac{A_1}{(x-x_1)^\alpha} + \frac{A_2}{(x-x_1)^{\alpha-1}} + \frac{A_3}{(x-x_1)^{\alpha-2}} + \cdots + \frac{A_\alpha}{x-x_1}$$

$$+ \frac{B_1}{(x-x_2)^\beta} + \frac{B_2}{(x-x_2)^{\beta-1}} + \frac{B_3}{(x-x_2)^{\beta-2}} + \cdots + \frac{B_\beta}{x-x_2}$$

$$+ \frac{C_1}{(x-x_3)^\gamma} + \frac{C_2}{(x-x_3)^{\gamma-1}} + \frac{C_3}{(x-x_3)^{\gamma-2}} + \cdots + \frac{C_\gamma}{x-x_3} + \cdots$$

The constants $A_1, A_2, \ldots, B_1, B_2, \ldots, C_1, C_2, \ldots$ are calculated according to the method of indefinite coefficients.

Example:

$$\int \frac{3x^3 + 10x^2 - x}{(x^2-1)^2}\,dx$$

$$(x^2-1)^2 = 0 \Rightarrow x_1 = x_1 = 1 \quad \text{and} \quad x_3 = x_4 = -1$$

$$\frac{3x^3 + 10x^2 - x}{(x^2 - 1)^2} = \frac{A_1}{(x-1)^2} + \frac{A_2}{x-1} + \frac{B_1}{(x+1)^2} + \frac{B_2}{(x+1)}$$

$$= \frac{A_1(x+1)^2 + A_2(x+1)^2(x-1) + B_1(x-1)^2 + B_2(x-1)^2(x+1)}{(x-1)^2(x+1)^2}$$

$$= \frac{(A_2 + B_2)x^3 + (A_1 + A_2 + B_1 - B_2)x^2 + (2A_1 - A_2 - 2B_1 - B_3)x}{(x-1)^2(x+1)^2}$$

$$+ \frac{(A_1 - A_2 + B_1 + B_2)}{(x_1 - 1)^2(x_2 + 1)^2}$$

The method of the indefinite coefficients leads to the system of equations

$$\left.\begin{aligned} A_2 + B_2 &= 3 \\ A_1 + A_2 + B_1 - B_2 &= 10 \\ 2A_1 - A_2 - 2B_1 - B_2 &= -1 \\ A_1 - A_2 + B_1 + B_2 &= 0 \end{aligned}\right\} \begin{aligned} &\Rightarrow A_1 = 3; \quad A_2 = 4; \\ &B_1 = 2; \quad B_2 = -1 \end{aligned}$$

Hence, the following holds:

$$\int \frac{3x^3 + 10x^2 - x}{(x^2 - 1)^2}\,dx$$

$$= 3\int \frac{dx}{(x-1)^2} + 4\int \frac{dx}{x-1} + 2\int \frac{dx}{(x+1)^2} - \int \frac{dx}{x+1}$$

$$= -\frac{3}{x-1} + 4\ln|x-1| - \frac{2}{x+1} - \ln|x+1| + C$$

Case 3

The equation $g(x) = 0$ has, besides real roots, *simple complex roots* which occur in the conjugate complex form. The above-mentioned procedures of decomposition into partial fractions may also be applied to this case, however, complex numerators may be involved. Calculating with complex quantities can be avoided by reducing the partial fractions that were obtained from complex roots to a common denominator. If, for example, x_1 and x_2 are two conjugate complex roots, the statement is

$$\frac{f(x)}{g(x)} = \frac{Px + Q}{(x - x_1)(x - x_2)}$$

where the coefficients are to be found by the method of the indefinite coefficients.
For the integration of the expression

$$\frac{Px + Q}{(x - x_1)(x - x_2)} = \frac{Px + Q}{x^2 + px + q},$$

a solution can be found by the following formula:

$$\int \frac{Px + Q}{x^2 + px + q}\, dx = \frac{P}{2} \ln |x^2 + px + q|$$

$$+ \frac{Q - \frac{Pp}{2}}{\sqrt{q - \frac{p^2}{4}}} \operatorname{arc\,tan} \frac{x + \frac{p}{2}}{\sqrt{q - \frac{p^2}{4}}} \quad \text{for} \quad q - \frac{p^2}{4} > 0$$

Example:

$$\int \frac{7x^2 - 10x + 37}{x^3 - 3x^2 + 9x + 13}\, dx$$

$$x^3 - 3x^2 + 9x + 13 = 0 \Rightarrow$$

$$\Rightarrow x_1 = -1; \quad x_2 = 2 + 3j; \quad x_3 = 2 - 3j$$

$$\frac{7x^2 - 10x + 37}{x^3 - 3x^2 + 9x + 13} = \frac{A}{x + 1} + \frac{Px + Q}{x^2 - 4x + 13}$$

$$= \frac{A(x^2 - 4x + 13) + (Px + Q)(x + 1)}{x^3 - 3x^2 + 9x + 13}$$

$$= \frac{(A + P)x^2 - (4A - Q - P)x + (13A + Q)}{x^3 - 3x^2 + 9x + 13}$$

The method of indefinite coefficients leads to the system of equations

$$\left.\begin{aligned} A + P &= 7 \\ 4A - Q - P &= 10 \\ 13A + Q &= 37 \end{aligned}\right\} \Rightarrow A = 3; \quad P = 4; \quad Q = -2$$

Hence, the following holds:

$$\int \frac{7x^2 - 10x + 37}{x^3 - 3x^2 + 9x + 13}\,dx$$

$$= 3\int \frac{dx}{x+1} + \int \frac{4x - 2}{x^2 - 4x + 13}\,dx$$

$$= 3 \ln |x + 1| + 2 \ln |x^2 - 4x + 13| + 2 \operatorname{arc\,tan} \frac{x-2}{3} + C$$

Case 4

The equation $g(x) = 0$ has *multiple complex roots* besides *real roots.* The best thing to do is to reduce the fractions obtained from conjugate complex roots to a common denominator, as in the example mentioned above. The decomposed form is:

$$\frac{f(x)}{g(x)} = \frac{A_1}{(x - x_1)^3} + \frac{A_2}{(x - x_1)^2} + \frac{A_3}{x - x_1} + \frac{P_1 x + Q_1}{(x^2 + px + q)^2} + \frac{P_2 x + Q_2}{(x^2 + px + q)}$$

In the example under consideration, x_1 appears as a triple *real* root and the conjugate complex roots appear as twofold roots.

7.4. A few special integrals

Note: To simplify the representation, the constants of integration have been omitted from the indefinite integrals.

7.4.1. Integrals of rational functions

1. $\int (ax + b)^n\,dx = \frac{(ax + b)^{n+1}}{a(n + 1)} \quad (n \neq -1)$

2. $\int \frac{dx}{ax + b} = \frac{1}{a} \ln |ax + b|$

3. $\int x(ax + b)^n\,dx = \frac{(ax + b)^{n+2}}{a^2(n + 2)} - \frac{b(ax + b)^{n+1}}{a^2(n + 1)} \quad (n \neq -1; -2)$

4. $\int \frac{x\,dx}{ax+b} = \frac{x}{a} - \frac{b}{a^2} \ln|ax+b|$

5. $\int \frac{x\,dx}{(ax+b)^2} = \frac{b}{a^2(ax+b)} + \frac{1}{a^2} \ln|ax+b|$

6. $\int \frac{x\,dx}{(ax+b)^n} = \frac{1}{a^2}\left[\frac{b}{(n-1)(ax+b)^{n-1}} - \frac{1}{(n-2)(ax+b)^{n-2}}\right]$

$(n \neq 1; 2)$

7. $\int \frac{x^2\,dx}{ax+b} = \frac{1}{a^3}\left[\frac{(ax+b)^2}{2} - 2b(ax+b) + b^2 \ln|ax+b|\right]$

8. $\int \frac{x^2\,dx}{(ax+b)^2} = \frac{1}{a^3}\left[ax+b - 2b\ln|ax+b| - \frac{b^2}{ax+b}\right]$

9. $\int \frac{x^2\,dx}{(ax+b)^3} = \frac{1}{a^3}\left[\ln|ax+b| + \frac{2b}{ax+b} - \frac{b^2}{2(ax+b)^2}\right]$

10. $\int \frac{x^2\,dx}{(ax+b)^n} = \frac{1}{a^3}\left[-\frac{1}{(n-3)(ax+b)^{n-3}}\right.$

$\left. + \frac{2b}{(n-2)(ax+b)^{n-2}} - \frac{b^2}{(n-1)(ax+b)^{n-1}}\right] \quad (n \neq 1; 2; 3)$

11. $\int \frac{dx}{x(ax+b)} = -\frac{1}{b} \ln\left|\frac{ax+b}{x}\right|$

12. $\int \frac{dx}{x^2(ax+b)} = -\frac{1}{bx} + \frac{a}{b^2} \ln\left|\frac{ax+b}{x}\right|$

13. $\int \frac{dx}{x^2(ax+b)^2} = -a\left[\frac{1}{b^2(ax+b)} + \frac{1}{ab^2x} - \frac{2}{b^2} \ln\left|\frac{ax+b}{x}\right|\right]$

14. $\int \frac{dx}{x^2+a^2} = \frac{1}{a} \operatorname{arc\,tan} \frac{x}{a}$

15. $\int \frac{dx}{x^2-a^2} = -\frac{1}{a} \tanh^{-1} \frac{x}{a} = \frac{1}{2a} \ln \frac{a-x}{a+x} \quad (|x| < a)$

16. $\int \frac{dx}{x^2-a^2} = -\frac{1}{a} \coth^{-1} \frac{x}{a} = \frac{1}{2a} \ln \frac{x-a}{x+a} \quad (|x| > a)$

17. $$\int \frac{dx}{ax^2 + bx + c}$$

$$= \frac{2}{\sqrt{4ac - b^2}} \arctan \frac{2ax + b}{\sqrt{4ac - b^2}} \quad \text{for} \quad 4ac - b^2 > 0$$

$$= -\frac{2}{\sqrt{b^2 - 4ac}} \tanh^{-1} \frac{2ax + b}{\sqrt{b^2 - 4ac}} \quad \text{for } 4ac - b^2 < 0$$

$$= \frac{1}{\sqrt{b^2 - 4ac}} \ln \left| \frac{2ax + b - \sqrt{b^2 - 4ac}}{2ax + b + \sqrt{b^2 - 4ac}} \right| \quad \text{for} \quad 4ac - b^2 < 0$$

18. $$\int \frac{x\,dx}{ax^2 + bx + c} = \frac{1}{2a} \ln |ax^2 + bx + c| - \frac{b}{2a} \int \frac{dx}{ax^2 + bx + c}$$

19. $$\int \frac{mx + n}{ax^2 + bx + c}\,dx = \frac{m}{2a} \ln |ax^2 + bx + c|$$

$$+ \frac{2an - bm}{a\sqrt{4ac - b^2}} \arctan \frac{2ax + b}{\sqrt{4ac - b^2}} \quad \text{for } 4ac - b^2 > 0$$

$$= \frac{m}{2a} \ln |ax^2 + bx + c|$$

$$- \frac{2an - bm}{a\sqrt{b^2 - 4ac}} \tanh^{-1} \frac{2ax + b}{\sqrt{b^2 - 4ac}} \quad \text{for } 4ac - b^2 < 0$$

20. $$\int \frac{dx}{(ax^2 + bx + c)^n} = \frac{2ax + b}{(n - 1)(4ac - b^2)(ax^2 + bx + c)^{n-1}} + \frac{(2n - 3)\,2a}{(n - 1)(4ac - b^2)} \int \frac{dx}{(ax^2 + bx + c)^{n-1}}$$

21. $$\int \frac{x\,dx}{(ax^2 + bx + c)^n} = -\frac{bx + 2c}{(n - 1)(4ac - b^2)(ax^2 + bx + c)^{n-1}} - \frac{-b(2n - 3)}{(n - 1)(4ac - b^2)} \int \frac{dx}{(ax^2 + bx + c)^{n-1}}$$

22. $$\int \frac{dx}{x(ax^2 + bx + c)} = \frac{1}{2c} \ln \left| \frac{x^2}{ax^2 + bx + c} \right| - \frac{b}{2c} \int \frac{dx}{ax^2 + bx + c}$$

7.4.2. Integrals of irrational functions

1. $\int \sqrt{a^2 - x^2}\, dx = \frac{1}{2}\left(x \sqrt{a^2 - x^2} + a^2 \arcsin \frac{x}{a}\right) \quad (|x| < a)$

2. $\int x \sqrt{a^2 - x^2}\, dx = -\frac{1}{3} \sqrt{(a^2 - x^2)^3}$

$$= -\frac{1}{3} (a^2 - x^2) \sqrt{a^2 - x^2} \quad (|x| < a)$$

3. $\int \frac{\sqrt{a^2 - x^2}\, dx}{x} = \sqrt{a^2 - x^2} - a \ln \left| \frac{a + \sqrt{a^2 - x^2}}{x} \right| \quad (|x| < a)$

4. $\int \frac{dx}{\sqrt{a^2 - x^2}} = \arcsin \frac{x}{a} \quad (|x| < a)$

5. $\int \frac{x^2\, dx}{\sqrt{a^2 - x^2}} = -\frac{x}{2} \sqrt{a^2 - x^2} + \frac{a^2}{2} \arcsin \frac{x}{a} \quad (|x| < a)$

6. $\int \sqrt{x^2 + a^2}\, dx = \frac{1}{2}\left(x \sqrt{x^2 + a^2} + a^2 \sinh^{-1} \frac{x}{a}\right)$

7. $\int x \sqrt{x^2 + a^2}\, dx = \frac{1}{3} \sqrt{(x^2 + a^2)^3} = \frac{1}{3} (x^2 + a^2) \sqrt{x^2 + a^2}$

8. $\int \frac{\sqrt{x^2 + a^2}\, dx}{x} = \sqrt{x^2 + a^2} - a \ln \left| \frac{a + \sqrt{x^2 + a^2}}{x} \right|$

9. $\int \frac{dx}{\sqrt{x^2 + a^2}} = \sinh^{-1} \frac{x}{a} = \ln \left| x + \sqrt{x^2 + a^2} \right|$

10. $\int \frac{x\, dx}{\sqrt{x^2 + a^2}} = \sqrt{x^2 + a^2}$

11. $\int \frac{x^2\, dx}{\sqrt{x^2 + a^2}} = \frac{x}{2} \sqrt{x^2 + a^2} - \frac{a^2}{2} \sin^{-1} \frac{x}{a}$

$$= \frac{x}{2} \sqrt{x^2 + a^2} - \frac{a^2}{2} \ln \left| x + \sqrt{x^2 + a^2} \right|$$

12. $\int \frac{dx}{x \sqrt{x^2 + a^2}} = -\frac{1}{a} \sinh^{-1} \frac{a}{x} = -\frac{1}{a} \ln \left| \frac{a + \sqrt{x^2 + a^2}}{x} \right|$

13. $$\int \frac{dx}{x^2\sqrt{x^2+a^2}} = -\frac{\sqrt{x^2+a^2}}{a^2 x}$$

14. $$\int \sqrt{x^2-a^2}\,dx = \frac{1}{2}\left(x\sqrt{x^2-a^2} \mp a^2 \cosh^{-1}\left|\frac{x}{a}\right|\right)$$

$$= \frac{1}{2}\left[x\sqrt{x^2-a^2} \mp a^2 \ln\left(|x| + \sqrt{x^2-a^2}\right)\right] \qquad (|x| \geqq |a|)$$

$-$ for $x > 0$, $+$ for $x < 0$

15. $$\int x\sqrt{x^2-a^2}\,dx = \frac{1}{3}\sqrt{(x^2-a^2)^3}$$

$$= \frac{1}{3}(x^2-a^2)\sqrt{x^2-a^2} \qquad (|x| \geqq |a|)$$

16. $$\int \frac{\sqrt{x^2-a^2}\,dx}{x} = \sqrt{x^2-a^2} - a \operatorname{arc\,cos}\frac{a}{x} \qquad (|x| \geqq |a|)$$

17. $$\int \frac{dx}{\sqrt{x^2-a^2}} = \cosh^{-1}\left|\frac{x}{a}\right| = \ln\left|\frac{x+\sqrt{x^2-a^2}}{a}\right| \qquad (|x| \geqq |a|)$$

18. $$\frac{x\,dx}{\sqrt{x^2-a^2}} = \sqrt{x^2-a^2} \qquad (|x| > a)$$

19. $$\int \frac{x^2\,dx}{\sqrt{x^2-a^2}} = \frac{x}{2}\sqrt{x^2-a^2} + \frac{a^2}{2}\cosh^{-1}\left|\frac{x}{a}\right|$$

$$= \frac{1}{2}\left[x\sqrt{x^2-a^2} + a^2 \ln\left(|x| + \sqrt{x^2-a_2}\right)\right] \qquad (|x| \geqq |a|)$$

20. $$\int \frac{dx}{\sqrt{ax^2+bx+c}} = \frac{1}{\sqrt{a}}\ln\left|2\sqrt{a(ax^2+bx+c)} + 2ax + b\right| \qquad (a > 0)$$

$$= \frac{1}{\sqrt{a}}\sinh^{-1}\frac{2ax+b}{\sqrt{4ac-b^2}} \qquad (a > 0;\ 4ac - b^2 > 0)$$

$$= \frac{1}{\sqrt{a}}\ln|2ax+b| \qquad (a > 0;\ 4ac - b^2 = 0)$$

$$= -\frac{1}{\sqrt{-a}}\operatorname{arc\,sin}\frac{2ax+b}{\sqrt{b^2-4ac}} \qquad (a < 0;\ 4ac - b^2 < 0)$$

21. $$\int \frac{x\,dx}{\sqrt{ax^2+bx+c}} = \frac{\sqrt{ax^2+bx+c}}{a} - \frac{b}{2a}\int \frac{dx}{\sqrt{ax^2+bx+c}}$$

7.4.3. Integrals of trigonometric functions

1. $$\int \sin cx\,dx = -\frac{1}{c}\cos cx$$

2. $$\int \sin^n cx\,dx = -\frac{\sin^{n-1} cx \cos cx}{nc} + \frac{n-1}{n}\int \sin^{n-2} cx\,dx \quad (n > 0)$$

3. $$\int x \sin cx\,dx = \frac{\sin cx}{c^2} - \frac{x \cos cx}{c}$$

4. $$\int x^n \sin cx\,dx = -\frac{x^n}{c}\cos cx + \frac{n}{c}\int x^{n-1}\cos cx\,dx \quad (n > 0)$$

5. $$\int \frac{\sin cx}{x}\,dx = cx - \frac{(cx)^3}{3\cdot 3!} + \frac{(cx)^5}{5\cdot 5!} - + \cdots$$

6. $$\int \frac{\sin cx}{x^n}\,dx = -\frac{1}{n-1}\,\frac{\sin cx}{x^{n-1}} + \frac{c}{n-1}\int \frac{\cos cx}{x^{n-1}}\,dx$$

7. $$\int \frac{dx}{\sin cx} = \frac{1}{c}\ln\left|\tan\frac{cx}{2}\right|$$

8. $$\int \frac{dx}{\sin^n cx} = -\frac{1}{c(n-1)}\,\frac{\cos cx}{\sin^{n-1} cx} + \frac{n-2}{n-1}\int \frac{dx}{\sin^{n-2} cx} \quad (n > 1)$$

9. $$\int \frac{dx}{1+\sin cx} = -\frac{1}{c}\tan\left(\frac{cx}{2} - \frac{\pi}{4}\right)$$

10. $$\int \frac{dx}{1-\sin cx} = \frac{1}{c}\tan\left(\frac{cx}{2} + \frac{\pi}{4}\right)$$

11. $$\int \frac{x\,dx}{1+\sin cx} = -\frac{x}{c}\tan\left(\frac{cx}{2} - \frac{\pi}{4}\right) + \frac{2}{c^2}\ln\left|\cos\left(\frac{cx}{2} - \frac{\pi}{4}\right)\right|$$

12. $$\int \frac{x\,dx}{1-\sin cx} = \frac{x}{c}\cot\left(\frac{\pi}{4} - \frac{cx}{2}\right) + \frac{2}{c^2}\ln\left|\sin\left(\frac{\pi}{4} - \frac{cx}{2}\right)\right|$$

13. $$\int \frac{\sin cx\,dx}{1\pm\sin cx} = \pm x + \frac{1}{c}\tan\left(\frac{\pi}{4} \mp \frac{cx}{2}\right)$$

14. $$\int \cos cx\,dx = \frac{1}{c}\sin cx$$

15. $$\int \cos^n cx\, dx = \frac{\cos^{n-1} cx \sin cx}{nc} + \frac{n-1}{n} \int \cos^{n-2} cx\, dx$$

16. $$\int x \cos cx\, dx = \frac{\cos cx}{c^2} + \frac{x \sin cx}{c}$$

17. $$\int x^n \cos cx\, dx = \frac{x^n \sin cx}{c} - \frac{n}{c} \int x^{n-1} \sin cx\, dx$$

18. $$\int \frac{\cos cx}{x}\, dx = \ln|cx| - \frac{(cx)^2}{2 \cdot 2!} + \frac{(cx)^4}{4 \cdot 4!} - + \cdots$$

19. $$\int \frac{\cos cx}{x^n}\, dx = -\frac{\cos cx}{(n-1)x^{n-1}} - \frac{c}{n-1} \int \frac{\sin cx\, dx}{x^{n-1}} \quad (n \neq 1)$$

20. $$\int \frac{dx}{\cos cx} = \frac{1}{c} \ln \left| \tan\left(\frac{cx}{2} + \frac{\pi}{4}\right) \right|$$

21. $$\int \frac{dx}{\cos^n cx} = \frac{1}{c(n-1)} \frac{\sin cx}{\cos^{n-1} cx} + \frac{n-2}{n-1} \int \frac{dx}{\cos^{n-2} cx} \quad (n > 1)$$

22. $$\int \frac{dx}{1 + \cos cx} = \frac{1}{c} \tan \frac{cx}{2}$$

23. $$\int \frac{dx}{1 - \cos cx} = -\frac{1}{c} \cot \frac{cx}{2}$$

24. $$\int \frac{x\, dx}{1 + \cos cx} = \frac{x}{c} \tan \frac{cx}{2} + \frac{2}{c^2} = \ln \left| \cos \frac{cx}{2} \right|$$

25. $$\int \frac{x\, dx}{1 - \cos cx} = -\frac{x}{c} \cot \frac{cx}{2} + \frac{2}{c^2} \ln \left| \sin \frac{cx}{2} \right|$$

26. $$\int \frac{\cos cx\, dx}{1 + \cos cx} = x - \frac{1}{c} \tan \frac{cx}{2}$$

27. $$\int \frac{\cos cx\, dx}{1 - \cos cx} = -x - \frac{1}{c} \cot \frac{cx}{2}$$

28. $$\int \frac{dx}{\cos cx + \sin cx} = \frac{1}{c\sqrt{2}} \ln \left| \tan\left(\frac{cx}{2} + \frac{\pi}{8}\right) \right|$$

29. $$\int \frac{dx}{\cos cx - \sin cx} = \frac{1}{c\sqrt{2}} \ln \left| \tan\left(\frac{cx}{2} - \frac{\pi}{8}\right) \right|$$

30. $$\int \frac{dx}{(\cos cx + \sin cx)^2} = \frac{1}{2c} \tan\left(cx - \frac{\pi}{4}\right)$$

31. $$\int \frac{dx}{(\cos cx - \sin cx)^2} = \frac{1}{2c} \tan\left(cx + \frac{\pi}{4}\right)$$

32. $$\int \frac{\cos cx\, dx}{\cos cx + \sin cx} = \frac{1}{2c}\left[cx + \ln\left|\sin\left(cx + \frac{\pi}{4}\right)\right|\right]$$
$$= \frac{1}{2c}(cx + \ln|\cos cx + \sin cx|)$$

33. $$\int \frac{\cos cx\, dx}{\cos cx - \sin cx} = \frac{1}{2c}\left[-cx + \ln\left|\cos\left(cx + \frac{\pi}{4}\right)\right|\right]$$
$$= \frac{1}{2c}(-cx + \ln|\cos cx - \sin cx|)$$

34. $$\int \frac{\sin cx\, dx}{\cos cx + \sin cx} = \frac{1}{2c}\left[cx - \ln\left|\sin\left(cx + \frac{\pi}{2}\right)\right|\right]$$
$$= \frac{1}{2c}(cx - \ln|\cos cx + \sin cx|)$$

35. $$\int \frac{\sin cx\, dx}{\cos cx - \sin cx} = -\frac{1}{2c}\left[cx + \ln\left|\cos\left(cx + \frac{\pi}{2}\right)\right|\right]$$
$$= -\frac{1}{2c}(cx + \ln|\cos cx - \sin cx|)$$

36. $$\int \frac{\cos cx\, dx}{\sin cx(1 + \cos cx)} = -\frac{1}{4c}\tan^2\frac{cx}{2} + \frac{1}{2c}\ln\left|\tan\frac{cx}{2}\right|$$

37. $$\int \frac{\cos cx\, dx}{\sin cx(1 - \cos cx)} = -\frac{1}{4c}\cot^2\frac{cx}{2} - \frac{1}{2c}\ln\left|\tan\frac{cx}{2}\right|$$

38. $$\int \frac{\sin cx\, dx}{\cos cx(1 + \sin cx)} = \frac{1}{4c}\cot^2\left(\frac{cx}{2} + \frac{\pi}{4}\right)$$
$$+ \frac{1}{2c}\ln\left|\tan\left(\frac{cx}{2} + \frac{\pi}{4}\right)\right|$$

39. $$\int \frac{\sin cx\, dx}{\cos cx(1-\sin cx)} = \frac{1}{4c}\tan^2\left(\frac{cx}{2}+\frac{\pi}{4}\right) - \frac{1}{2c}\ln\left|\tan\left(\frac{cx}{2}+\frac{\pi}{4}\right)\right|$$

40. $$\int \sin cx \cos cx\, dx = \frac{1}{2c}\sin^2 cx$$

41. $$\int \sin^n cx \cos cx\, dx = \frac{1}{c(n+1)}\sin^{n+1} cx \quad (n \neq -1)$$

42. $$\int \sin cx \cos^n cx\, dx = -\frac{1}{c(n+1)}\cos^{n+1} cx \quad (n \neq -1)$$

43. $$\int \sin^n cx \cos^m cx\, dx = -\frac{\sin^{n-1} cx \cos^{m+1} cx}{c(n+m)} + \frac{n-1}{n+m}\int \sin^{n-2} cx \cos^m cx\, dx$$
$$= \frac{\sin^{n+1} cx \cos^{m-1} cx}{c(n+m)} + \frac{m-1}{n+m}\int \sin^n cx \cos^{m-2} cx\, dx$$
$$(m; n > 0)$$

44. $$\int \frac{dx}{\sin cx \cos cx} = \frac{1}{c}\ln|\tan cx|$$

45. $$\int \frac{dx}{\sin cx \cos^n cx} = \frac{1}{c(n-1)\cos^{n-1} cx} + \int \frac{dx}{\sin cx \cos^{n-2} cx} \quad (n \neq 1)$$

46. $$\int \frac{dx}{\sin^n cx \cos cx} = -\frac{1}{c(n-1)\sin^{n-1} cx} + \int \frac{dx}{\sin^{n-2} cx \cos cx} \quad (n \neq 1)$$

47. $$\int \frac{\sin cx\, dx}{\cos^n cx} = \frac{1}{c(n-1)\cos^{n-1} cx} \quad (n \neq 1)$$

48. $$\int \frac{\sin^2 cx\, dx}{\cos cx} = -\frac{1}{c}\sin cx + \frac{1}{c}\ln\left|\tan\left(\frac{\pi}{4}+\frac{cx}{2}\right)\right|$$

49. $$\int \frac{\sin^2 cx\,dx}{\cos^n cx} = \frac{\sin cx}{c(n-1)\cos^{n-1} cx} - \frac{1}{n-1}\int \frac{dx}{\cos^{n-2} cx} \quad (n \neq 1)$$

50. $$\int \frac{\sin^n cx\,dx}{\cos cx} = -\frac{\sin^{n-1} cx}{c(n-1)} + \int \frac{\sin^{n-2} cx\,dx}{\cos cx} \quad (n \neq 1)$$

51. $$\int \frac{\sin^n cx\,dx}{\cos^m cx} = \frac{\sin^{n+1} cx}{c(m-1)\cos^{m-1} cx} - \frac{n-m+2}{m-1}\int \frac{\sin^n cx\,dx}{\cos^{m-2} cx} \quad (m \neq 1)$$

$$= -\frac{\sin^{n-1} cx}{c(n-m)\cos^{m-1} cx} + \frac{n-1}{n-m}\int \frac{\sin^{n-2} cx\,dx}{\cos^m cx} \quad (m \neq n)$$

$$= \frac{\sin^{n-1} cx}{c(m-1)\cos^{m-1} cx} - \frac{n-1}{m-1}\int \frac{\sin^{n-1} cx\,dx}{\cos^{m-2} cx} \quad (m \neq 1)$$

52. $$\int \frac{\cos cx\,dx}{\sin^n cx} = -\frac{1}{c(n-1)\sin^{n-1} cx} \quad (n \neq 1)$$

53. $$\int \frac{\cos^2 cx\,dx}{\sin cx} = \frac{1}{c}\left(\cos cx + \ln\left|\tan\frac{cx}{2}\right|\right)$$

54. $$\int \frac{\cos^2 cx\,dx}{\sin^n cx} = -\frac{1}{n-1}\left(\frac{\cos cx}{c\sin^{n-1} cx} + \int \frac{dx}{\sin^{n-2} cx}\right) \quad (n \neq 1)$$

55. $$\int \frac{\cos^n cx\,dx}{\sin cx} = \frac{\cos^{n-1} cx}{c(n-1)} + \int \frac{\cos^{n-2} cx\,dx}{\sin cx} \quad (n \neq 1)$$

56. $$\int \frac{\cos^n cx\,dx}{\sin^m cx} = -\frac{\cos^{n+1} cx}{c(m-1)\sin^{m-1} cx} - \frac{n-m+2}{m-1}\int \frac{\cos^n cx\,dx}{\sin^{m-2} cx} \quad (m \neq 1)$$

$$= \frac{\cos^{n-1} cx}{c(n-m)\sin^{m-1} cx} + \frac{n-1}{n-m}\int \frac{\cos^{n-2} cx\,dx}{\sin^m cx} \quad (m \neq n)$$

$$= -\frac{\cos^{n-1} cx}{c(m-1)\sin^{m-1} cx} - \frac{n-1}{m-1}\int \frac{\cos^{n-2} cx\,dx}{\sin^{m-2} cx} \quad (m \neq 1)$$

57. $\int \tan cx\, dx = -\frac{1}{c} \ln |\cos cx|$

58. $\int \tan^n cx\, dx = \frac{1}{c(n-1)} \tan^{n-1} cx - \int \tan^{n-2} cx\, dx \quad (n \neq 1)$

59. $\int \frac{\tan^n cx\, dx}{\cos^2 cx} = \frac{1}{c(n+1)} \tan^{n+1} cx \quad (n \neq -1)$

60. $\int \frac{dx}{\tan cx + 1} = \frac{x}{2} + \frac{1}{2c} \ln |\sin cx + \cos cx|$

61. $\int \frac{dx}{\tan cx - 1} = -\frac{x}{2} + \frac{1}{2c} \ln |\sin cx - \cos cx|$

62. $\int \frac{\tan cx\, dx}{\tan cx + 1} = \frac{x}{2} - \frac{1}{2c} \ln |\sin cx + \cos cx|$

63. $\int \frac{\tan cx\, dx}{\tan cx - 1} = \frac{x}{2} + \frac{1}{2c} \ln |\sin cx - \cos cx|$

64. $\int \cot cx\, dx = \frac{1}{c} \ln |\sin cx|$

65. $\int \cot^n cx\, dx = -\frac{1}{c(n-1)} \cot^{n-1} cx - \int \cot^{n-2} cx\, dx \quad (n \neq 1)$

66. $\int \frac{\cot^n cx\, dx}{\sin^2 cx} = -\frac{1}{c(n+1)} \cot^{n+1} cx \quad (n \neq -1)$

67. $\int \frac{dx}{1 + \cot cx} = \int \frac{\tan cx\, dx}{\tan cx + 1}$

68. $\int \frac{dx}{1 - \cot cx} = \int \frac{\tan cx\, dx}{\tan cx - 1}$

7.4.4. Integrals of the hyperbolic functions

1. $\int \sinh cx\, dx = \frac{1}{c} \cosh cx$

2. $\int \cosh cx\, dx = \frac{1}{c} \sinh cx$

3. $\int \sinh^2 cx\,dx = \frac{1}{4c}\sinh 2cx - \frac{x}{2}$

4. $\int \cosh^2 cx\,dx = \frac{1}{4c}\sinh 2cx + \frac{x}{2}$

5. $$\int \sinh^n cx\,dx = \frac{1}{cn}\sinh^{n-1} cx\cosh cx - \frac{n-1}{n}\int \sinh^{n-2} cx\,dx \quad (n > 0)$$
$$= \frac{1}{c(n+1)}\sinh^{n+1} cx\cosh cx - \frac{n+2}{n+1}\int \sinh^{n+2} cx\,dx \quad (n < 0;\ n \neq -1)$$

6. $$\int \cosh^n cx\,dx = \frac{1}{cn}\sinh cx\cosh^{n-1} cx + \frac{n-1}{n}\int \cosh^{n-2} cx\,dx \quad (n > 0)$$
$$= -\frac{1}{c(n+1)}\sinh cx\cosh^{n+1} cx + \frac{n+2}{n+1}\int \cosh^{n+2} cx\,dx \quad (n < 0;\ n \neq -1)$$

7. $$\int \frac{dx}{\sinh cx} = \frac{1}{c}\ln\left|\tanh\frac{cx}{2}\right| = \frac{1}{c}\ln\left|\frac{\cosh cx - 1}{\sinh cx}\right|$$
$$= \frac{1}{c}\ln\left|\frac{\sinh cx}{\cosh cx + 1}\right| = \frac{1}{2c}\ln\left|\frac{\cosh cx - 1}{\cosh cx + 1}\right|$$

8. $$\int \frac{dx}{\cosh cx} = \frac{2}{c}\arctan e^{cx}$$

9. $$\int \frac{dx}{\sinh^n cx} = -\frac{1}{c(n-1)}\frac{\cosh cx}{\sinh^{n-1} cx} - \frac{n-2}{n-1}\int \frac{dx}{\sinh^{n-2} cx} \quad (n \neq 1)$$

10. $$\int \frac{dx}{\cosh^n cx} = \frac{1}{c(n-1)}\frac{\sinh cx}{\cosh^{n-1} cx} + \frac{n-2}{n-1}\int \frac{dx}{\cosh^{n-2} cx} \quad (n \neq 1)$$

11. $$\int \frac{\cosh^n cx}{\sinh^m cx}\,dx = \frac{1}{c(n-m)}\,\frac{\cosh^{n-1} cx}{\sinh^{m-1} cx} + \frac{n-1}{n-m}\int \frac{\cosh^{n-2} cx}{\sinh^m cx}\,dx \quad (m \neq n)$$

$$= -\frac{1}{c(m-1)}\,\frac{\cosh^{n+1} cx}{\sinh^{m-1} cx} + \frac{n-m+2}{m-1}\int \frac{\cosh^n cx}{\sinh^{m-2} cx}\,dx \quad (m \neq 1)$$

$$= -\frac{1}{c(m-1)}\,\frac{\cosh^{n-1} cx}{\sinh^{m-1} cx} + \frac{n-1}{m-1}\int \frac{\cosh^{n-2} cx}{\sinh^{m-2} cx}\,dx \quad (m \neq 1)$$

12. $$\int \frac{\sinh^m cx}{\cosh^n cx}\,dx = \frac{1}{c(m-n)}\,\frac{\sinh^{m-1} cx}{\cosh^{n-1} cx} - \frac{m-1}{m-n}\int \frac{\sinh^{m-2} cx}{\cosh^n cx}\,dx \quad (m \neq n)$$

$$= \frac{1}{c(n-1)}\,\frac{\sinh^{m+1} cx}{\cosh^{n-1} cx} - \frac{m-n+2}{n-1}\int \frac{\sinh^m cx}{\cosh^{n-2} cx}\,dx \quad (n \neq 1)$$

$$= -\frac{1}{c(n-1)}\,\frac{\sinh^{m-1} cx}{\cosh^{n-1} cx} + \frac{m-1}{n-1}\int \frac{\sinh^{m-2} cx}{\cosh^{n-2} cx}\,dx \quad (n \neq 1)$$

13. $$\int x \sinh cx\,dx = \frac{1}{c}\,x \cosh cx - \frac{1}{c^2} \sinh cx$$

14. $$\int x \cosh cx\,dx = \frac{1}{c}\,x \sinh cx - \frac{1}{c^2} \cosh cx$$

15. $$\int \tanh cx\,dx = \frac{1}{c} \ln |\cosh cx|$$

16. $$\int \coth cx\,dx = \frac{1}{c} \ln |\sinh cx|$$

17. $$\int \tanh^n cx\,dx = -\frac{1}{c(n-1)}\tanh^{n-1} cx + \int \tanh^{n-2} cx\,dx \qquad (n \neq 1)$$

18. $$\int \coth^n cx\,dx = -\frac{1}{c(n-1)}\coth^{n-1} cx + \int \coth^{n-2} cx\,dx \qquad (n \neq 1)$$

19. $$\int \sinh bx \sinh cx\,dx = \frac{1}{b^2 - c^2}(b \sinh cx \cosh bx - c \cosh cx \sinh bx) \quad (b^2 \neq c^2)$$

20. $$\int \cosh bx \cosh cx\,dx = \frac{1}{b^2 - c^2}(b \sinh bx \cosh cx - c \sinh cx \cosh bx) \quad (b^2 \neq c^2)$$

21. $$\int \cosh bx \sinh cx\,dx = \frac{1}{b^2 - c^2}(b \sinh bx \sinh cx - c \cosh bx \cosh cx) \quad (b^2 \neq c^2)$$

22. $$\int \sinh(ax+b)\sin(cx+d)\,dx = \frac{a}{a^2+c^2}\cosh(ax+b)\sin(cx+d) - \frac{c}{a^2+c^2}\sinh(ax+b)\cos(cx+d)$$

23. $$\int \sinh(ax+b)\cos(cx+d)\,dx = \frac{a}{a^2+c^2}\cosh(ax+b)\cos(cx+d) + \frac{c}{a^2+c^2}\sinh(ax+b)\sin(cx+d)$$

24. $$\int \cosh(ax+b)\sin(cx+d)\,dx = \frac{a}{a^2+c^2}\sinh(ax+b)\sin(cx+d) - \frac{c}{a^2+c^2}\cosh(ax+b)\cos(cx+d)$$

25. $$\int \cosh(ax+b)\cos(cx+d)\,dx = \frac{a}{a^2+c^2}\sinh(ax+b)\cos(cx+d) + \frac{c}{a^2+c^2}\cosh(ax+b)\sin(cx+d)$$

7.4.5. Integrals of exponential functions

1. $$\int e^{cx}\,dx = \frac{1}{c}\,e^{cx}$$

2. $$\int x\,e^{cx}\,dx = \frac{e^{cx}}{c^2}\,(cx-1)$$

3. $$\int x^2\,e^{cx}\,dx = e^{cx}\left(\frac{x^2}{c}-\frac{2x}{c^2}+\frac{2}{c^3}\right)$$

4. $$\int x^n\,e^{cx}\,dx = \frac{1}{c}\,x^n\,e^{cx} - \frac{n}{c}\int x^{n-1}\,e^{cx}\,dx$$

5. $$\int \frac{e^{cx}\,dx}{x} = \ln|x| + \frac{cx}{1\cdot 1!} + \frac{(cx)^2}{2\cdot 2!} + \cdots$$

6. $$\int \frac{e^{cx}\,dx}{x^n} = \frac{1}{n-1}\left(-\frac{e^{cx}}{x^{n-1}} + c\int\frac{e^{cx}\,dx}{x^{n-1}}\right) \quad (n \neq 1)$$

7. $$\int e^{cx}\ln x\,dx = \frac{1}{c}\left(e^{cx}\ln|x| - \int\frac{e^{cx}\,dx}{x}\right)$$

8. $$\int e^{cx}\sin bx\,dx = \frac{e^{cx}}{c^2+b^2}\,(c\sin bx - b\cos bx)$$

9. $$\int e^{cx}\cos bx\,dx = \frac{e^{cx}}{c^2+b^2}\,(c\cos bx + b\sin bx)$$

10. $$\int e^{cx}\sin^n x\,dx = \frac{e^{cx}\sin^{n-1}x}{c^2+n^2}\,(c\sin x - n\cos x) + \frac{n(n-1)}{c^2+n^2}\int e^{cx}\sin^{n-2}x\,dx$$

11. $$\int e^{cx}\cos^n x\,dx = \frac{e^{cx}\cos^{n-1}x}{c^2+n^2}\,(c\cos x + n\sin x) + \frac{n(n-1)}{c^2+n^2}\int e^{cx}\cos^{n-2}x\,dx$$

7.4.6. Integrals of the logarithmic functions

For $x > 0$, the following holds:

1. $\int \ln x \, dx = x \ln x - x$

2. $\int (\ln x)^2 \, dx = x(\ln x)^2 - 2x \ln x + 2x$

3. $\int (\ln x)^n \, dx = x(\ln x)^n - n \int (\ln x)^{n-1} \, dx \quad (n \neq -1)$

4. $$\int \frac{dx}{\ln x} = \ln |\ln x| + \ln x + \frac{(\ln x)^2}{2 \cdot 2!} + \frac{(\ln x)^3}{3 \cdot 3!} + \cdots$$

5. $$\int \frac{dx}{(\ln x)^n} = -\frac{x}{(n-1)(\ln x)^{n-1}} + \frac{1}{n-1} \int \frac{dx}{(\ln x)^{n-1}} \quad (n \neq 1)$$

6. $$\int x^m \ln x \, dx = x^{m+1} \left(\frac{\ln x}{m+1} - \frac{1}{(m+1)^2} \right) \quad (m \neq -1)$$

7. $$\int x^m (\ln x)^n \, dx = \frac{x^{m+1}(\ln x)^n}{m+1} - \frac{n}{m+1} \int x^m (\ln x)^{n-1} \, dx \quad (m;\, n \neq -1)$$

8. $$\int \frac{(\ln x)^n \, dx}{x} = \frac{(\ln x)^{n+1}}{n+1} \quad (n \neq -1)$$

9. $$\int \frac{\ln x \, dx}{x^m} = -\frac{\ln x}{(m-1)x^{m-1}} - \frac{1}{(m-1)^2 x^{m-1}} \quad (m \neq 1)$$

10. $$\int \frac{(\ln x)^n \, dx}{x^m} = -\frac{(\ln x)^n}{(m-1)x^{m-1}} + \frac{n}{m-1} \int \frac{(\ln x)^{n-1} \, dx}{x^m} \quad \begin{matrix} (m \neq 1) \\ (n \neq -1) \end{matrix}$$

11. $$\int \frac{x^m \, dx}{(\ln x)^n} = -\frac{x^{m+1}}{(n-1)(\ln x)^{n-1}} + \frac{m+1}{n-1} \int \frac{x^m \, dx}{(\ln x)^{n-1}} \quad (n \neq 1)$$

12. $$\int \frac{dx}{x \ln x} = \ln |\ln x|$$

13. $$\int \frac{dx}{x^n \ln x} = \ln |\ln x| - (n-1) \ln x + \frac{(n-1)^2 (\ln x)^2}{2 \cdot 2!} - \frac{(n-1)^3 (\ln x)^3}{3 \cdot 3!} + - \cdots$$

14. $$\int \frac{dx}{x (\ln x)^n} = -\frac{1}{(n-1)(\ln x)^{n-1}} \quad (n \neq 1)$$

15. $\int \sin(\ln x)\,dx = \frac{x}{2}[\sin(\ln x) - \cos(\ln x)]$

16. $\int \cos(\ln x)\,dx = \frac{x}{2}[\sin(\ln x) + \cos(\ln x)]$

17. $\int e^{cx} \ln x\,dx = \frac{1}{c}\left(e^{cx} \ln x - \int \frac{e^{cx}\,dx}{x}\right)$

7.4.7. Integrals of the inverse trigonometric functions (arc functions)

1. $\int \operatorname{arc\,sin} \frac{x}{c}\,dx = x \operatorname{arc\,sin} \frac{x}{c} + \sqrt{c^2 - x^2}$

2. $\int x \operatorname{arc\,sin} \frac{x}{c}\,dx = \left(\frac{x^2}{2} - \frac{c^2}{4}\right) \operatorname{arc\,sin} \frac{x}{c} + \frac{x}{4}\sqrt{c^2 - x^2}$

3. $\int x^2 \operatorname{arc\,sin} \frac{x}{c}\,dx = \frac{x^3}{3} \operatorname{arc\,sin} \frac{x}{c} + \frac{x^2 + 2c^2}{9}\sqrt{c^2 - x^2}$

4. $\int \operatorname{arc\,cos} \frac{x}{c}\,dx = x \operatorname{arc\,cos} \frac{x}{c} - \sqrt{c^2 - x^2}$

5. $\int x \operatorname{arc\,cos} \frac{x}{c}\,dx = \left(\frac{x^2}{2} - \frac{c^2}{4}\right) \operatorname{arc\,cos} \frac{x}{c} - \frac{x}{4}\sqrt{c^2 - x^2}$

6. $\int x^2 \operatorname{arc\,cos} \frac{x}{c}\,dx = \frac{x^3}{3} \operatorname{arc\,cos} \frac{x}{c} - \frac{x^2 + 2c^2}{9}\sqrt{c^2 - x^2}$

7. $\int \operatorname{arc\,tan} \frac{x}{c}\,dx = x \operatorname{arc\,tan} \frac{x}{c} - \frac{a}{2}\ln(c^2 + x^2)$

8. $\int x \operatorname{arc\,tan} \frac{x}{c}\,dx = \frac{c^2 + x^2}{2} \operatorname{arc\,tan} \frac{x}{c} - \frac{cx}{2}$

9. $\int x^2 \operatorname{arc\,tan} \frac{x}{c}\,dx = \frac{x^3}{3} \operatorname{arc\,tan} \frac{x}{c} - \frac{cx^2}{6} + \frac{c^3}{6}\ln(c^2 + x^2)$

10. $\int x^n \operatorname{arc\,tan} \frac{x}{c}\,dx = \frac{x^{n+1}}{n+1} \operatorname{arc\,tan} \frac{x}{c} - \frac{c}{n+1}\int \frac{x^{n+1}\,dx}{c^2 + x^2}$

$(n \neq -1)$

11. $\int \operatorname{arc\,cot} \frac{x}{c}\, dx = x \operatorname{arc\,cot} \frac{x}{c} + \frac{c}{2} \ln (c^2 + x^2)$

12. $\int x \operatorname{arc\,cot} \frac{x}{c}\, dx = \frac{c^2 + x^2}{2} \operatorname{arc\,cot} \frac{x}{c} + \frac{cx}{2}$

13. $\int x^2 \operatorname{arc\,cot} \frac{x}{c}\, dx = \frac{x^3}{3} \operatorname{arc\,cot} \frac{x}{c} + \frac{cx^2}{6} - \frac{c^3}{6} \ln (c^2 + x^2)$

14. $\int x^n \operatorname{arc\,cot} \frac{x}{c}\, dx = \frac{x^{n+1}}{n+1} \operatorname{arc\,cot} \frac{x}{c} + \frac{c}{n+1} \int \frac{x^{n+1}\, dx}{c^2 + x^2}$ $(n \neq -1)$

7.4.8. Integrals of the inverse hyperbolic functions (area functions)

1. $\int \sinh^{-1} \frac{x}{c}\, dx = x \sinh^{-1} \frac{x}{c} - \sqrt{x^2 + c^2}$

2. $\int \cosh^{-1} \frac{x}{c}\, dx = x \cosh^{-1} \frac{x}{c} - \sqrt{x^2 - c^2}$

3. $\int \tanh^{-1} \frac{x}{c}\, dx = x \tanh^{-1} \frac{x}{c} + \frac{c}{2} \ln |c^2 - x^2|$ $(|x| < |c|)$

4. $\int \coth^{-1} \frac{x}{c}\, dx = x \coth^{-1} \frac{x}{c} + \frac{c}{2} \ln |x^2 - c^2|$ $(|x| > |c|)$

7.5. Definite integral

7.5.1. General

Definition

For $x = a$ and $x = b$, the indefinite integral $\int f(x)\, dx = F(x) + C$ takes on nonunique values $F(a) + C$ and $F(b) + C$, respectively. If we find the difference, the indefinite constant of integration is eliminated and we obtain a definite value, the so-called definite integral between the lower limit a and the upper limit b:

$$\int_a^b f(x)\, dx = \int f(x)\, dx \Big|_a^b = F(x) \Big|_a^b = F(b) - F(a)$$

Calculating rules for definite integrals

$$\int_a^b f(x)\,dx = -\int_b^a f(x)\,dx$$

$$\int_a^a f(x)\,dx = 0$$

$$\int_a^b f(x)\,dx = \int_a^c f(x)\,dx + \int_c^b f(x)\,dx$$

$$\int_a^b [f(x) + g(x) + h(x)]\,dx = \int_a^b f(x)\,dx + \int_a^b g(x)\,dx + \int_a^b h(x)\,dx$$

$$\int_a^b cf(x)\,dx = c\int_a^b f(x)\,dx$$

Example:

$$\int_1^3 (2x + 3x^2)\,dx = \int (2x + 3x^2)\,dx \,\Big|_1^3 = (x^2 + x^3)\,\Big|_1^3$$

$$= (9 + 27) - (1 + 1) = \underline{\underline{34}}$$

7.5.2. Mean-value theorems for integral calculus

If $y = f(x)$ is continuous in the interval $[a, b]$, at least one value ξ exists within the interval for which

$$\int_a^b f(x)\,dx = (b - a)\,f(\xi)$$

This may also be written as:

$$\int_a^b f(x)\,dx = (b - a)\,f[a + \vartheta(b - a)] \quad \text{for} \quad 0 < \vartheta < 1$$

Extended mean-value theorem for integrals

If $f(x)$ and $g(x)$ are continuous in the interval $[a, b]$ and if $g(x)$ retains its sign within the interval, then

$$\int_a^b f(x)\, g(x)\, dx = f(\xi) \int_a^b g(x)\, dx$$

ξ is a value in the interval (a, b)

$$\int_a^b f(x)\, g(x)\, dx = f[a + \vartheta(b - a)] \int_a^b g(x)\, dx \quad \text{for} \quad 0 < \vartheta < 1$$

Arithmetic mean

of the function $f(x)$ between the limits a and b

$$\mathrm{AM} = \frac{1}{b - a} \int_a^b f(x)\, dx$$

Root-mean-square

of the function $f(x)$ between the limits a and b

$$\mathrm{RMS} = \sqrt{\frac{1}{b - a} \int_a^b f(x)^2\, dx}$$

Definite integral as a function of the upper limit

The definite integral with variable upper limit is a continuous function of this limit of integration.

$$\int_a^x f(x)\, dx = F(x) - F(a)$$

7.5.3. Geometrical interpretation of the definite integral

The definite integral $\int_a^b f(x)\, dx$ numerically states the area which is bounded by the curve $y = f(x)$, the x-axis and the ordinates $f(a)$ and $f(b)$. Areas below the x-axis take a negative sign in calculation, hence they must be taken in absolute value. If the function has one or several zeros within the limits, we calculate the area as a sum of the individual subareas above and/or below the x-axis, taking the subareas below the x-axis in absolute value.

7.5.4. Methods of approximation for definite integrals

Rectangle formula

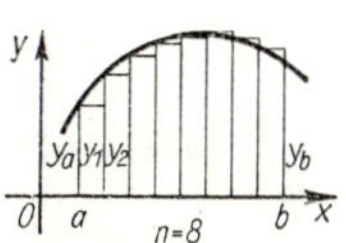

$$\int_a^b y\,dx \approx \frac{b-a}{n}(y_a + y_1 + y_2 + \cdots + y_{n-1})$$

n number of equal parts of the interval

Trapezoidal formula

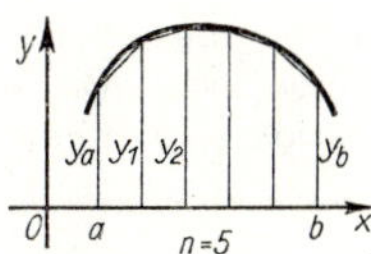

$$\int_a^b y\,dx \approx \frac{b-a}{2n}(y_a + 2y_1 + 2y_2 + \cdots + 2y_{n-1} + y_b)$$

n see above

Tangent formula

$$\int_a^b y\,dx \approx \frac{2(b-a)}{n}(y_1 + y_3 + y_5 + \cdots + y_{n-1}) \quad (n \text{ is even})$$

Simpson's rule

$$\int_a^b y\,dx \approx \frac{b-a}{3n}(y_a + 4y_1 + 2y_2 + 4y_3 + 2y_4 + \cdots + 2y_{n-2} + 4y_{n-1} + y_b)$$

Kepler's barrel rule

$$\int_a^b y\,dx \approx \frac{b-a}{6}(y_a + 4y_1 + y_b) \quad (n = 2)$$

Example:

$$\int_{-\frac{\pi}{2}}^{+\frac{\pi}{2}} \cos x \, dx = [\sin x]_{-\frac{\pi}{2}}^{+\frac{\pi}{2}} = \underline{\underline{2}}$$

By Kepler's barrel rule, we obtain

$$\int_{-\frac{\pi}{2}}^{+\frac{\pi}{2}} \cos x \, dx \approx \frac{\pi}{6} (0 + 4 \cdot 1 + 0) = \frac{2\pi}{3} \approx \underline{\underline{2.094}}$$

Integration by expansion in series

If the function to be integrated can be expanded in a convergent power series $f(x) = a_0 + a_1 x + a_2 x^2 + \cdots$ and if the limits of integration lie within the range of convergence, then:

$$\int_a^b f(x) \, dx = a_0 \int_a^b dx + a_1 \int_a^b x \, dx + a_2 \int_a^b x^2 \, dx + \cdots,$$

where $|a| < r$; $|b| < r$ (r radius of convergence)

Applications

Sine integral

$$\mathrm{Si}(x) = \int_0^x \frac{\sin t}{t} \, dt = x - \frac{x^3}{3 \cdot 3!} + \frac{x^5}{5 \cdot 5!} - \frac{x^7}{7 \cdot 7!} + - \cdots$$

for $|x| < \infty$

Cosine integral

$$\mathrm{Ci}(x) = \int_x^\infty \frac{\cos t}{t} \, dt = C - \ln |x| - \frac{x^2}{2 \cdot 2!} + \frac{x^4}{4 \cdot 4!} - \frac{x^6}{6 \cdot 6!}$$

$$+ - \cdots \quad \text{for} \quad |x| < \infty$$

C **(Euler's constant)** $= 0.5772\ldots$

Exponential integral

$$\operatorname{Ei}(x) = \int_{-\infty}^{x} \frac{e^t}{t}\,dt = C + \ln|x| + \frac{x}{1\cdot 1!} + \frac{x^2}{2\cdot 2!} + \frac{x^3}{3\cdot 3!} + \cdots$$

for $x < 0$ $\quad C$ see above

Logarithmic integral

$$\operatorname{Li}(x) = \int_{0}^{x} \frac{dt}{\ln t} = C + \ln|\ln x| + \frac{\ln x}{1\cdot 1!} + \frac{(\ln x)^2}{2\cdot 2!} + \frac{(\ln x)^3}{3\cdot 3!} + \cdots$$

for $0 < x < \infty$ $\quad C$ see above

Gaussian error integral

$$\Phi(x) = \frac{1}{\sqrt{2\pi}} \int_{-\infty}^{x} e^{-\frac{t^2}{2}}\,dt$$

$$= \frac{1}{2} + \frac{1}{\sqrt{2\pi}} \left(\frac{x}{1} - \frac{x^3}{2\cdot 3\cdot 1!} + \frac{x^5}{2^2\cdot 5\cdot 2!} - \frac{x^7}{2^3\cdot 7\cdot 3!} + - \cdots\right)$$

for $|x| < \infty$

7.5.5. Graphical integration

To construct the integral curve $F(x) = \int_a^b f(x)\,dx$ belonging to a given curve $y = f(x)$, the curve $y = f(x)$ is replaced by a step function with steps parallel to the abscissa in such a way that the two hatched corners associated with each step have the same area. The ordinates of the steps are marked off on the y-axis, $\overline{OB_1}$, $\overline{OB_2}$, etc., the points B_1, B_2, etc. are connected by lines with the pole $P(-1, 0)$. We then draw lines parallel to these joining lines, starting from C_0 in such a way that $\overline{C_0C_1} \parallel \overline{PB_1}$, $\overline{C_1C_2} \parallel \overline{PB_2}$, $\overline{C_2C_3} \parallel \overline{PB_3}$, etc. The open polygon obtained in this way represents a number of tangents of the desired integral curve which contact the curve at the points C_0, D_1, D_2, D_3, etc. The integral curve is then drawn with the help of a French curve.

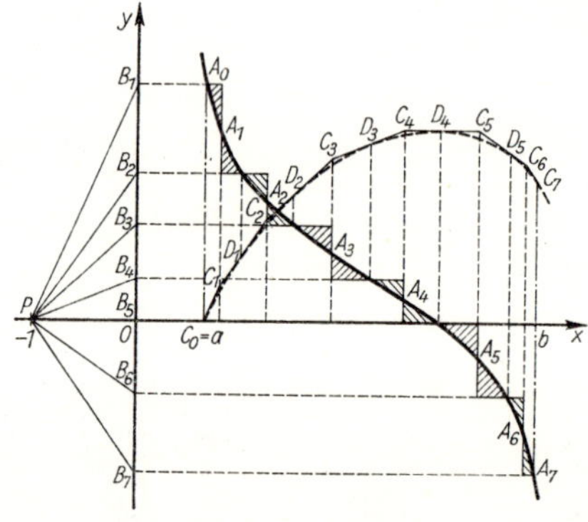

7.5.6. Improper integrals

Explanation

Integrals with infinite limits and integrals which become infinite within the interval of integration are called *improper integrals.*

$$\int_a^{+\infty} f(x)\,dx = \lim_{b\to+\infty} \int_a^b f(x)\,dx$$

$$\int_{-\infty}^{b} f(x)\,dx = \lim_{a\to-\infty} \int_a^b f(x)\,dx$$

$$\int_{-\infty}^{+\infty} f(x)\,dx = \lim_{\substack{a\to-\infty\\ a\to+\infty}} \int_a^b f(x)\,dx$$

If the limit exists, it is taken as the value of the improper integral. For integrals of dicsontinuous functions, the following limit is to be found:

$$\int_a^b f(x)\,dx = \lim_{\varepsilon\to 0} \int_a^{b-\varepsilon} f(x)\,dx \quad \text{for} \quad \lim_{x\to b} f(x) = \infty$$

If this limit exists, it is referred to as the value of the improper integral.

Examples:

1. $$\int_0^1 \frac{dx}{x^n} = \lim_{\varepsilon \to 0} \int_\varepsilon^1 \frac{dx}{x^n} = \lim_{\varepsilon \to 0} \left\{\frac{x^{1-n}}{1-n}\right\}_\varepsilon^1 = \frac{1}{1-n}\left(1 - \lim_{\varepsilon \to 0} \varepsilon^{1-n}\right)$$

The last limit exists for $1 - n > 0 \Rightarrow n < 1$ and has value 0.

$$\int_0^1 \frac{dx}{x^n} = \frac{1}{1-n} \qquad (n < 1)$$

2. $$\int_1^\infty \frac{dx}{x^n} = \lim_{b \to \infty} \int_1^b \frac{dx}{x^n} = \lim_{b \to \infty} \left\{\frac{x^{1-n}}{1-n}\right\}_1^b = \lim_{b \to \infty} \frac{\dfrac{1}{b^{n-1}} - 1}{1-n}$$

This limit exists for $n - 1 > 0 \Rightarrow n > 1$.

$$\int_1^\infty \frac{dx}{x^n} = \frac{1}{n-1} \qquad (n > 1)$$

3. $\displaystyle\int_0^1 \frac{dx}{x}$ does not exist because ln 0 does not exist.

7.5.7. A few definite integrals

1. $$\int_0^1 \frac{dx}{\sqrt{1-x^2}} = \frac{\pi}{2}$$

2. $$\int_0^\infty \frac{dx}{(1+x)\sqrt{x}} = \pi$$

3. $$\int_a^b \frac{dx}{\sqrt{(x-a)(b-x)}} = \pi$$

4. $$\int_0^a \frac{dx}{\sqrt{a^2 - x^2}} = \frac{\pi}{2}$$

5. $$\int_0^\infty \frac{dx}{a^2 + x^2} = \frac{\pi}{2a}$$

6. $$\int_a^b \frac{dx}{x^2 - a^2} = -\infty$$

7. $$\int_0^1 \frac{x\,dx}{\sqrt{1 - x^2}} = 1$$

8. $$\int_0^a \frac{x^2\,dx}{\sqrt{ax - x^2}} = \frac{3\pi a^2}{8}$$

9. $$\int_0^\infty \frac{dx}{(1 - x)\sqrt{x}} = 0$$

10. $$\int_0^{2b} \sqrt{2bx - x^2}\,dx = -\frac{\pi b^2}{2}$$

11. $$\int_{-1}^{+1} a^x\,dx = \frac{a^2 - 1}{a \ln a} \quad (a > 0)$$

12. $$\int_0^\infty e^{-x} x^n\,dx = n! \quad (n \in N \setminus \{0\})$$

13. $$\int_0^\infty e^{-x^2}\,dx = \frac{1}{2}\sqrt{\pi}$$

14. $$\int_0^\infty \frac{x\,dx}{e^x + 1} = \frac{\pi^2}{12}$$

15. $$\int_0^\infty \frac{x\,dx}{e^x - 1} = \frac{\pi^2}{6}$$

16. $$\int_0^\infty \frac{\sin ax\,dx}{x} = \begin{cases} \dfrac{\pi}{2} & \text{for} \quad a > 0 \\ -\dfrac{\pi}{2} & \text{for} \quad a < 0 \end{cases}$$

17. $$\int_0^\infty \frac{\cos ax\,dx}{x} = \infty$$

18. $$\int_0^\pi \sin ax\,dx = \frac{1 - \cos a\pi}{a}$$

19. $$\int_0^\pi \cos ax\,dx = \frac{\sin a\pi}{a}$$

20. $$\int_0^{\frac{\pi}{2}} \sin^{2n} x\,dx = \int_0^{\frac{\pi}{2}} \cos^{2n} x\,dx = \frac{1 \cdot 3 \cdot 5 \cdots (2n-1)}{2 \cdot 4 \cdot 6 \cdots 2n} \cdot \frac{\pi}{2}$$

$(n \in N \setminus \{0\})$

21. $$\int_0^{\frac{\pi}{2}} \sin^{2n+1} x\,dx = \int_0^{\frac{\pi}{2}} \cos^{2n+1} x\,dx = \frac{2 \cdot 4 \cdot 6 \cdots 2n}{1 \cdot 3 \cdot 5 \cdots (2n+1)}$$

$(n \in N \setminus \{0\})$

22. $$\int_0^{\frac{\pi}{2}} \frac{dx}{1 + \cos x} = 1$$

23. $$\int_0^\infty \frac{\tan ax\,dx}{x} = \begin{cases} \dfrac{\pi}{2} & \text{for} \quad a > 0 \\[2ex] -\dfrac{\pi}{2} & \text{for} \quad a < 0 \end{cases}$$

24. $$\int_0^{\frac{\pi}{4}} \tan x\,dx = \frac{1}{2} \ln 2$$

25. $$\int_0^\infty \frac{\sin x\,dx}{\sqrt{x}} = \int_0^\infty \frac{\cos x\,dx}{\sqrt{x}} = \sqrt{\frac{\pi}{2}}$$

26. $$\int_0^1 \frac{\ln x}{x+1}\,dx = -\frac{\pi^2}{12}$$

27. $$\int_0^1 \frac{\ln x}{x-1}\,dx = \frac{\pi^2}{6}$$

28. $$\int_0^1 \frac{\ln (x+1)}{x^2+1}\,dx = \frac{\pi}{8} \ln 2$$

29. $$\int_0^1 \frac{\ln x}{x^2-1}\,dx = \frac{\pi^2}{8}$$

7.5.8. Applications of the definite integral

Geometrical applications

Areas

1. An area bounded by the curve $y = f(x)$, the x-axis and the straight lines $x = x_1$ and $x = x_2$

$$A = \int_{x_1}^{x_2} f(x)\,dx$$ (see page 361)

For the determination of the *absolute area*, the dimensions of the parts of the area below the abscissa which have negative sign must be taken in absolute value, otherwise we will obtain the so-called *relative area*.

Example:

Find the area of the surface bounded by the curve

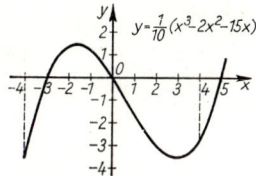

$$y = \frac{1}{10}(x^3 - 2x^2 - 15x),$$

the x-axis and the parallel lines $x = -4$ and $x = 4$.

Zeros: $x_1 = -3$; $x_2 = 0$; $x_3 = 5$; $\quad x_2; x_1 \in X$

$$A = \int_{-4}^{4} f(x)\,dx = \left|\int_{-4}^{-3} f(x)\,dx\right| + \int_{-3}^{0} f(x)\,dx + \left|\int_{0}^{4} f(x)\,dx\right|$$

$$= \left|\frac{1}{10}\int_{-4}^{-3}(x^3 - 2x^2 + 15x)\,dx\right|$$

$$+ \frac{1}{10}\int_{-3}^{0}(x^3 - 2x^2 - 15x)\,dx$$

$$+ \left|\frac{1}{10}\int_{0}^{4}(x^3 - 2x^2 - 15x)\,dx\right|$$

$$= \left|\frac{1}{10}\left[\frac{x^4}{4} - \frac{2x^3}{3} - \frac{15x^2}{2}\right]_{-4}^{-3}\right|$$

$$+ \frac{1}{10}\left[\frac{x^4}{4} - \frac{2x^3}{3} - \frac{15x^2}{2}\right]_{-3}^{0}$$

$$+ \left|\frac{1}{10}\left[\frac{x^4}{4} - \frac{2x^3}{3} - \frac{15x^2}{2}\right]_{0}^{4}\right|$$

$$= \left|\frac{1}{10}\left(-\frac{117}{4} + \frac{40}{3}\right)\right| + \frac{1}{10}\cdot\frac{117}{4} + \left|\frac{1}{10}\left(-\frac{296}{3}\right)\right|$$

$$= 14\,\frac{23}{60} \text{ units of area}$$

2. An area bounded by the curve $x = \varphi(t)$; $y = \psi(t)$, the x-axis and the ordinates $\psi(t_1)$ and $\psi(t_2)$

$$A = \int_{t_1}^{t_2} \psi(t)\, \dot{\varphi}(t)\, dx$$

3. An area bounded from above by the curve $y = f(x)$ and from below by the curve $y = g(x)$ and lying between the parallel lines $x = x_1$ and $x = x_2$

$$A = \int_{x_1}^{x_2} [f(x) - g(x)]\, dx$$

4. An area bounded by the curve $r = f(\varphi)$ and the radius vectors

$$r_1 = f(\varphi_1) \quad \text{and} \quad r_2 = f(\varphi_2)$$

$$A = \frac{1}{2} \int_{\varphi_1}^{\varphi_2} r^2\, d\varphi$$ **(Leibniz's sector formula)**

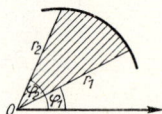

5. An area bounded by the curve $x = \varphi(t)$; $y = \psi(t)$ and the radius vectors OP_1 and OP_2

$$A = \frac{1}{2} \int_{t_1}^{t_2} [\varphi\dot{\psi} - \dot{\varphi}\psi]\, dt$$ **(Leibniz's sector formula)**

Arc length

Length s of a curve element between the points P_1 and P_2

$$s = \int_{x_1}^{x_2} \sqrt{1 + y'^2}\, dx \quad \text{for curve} \quad y = f(x)$$

$$s = \int_{t_1}^{t_2} \sqrt{\dot{\varphi}^2 + \dot{\psi}^2}\, d \quad \text{for curve} \quad x = \varphi(t);\ y = \psi(t)t$$

$$s = \int_{\varphi_1}^{\varphi_2} \sqrt{r^2 + \left(\frac{dr}{d\varphi}\right)^2}\, d\varphi = \int_{r_1}^{r_2} \sqrt{1 + r^2 \left(\frac{d\varphi}{dr}\right)^2}\, dr$$

for curve $r = f(\varphi)$

Surface-areas of solids of revolution (complanation)

$$A_{Sx} = 2\pi \int_{x_1}^{x_2} y \sqrt{1 + y'^2}\, dx$$ when curve $y = f(x)$ rotates about the x-axis

$$A_{Sy} = 2\pi \int_{y_1}^{y_2} x \sqrt{1 + \left(\frac{dx}{dy}\right)^2}\, dy$$ when curve $y = f(x)$ rotates about the y-axis

$$A_{Sx} = 2\pi \int_{t_1}^{t_2} \psi \sqrt{\dot{\varphi}^2 + \dot{\psi}^2}\, dt$$ when curve $x = \varphi(t)$; $y = \psi(t)$ rotates about the x-axis

$$A_{Sy} = 2\pi \int_{t_1}^{t_2} \varphi \sqrt{\dot{\varphi}^2 + \dot{\psi}^2}\, dt$$ about the y-axis

$$A_{Sx} = 2\pi \int_{\varphi_1}^{\varphi_2} r \sin\varphi \sqrt{r^2 + \left(\frac{dr}{d\varphi}\right)^2}\, d\varphi$$ when curve $r = r(\varphi)$ rotates about the x-axis

$$A_{Sy} = 2\pi \int_{\varphi_1}^{\varphi_2} r \cos\varphi \sqrt{r^2 + \left(\frac{dr}{d\varphi}\right)^2}\, d\varphi$$ about the y-axis

Volume of solids of revolution (cubature)

$$V_x = \pi \int_{x_1}^{x_2} y^2\, dx$$ when $y = f(x)$ rotates about the x-axis

$$V_y = \pi \int_{y_1}^{y_2} [g(y)]^2\, dx$$ when $y = f(x)$ rotates about the y-axis

$$= \pi \int_{x_1}^{x_2} x^2 y'\, dx$$ $y = f(x) \Leftrightarrow x = g(y)$

$$V_x = \pi \int_{t_1}^{t_2} \psi^2 \dot{\varphi}\, dt$$ when curve $x = \varphi(t)$; $y = \psi(t)$ rotates about the x-axis

$$V_y = \pi \int_{t_1}^{t_2} \varphi^2 \dot{\psi}\, dt$$ about the y-axis

$$V_x = \pi \int_{\varphi_1}^{\varphi_2} r^2 \sin^2 \varphi \left(\frac{dr}{d\varphi} \cos \varphi - r \sin \varphi\right) d\varphi$$

when curve $r = f(\varphi)$ rotates about the x-axis

$$V_y = \pi \int_{\varphi_1}^{\varphi_2} r^2 \cos^2 \varphi \left(\frac{dr}{d\varphi} \sin \varphi + r \cos \varphi\right) d\varphi$$

about the y-axis

Use in engineering

Work

$$W = \int_{s_1}^{s_2} F\, ds$$

Static moments

1. Static moment of a homogeneous plane element of a curve (density $\varrho = 1$)

$$M_x = \int_{x_1}^{x_2} y \sqrt{1 + y'^2}\, dx$$

(with respect to the x-axis) for curve $y = f(x)$

$$M_y = \int_{x_1}^{x_2} x \sqrt{1 + y'^2}\, dx$$

(with respect to the y-axis)

$$M_x = \int_{t_1}^{t_2} \psi \sqrt{\dot{\varphi} + \dot{\psi}}\, dt$$

(with respect to the x-axis) for curve $x = \varphi(t)$; $y = \psi(t)$

$$M_y = \int_{t_1}^{t_2} \varphi \sqrt{\dot{\varphi} + \dot{\psi}}\, dt$$

(with respect to the y-axis)

$$M_x = \int_{\varphi_1}^{\varphi_2} r \sqrt{r^2 + \frac{dr}{d\varphi}} \sin \varphi\, d\varphi$$

(with respect to the x-axis) for curve $r = f(\varphi)$

$$M_y = \int_{\varphi_1}^{\varphi_2} r \sqrt{r^2 + \left(\frac{dr}{d\varphi}\right)^2} \cos \varphi\, d\varphi$$

(with respect to the y-axis)

2. Static moment of a homogeneous plane element of a surface which is bounded by the curve $y = f(x)$, the x-axis and the straight lines $x = x_1$ and $x = x_2$

$$M_x = \frac{1}{2}\int_{x_1}^{x_2} y^2\,dx \qquad \text{(with respect to the } x\text{-axis)}$$

$$M_y = \int_{x_1}^{x_2} xy\,dx \qquad \text{(with respect to the } y\text{-axis)}$$

3. Static moment of a homogeneous plane element of a surface which is bounded from above by the curve $y = f(x)$, from below by the curve $y = g(x)$ and by the straight lines $x = x_1$ and $x = x_2$ ($\varrho = 1$)

$$M_x = \frac{1}{2}\int_{x_1}^{x_2} [f(x)^2 - g(x)^2]\,dx \qquad \text{(with respect to the } x\text{-axis)}$$

$$M_y = \int_{x_1}^{x_2} x[f(x) - g(x)]\,dx \qquad \text{(with respect to the } y\text{-axis)}$$

4. Static moment of a homogeneous solid of revolution ($\varrho = 1$)

$$M_{yz} = \pi\int_{x_1}^{x_2} xy^2\,dx$$

(with respect to the y,z-plane perpendicular to the axis of revolution x at the origin)

Centroid

1. Centroid of a homogeneous plane element of the curve $y = f(x)$ between points P_1 and P_2

$$x_C = \frac{\int_{x_1}^{x_2} x\sqrt{1 + y'^2}\,dx}{\int_{x_1}^{x_2} \sqrt{1 + y'^2}\,dx} = \frac{M_y}{s};$$

$$y_C = \frac{\int_{x_1}^{x_2} y\sqrt{1 + y'^2}\,dx}{\int_{x_1}^{x_2} \sqrt{1 + y'^2}\,dx} = \frac{M_x}{s}$$

2. Centroid of a homogeneous plane element of a surface that is bounded by the curve $y = f(x)$, the x-axis and the straight lines $x = x_1$ and $x = x_2$

$$x_C = \frac{\int_{x_1}^{x_2} xy\,dx}{\int_{x_1}^{x_2} y\,dx} = \frac{M_y}{A}; \qquad y_C = \frac{\int_{x_1}^{x_2} y^2\,dx}{2\int_{x_1}^{x_2} y\,dx} = \frac{M_x}{A}$$

3. Centroid of a homogeneous plane surface that is bounded from above by the curve $y = f(x)$ and from below by the curve $y = g(x)$

$$x_C = \frac{\int_{x_1}^{x_2} x[f(x) - g(x)]\,dx}{\int_{x_1}^{x_2} [f(x) - g(x)]\,dx} = \frac{M_y}{A}$$

$$y_C = \frac{\int_{x_1}^{x_2} [f(x)^2 - g(x)^2]\,dx}{2\int_{x_1}^{x_2} [f(x) - g(x)]\,dx} = \frac{M_x}{A}$$

4. Center of gravity of a homogeneous solid of revolution generated by the curve $y = f(x)$ rotating about the x-axis

$$x_C = \frac{\int_{x_1}^{x_2} xy^2\,dx}{\int_{x_1}^{x_2} y^2\,dx} = \frac{M_{yz}}{V}; \quad y_C = 0; \quad z_C = 0$$

5. Center of gravity of a solid

$$x_C = \frac{M_{yz}}{V} = \frac{\int_V x\,dV}{V}; \qquad y_C = \frac{M_{xz}}{V} = \frac{\int_V y\,dV}{V}$$

$$z_C = \frac{M_{xy}}{V} = \frac{\int_V z\,dV}{V}$$

For curves in parametric representation or in polar coordinates, the centroid is found on the basis of the moment and the arc or area.

Areal moments of inertia (theory of the strength of materials)

1. *Equatorial moment of inertia* of a plane arc of a curve s

$$I_x = \int_{x_1}^{x_2} y^2 \sqrt{1 + y'^2}\, dx$$ (with respect to the x-axis) for curve $y = f(x)$

$$I_y = \int_{x_1}^{x_2} x^2 \sqrt{1 + y'^2}\, dx$$ (with respect to the y-axis)

$$I_x = \int_{t_1}^{t_2} \psi^2 \sqrt{\dot{\varphi}^2 + \dot{\psi}^2}\, dt$$ (with respect to the x-axis) for curve $x = \varphi(t); y = \psi(t)$

$$I_y = \int_{t_1}^{t_2} \varphi^2 \sqrt{\dot{\varphi}^2 + \dot{\psi}^2}\, dt$$ (with respect to the y-axis)

$$I_x = \int_{\varphi_1}^{\varphi_2} r^2 \sin^2 \varphi \sqrt{r^2 + \left(\frac{dr}{d\varphi}\right)^2}\, d\varphi$$ (with respect to the x-axis) for curve $r = f(\varphi)$

$$I_y = \int_{\varphi_1}^{\varphi_2} r^2 \cos^2 \varphi \sqrt{r^2 + \left(\frac{dr}{d\varphi}\right)^2}\, d\varphi$$ (with respect to the y-axis)

2. *Equatorial moments of inertia* of the surface A, general

$$I_x = \int_A y^2\, dA; \quad I_y = \int_A x^2\, dA$$ dA element of surface

Theorem of STEINER $I = I_C + a^2 A$

I_C moment of intertia with respect to centroid
a distance between reference axis and centroid

3. *Equatorial moment of inertia* of a homogeneous plane surface bounded by the curve $y = f(x)$, the x-axis and the straight lines $x = x_1$ and $x = x_2$

$$I_x = \frac{1}{3} \int_{x_1}^{x_2} y^3\, dx$$ (with respect to the x-axis)

$$I_y = \int_{x_1}^{x_2} x^2 y \, dx \qquad \text{(with respect to the } y\text{-axis)}$$

See also page 384

4. *Equatorial moment of inertia* of a homogeneous plane surface bounded from above by the curve $y = f(x)$, from below by the curve $y = g(x)$ and by the straight lines $x = x_1$ and $x = x_2$

$$I_x = \frac{1}{3} \int_{x_1}^{x_2} \{[f(x)]^3 - [g(x)]^3\} \, dx \qquad \text{(with respect to the } x\text{-axis)}$$

$$I_y = \int_{x_1}^{x_2} x^2 [f(x) - g(x)] \, dx \qquad \text{(with respect to the } y\text{-axis)}$$

5. *Polar moment of inertia*

$$I_{\mathrm{p}} = \int_A r^2 \, dA = I_x + I_y \qquad \text{(with respect to the origin)}$$

See also page 385.

6. *Centrifugal moment of inertia*

$$I_{xy} = \int_A xy \, dA$$ dA element of surface

Mass moment of inertia (dynamics)

$$J = \int_m r^2 \, dm$$

dm mass element $= \varrho \, dV$
dV volume element
r distance from center of revolution

Mass moment of inertia of a homogeneous solid of density ϱ that is generated by revolution of the plane surface bounded by the curve $y = f(x)$, the x-axis, and the straight lines $x = x_1$ and $x = x_2$ about the x-axis

$$J_x = \frac{\pi \varrho}{2} \int_{x_1}^{x_2} y^4 \, dx$$

Mass moment of inertia of a homogeneous soild of density ϱ that is generated by revolution of a plane surface bounded by the curves $x = g(y)$ and the straight lines $y = y_1$ and $y = y_2$ about the y-axis

$$J_y = \frac{\pi \varrho}{2} \int_{y_1}^{y_2} x^4 \, dy$$

7.6. Line integral

7.6.1. Line integrals in the plane

A line integral is defined as a definite integral whose path of integration is not established by two points of the x-axis but by an element of a curve $C[x = f(t), y = g(t)]$ given in the form of an equation. The limits of integration are A and B with $t = t_A$ and $t = t_B$.

$$L = \int_A^B [P(x, y) \, dx \underset{(C)}{+} Q(x, y) \, dy]$$

In this case we speak of a line integral taken over the curve C between the points A and B.
For calculating, we use the definite integral

$$L = \int_{t_A}^{t_B} \{P[f(t), g(t)] \, \dot{f}(t) + Q[f(t), g(t)] \, \dot{g}(t)\} \, dt$$

The value of the line interval is independent of both the position of the coordinate system and the choice of the parameter.
By exchanging the limits A and B, we obtain

$$\int_A^B [P(x, y) \, dx \underset{(C)}{+} Q(x, y) \, dy]$$

$$= - \int_B^A [P(x, y) \, dx \underset{(C)}{+} Q(x, y) \, dy]$$

If the line integral is taken over a closed path, a contour, with the direction chosen in such a way that the interior of the closed curve is to the left, we write $\oint$ (*contour integral*).

Area of a plane figure

$$A = \frac{1}{2} \oint_{(C)} (x\,dy - y\,dx)$$

C curved boundary of the plane figure

Line integral of a complete differential

If the *condition of integrability*

$$\frac{\partial P}{\partial y} - \frac{\partial Q}{\partial x} = 0$$

is satisfied, i.e. if $P\,dx + Q\,dy$ is a complete differential of a function $\varphi(x, y)$, the line integral is *independent of the path of integration* taken. Then the line integral only depends on the limits A and B.
From this follows: The value of the line integral of a complete differential taken over a closed path of integration is zero.
Conversely: If the value of a line integral taken over a closed path of integration in a simply connected range is equal to zero, the integrand is a complete differential.
If the integrability condition for the vector $\mathbf{F} = P\mathbf{i} + Q\mathbf{j}$ is given, the field is called a potential field.

Example:

Find the line integral $\int_{A\,(C)}^{B} [(xy + y^2)\,dx + x\,dy]$ over the parabola $y = 2x^2$ between the limits $A(0, 0)$ and $B(2, 8)$. If the curve is given in explicit form, one of the unknowns is chosen as parameter. We choose x, hence $y = 2x^2$; $dy = 4x\,dx$.

$$L = \int_0^2 (x \cdot 2x^2 + 4x^4 + x \cdot 4x)\,dx = \underline{\underline{44\frac{4}{15}}}$$

7.6.2. Line integrals in space

The theorems given in section "Line integrals in a plane" can directly be applied to problems of space.

$$L = \int_{A\,(C)}^{B} [P(x, y, z)\,dx + Q(x, y, z)\,dy + R(x, y, z)\,dz]$$

where C is given in the form:

$$x = f(t); \quad y = g(t); \quad z = h(t); \quad A(x_A, y_A, z_A); \quad B(x_B, y_B, z_B).$$

$$L = \int_{t_A}^{t_B} \{P[f(t), g(t), h(t)]\, \dot{f}(t) + Q[f(t), g(t), h(t)]\, \dot{g}(t)$$

$$+ R[f(t), g(t), h(t)]\, \dot{h}(t)\}\, dt$$

The line integral becomes *independent of the path of integration* if the condition of integrability

$$\frac{\partial R}{\partial y} - \frac{\partial Q}{\partial z} = 0; \quad \frac{\partial P}{\partial z} - \frac{\partial R}{\partial x} = 0; \quad \frac{\partial Q}{\partial x} - \frac{\partial P}{\partial y} = 0$$

holds.

7.6.3. Line integral of a vector

$$\mathbf{F} = P(x, y, z)\, \mathbf{i} + Q(x, y, z)\, \mathbf{j} + R(x, y, z)\, \mathbf{k}$$

The vector is called *field vector*, the associated space *field of the vector* **F**.
The space curve C over which the line integral is taken is given in the vector form $\mathbf{r} = \mathbf{r}(t)$.
The scalar product

$$\mathbf{F} \cdot d\mathbf{r} = (P\mathbf{i} + Q\mathbf{j} + R\mathbf{k}) \cdot (dx\mathbf{i} + dy\mathbf{j} + dz\mathbf{k})$$

$$= P\, dx + Q\, dy + R\, dz$$

turns out to be the integrand of the line integral (see 7.6.1.).
Hence:

$$L = \int_{A \ (C)}^{B} \mathbf{F}\, d\mathbf{r}$$

This line integral is referred to as the line integral of vector **F** over the curve C within the limits A and B.

Example:

Given the field of force $\mathbf{F} = -y\mathbf{i} + x\mathbf{j} + \frac{1}{z+1}\mathbf{k}$. What is the work $W = \int \mathbf{F}\, d\mathbf{r}$ to be done to move a point particle in the field of force along the helix $\mathbf{r} = (a \cos t)\, \mathbf{i} + (a \sin t)\, \mathbf{j} + ct\mathbf{k}$ from $P_1(a, 0, 0)$ to $P_2(a, 0, 2\pi c)$ $(c \in N)$?

From the equation of the *helix* follows that

$$x = a\cos t; \qquad y = a\sin t; \qquad z = ct$$

$$dx = -a\sin t\,dt; \quad dy = a\cos t\,dt; \quad dz = c\,dt$$

Hence

$$d\mathbf{r} = [(-a\sin t)\,\mathbf{i} + (a\cos t)\,\mathbf{j} + c\mathbf{k}]\,dt$$

Limits:

$$\frac{y}{x} = \frac{\sin t}{\cos t} = \tan t; \quad t = \arctan\frac{y}{x}$$

For P_1 we have $t_1 = \arctan\dfrac{0}{a} = \arctan 0 = 0; \pi; 2\pi; \ldots;$

with $z_1 = ct_1 = 0$ only $t_1 = 0$ is possible.

For P_2 we have $t_2 = \arctan\dfrac{0}{a} = 0; \pi; 2\pi; \ldots;$

with $z_2 = ct_2 = 2\pi c$ we obtain $t_2 = 2\pi$

Here, $\arctan x$ must be taken as an ambiguous function because c turns of the helix are given.
Thus, the integral is written as

$$W = \int_0^{2\pi}\left[(-a\sin t)\,\mathbf{i} + (a\cos t)\,\mathbf{j} + \frac{1}{ct+1}\mathbf{k}\right][(-a\sin t)\,\mathbf{i}$$

$$+ (a\cos t)\,\mathbf{j} + c\mathbf{k}]\,dt = \int_0^{2\pi}\left(a^2\sin^2 t + a^2\cos^2 t + \frac{c}{ct+1}\right)dt$$

$$= \underline{\underline{2\pi a^2 + \ln(2\pi c + 1)}}$$

7.7. Multiple integrals

7.7.1. Double integral

Double integrals are obtained from functions of two variables $z = f(x, y)$ [or $f(r, \varphi)$] and are extended over a surface A in the x,y-plane [or r,φ-plane]. They yield a number as the sum of the limit

$\lim_{n \to \infty} \sum_{i}^{n} f(x_i, y_i) \Delta A_i$, all $\Delta A_i \to 0$ where $f(x_i, y_i)$ is the value of the function z for an arbitrary point P_i (P_i lies in the surface A or on its contour) and ΔA_i is the associated surface element.

Geometrically, the double integral represents the coefficient of measure of the volume of the cylindrical body that is bounded by the surface A in the x,y-plane, the perpendiculars erected on the contour of A parallel to the z-axis, and a part of the surface $z = f(x, y)$.

The volume found is positive if z is positive, otherwise it is negative. If the surface $z = f(x, y)$ intersects the x,y-plane within A, the operation must be based on the respective number of parts whose coefficient of measure must be taken in absolute value.

For $z = f(x, y) = 1$, the calculation of the double integral comes to the calculation of an area:

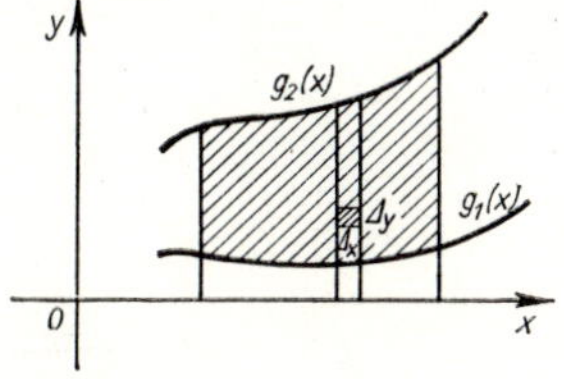

$$A = \int_{x=x_1}^{x_2} \int_{y=g_1(x)}^{g_2(x)} dy\, dx,$$

$dA = dx\, dy$ surface differential,

where $g_1(x)$ and $g_2(x)$ are the variable limits of the variable y.

Calculation of the double integral

In Cartesian coordinates:

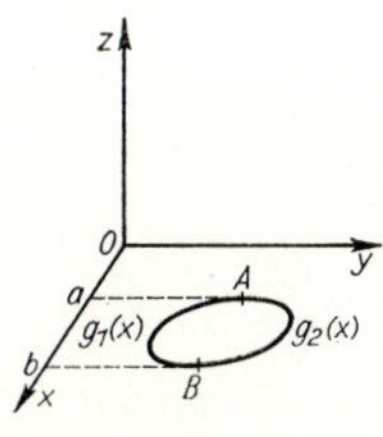

$$\int_A f(x, y)\, dA = \int_a^b \left[\int_{g_1(x)}^{g_2(x)} f(x, y)\, dy \right] dx$$

$$= \int_a^b \int_{g_1(x)}^{g_2(x)} f(x, y)\, dy\, dx$$

$g_1(x)$ and $g_2(x)$ are the variable limits of the variable y. We always integrate over the variable with *fixed limits last*. With $h_1(y)$ and $h_2(y)$ as variable limits, the above integral becomes

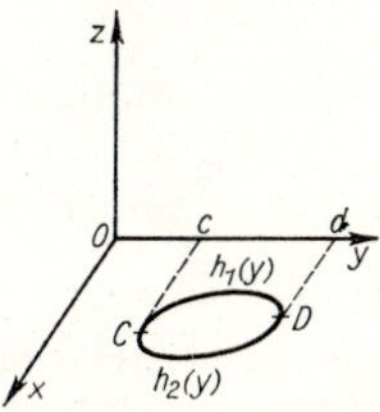

$$\int_A f(x, y)\, dA = \int_c^d \int_{h_1(y)}^{h_2(y)} f(x, y)\, dx\, dy$$

In polar coordinates:

$$\int_A f(r, \varphi)\, dA = \int_{\varphi_1}^{\varphi_2} \int_{g_1(\varphi)}^{g_2(\varphi)} f(r, \varphi)\, r\, dr\, d\varphi$$

Example:

Find the volume of the sphere $x^2 + y^2 + z^2 = r^2$.
It will suffice to determine the first octant $x \geqq 0$; $y \geqq 0$; $z \geqq 0$ because a symmetrical solid is given.

$$\frac{V}{8} = \int_A f(x, y)\, dA$$

where $f(x, y) = z = \sqrt{r^2 - x^2 - y^2}$

$$dA = dx\, dy$$

Limits:

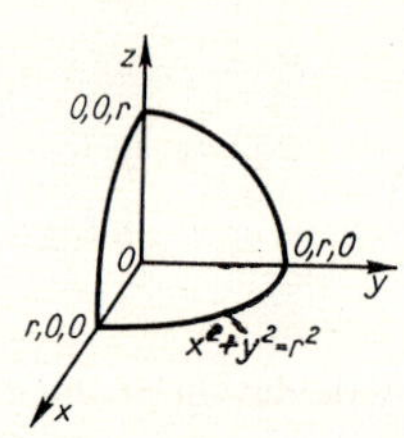

x extends from 0 to $\sqrt{r^2 - y^2}$.
(From the equation of the sphere for $z = 0$, i.e. in the x,y-plane)
y extends from 0 to r.
With this we have

$$\frac{V}{8} = \int_0^r \int_0^{\sqrt{r^2-y^2}} \sqrt{r^2 - x^2 - y^2}\, dx\, dy \qquad x \geqq 0;\ y \geqq 0$$

Substitution: $x = \sqrt{r^2 - y^2} \sin \varphi$ with $dx = \sqrt{r^2 - y^2} \cos \varphi\, d\varphi$
New limits:

For $x = 0$, we have $\varphi = 0$; for $x = \sqrt{r^2 - y^2}$, we have $\varphi = \dfrac{\pi}{2}$

$$\frac{V}{8} = \int_0^r \int_0^{\pi/2} \sqrt{r^2 - (r^2 - y^2) \sin^2 \varphi - y^2}\, \sqrt{r^2 - y^2} \cos \varphi\, d\varphi\, dy$$

$$= \int_0^r \int_0^{\pi/2} \sqrt{(r^2 - y^2)(1 - \sin^2 \varphi)}\, \sqrt{r^2 - y^2} \cos \varphi\, d\varphi\, dy$$

$$= \int_0^r \int_0^{\pi/2} (r^2 - y^2) \cos^2 \varphi \, d\varphi \, dy$$

$$= \int_0^r \left[(r^2 - y^2) \left(\frac{\varphi}{2} + \frac{1}{4} \sin 2\varphi \right) \right]_0^{\pi/2} dy = \int_0^r (r^2 - y^2) \frac{\pi}{4} \, dy$$

$$= \frac{\pi}{4} \left[r^2 y - \frac{y^3}{3} \right]_0^r = \frac{\pi r^3}{6};$$

hence $V = \dfrac{4}{3} \pi r^3$

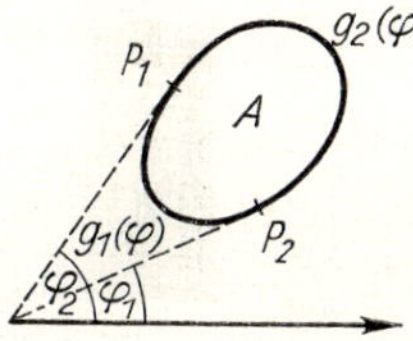

Applications of double integrals

1. *Area of a plane surface* in the x,y-plane general

$$A = \int_A dA$$

in Cartesian coordinates

$$A = \int_{x_1}^{x_2} \int_{g_1(x)}^{g_2(x)} dy \, dx$$

in polar coordinates

$$A = \int_{\varphi_1}^{\varphi_2} \int_{g_1(\varphi)}^{g_2(\varphi)} r \, dr \, d\varphi$$

2. *Area of a part of a surface* $z = f(x, y)$ whose projection onto the x,y-plane is
in general

$$A_S = \int_A \frac{dA}{\cos \gamma}$$ where γ is the angle between the tangent to the surface element and the x,y-plane,
in Cartesian coordinates

$$A_S = \int_{x_1}^{x_2} \int_{g_1(x)}^{g_2(x)} \sqrt{f_x^2 + f_y^2 + 1} \, dy \, dx$$

in polar coordinates

$$A_S = \int_{\varphi_1}^{\varphi_2} \int_{g_1(\varphi)}^{g_2(\varphi)} \sqrt{f_\varphi^2 + r^2 f_r^2 + r^2}\, dr\, d\varphi$$

3. *Static moment*
general

$$M_x = \int_A x\, dA$$

in Cartesian coordinates

$$M_x = \int_{x_1}^{x_2} \int_{g_1(x)}^{g_2(x)} y\, dy\, dx; \qquad M_y = \int_{x_1}^{x_2} \int_{g_1(x)}^{g_2(x)} x\, dy\, dx$$

in polar coordinates

$$M_x = \int_{\varphi_1}^{\varphi_2} \int_{g_1(\varphi)}^{g_2(\varphi)} r^2 \sin\varphi\, dr\, d\varphi$$

$$M_y = \int_{\varphi_1}^{\varphi_2} \int_{g_1(\varphi)}^{g_2(\varphi)} r^2 \cos\varphi\, dr\, d\varphi$$

in parametric representation:
It is advisable to convert this representation into one of the above representations.

4. *Centroid* of a plane homogeneous surface
general

$$x_S = \frac{M_y}{A}; \qquad y_S = \frac{M_x}{A}$$

5. *Axial areal moments of inertia*
general

$$I_x = \int_A y^2\, dA$$

$$I_y = \int_A x^2\, dA$$

in Cartesian coordinates

$$I_x = \int_{x_1}^{x_2} \int_{g_1(x)}^{g_2(x)} y^2 \, dy \, dx$$

$$I_y = \int_{x_1}^{x_2} \int_{g_1(x)}^{g_2(x)} x^2 \, dy \, dx$$

in polar coordinates

$$I_x = \int_{\varphi_1}^{\varphi_2} \int_{g_1(\varphi)}^{g_2(\varphi)} r^3 \sin^2 \varphi \, dr \, d\varphi$$

$$I_y = \int_{\varphi_1}^{\varphi_2} \int_{g_1(\varphi)}^{g_2(\varphi)} r^3 \cos^2 \varphi \, dr \, d\varphi$$

in parametric representation
See note under 3.

6. *Polar moment of inertia*
general

$$I_p = \int_A r^2 \, dA$$

in Cartesian coordinates

$$I_p = \int_{x_1}^{x_2} \int_{g_1(x)}^{g_2(x)} (x^2 + y^2) \, dy \, dx$$

in polar coordinates

$$I_p = \int_{\varphi_1}^{\varphi_2} \int_{g_1(\varphi)}^{g_2(\varphi)} r^3 \, dr \, d\varphi$$

in parametric representation.
See note under 3.

7. *Volume of a cylinder*
general

$$V = \int_A z \, dA$$

in Cartesian coordinates

$$V = \int_{x_1}^{x_2} \int_{g_1(x)}^{g_2(x)} z \, dy \, dx$$

in polar coordinates

$$V = \int_{\varphi_1}^{\varphi_2} \int_{g_1(\varphi)}^{g_2(\varphi)} zr \, dr \, d\varphi$$

7.7.2. Triple integrals

Triple integrals are obtained from functions of three variables $u = f(x, y, z)$ in cylindrical coordinates $u = f(\varrho, \varphi, z)$ or in spherical coordinates $u = f(r, \varphi, \vartheta)$, taken over a volume in space.

Calculation of triple integrals

In Cartesian coordinates:

$$\int_V f(x, y, z) \, dV = \int_a^b \left\{ \int_{g_1(x)}^{g_2(x)} \left[\int_{h_1(x,y)}^{h_2(x,y)} f(x, y, z) \, dz \right] dy \right\} dx$$

$$= \int_a^b \int_{g_1(x)}^{g_2(x)} \int_{h_1(x,y)}^{h_2(x,y)} f(x, y, z) \, dz \, dy \, dx$$

The limits $h_1(x, y)$ and $h_2(x, y)$ are the lower and upper bounding surfaces of the volume V which are separated by the boundary curve of the volume (line connecting the points of contact of all tangential planes to the volume parallel to the z-axis). The limits $g_1(x)$ and $g_2(x)$ are the lower and upper parts of the curve in the x,y-plane which is obtained by projecting the boundary curve onto the x,y-plane. The limits $x = a$ and $x = b$ separate the two curves $g_1(x)$ and $g_2(x)$.

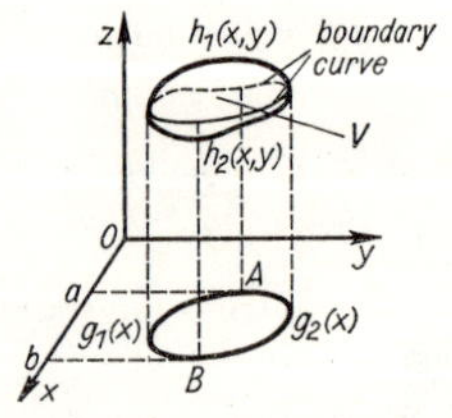

Similarly to the double integral, the triple integral can be solved by integrating in any convenient succession; it should be noted that the limits are changed, however. Again we integrate over the variable with fixed limits last.

In cylindrical coordinates we have

$$\int_V f(\varrho, \varphi, z)\, dV = \int_{\varphi_1}^{\varphi_2} \int_{g_1(\varphi)}^{g_2(\varphi)} \int_{h_1(\varrho,\varphi)}^{h_2(\varrho,\varphi)} f(\varrho, \varphi, z)\, \varrho\, dz\, d\varrho\, d\varphi$$

In spherical coordinates

$$\int_V f(r, \varphi, \vartheta)\, dV = \int_{\varphi_1}^{\varphi_2} \int_{g_1(\varphi)}^{g_2(\varphi)} \int_{h_1(\vartheta,\varphi)}^{h_2(\vartheta,\varphi)} f(r, \varphi, \vartheta)\, r^2 \sin\vartheta\, dr\, d\vartheta\, d\varphi$$

Applications of triple integrals

1. *Volume* of a solid
general

$$V = \int_V dV$$

in Cartesian coordinates

$$V = \int_{x_1}^{x_2} \int_{g_1(x)}^{g_2(x)} \int_{h_1(x,y)}^{h_2(x,y)} dz\, dy\, dx$$

in cylindrical coordinates

$$V = \int_{\varphi_1}^{\varphi_2} \int_{g_1(\varphi)}^{g_2(\varphi)} \int_{h_1(\varrho,\varphi)}^{h_2(\varrho,\varphi)} \varrho\, dz\, d\varrho\, d\varphi$$

in spherical coordinates

$$V = \int_{\varphi_1}^{\varphi_2} \int_{g_1(\varphi)}^{g_2(\varphi)} \int_{h_1(\vartheta,\varphi)}^{h_2(\vartheta,\varphi)} r^2 \sin\vartheta\, dr\, d\vartheta\, d\varphi$$

2. *Center of gravity* of a homogeneous solid
general

$$x_C = \frac{\int_V x\, dV}{V}; \qquad y_C = \frac{\int_V y\, dV}{V}; \qquad z_C = \frac{\int_V z\, dV}{V}$$

$$x_C = \frac{1}{V} \int_{x_1}^{x_2} \int_{g_1(x)}^{g_2(x)} \int_{h_1(x,y)}^{h_2(x,y)} x\, dz\, dy\, dx$$

$$y_C = \frac{1}{V} \int_{x_1}^{x_2} \int_{g_1(x)}^{g_2(x)} \int_{h_1(x,y)}^{h_2(x,y)} y \, dz \, dy \, dx$$

$$z_C = \frac{1}{V} \int_{x_1}^{x_2} \int_{g_1(x)}^{g_2(x)} \int_{h_1(x,y)}^{h_1(x,y)} z \, dz \, dy \, dx$$

Example:

Find the volume of the solid bounded by the following surfaces:

$$z = 2x^2y, \quad (x-2)^2 + y^2 = 4, \quad z = 0, \quad y \geqq 0.$$

Limits:

z from 0 to $2x^2y$

y from 0 to $\sqrt{4x - x^2}$ because $(x-2)^2 + y^2 = 4;$

$$x^2 - 4x + 4 + y^2 = 4;$$

$$y = \sqrt{4x - x^2}$$

x from 0 to 4; this follows from the equation of the circle.

$$V = \int_0^4 \int_0^{\sqrt{4x-x^2}} \int_0^{2x^2y} dz \, dy \, dx = \int_0^4 \int_0^{\sqrt{4x-x^2}} 2x^2y \, dy \, dx$$

$$= \int_0^4 \left[x^2y^2\right]_0^{\sqrt{4x-x^2}} dx = \int_0^4 (4x^3 - x^4) \, dx$$

$$= \left(x^4 - \frac{x^5}{5}\right)\Bigg|_0^4 = \underline{\underline{51.2 \text{ units of volume}}}$$

8. Differential equations

8.1. General

Definition of the differential equation

Equations involving, besides one or several variables, functions of these variables and their derivatives are called differential equations.

Ordinary differential equations are conditional equations for a function of an independent variable which contain at least one derivative of the unknown function with respect to this variable.

$$f(x, y, y', y'', \ldots y^{(k)}) = 0$$

Partial differential equations are conditional equations for a function of several independent variables which contain at least one derivative of the unknown function with respect to one of the independent variables.

For example for $z(x, y)$: $f(x, y, z, z_x, z_y, z_{xx}, z_{yy}, z_{xy}, \ldots) = 0$

Definition of the solution of a differential equation

The solution (integral) of a differential equation is defined as the set of all functions whose derivatives identically satisfy the differential equation. The *general solution* of a differential equation of nth order is the set of all functions of solution which contain exactly n arbitrary parameters (constants).

A *particular solution* of a differential equation is obtained if, by imposition of additional *initial* or *boundary conditions*, the values of the parameters are specified.

A solution of a differential equation is called *singular* if it cannot be obtained from the general solution by the choice of a special parameter.

Order, degree of a differential equation

The order of the highest derivative of the function to be found occurring in a differential equation is its order (nth derivative $\Rightarrow$ differential equation of the nth order).

The highest power of the function to be found or its derivatives is the degree of the differential equation.

Geometrical interpretation of the differential equation

The graphical representation of a general solution of a differential equation of the nth order is a family of curves with n parameters. Conversely, every family of curves has its differential equations.
The particular solution corresponds to a certain curve of the family of curves (curve of solution, integral curve).
Differential equations of the first order determine, for each point (x, y) of the domain of definition of the function, the direction $y' = \tan \alpha$ of the curve through this point and included in the family of curves of the general solution of the differential equation $f(x, y, y') = 0$ or $y' = f(x, y)$.

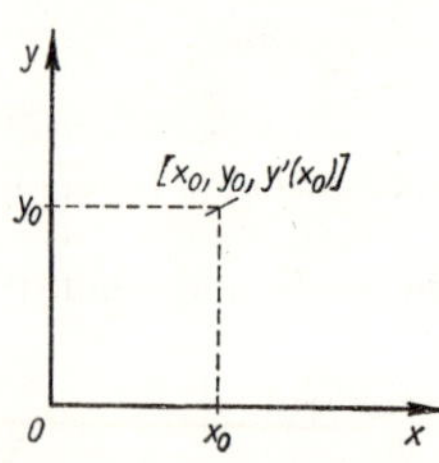

By the set of three values (x, y, y'), one line element of the family of curves of the solution set is fixed; all linie elements yield the *directional field* in the Cartesian system of coordinates. The family of curves of the solution set of a differential equation includes all curves whose directions at each point correspond to the directional field.
The lines joining all points with the same direction of the line elements are called *isoclinic lines* ($y' =$ const.).
If the isoclinic lines are known, solution curves of differential equations can be derived and plotted on a graph, giving good approximations.

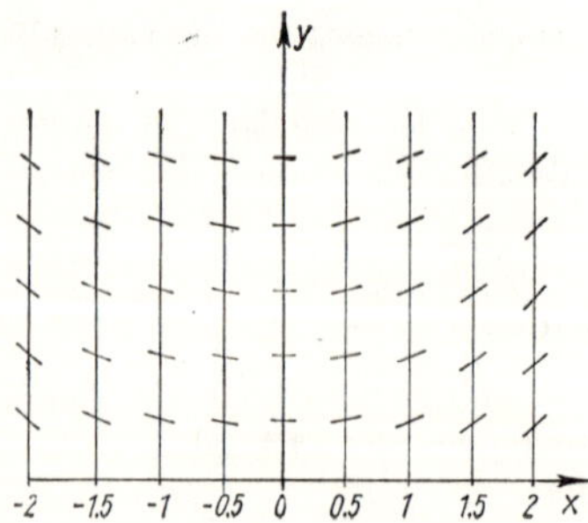

Examples:

1. $y' = f(x) = \frac{1}{2} x$

   for

   $y > 0$

   $x \in [-2, 2]$

   Equation of isoclinics:

   $y' = C \Rightarrow x = 2c$

2. $y' = -y$

Equation of isoclinics:

$$y' = C \Rightarrow y = -C$$

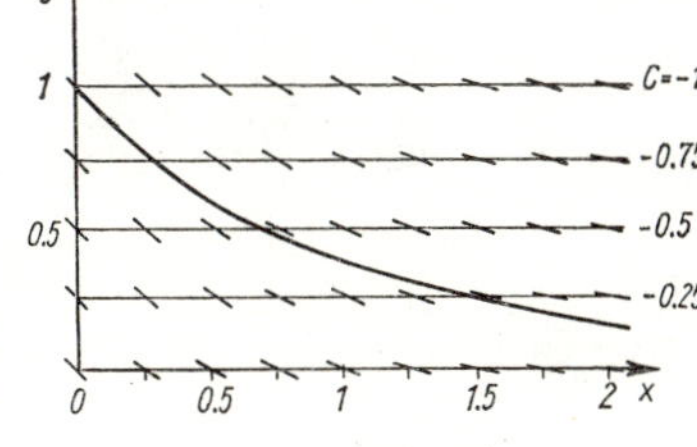

The decreasing exponential function $y = e^{-x}$ is clearly represented.

Differential equations of the second order determine direction and curvature of the arc elements for each point of the domain of definition. The isoclinic method is to be used by putting $y' = f(t) = z$ (t is a parameter) and converting the differential equation in x of the second order into a differential equation in z of the first order.

Curves that cut each curve of a family of curves exactly once are called *trajectories* of this family of curves.

Curves that cut a given family of curves at a constant angle are called *isogonal trajectories*. If the intersection occurs at the angle of 90°, they are called *orthogonal trajectories*.

Application: Determination of the potential surfaces or potential lines of a given field of lines at a given behavior.

For differential equations of the first order, the following relations hold:

Given family of curves	$F(x, y, c) = 0$	$y = f(x, c)$
Differential equation of the family of curves	$\dfrac{\partial F}{\partial x} + \dfrac{\partial F}{\partial y} y' = 0$	$y' = g(x, y)$
Direction of the curves	$y' = \dfrac{-F_x}{F_y}$	$y' = g(x, y)$
Orthogonal trajectories	$\dfrac{\partial F}{\partial y} - \dfrac{\partial F}{\partial x} y' = 0$	$y' = -\dfrac{1}{g(x, y)}$
Isogonal trajectories (angle of intersection φ)	$y' = \dfrac{-F_x/F_y + \tan\varphi}{1 + F_y/F_x \tan\varphi}$	$y' = \dfrac{g(x, y) + \tan\varphi}{1 - g(x, y)\tan\varphi}$

Example:

Given family of curves $4x^2 + 5y + c = 0$

Differential equation of the family of curves

$$8x + 5y' = 0$$

Orthogonal trajectories $5 - 8xy' = 0$

Isogonal trajectories
(angle of intersection $\varphi = 30°$)

$$y' = \frac{\frac{-4x}{5} + \tan 30°}{1 + \frac{5}{4x} \tan 30°}$$

$$48x^2 - 20x\sqrt{3} - \left(60x + 25\sqrt{3}\right)y' = 0$$

Setting up differential equations

We differentiate the equation of the family of curves repeatedly until the parameters can be eliminated.

Example:

Find the differential equation of all paraboles which are open to the right.

Solution:

Statement of the equation for the family of curves, using the parameters c, d and p:

$$(y - d)^2 = 2p(x - c)$$

$$2(y - d)y' = 2p$$

$$(y - d)y'' + y'^2 = 0$$

$$y'y'' + (y - d)y''' + 2y'y'' = 0$$

The differential equation of all parabolas follows from the last two equations

$$y'''y'^2 - 3y'y''^2 = 0$$

8.2. Ordinary differential equations of the first order

8.2.1. Separation of variables

Differential equations of the form $y' = \dfrac{\varphi(x)}{\psi(y)}$

$$\frac{dy}{dx} = \frac{\varphi(x)}{\psi(y)} \quad \text{yields} \quad \psi(y)\,dy = \varphi(x)\,dx$$

Solution:

$$\int \psi(y)\,dy = \int \varphi(x)\,dx + C$$

Examples:

1. $xe^{x+y} = yy' \quad \text{or} \quad xe^{x+y}\,dx = y\,dy$

$$xe^x\,dx = ye^{-y}\,dy$$

$$\int xe^x\,dx = \int ye^{-y}\,dy$$

$$xe^x - \int e^x\,dx = -ye^{-y} + \int e^{-y}\,dy \quad \text{(integration by parts)}$$

$$e^x(x - 1) = -e^{-y}(1 + y) + C$$

2. $y'(2x - 7) + y(2x^2 - 3x - 14) = 0$

$$\frac{dy}{dx}(2x - 7) = -y(2x^2 - 3x - 14)$$

$$\frac{dy}{y} = -\frac{2x^2 - 3x - 14}{2x - 7}\,dx = -(x + 2)\,dx$$

$$\int \frac{dy}{y} = -\int (x + 2)\,dx$$

$$\ln y = -\frac{x^2}{2} - 2x + C$$

$$y = e^{-\frac{x^2}{2} - 2x + c} = e^C \cdot e^{-\frac{x^2}{2} - 2x} = C_1 \cdot e^{-\frac{x^2}{2} - 2x}$$

8.2.2. Homogeneous differential equations of the first order

$$y' = \frac{\varphi(x, y)}{\psi(x, y)} \quad \text{or} \quad \varphi(x, y)\, dx - \psi(x, y)\, dy = 0$$

where $\varphi(x, y)$ and $\psi(x, y)$ are homogeneous functions of the same degree (for homogeneous functions see page 106).

Solution: Substituting $\frac{y}{x} = z$ leads to a differential equation to which the method of separation of variables can be applied.

Example:

$$(3x - 2y)\, dx - x\, dy = 0$$

Substituting $\frac{y}{x} = z; \quad y = zx; \quad y' = xz' + z;$

$$dy = x\, dz + z\, dx$$

we get:

$$(3x - 2zx)\, dx - x(x\, dz + z\, dx) = 0$$

$$\frac{3}{x}\, dx = \frac{dz}{1 - z} \qquad \text{(separation of variables)}$$

$$3 \int \frac{dx}{x} = \int \frac{dz}{1 - z}$$

$$3 \ln x = -\ln (1 - z) + C$$

$$\ln x^3 = -\ln \left(1 - \frac{y}{x}\right) + C$$

$$\ln \left(x^3 \frac{x - y}{x}\right) = C; \quad \ln (x^3 - x^2 y) = C$$

$$x^3 - x^2 y = e^C = C_1$$

8.2.3. Inhomogeneous differential equations of the first order

$$y'\varphi(x) + y\psi(x) + \omega(x) = 0$$

Solution 1:

Integration by substitution

$$y' + y\frac{\psi(x)}{\varphi(x)} + \frac{\omega(x)}{\varphi(x)} = 0$$

$$y' + yP(x) + Q(x) = 0$$

where

$$P(x) = \frac{\psi(x)}{\varphi(x)}, \quad Q(x) = \frac{\omega(x)}{\varphi(x)}$$

Substitution $y = uv$; $\quad y' = u'v + uv'$

$$u'v + uv' + uvP(x) + Q(x) = 0$$

$$u'v + u[v' + vP(x)] + Q(x) = 0 \tag{1}$$

v is chosen in such a way that the factor of u in (1) vanishes.

$$v' + vP(x) = 0;$$

$$v = e^{-\int P(x)dx}$$

Thus, equation (1) is written as

$$u'e^{-\int P(x)dx} = -Q(x); \quad u' = -Q(x)e^{\int P(x)dx}$$

$$u = -\int Q(x)e^{\int P(x)dx}\, dx + C$$

$$y = uv = e^{-\int P(x)dx}\left[-\int Q(x)e^{\int P(x)dx}\, dx + C\right]$$

(general solution)

Example:

$$(4 + x)y' + y = 6 + 2x$$

$$y' + y\frac{1}{4 + x} - \frac{6 + 2x}{4 + x} = 0$$

$$P(x) = \frac{1}{4 + x};$$

$$\int P(x)\,dx = \int \frac{1}{4+x}\,dx = \ln|4+x| + C_1$$

$$= \ln|4+x| + \ln C_2 = \ln|C_2(4+x)|$$

$$Q(x) = -\frac{6+2x}{4+x};$$

$$Q(x)e^{\int P(x)dx}\,dx = -\int \frac{6+2x}{4+x}\,e^{\ln|C_2(4+x)|}\,dx$$

$$= -\int C_2(6+2x)\,dx = -C_2(6x+x^2) + C_3$$

$$y = e^{-\ln|C_2(4+x)|}\,[C_2(6x+x^2) - C_3]$$

$$= \frac{1}{C_2(4+x)}\,[C_2(6x+x^2) - C_3] = \frac{6x+x^2+C_4}{4+x}$$

Solution 2:

Integration by variation of the constants

$$y'\varphi(x) + y\psi(x) + \omega(x) = 0 \tag{1}$$

At first we solve the homogeneous equation

$$y'\varphi(x) + y\psi(x) = 0$$

(Here, the word homogeneous refers to the fact that the term free of y is missing.)
Separation of variables leads to the result

$$y = Ce^{-\int \frac{\psi(x)}{\varphi(x)}dx}$$

We replace the constant C by a function $z(x)$.

$$y = z(x)e^{-\int \frac{\psi(x)}{\varphi(x)}dx}$$

$$y' = z'e^{-\int \frac{\psi(x)}{\varphi(x)}dx} + ze^{-\int \frac{\psi(x)}{\varphi(x)}dx}\left[-\frac{\psi(x)}{\varphi(x)}\right] \tag{2}$$

From (1) and (2) we obtain z and finally y.

Example:

$$(4 + x)y' + y = 6 + 2x \tag{3}$$

Homogeneous equation $(4 + x)y' + y = 0$

$$y = Ce^{-\int \frac{1}{4+x}dx} = Ce^{-\ln|4+x|+C_1}$$

$$= Ce^{-\ln|C_2(4+x)|} = \frac{C}{C_2(4+x)} = \frac{C_3}{4+x}$$

Variation of constants yields

$$y = z(x)\frac{1}{4+x}; \qquad y' = z'\frac{1}{4+x} - z\frac{1}{(4+x)^2}$$

$$(4 + x)\left[z'\frac{1}{4+x} - z\frac{1}{(4+x)^2}\right] + z\frac{1}{4+x} = 6 + 2x$$

$$z' = 6 + 2x$$

$$z = 6x + 2\frac{x^2}{2} + C_4 = 6x + x^2 + C_4$$

$$y = (6x + x^2 + C_4)\frac{1}{4+x}$$

8.2.4. Total (exact) differential equations of the first order

$$\varphi(x, y)\, dx + \psi(x, y)\, dy = 0$$

on condition that the left-hand side represents a *complete differential*:

$$\frac{\partial\varphi(x, y)}{\partial y} = \frac{\partial\psi(x, y)}{\partial x} \qquad \textit{(integrability condition)}$$

Direct integration leads to the solution

$$\int \varphi(x, y)\, dx + \int \left[\psi(x, y) - \int \frac{\partial\varphi(x, y)}{\partial y}\, dx\right] dy = C \tag{I}$$

or

$$\int \psi(x, y)\, dy + \int \left[\varphi(x, y) - \int \frac{\partial \psi(x, y)}{\partial x}\, dy\right] dx = C \qquad \text{(II)}$$

Example:

$$(3x^2 + 8ax + 2by^2 + 3y)\, dx + (4bxy + 3x + 5)\, dy = 0$$

$$\frac{\partial(3x^2 + 8ax + 2by^2 + 3y)}{\partial y} = 4by + 3$$

$$\frac{\partial(4bxy + 3x + 5)}{\partial x} = 4by + 3$$

Hence, the left-hand side of the equation represents a complete differential.
Application of the solution formula (I)

$$\int (3x^2 + 8ax + 2by^2 + 3y)\, dx + \int [4bxy + 3x + 5 - \int (4by + 3)\, dx]\, dy = C$$

$$x^3 + 4ax^2 + 2bxy^2 + 3xy + \int [4bxy + 3x + 5 - (4bxy + 3x + C_1)]\, dy = C$$

$$x^3 + 4ax^2 + 2bxy^2 + 3xy + (5 + C_1)y = C_3$$

8.2.5. Integrating factors

A function $\mu(x, y)$ is called an *integrating factor* of the differential equation $\varphi(x, y)\, dx + \psi(x, y)\, dy = 0$ if the left-hand side of the equation is converted into a complete differential by multiplication by $\mu(x, y)$:

$$\frac{\partial[\mu(x, y)\varphi(x, y)]}{\partial y} = \frac{\partial[\mu(x, y)\psi(x, y)]}{\partial x}$$

Often the solution can be simplified by treating $\mu(x, y)$ as depending only on x or y or by setting up special combinations of the two such as $x^2 + y^2$, $\dfrac{x}{y}$, etc.

Special integrating factors

The integrating factor is a function of x only:

$$\mu = e^{-\int \frac{1}{\varphi}\left(\frac{\partial \psi}{\partial x} - \frac{\partial \varphi}{\partial y}\right) dx}$$

The integrating factor is a function of y only:

$$\mu = e^{\int \frac{1}{\psi}\left(\frac{\partial \psi}{\partial x} - \frac{\partial \varphi}{\partial y}\right) dy}$$

The integrating factor is a function of xy only:

$$\mu = e^{\int \frac{1}{x\varphi - y\psi}\left(\frac{\partial \psi}{\partial x} - \frac{\partial \varphi}{\partial y}\right) dz}; \quad z = xy$$

The integrating factor is a function of $\frac{y}{x}$ only:

$$\mu = e^{\int \frac{x^2}{x\varphi + y\psi}\left(\frac{\partial \psi}{\partial x} - \frac{\partial \varphi}{\partial x}\right) dz}; \quad z = \frac{y}{x}$$

The integrating factor is a function of $x^2 + y^2$ only:

$$\mu = e^{\int \frac{1}{2(y\varphi - x\psi)}\left(\frac{\partial \psi}{\partial x} - \frac{\partial \varphi}{\partial y}\right) dz}; \quad z = x^2 + y^2$$

Note: For brevity, φ takes the place of $\varphi(x, y)$ and ψ the place of $\psi(x, y)$ in these formulas.

Example:

$$(3x - 2y)\, dx - x\, dy = 0$$

Proof of the integrability condition

$$\frac{\partial(3x - 2y)}{\partial y} = -2 \qquad \frac{\partial(-x)}{\partial x} = -1$$

The integrability condition is not satisfied.
Assumption: The integrating factor is a function of x only.

$$\mu = e^{-\int \frac{1}{-x}(-1+2)dx} = e^{\int \frac{dx}{x}} = e^{\ln x} = x$$

Multiplication of the initial equation by $\mu = x$ yields

$$(3x^2 - 2xy)\,dx - x^2\,dy = 0 \quad \text{(total differential equation)}$$

Solution: $x^3 - x^2y = C$

8.2.6. Bernoulli differential equation

$$y' + \varphi(x)y = \psi(x)y^n \qquad (n \neq 1)$$

Substitution:

$$y = z^{\frac{1}{1-n}}; \quad y' = \frac{1}{1-n} z^{\frac{n}{1-n}} \cdot z'$$

$$z' + (1-n)\varphi(x)z = (1-n)\psi(x)$$

Example:

$$y' + \frac{y}{x} - x^2y^3 = 0$$

Substitution: $y = z^{\frac{1}{1-3}} = z^{-\frac{1}{2}}; \quad y' = -\frac{1}{2} z^{-\frac{3}{2}} z'$

$$-\frac{1}{2} z^{-\frac{3}{2}} z' + \frac{1}{x} z^{-\frac{1}{2}} = x^2 z^{-\frac{3}{2}}$$

$$z'x - 2z = -2x^3$$

By variation of the constants, this differential equation yields the solution $z = x^2(C - 2x)$.

$$x^2y^2(C - 2x) - 1 = 0$$

8.2.7. Clairaut differential equation

$$y = xy' + \varphi(y')$$

By differentiation with respect to x we obtain

$$0 = y''[x + \varphi'(y')]$$

This equation is satisfied either by $y'' = 0$ with the solution $y = C_1x + C_2$ (general integral) or by $x + \varphi'(y') = 0$. When we eliminate y' from the last equation and from the initial equation, we obtain $y = g(x)$ (singular solution).
Geometrically interpreted, the general integral is a *family of straight lines*, whereas the singular solution is the *envelope of this family of straight lines.*

Example:

$$y = xy' - 2y'^2 + y'$$

$$0 = xy'' - 4y'y'' + y'' = y''(x - 4y' + 1)$$

$$y'' = 0 \Rightarrow y = C_1x + C_2 \quad \text{(general integral)}$$

$$x - 4y' + 1 = 0; \quad y' = \frac{x+1}{4}$$

Solution:

$$y = x\frac{x+1}{4} - 2\frac{(x+1)^2}{16} + \frac{x+1}{4} = \frac{x^2+2x+1}{8}$$

8.2.8. Riccati differential equation

$$y' = \varphi(x)y^2 + \psi(x)y + \omega(x)$$

A solution is only possible if a *particular integral* y_1 can be found.

Substitution: $y - y_1 = \dfrac{1}{z}$

Example:

$$x^2y' + xy - x^2y^2 + 1 = 0 \tag{1}$$

$$y' = y^2 - \frac{1}{x}y - \frac{1}{x^2} \tag{2}$$

Due to the given form of the equation, a trial can be made with $y = \dfrac{A}{x}$ to find a particular integral:

$$y = \frac{A}{x}; \quad y' = -\frac{A}{x^2}$$

$$-A + A - A^2 + 1 = 0$$

$$A^2 = 1; \quad A_1 = 1, \; A_2 = -1 \Rightarrow y_1 = \frac{1}{x}$$

Substitution: $y - \frac{1}{x} = \frac{1}{z}$; $\quad y' = -\frac{1}{z^2}z' - \frac{1}{x^2}$

$$-\frac{1}{z^2}z' - \frac{1}{x^2} = \left(\frac{1}{z} + \frac{1}{x}\right)^2 - \frac{1}{x}\left(\frac{1}{z} + \frac{1}{x}\right) - \frac{1}{x^2}$$

$$z' + \frac{z}{x} + 1 = 0$$

According to the method of variation of constants we obtain

$$z = -\frac{x}{2} + \frac{C}{x} \Rightarrow y = \frac{1}{x} + \frac{2x}{C_1 - x^2}$$

8.3. Ordinary differential equations of the second order

8.3.1. Special cases

Differential equation without y and y'

$$y'' = \varphi(x)$$

$$y = \int \left[\int \varphi(x)\,dx\right] dx + C_1x + C_2$$

Example:

$$y'' = 4x^2 + 5x$$

$$y' = \int (4x^2 + 5x)\,dx = \frac{4x^3}{3} + \frac{5x^2}{2} + C_1$$

$$y = \int \left(\frac{4x^3}{3} + \frac{5x^2}{2} + C_1\right) dx = \frac{x^4}{3} + \frac{5x^3}{6} + C_1x + C_2$$

Differential equation without x and y'

$$y'' = \varphi(y)$$

Substitution: $y' = p \Rightarrow y'' = p' = \varphi(y)$.

$$p' = \frac{dp}{dy}y' = \frac{dp}{dy}p \Rightarrow \frac{dp}{dy}p = \varphi(y)$$

Solution by separation of variables yields an equation for y' whose quadrature leads to y.

Example:

$$y'' = \frac{y}{a^2}$$

$$\frac{dp}{dy} p = \frac{y}{a^2}; \qquad p = y'$$

$$p\,dp = \frac{y}{a^2}\,dy$$

$$\int p\,dp = \frac{1}{a^2}\int y\,dy$$

$$\frac{p^2}{2} = \frac{y^2}{2a^2} + C_1$$

$$p = y' = \sqrt{\frac{y^2 + 2C_1a^2}{a^2}}$$

$$\frac{a\,dy}{\sqrt{C_2 + y^2}} = dx$$

$$a\int \frac{dy}{\sqrt{C_2 + y^2}} = x$$

$$x = a \sinh^{-1} \frac{y}{\sqrt{C_2}} + C_3$$

$$y = \sqrt{C_2} \sinh \frac{x - C_3}{a}$$

Differential equation without x and y

$$y'' = f(y')$$

Substitution: $y' = p$

Example:

$$y'' = 2y'^2; \quad y' = p; \quad y'' = p'$$

$$p' = 2p^2$$

$$\int \frac{dp}{p^2} = \int 2dx$$

$$-\frac{1}{p} = 2x + C_1; \quad p = -\frac{1}{2x + C_1} = y'$$

$$\int dy = -\int \frac{dx}{2x + C_1}$$

$$y = -\frac{1}{2} \ln |2x + C_1| + C_2 = \ln \left| \frac{1}{\sqrt{2x + C_1}} \right| + \ln C_3$$

$$y = \ln \left| \frac{C_3}{\sqrt{2x + C_1}} \right|$$

Differential equation without y

$$y'' = f(y', x)$$

Substitution: $y' = p$

Example:

$$xy'' + y' - 1 = 0$$

$$y'' = -\frac{y'}{x} + \frac{1}{x}$$

$$p' = -\frac{p}{x} + \frac{1}{x}$$

$$\frac{dp}{1 - p} = \frac{dx}{x}$$

$$-\ln (p - 1) = \ln x + \ln C_1 = \ln C_1 x$$

$$\frac{1}{p - 1} = C_1 x$$

$$p = \frac{1}{C_1 x} + 1 = y'$$

$$\int dy = \int \frac{dx}{C_1 x} + \int dx$$

$$y = \frac{1}{C_1} \ln |x| + x + C_2 = x + C_3 \ln |x| + C_2$$

Differential equation without x

$$y'' = f(y, y')$$

Substitution: $y' = p$

Example 1:

$$y'' = y'^2 y + 3y; \quad y' = p; \quad y'' = \frac{dp}{dy} p$$

$$p \frac{dp}{dy} = p^2 y + 3y$$

$$\frac{dp}{dy} - py = \frac{3y}{p}$$ BERNOULLI equation, procedure for solution see page 400

Example 2:

$$y'' = y'^2 \frac{1}{y}; \quad y' = p; \quad y'' = \frac{dp}{dy} p$$

$$p \frac{dp}{dy} = p^2 \frac{1}{y}$$

$$\frac{dp}{p} = \frac{dy}{y}$$

$$\ln p = \ln y + \ln C_1 = \ln C_1 y$$

$$p = \frac{dy}{dx} = C_1 y$$

$$\frac{dy}{y} = C_1\, dx$$

$$\ln y = C_1 x + C_2 \Rightarrow y = e^{C_1 x + C_2} = C_3 e^{C_1 x}$$

Differential equation without y'

$y'' = f(y, x)$ see the following Sections

8.3.2. Linear homogeneous differential equation of the second order with constant coefficients

$$ay'' + by' + cy = 0 \tag{1}$$

Statement $y = e^{rx} \Rightarrow y' = re^{rx}$ and $y'' = r^2 e^{rx}$

$$ar^2 e^{rx} + bre^{rx} + ce^{rx} = 0$$

Characteristic equation

$$ar^2 + br + c = 0$$

$$r_{1;2} = -\frac{b}{2a} \pm \sqrt{\frac{b^2}{4a^2} - \frac{c}{a}}$$

Case 1: $r_1 \neq r_2 \quad r_1; r_2 \in R$

Solution of the differential equation

$$y = C_1 e^{r_1 x} + C_2 e^{r_2 x}$$

Case 2: $r_1 = r_2 = r; \quad r_1, r_2, r \in R$

Solution of the differential equation

$$y = e^{rx}(C_1 x + C_2)$$

Case 3: $r_{1;2} = a \pm bj$

Solution of the differential equation

$$y = e^{ax}(C_1 \cos bx + C_2 \sin bx)$$

Example 1:

$$2y'' - 8y' + 6y = 0$$

$$2r^2 - 8r + 6 = 0$$

(Case 1)

$$r^2 - 4r + 3 = 0 \quad \text{with} \quad r_1 = 3; \quad r_2 = 1$$

$$y = C_1 e^{3x} + C_2 e^x$$

Example 2:

$$3y'' + 18y' + 27y = 0$$

$$3r^2 + 18r + 27 = 0; \quad r_{1;2} = -3$$ (Case 2)

$$y = e^{-3x}(C_1 x + C_2)$$

Example 3:

$$y'' + 2y' + 5y = 0$$

$$r^2 + 2r + 5 = 0; \quad r_{1;2} = -1 \pm 2j$$ (Case 3)

$$y = e^{-x}(C_1 \cos 2x + C_2 \sin 2x)$$

8.3.3. Linear homogeneous differential equation of the second order with variable coefficients

$$y''\varphi(x) + y'\psi(x) + y\omega(x) = 0$$

A *particular integral* y_1 must be found for its solution.

Statement for solution:

$$y = y_1 z$$

By further substitution $z' = u$, a differential equation of the first order (*reduction of the order*) is obtained.

Example:

$$x^2(\ln x - 1)y'' - xy' + y = 0$$

$$y_1 = x$$

$$y = y_1 z = xz$$

$$y' = xz' + z$$

$$y'' = xz'' + 2z'$$

$$x^3(\ln x - 1)z'' + x^2(2 \ln x - 3)z' = 0; \quad z' = u; \quad z'' = u'$$

$$xu'(\ln x - 1) = u(3 - 2 \ln x)$$

$$\frac{du}{u} = \frac{3 - 2 \ln x}{x(\ln x - 1)}\, dx$$

Substitution: $\ln x = v; \quad \dfrac{dx}{x} = dv$

$$\int \frac{du}{u} = 3\int \frac{dv}{v-1} - 2\int \frac{vdv}{v-1}$$

$$\ln u = 3 \ln (v - 1) - 2\int \left(1 + \frac{1}{v-1}\right) dv$$

$$\ln u = 3 \ln (v - 1) - 2v - 2 \ln (v - 1) + \ln C_1$$

$$u = C_1 \frac{\ln x - 1}{x^2}$$

$$dz = C_1 \frac{\ln x - 1}{x^2} dx$$

$$z = C_2 - \frac{C_1}{x} \ln x$$

$$y = C_2 x - C_1 \ln |x|$$

8.3.4. Euler differential equation

Euler differential equation of the second order without disturbance function (homogeneous Euler differential equation)

$$ax^2 y'' + bxy' + cy = 0$$

Statement:

$$y = x^r; \quad y' = rx^{r-1}; \quad y'' = r(r-1)x^{r-2}$$

$$ax^2 r(r-1)x^{r-2} + bxrx^{r-1} + cx^r = 0$$

$$x^r [ar(r-1) + br + c] = 0$$

$x^r = 0$ yields the trivial solution $y = 0$.

Characteristic equation for r

$$ar(r-1) + br + c = 0$$

$$r_{1;2} = \frac{a-b}{2a} \pm \sqrt{\frac{(a-b)^2}{4a^2} - \frac{c}{a}}$$

Case 1: $r_1 \neq r_2$; $r_1, r_2 \in R$

Solution of the differential equation $y = C_1 x^{r_1} + C_2 x^{r_2}$

Case 2: $r_1 = r_2 = r$; $r_1, r_2, r \in R$

Solution of the differential equation $y = x^r (C_1 \ln |x| + C_2)$

Case 3: $r_{1;2} = a_1 \pm b_1 j$

Solution of the differential equation

$$y = x^{a_1} [C_1 \cos (b_1 \ln |x|) + C_2 \sin (b_1 \ln |x|)]$$

Example 1:

$$3x^2y'' + 15xy' - 36y = 0$$

$$x^2y'' + 5xy' - 12y = 0$$

Characteristic equation

$$r^2 + (5 - 1)r - 12 = 0 \quad \text{with} \quad r_1 = 2, \quad r_2 = -6 \quad \text{(Case 1)}$$

Solution of the differential equation

$$y = C_1x^2 + C_2x^{-6} = C_1x^2 + \frac{C_2}{x^6}$$

Example 2:

$$4x^2y'' - 16xy' + 25y = 0$$

Characteristic equation

$$4r^2 + (-16 - 4)r + 25 = 0$$

$$4r^2 - 20r + 25 = 0 \quad \text{with} \quad r_1 = r_2 = \frac{5}{2} \qquad \text{(Case 2)}$$

Solution of the differential equation

$$y = x^{\frac{5}{2}} (C_1 \ln |x| + C_2) = \sqrt{x^5} (C_1 \ln |x| + C_2)$$

$$= x^2 \sqrt{x} (C_1 \ln |x| + C_2)$$

Example 3:

$$x^2y'' - 7xy' + 20y = 0$$

Characteristic equation

$$r^2 + (-7 - 1)r + 20 = 0$$

$$r^2 - 8r + 20 = 0 \quad \text{with} \quad r_{1;2} = 4 \pm 2j \qquad \text{(Case 3)}$$

Solution of the differential equation

$$y = x^4[C_1 \cos (2 \ln |x|) + C_2 \sin (2 \ln |x|)]$$

Euler differential equation of the second order with disturbance function (complete Euler differential equation)

$$ax^2y'' + bxy' + cy = \varphi(x)$$

We first solve the homogeneous Euler differential equation

$$ax^2y'' + bxy' + cy = 0$$

For solution see previous Section.
By choosing a statement which is appropriate for the degree and the form of the disturbance term $\varphi(x)$, frequently a particular integral of the complete differential equation can additionally be found. Then the solution of the differential equation is the sum obtained from the solution of the homogeneous equation and the particular integral.

Statements for the finding of a particular integral

a) The disturbance term $\varphi(x)$ is a rational integral function of the n th degree.
Statement: $y = Ax^n + Bx^{n-1} + Cx^{n-2} + \cdots + K$

b) The disturbance term is an exponential function.
Statement: $y = Ae^{nx}$ where e^{nx} is the exponential function occurring in the disturbance term.

c) The disturbance term is a function of $\sin mx$ or of $\cos mx$ or $\sin mx$ and $\cos mx$.
Statement: $y = A \sin mx + B \cos mx$

d) The disturbance term is a function of $\sinh mx$ or of $\cosh mx$ or of $\sinh mx$ and $\cosh mx$.
Statement: $y = A \sinh mx + B \cosh mx$

e) The disturbance term is an algebraic sum of the functions individually given above.
Statement: Then the statement also is an algebraic sum of the individual statements.

Example:

$$x^2y'' - 2xy' - 10y = 2x^2 - 3x + 10$$

Solution of the homogeneous differential equation

$$x^2y'' - 2xy' - 10y = 0$$

Characteristic equation

$$r^2 + (-2 - 1)r - 10 = 0$$

$$r^2 - 3r - 10 = 0 \quad \text{yields} \quad r_1 = 5; \quad r_2 = -2$$

Thus, for the homogeneous differential equation

$$y = C_1x^5 + C_2x^{-2}$$

Determination of a particular integral:
Statement:

$$y = Ax^2 + Bx + C$$

$$y' = 2Ax + B$$

$$y'' = 2A$$

By insertion into the initial equation

$$x^2 \cdot 2A - 2x(2Ax + B) - 10(Ax^2 + Bx + C) = 2x^2 - 3x + 10$$

By *comparison of coefficients* follows

$$A = -\frac{1}{6}, \quad B = \frac{1}{4}, \quad C = -1$$

Particular integral $y_1 = -\frac{1}{6}x^2 + \frac{1}{4}x - 1$

Hence, solution of the complete differential equation:

$$y = C_1x^5 + C_2x^{-2} + y_1 = C_1x^5 + \frac{C_2}{x^2} - \frac{1}{6}x^2 + \frac{1}{4}x - 1$$

8.3.5. Linear inhomogeneous differential equation of the second order with constant coefficients

$$ay'' + by' + cy = \varphi(x) \qquad \varphi(x) \not\equiv 0$$

or by division by a

$$y'' + dy' + ey = \omega(x) \quad \text{with} \quad \omega(x) \not\equiv 0 \quad \text{(normal form)}$$

First we solve the homogeneous equation for $\omega(x) = 0$ in the same way as given in the above Sections dealing with homogeneous equations.
Then we apply the method of variation of constants to the inhomogeneous differential equation (with disturbance function). In individual cases we may arrive at a solution more readily when using one of the following statements:

$$y = A_0 + A_1x + \cdots + A_nx^n$$

or

$$y = Ae^{nx}$$

or

$$y = A \sin mx + B \cos nx$$

or

$$y = A \sinh mx + B \cosh nx$$

or an algebraic sum of the above functions.
Then the solution of the differential equation is composed of the solution of the associated homogeneous differential equation and the particular integral.
If the disturbance function or a term of it also is a solution of the homogeneous differential equation (*case of resonance*), then the variation of constants will give a result though this procedure may be time-consuming.

Example 1:

$$y'' - 2y' - 8y = 3 \sin x + 4$$

Solution of the homogeneous differential equation

$$y'' - 2y' - 8y = 0$$

Characteristic equation

$$r^2 - 2r - 8 = 0 \quad \text{with} \quad r_1 = 4; \quad r_2 = -2$$

Solution of the homogeneous differential equation:

$$y = C_1e^{4x} + C_2e^{-2x}$$

Statement for the determination of a particular integral:

$$y = A \sin x + B \cos x + C$$

$$y' = A \cos x - B \sin x$$

$$y'' = -A \sin x - B \cos x$$

Insertion into the initial equation

$$-A \sin x - B \cos x - 2A \cos x + 2B \sin x$$
$$- 8A \sin x - 8B \cos x - 8C = 3 \sin x + 4$$

By comparison of coefficients we obtain

$$A = -\frac{27}{85}; \quad B = \frac{6}{85}; \quad C = -\frac{1}{2}$$

Hence, solution of the differential equation

$$y = C_1 e^{4x} + \frac{C_2}{e^{2x}} - \frac{27}{85} \sin x + \frac{6}{85} \cos x - \frac{1}{2}$$

Example 2 (case of resonance):

$$y'' + y' - 2y = \cosh x$$

Characteristic equation of the homogeneous differential equation

$$r^2 + r - 2 = 0 \quad \text{with} \quad r_1 = 1; \quad r_2 = -2$$

Solution of the homogeneous differential equation

$$y = C_1 e^x + C_2 e^{-2x}$$

For $\cosh x$ we may put $\cosh x = \frac{1}{2}(e^x + e^{-x})$.

With $C_1 = \frac{1}{2}$ and $C_2 = 0$, the solution of the homogeneous equation becomes a term of the disturbance function and thus we have a case of resonance. We proceed by variation of constants:

$$y = z_1 e^x + z_2 e^{-2x}$$

$$y' = z_1' e^x + z_1 e^x + z_2' e^{-2x} - 2z_2 e^{-2x}$$

Additional condition: $z_1' e^x + z_2' e^{-2x} = 0$

$$y'' = z_1' e^x + z_1 e^x - 2z_2' e^{-2x} + 4z_2 e^{-2x}$$

Inserted into the initial equation, we have

$$z_1'e^x + z_1e^x - 2z_2'e^{-2x} + 4z_2e^{-2x} + z_1e^x - 2z_2e^{-2x} - 2z_1e^x - 2z_2e^{-2x} = \cosh x$$

From this follows that

$$z_1'e^x - 2z_2'e^{-2x} = \cosh x$$ (the terms with z_1 and z_2 will always vanish)

In connection with the above statement, a system of equations with the two unknowns z_1' and z_2' is obtained:

$$\begin{aligned} z_1'e^x - 2z_2'e^{-2x} &= \cosh x \\ z_1'e^x + z_2'e^{-2x} &= 0 \end{aligned}$$

From this $z_1' = \dfrac{\cosh x}{3e^x}$ and $z_2' = -\dfrac{\cosh x}{3e^{-2x}}$

$$z_1' = \frac{e^x + e^{-x}}{6e^x}$$

Separation of the variables

$$dz_1 = \frac{1}{6}(1 + e^{-2x})\,dx$$

$$z_1 = \frac{1}{6}x - \frac{1}{12}e^{-2x} + K_1$$

$$z_2 = -\frac{1}{18}e^{3x} - \frac{1}{6}e^x + K_2$$

Solution of the differential equation

$$y = \left(\frac{1}{6}x - \frac{1}{12}e^{-2x} + K_1\right)e^x - \left(\frac{1}{18}e^{3x} + \frac{1}{6}e^x - K_2\right)e^{-2x}$$

$$y = e^x\left(K_3 + \frac{x}{6}\right) - \frac{1}{4}e^{-x} + K_2e^{-2x}$$

8.3.6. Linear inhomogeneous differential equation of the second order with variable coefficients

$$y''\varphi_1(x) + y'\varphi_2(x) + y\varphi_3(x) = \varphi_4(x) \qquad \varphi_1(x) \not\equiv 0$$

$$\varphi_4(x) \not\equiv 0$$

or after division by $\varphi_1(x)$

$$y'' + y'\varphi(x) + y\psi(x) = \omega(x) \quad \text{with} \quad \omega(x) \not\equiv 0 \quad \text{(normal form)}$$

Solution:

Let us assume $y_1 = y_1(x)$ to be a nonidentically vanishing solution of the homogeneous equation $y'' + y'\varphi(x) + y\psi(x) = 0$. If we put $z = z(x) = \frac{d}{dx}\left(\frac{y}{y_1}\right)$, this equation changes into the linear homogeneous differential equation of the first order

$$z' + \left(\varphi + \frac{2y_1'}{y_1}\right) z = 0, \quad \varphi = \varphi(x)$$

which can be integrated according to the method of separation of variables. If z is a particular integral, then $y_2 = y_2(x) = y_1 \int z\, dx$ is a second particular integral, linearly independent of y_1, of

$$y'' + y'\varphi + y\psi = 0, \quad \psi = \psi(x)$$

Hence, $y = C_1 y_1 + C_2 y_2$ is the general integral of the homogeneous differential equation.
We find the general integral of the inhomogeneous differential equation

$$y'' + y'\varphi + y\psi = \omega(x)$$

by variation of the constants of integration.

Example:

$$x^2 y'' - 2xy' + (x^2 + 2)y = x^4$$

Division by x^2 leads to the normal form

$$y'' - \frac{2}{x} y' + \frac{x^2 + 2}{x^2} y = x^2$$

We first solve the homogeneous differential equation

$$y'' - \frac{2}{x} y' + \frac{x^2 + 2}{x^2} y = 0$$

which has the particular integral $y_1 = x \sin x$.
If we put $z = \frac{d}{dx}\left(\frac{y}{x \sin x}\right)$ then follows that $y = x \sin x \int z \, dx$ and

$$y' = (\sin x + x \cos x) \int z \, dx + xz \sin x$$

$$y'' = (2 \cos x - x \sin x) \int z \, dx + 2z(\sin x + x \cos x) + xz' \sin x$$

By inserting into the differential equation we obtain

$$(2 \cos x - x \sin x) \int z \, dx + 2z(\sin x + x \cos x) + xz' \sin x$$

$$- \left(\frac{2}{x} \sin x + 2 \cos x\right) \int z \, dx - 2z \sin x + x \sin x \int z \, dx$$

$$+ \frac{2}{x} \sin x \int z \, dx = 0 \quad \text{or} \quad xz' \sin x + 2xz \cos x = 0;$$

$$\frac{z'}{z} = - \frac{2 \cos x}{\sin x}; \qquad \ln |z| = -2 \ln |\sin x|$$

Hence $z = \dfrac{1}{\sin^2 x}$.
From this follows for the second integral linearly independent of y_1

$$y_2 = x \sin x \int \frac{dx}{\sin^2 x} = -x \sin x \cot x = -x \cos x$$

Thus, the general integral of the homogeneous differential equation to be found is

$$y = x(C_1 \sin x + C_2 \cos x) = C_1 y_1 + C_2 y_2$$

We find the general integral of the inhomogeneous differential equation by variation of the constants.
If we replace C_1 and C_2 by the functions $z_1 = z_1(x)$ and $z_2 = z_2(x)$, then $y = z_1 y_1 + z_2 y_2$.
If we insert this into the inhomogeneous differential equation,

another condition must be imposed upon the functions z_1 and z_2 in order to determine them:

Additional condition: $z_1'y_1 + z_2'y_2 = 0$

Then the following relations hold:

$$y' = z_1y_1' + z_2y_2'$$

$$y'' = z_1'y_1' + z_2'y_2' + z_1y_1'' + z_2y_2''$$

By inserting them into the normal form we obtain

$$z_1'y_1' + z_2'y_2' + z_1y_1'' + z_2y_2'' - \frac{2}{x} z_1y_1' - \frac{2}{x} z_2y_2' + z_1y_1$$
$$+ z_2y_2 + \frac{2}{x^2} z_1y_1 + \frac{2}{x^2} z_2y_2 = x^2$$

$$z_1'y_1' + z_2'y_2' + z_1\left(y_1'' - \frac{2}{x} y_1' + y_1 + \frac{2}{x^2} y_1\right)$$
$$+ z_2\left(y_2'' - \frac{2}{x} y_2' + y_2 + \frac{2}{x^2} y_2\right) = x^2$$

Since y_1 and y_2 are particular integrals of the homogeneous differential equation, the expressions in parantheses are equal to zero. Thus, together with the additional condition, the following system of equations holds:

$$\begin{array}{l|} z_1'y_1' + z_2'y_2' = x^2 \\ z_1'y_1 + z_2'y_2 = 0 \\ \hline \end{array}$$

The determinant of this system is

$$\Delta = \begin{vmatrix} y_1' & y_2' \\ y_1 & y_2 \end{vmatrix} = y_1'y_2 - y_2'y_1$$
$$= (\sin x + x\cos x)(-x\cos x)$$
$$-(-\cos x + x\sin x)x\sin x = -x^2 \not\equiv 0$$

The above system of equations yields

$$z_1' = \frac{\begin{vmatrix} x^2 & y_2' \\ 0 & y^2 \end{vmatrix}}{\Delta} \quad \text{and} \quad z_2' = \frac{\begin{vmatrix} y_1' & x^2 \\ y_1 & 0 \end{vmatrix}}{\Delta}$$

By integration we obtain $z_1 = \int \frac{x^2 y_2\, dx}{\Delta}$ and $z_2 = -\int \frac{x^2 y_1\, dx}{\Delta}$.

$$z_1 = \int \frac{x^2(-x \cos x)}{-x^2}\, dx = \int x \cos x\, dx = \cos x + x \sin x + C_3$$

$$z_2 = -\int \frac{x^2 \cdot x \sin x}{-x^2}\, dx = \int x \sin x\, dx = \sin x - x \cos x + C_4$$

The general integral of the inhomogeneous differential equation

$$\begin{aligned} y &= (\cos x + x \sin x + C_3) x \sin x \\ &\quad + (\sin x - x \cos x + C_4)(-x \cos x) \\ &= x^2 + C_3 x \sin x - C_4 x \cos x \end{aligned}$$

8.4. Ordinary differential equations of the third order

8.4.1. Linear homogeneous differential equation of the third order with constant coefficients

$$y''' + ay'' + by' + cy = 0$$

According to the explanations given on page 405, the statement $y = e^{rx}$ leads to the characteristic equation

$$r^3 + ar^2 + br + c = 0$$

Case 1:

$$r_1 \neq r_2 \neq r_3; \quad r_1, r_2, r_3 \in R$$

Solution of the differential equation

$$y = C_1 e^{r_1 x} + C_2 e^{r_2 x} + C_3 e^{r_3 x}$$

Case 2:

$$r_1 = r_2 = r; \quad r_3 \neq r; \quad r, r_3 \in R$$

Solution of the differential equation

$$y = (C_1 x + C_2) e^{rx} + C_3 e^{r_3 x}$$

Case 3:

$$r_1 = r_2 = r_3 = r; \qquad r \in R$$

Solution of the differential equation

$$y = (C_1x^2 + C_2x + C_3)e^{rx}$$

Case 4:

$$r_1 \in R; \quad r_{2;3} = a_1 \pm b_1j$$

Solution of the differential equation

$$y = C_1e^{r_1x} + (C_2 \cos b_1x + C_3 \sin b_1x)e^{a_1x}$$

This procedure can easily be generalized to be applicable to a linear homogeneous differential equation of nth order with constant coefficients.

8.4.2. Linear inhomogeneous differential equation of the third order with constant coefficients

$$y''' + ay'' + by' + cy = \varphi(x)$$

This differential equation is treated in practically the same way as the inhomogeneous differential equation of the second order (see page 411); that is to say, we first solve the associated homogeneous equation and then find a particular integral by means of a suitable statement. By addition of these results, we obtain the general solution of the differential equation.
Analogously this procedure can also be applied to differential equations of higher orders.

8.5. Integration of differential equations by power series

This method yields approximate solutions. It is used if a differential equation cannot be brought into one of the forms explained above.
We insert

$$y = a_0 + a_1x + a_2x^2 + \cdots + a_nx^n$$

together with the associated derivatives into the differential equation and compare the coefficients of the same powers of x. The initial conditions give a_0 by means of which all other coefficients can be determined (*method of indefinite coefficients, comparison of coefficients*).

We obtain the same result by differentiating several times both the differential equation and the statement in the form of a series and put x equal to zero; in this way the coefficients are obtained when equating the respective derivatives.

Example:

$$y' = y^2 + x^3 \text{ with the initial condition } x = 0,\ y = -1$$

Statement:

$$y = a_0 + a_1x + a_2x^2 + a_3x^3 + \cdots + a_nx^n$$

with

$$y' = a_1 + 2a_2x + 3a_3x^2 + \cdots$$

inserted into the differential equation

$$\begin{aligned} &a_1 + 2a_2x + 3a_3x^2 + 4a_4x^3 + \cdots \\ &= (a_0 + a_1x + a_2x^2 + a_3x^3 + \cdots)^2 + x^3 \\ &= a_0^2 + 2a_0a_1x + (a_1^2 + 2a_0a_2)x^2 \\ &\quad + (1 + 2a_0a_3 + 2a_1a_2)x^3 + \cdots \end{aligned}$$

By comparison of coefficients

$$\begin{aligned} a_1 &= a_0^2 \\ 2a_2 &= 2a_0a_1 \\ 3a_3 &= a_1^2 + 2a_0a_2 \\ 4a_4 &= 1 + 2a_0a_3 + 2a_1a_2 \end{aligned}$$

etc.

From the initial conditions $x = 0$, $y = -1$ it follows that

$$a_0 = -1; \quad a_1 = 1; \quad a_2 = -1; \quad a_3 = 1; \quad a_4 = \frac{1}{4}$$

Thus, the approximate solution of the differential equation reads

$$y \approx -1 + x - x^2 + x^3 + \frac{1}{4}x^4$$

8.6. Partial differential equations

General form: $f(x, y, z, z_x, z_y, z_{xx}, z_{yy}, z_{xy}, \ldots)$ for a function $z(x, y)$.
A partial differential equation is said to be linear if it is linear with respect to the quantities $z, z_x, z_y, z_{xx}, z_{yy}, z_{xy}, \ldots$
A partial differential equation of the first order is called homogeneous if no term free of z and its derivatives occurs, otherwise the equation is called *inhomogeneous*.
The order of a partial differential equation is determined by the order of the highest partial derivative occurring in it.
The general solution of a partial differential equation is distinguished from that of an ordinary differential equation by the fact that arbitrary functions of the independent variables occur in the place of arbitrary constants.

8.6.1. Simple partial differential equations

	Solution:
$z_x = 0$	$z = w(y)$
$z_y = 0$	$z = w(x)$
$z_{xx} = 0$	$z = xw_1(y) + w_2(y)$
$z_{yy} = 0$	$z = yw_1(x) + w_2(x)$
$z_{xy} = 0$	$z = w_1(x) + w_2(y)$
$z_x - z_y = 0$	$z = w(x + y)$
$z_{xy} = f(x, y)$	$z = \iint f(x, y)\,dx\,dy + w_1(x) + w_2(y)$
$z_{xx} - z_{yy} = 0$	$z = w_1(x + y) + w_2(x - y)$
$z_x + z_y = 0$	$z = w(x - y)$
$z_{xx} - \dfrac{z_{yy}}{t^2} = 0$	$z = w_1(x + ty) + w_2(x - ty)$
(t^2 real constant)	
$az_x + bz_y = 0$	$z = w(ay - bx)$

$$z_{xx} + z_{yy} = 0 \qquad z = w_1(x + jy) + w_2(x - jy)$$

$$z_x g_y - z_y g_x = 0 \qquad z = w[g(x, y)]$$

(g given function of x and y)

$$x z_x - y z_y = 0 \qquad z = w(xy)$$

$$y z_x - x z_y = 0 \qquad z = w(x^2 + y^2)$$

8.6.2. Linear partial differential equation of the first order for $z = f(x, y)$

$$P z_x + Q z_y = R$$

where P, Q, R are given functions of x, y, z.

$$dx : dy : dz = P : Q : R$$

or written as

$$\frac{dx}{dy} = \frac{P}{Q}; \quad \frac{dx}{dz} = \frac{P}{R}; \quad \frac{dy}{dz} = \frac{Q}{R}$$

Two of these ordinary differential equations at a time give the solutions

$u(x, y, z) = C_1$ and

$v(x, y, z) = C_2$

from which the *general solution* of the partial differential equation is obtained:

$$w(u, v) = 0$$

The great number of solutions is due to the arbitrary function w; we have to find *particular solutions* by determining the arbitrary function w with the help of *additional boundary conditions.*

Example:

$$2xy z_x + 4y^2 z_y = x^2 y$$

$$dx : dy : dz = 2xy : 4y^2 : x^2 y$$

Differential equation for the determination of u

$$\frac{dx}{dy} = \frac{2xy}{4y^2} = \frac{x}{2y}$$

$$\frac{dx}{x} = \frac{dy}{2y}$$

$$\ln x = \frac{1}{2} \ln y + C_1'$$

$$2 \ln x - \ln y = C_1''$$

$$\ln \frac{x^2}{y} = \ln C_1; \quad C_1 = \frac{x^2}{y} = u$$

Differential equation for the determination of v

$$\frac{dz}{dx} = \frac{x^2 y}{2xy} = \frac{x}{2}$$

$$dz = \frac{x}{2} dx$$

$$z = \frac{x^2}{4} + C_2; \quad C_2 = z - \frac{x^2}{4} = v$$

General solution

$$w\left(\frac{x^2}{y}, z - \frac{x^2}{4}\right) = 0$$

Further, we have to find the particular solution that takes the values $z(x, 4) = \frac{5}{4} x^2$ for $y = 4$.

$$C_1 = \frac{x^2}{4}$$

$$C_2 = \frac{5}{4} x^2 - \frac{1}{4} x^2 = x^2$$

From these two equations it follows that

$$C_1 = \frac{1}{4} C_2$$

Particular solution

$$\frac{x^2}{y} = \frac{1}{4}\left(z - \frac{x^2}{4}\right)$$

$$4x^2 = yz - \frac{x^2 y}{4}$$

$$z = \frac{4x^2}{y} + \frac{x^2}{4} = x^2\left(\frac{4}{y} + \frac{1}{4}\right)$$

9. Infinite series, Fourier series, Fourier integral, Laplace transformation

9.1. Infinite series

9.1.1. General

The sequence $\{a_k\}_{\text{ordered}} = a_1, a_2, a_3, \ldots$ with ordered terms has the *partial sums* $s_n = \sum_{k=1}^{n} a_k = a_1 + a_2 + a_3 + \cdots + a_n$. *Partial sum series* $\{s_k\}_{\text{ordered}} = s_1, s_2, s_3, \ldots = a_1, a_1 + a_2, a_1 + a_2 + a_3, \ldots$

The infinite series $\sum_{k=1}^{\infty} a_k$ is another representation of the partial sum sequence.

Example:

$$\{a_k\}_{\text{ordered}} = 1;\quad \frac{1}{2};\quad \frac{1}{4};\quad \frac{1}{8};\ \ldots$$

$$\{s_k\}_{\text{ordered}} = 1;\ 1 + \frac{1}{2};\quad 1 + \frac{1}{2} + \frac{1}{4};\quad 1 + \frac{1}{2} + \frac{1}{4} + \frac{1}{8};\ \ldots = 1;\ 1\frac{1}{2};\quad 1\frac{3}{4};\quad 1\frac{7}{8};\ \ldots$$

$$\sum_{k=1}^{\infty} a_k = 1 + \frac{1}{2} + \frac{1}{4} + \frac{1}{8} + \cdots$$

9.1.2. Convergence criteria

Convergent and divergent infinite series

The infinite series $\sum_{k=1}^{\infty} a_k$ *converges* if for the partial sums $s_n = \sum_{k=1}^{n} a_k$ the limit value $\lim_{n\to\infty} s_n$ exists and has a finite value s. s is called the sum of the convergent series.

An infinite series is termed "*definitely divergent*" if $\lim_{n\to\infty} s_n = \infty$

(improper limit value) or "*indefinitely divergent*" if the limit value $\lim_{n\to\infty} s_n$ does not exist.

The series $\sum_{k=1}^{\infty} a_k$ is called "*absolutely convergent*" if the series of the absolute values $\sum_{k=1}^{\infty} |a_k|$ converges.

The series $\sum_{k=1}^{\infty} a_k$ is called "*unconditionally convergent*" if its sum is independent of the sequence of the terms, otherwise it is called "*conditionally convergent*".

The remainder R_n is the difference between the sum s of the series and the partial sum s_n:

$$R_n = s - s_n$$

For convergent series we have

$$\lim_{n\to\infty} R_n = 0$$

Main criterion for series with arbitrary terms

$$\lim_{k\to\infty} (s_{k+p} - s_k) = 0 \qquad p \in N$$

This criterion is *necessary* and also *sufficient*.

$$\lim_{k\to\infty} a_k = 0$$

is a *necessary* but not *sufficient* convergence criterion. *Sufficient* but not *necessary* convergence criteria for series with positive terms are:

Quotient criterion: $\left|\dfrac{a_{k+1}}{a_k}\right| \leqq q < 1$ (D'ALEMBERT)

Root criterion: $\sqrt[k]{|a_k|} \leqq q < 1$ (CAUCHY)

The two criteria written in a different manner:

$$\lim_{k\to\infty} \left|\frac{a_{k+1}}{a_k}\right| < 1; \quad \lim_{k\to\infty} \sqrt[k]{|a_k|} < 1$$

If the limit values $> 1 \Rightarrow$ divergence
$= 1$, no direct decision can be made with regard to convergence or divergence.

Method of series comparison (for series with positive terms)

If $b_k \leqq a_k$ is always valid in the series $\sum_{k=1}^{\infty} a_k$ and $\sum_{k=1}^{\infty} b_k$, the convergence of the series $\sum_{k=1}^{\infty} b_k$ is a consequence of the convergence of the series $\sum_{k=1}^{\infty} a_k$. In this case, the series $\sum_{k=1}^{\infty} a_k$ is called "*convergent majorant*" or *superseries* of the series $\sum_{k=1}^{\infty} b_k$.

If in the series $\sum_{k=1}^{\infty} a_k$ and $\sum_{k=1}^{\infty} b_k$ always $b_k \geqq a_k$, the divergence of the series $\sum_{k=1}^{\infty} b_k$ is a consequence of the divergence of the series $\sum_{k=1}^{\infty} a_k$. In this case, the series $\sum_{k=1}^{\infty} a_k$ is called "*divergent minorant*" or *subseries* of the series $\sum_{k=1}^{\infty} b_k$.

Alternating series

$$a_1 - a_2 + a_3 - a_4 + - \cdots$$

An alternating series converges if the series of the absolute values converges.
A *sufficient* convergence criterion for alternating series: $\lim_{k\to\infty} a_k = 0$ with the sequence $\{a_k\}_{\text{ordered}}$ decreasing monotonously (LEIBNIZ' *convergence criterion*).
Convergent series can be added or subtracted term by term.
Absolutely convergent series can be multiplied by each other in the same way as polynominals.

9.1.3. Some infinite convergent series

$$1 + \frac{1}{1!} + \frac{1}{2!} + \frac{1}{3!} + \cdots = e \qquad \left(a_k = \frac{1}{(k-1)!}\right)$$

$$1 - \frac{1}{2} + \frac{1}{3} - \frac{1}{4} + - \cdots = \ln 2 \qquad \left(a_k = \frac{(-1)^{k-1}}{k}\right)$$

$$1 + \frac{1}{2} + \frac{1}{4} + \frac{1}{8} + \cdots = 2 \qquad \left(a_k = \frac{1}{2^{k-1}}\right)$$

$$1 + \frac{1}{2^2} + \frac{1}{3^2} + \frac{1}{4^2} + \cdots = \frac{\pi^2}{6} \qquad \left(a_k = \frac{1}{k^2}\right)$$

$$1 - \frac{1}{2^2} + \frac{1}{3^2} - \frac{1}{4^2} + - \cdots = \frac{\pi^2}{12} \qquad \left(a_k = \frac{(-1)^{k-1}}{k^2}\right)$$

$$\frac{1}{1^2} + \frac{1}{3^2} + \frac{1}{5^2} + \cdots = \frac{\pi^2}{8} \qquad \left(a_k = \frac{1}{(2k-1)^2}\right)$$

$$\frac{1}{1 \cdot 2} + \frac{1}{2 \cdot 3} + \frac{1}{3 \cdot 4} + \cdots = 1 \qquad \left(a_k = \frac{1}{k(k+1)}\right)$$

$$\frac{1}{1 \cdot 3} + \frac{1}{3 \cdot 5} + \frac{1}{5 \cdot 7} + \cdots = \frac{1}{2} \qquad \left(a_k = \frac{1}{(2k-1)(2k+1)}\right)$$

$$\frac{1}{1 \cdot 2 \cdot 3} + \frac{1}{2 \cdot 3 \cdot 4} + \frac{1}{3 \cdot 4 \cdot 5} + \cdots = \frac{1}{4}$$

$$\left(a_k = \frac{1}{k(k+1)(k+2)}\right)$$

Further infinite series for the number π are obtained from the series expansions of the cyclometric functions (cf. page 434).

$$\frac{\pi}{4} = \operatorname{arc\,tan} 1 = 1 - \frac{1}{3} + \frac{1}{5} - \frac{1}{7} + - \cdots \qquad \text{(Leibniz)}$$

$$\frac{\pi}{4} = \operatorname{arc\,tan} \frac{1}{2} + \operatorname{arc\,tan} \frac{1}{3} = \left(\frac{1}{2} + \frac{1}{3}\right)$$

$$- \frac{1}{3}\left(\frac{1}{2^3} + \frac{1}{3^3}\right) + \frac{1}{5}\left(\frac{1}{2^5} + \frac{1}{3^5}\right) - + \cdots \qquad \text{(Euler)}$$

$$\frac{\pi}{4} = 4 \operatorname{arc\,tan} \frac{1}{5} - \operatorname{arc\,tan} \frac{1}{239}$$

$$= 4\left(\frac{1}{5} - \frac{1}{3 \cdot 5^3} + \frac{1}{5 \cdot 5^5} - + \cdots\right)$$

$$- \left(\frac{1}{239} - \frac{1}{3 \cdot 239^3} + \frac{1}{5 \cdot 239^5} - + \cdots\right)$$

$$\frac{\pi}{4} = \operatorname{arc\,tan} \frac{1}{2} + \operatorname{arc\,tan} \frac{1}{5} + \operatorname{arc\,tan} \frac{1}{8}$$

$$\frac{\pi}{4} = 8 \operatorname{arc\,tan} \frac{1}{10} - \operatorname{arc\,tan} \frac{1}{239} - 4 \operatorname{arc\,tan} \frac{1}{515}$$

9.1.4. Power series

Definition

Power series are infinite series of the form

$$a_0 + a_1x + a_2x^2 + a_3x^3 + \cdots = \sum_{k=0}^{\infty} a_kx^k$$

The interval within which the power series converges is called *convergence range*. Its limit is called *convergence radius r*. The power series converges for all $|x| < r$, it diverges for all $|x| > r$.
The following theorem holds for the convergence radius r:
If α is the largest accumulation point of the series

$$|a_1|;\ \sqrt{|a_2|};\ \sqrt[3]{|a_3|};\ \ldots;\ \sqrt[k]{|a_k|}\ \ldots,$$

$$\text{thus}\quad \limsup_{k\to\infty} \sqrt[k]{|a_k|} = \overline{\lim_{k\to\infty}}\, \sqrt[k]{|a_k|},$$

$$\text{then}\quad r = \frac{1}{\alpha} = \frac{1}{\overline{\lim\limits_{k\to\infty}}\, \sqrt[k]{|a_k|}}$$

For $\alpha = 0$ this means that $r = \infty$, i.e. the power series is *continuously convergent*. For $\alpha = \infty$, this means that $r = 0$, i.e. the power series is only convergent for $x = 0$.
Radius of convergence:

$$r = \lim_{k\to\infty} \left|\frac{a_k}{a_{k+1}}\right| \quad \text{or} \quad r = \left(\lim_{k\to\infty} \sqrt[k]{|a_k|}\right)^{-1}$$

Within the radius of convergence, the power series converges in an absolute manner.
Within its radius of convergence, every power series may be differentiated or integrated term by term.

Methods for the expansion of functions in power series

Taylor series

$$f(x_0 + h) = f(x_0) + \frac{h}{1!} f'(x_0) + \frac{h^2}{2!} f''(x_0)$$

$$+ \frac{h^3}{3!} f'''(x_0) + \cdots + \frac{h^n}{n!} f^{(n)}(x_0) + R_n$$

with the series converging for $\lim\limits_{n\to\infty} R_n = 0$.

General form of the remainder:

$$R_n = \frac{h^{n+1}}{n!p}(1 - \vartheta)^{n+1-p} f^{(n+1)}(x_0 + \vartheta h) \quad \text{for} \quad 0 < \vartheta < 1$$

$p \in N$ (LAGRANGE)

Special cases

(1) $p = n + 1$

Remainder $R_n = \frac{h^{n+1}}{(n+1)!} f^{(n+1)}(x_0 + \vartheta h) \quad \text{for} \quad 0 < \vartheta < 1$

(LAGRANGE)

(2) $p = 1$

Remainder $R_n = \frac{h^{n+1}}{n!}(1 - \vartheta)^n f^{(n+1)}(x_0 + \vartheta h) \text{ for } 0 < \vartheta < 0$

(CAUCHY)

Another form of the TAYLOR series:

$$f(x) = f(a) + \frac{x-a}{1!} f'(a) + \frac{(x-a)^2}{2!} f''(a) + \cdots$$

$$+ \frac{(x-a)^n}{n!} f^{(n)}(a) + R_n$$

Remainder $R_n = \frac{(x-a)^{n+1}}{(n+1)!} f^{(n+1)}[a + \vartheta(x-a)]$

for $0 < \vartheta < 1$ (LAGRANGE)

Remainder $R_n = \frac{(x-a)^{n+1}}{n!}(1-\vartheta)^n f^{(n+1)}[a + \vartheta(x-a)]$

for $0 < \vartheta < 1$ (CAUCHY)

Condition of validity of the TAYLOR series:
It must be possible to differentiate the function $f(x)$ infinitely at the point a.

MacLaurin series

$$f(x) = f(0) + \frac{f'(0)}{1!}x + \frac{f''(0)}{2!}x^2 + \cdots + \frac{f^{(n)}(0)}{n!} + R_n$$

Remainder $R_n = \frac{x^{n+1}}{(n+1)!} f^{(n+1)}(\vartheta x)$ for $0 < \vartheta < 1$ (LAGRANGE)

Remainder $R_n = \frac{x^{n+1}}{n!}(1-\vartheta)^n f^{(n+1)}(\vartheta x)$ for $0 < \vartheta < 1$ (CAUCHY)

$$\lim_{n\to\infty} R_n = 0$$

Condition for the validity of the MacLaurin series:
It must be possible to differentiate the function $f(x)$ infinitely at the point 0.

Expansion of series by integration

A power series may be integrated term by term over an interval if it is uniformly convergent within this interval.

Example:

$\arctan x = \int_0^x \frac{dz}{1+z^2}$. In the interval $0 \leqq z \leqq x$, the series $\frac{1}{1+z^2} = 1 - z^2 + z^4 - z^6 + - \cdots$ converges uniformly for every $|z| < 1$. By integration we obtain $\arctan x = x - \frac{x^3}{3} + \frac{x^5}{5} - + \cdots$ for $|x| < 1$. This series moreover converges for $x = \pm 1$ according to the LEIBNIZ convergence criterion for alternating series (cf. page 427).

SURVEY OF ALREADY EXPANDED SERIES

Binomial series

$$(1 \pm x)^n = 1 \pm \binom{n}{1} x + \binom{n}{2} x^2 \pm \binom{n}{3} x^3 + \pm \cdots$$

for $|x| \leqq 1$; $n \in R$

When n is a positive integer, the series breaks off at the $(n+1)$th term.
$(a \pm x^n) = a^n \left(1 \pm \frac{x}{a}\right)^n$ can be expanded by means of the above

series when substituting $\frac{x}{a}$ for x.

$$(1 \pm x)^{\frac{1}{2}} = 1 \pm \frac{1}{2}x - \frac{1 \cdot 1}{2 \cdot 4}x^2 \pm \frac{1 \cdot 1 \cdot 3}{2 \cdot 4 \cdot 6}x^3 - \frac{1 \cdot 1 \cdot 3 \cdot 5}{2 \cdot 4 \cdot 6 \cdot 8}x^4 \pm - \cdots \quad \text{for } |x| \leqq 1$$

$$(1 \pm x)^{\frac{1}{3}} = 1 \pm \frac{1}{3}x - \frac{1 \cdot 2}{3 \cdot 6}x^2 \pm \frac{1 \cdot 2 \cdot 5}{3 \cdot 6 \cdot 9}x^3 - \frac{1 \cdot 2 \cdot 5 \cdot 8}{3 \cdot 6 \cdot 9 \cdot 12}x^4 \pm - \cdots \quad \text{for } |x| \leqq 1$$

$$(1 \pm x)^{\frac{1}{4}} = 1 \pm \frac{1}{4}x - \frac{1 \cdot 3}{4 \cdot 8}x^2 \pm \frac{1 \cdot 3 \cdot 7}{4 \cdot 8 \cdot 12}x^3 - \frac{1 \cdot 3 \cdot 7 \cdot 11}{4 \cdot 8 \cdot 12 \cdot 16}x^4 \pm - \cdots \quad \text{for } |x| \leqq 1$$

$$(1 \pm x)^{-1} = 1 \mp x + x^2 \mp x^3 + x^4 \mp + \cdots \quad \text{for} \quad |x| < 1$$

$$(1 \pm x)^{-\frac{1}{2}} = 1 \mp \frac{1}{2}x + \frac{1 \cdot 3}{2 \cdot 4}x^2 \mp \frac{1 \cdot 3 \cdot 5}{2 \cdot 4 \cdot 6}x^3 + \frac{1 \cdot 3 \cdot 5 \cdot 7}{2 \cdot 4 \cdot 6 \cdot 8}x^4 \mp + \cdots \quad \text{for } |x| < 1$$

$$(1 \pm x)^{-\frac{1}{3}} = 1 \mp \frac{1}{3}x + \frac{1 \cdot 4}{3 \cdot 6}x^2 \mp \frac{1 \cdot 4 \cdot 7}{3 \cdot 6 \cdot 9}x^3 + \frac{1 \cdot 4 \cdot 7 \cdot 10}{3 \cdot 6 \cdot 9 \cdot 12}x^4 \mp + \cdots \quad \text{for} \quad |x| < 1$$

$$(1 \pm x)^{-\frac{1}{4}} = 1 \mp \frac{1}{4}x + \frac{1 \cdot 5}{4 \cdot 8}x^2 \mp \frac{1 \cdot 5 \cdot 9}{4 \cdot 8 \cdot 12}x^3 + \frac{1 \cdot 5 \cdot 9 \cdot 13}{4 \cdot 8 \cdot 12 \cdot 16}x^4 \mp + \cdots \quad \text{for} \quad |x| < 1$$

Exponential series

$$e^x = 1 + \frac{x}{1!} + \frac{x^2}{2!} + \frac{x^3}{3!} + \cdots \quad \text{for } |x| < \infty$$

$$a^x = e^{x \ln a} = 1 + \frac{x \ln a}{1!} + \frac{x^2 \ln^2 a}{2!} + \frac{x^3 \ln^3 a}{3!} + \cdots$$

$$\text{for } |x| < \infty, \ a > 0$$

Logarithmic series

$$\ln x = \frac{x-1}{1} - \frac{(x-1)^2}{2} + \frac{(x-1)^3}{3} - + \cdots \quad \text{for } 0 < x \leqq 2$$

$$\ln x = \frac{x-1}{x} + \frac{(x-1)^2}{2x^2} + \frac{(x-1)^3}{3x^3} + \cdots \quad \text{for } x > \frac{1}{2}$$

$$\ln x = 2\left[\frac{x-1}{x+1} + \frac{(x-1)^3}{3(x+1)^3} + \frac{(x-1)^5}{5(x+1)^5} + \cdots\right]$$

$$\text{for } x > 0$$

$$\ln(1+x) = x - \frac{x^2}{2} + \frac{x^3}{3} - \frac{x^4}{4} + - \cdots \quad \text{for } -1 < x \leqq 1$$

$$\ln(1-x) = -\left(x + \frac{x^2}{2} + \frac{x^3}{3} + \frac{x^4}{4} + \cdots\right) \quad \text{for } -1 \leqq x < 1$$

$$\ln\frac{1+x}{1-x} = 2 \tanh^{-1} x = 2\left[x + \frac{x^3}{3} + \frac{x^5}{5} + \frac{x^7}{7} + \cdots\right]$$

$$\text{for } |x| < 1$$

$$\ln\frac{x+1}{x-1} = 2 \coth^{-1} x = 2\left[\frac{1}{x} + \frac{1}{3x^3} + \frac{1}{5x^5} + \cdots\right]$$

$$\text{for } |x| > 1$$

Trigonometric series

$$\sin x = x - \frac{x^3}{3!} + \frac{x^5}{5!} - \frac{x^7}{7!} + - \cdots \quad \text{for } |x| < \infty$$

$$\cos x = 1 - \frac{x^2}{2!} + \frac{x^4}{4!} - \frac{x^6}{6!} + - \cdots \quad \text{for } |x| < \infty$$

$$\tan x = x + \frac{1}{3}x^3 + \frac{2}{15}x^5 + \frac{17}{315}x^7 + \cdots \quad \text{for} \quad |x| < \frac{\pi}{2}$$

$$\cot x = \frac{1}{x} - \frac{1}{3}x - \frac{1}{45}x^3 - \frac{2}{945}x^5 - \cdots \quad \text{for } 0 < |x| < \pi$$

Series for inverse trigonometric functions

$$\text{arc}\sin x = x + \frac{1}{2}\frac{x^3}{3} + \frac{1\cdot 3}{2\cdot 4}\frac{x^5}{5} + \frac{1\cdot 3\cdot 5}{2\cdot 4\cdot 6}\frac{x^7}{7} + \cdots \quad \text{for} \quad |x| < 1$$

$$\text{arc}\cos x = \frac{\pi}{2} - x - \frac{1}{2}\frac{x^3}{3} - \frac{1\cdot 3}{2\cdot 4}\frac{x^5}{5} - \frac{1\cdot 3\cdot 5}{2\cdot 4\cdot 6}\frac{x^7}{7} - \cdots \quad \text{for} \quad |x| < 1$$

$$\text{arc}\tan x = x - \frac{x^3}{3} + \frac{x^5}{5} - \frac{x^7}{7} + - \cdots \quad \text{for} \quad |x| < 1$$

$$\text{arc}\cot x = \frac{\pi}{2} - x + \frac{x^3}{3} - \frac{x^5}{5} + \frac{x^7}{7} - + \cdots \quad \text{for} \quad |x| < 1$$

Series for hyperbolic functions

$$\sinh x = x + \frac{1}{3!}x^3 + \frac{1}{5!}x^5 + \frac{1}{7!}x^7 + \cdots \quad \text{for} \quad |x| < \infty$$

$$\cosh x = 1 + \frac{1}{2!}x^2 + \frac{1}{4!}x^4 + \frac{1}{6!}x^6 + \cdots \quad \text{for} \quad |x| < \infty$$

$$\tanh x = x - \frac{1}{3}x^3 + \frac{2}{15}x^5 - \frac{17}{315}x^7 + - \cdots \quad \text{for} \quad |x| < \frac{\pi}{2}$$

$$\coth x = \frac{1}{x} + \frac{x}{3} - \frac{x^3}{45} + \frac{2x^5}{945} - + \cdots \quad \text{for} \quad < |x| < \pi$$

Series for inverse hyperbolic functions

$$\sinh^{-1} x = x - \frac{1}{2}\frac{x^3}{3} + \frac{1\cdot 3}{2\cdot 4}\frac{x^5}{5} - \frac{1\cdot 3\cdot 5}{2\cdot 4\cdot 6}\frac{x^7}{7} + - \cdots \quad \text{for} \quad |x| < 1$$

$$\cosh^{-1} x = \pm\left[\ln(2x) - \frac{1}{2 \cdot 2x^2} - \frac{1 \cdot 3}{2 \cdot 4 \cdot 4x^4} - \frac{1 \cdot 3 \cdot 5}{2 \cdot 4 \cdot 6 \cdot 6x^6} - \cdots\right] \qquad \text{for} \quad x > 1$$

$$\tanh^{-1} x = x + \frac{x^3}{3} + \frac{x^5}{5} + \frac{x^7}{7} + \cdots \qquad \text{for} \quad |x| < 1$$

$$\coth^{-1} x = \frac{1}{x} + \frac{1}{3x^3} + \frac{1}{5x^5} + \frac{1}{7x^7} + \cdots \qquad \text{for} \quad |x| > 1$$

9.1.5. Approximation formulas

For very small values of ε, approximation formulas are obtained from the power series; these are frequently used in practice.

$$(1 \pm \varepsilon)^n \approx 1 \pm n\varepsilon \qquad \varepsilon \in R; \quad |\varepsilon| \ll 1$$

$$(a \pm \varepsilon)^n \approx a^n \left(1 \pm n\frac{\varepsilon}{a}\right) \qquad \text{for} \quad \varepsilon \ll a$$

Special cases:

$$(1 \pm \varepsilon)^2 \approx 1 \pm 2\varepsilon \qquad (a \pm \varepsilon)^2 \approx a^2 \left(1 \pm \frac{2\varepsilon}{a}\right)$$

$$(1 \pm \varepsilon)^3 \approx 1 \pm 3\varepsilon \qquad (a \pm \varepsilon)^3 \approx a^3 \left(1 \pm \frac{3\varepsilon}{a}\right)$$

$$\sqrt{1 \pm \varepsilon} \approx 1 \pm \frac{1}{2}\varepsilon \qquad \sqrt{a \pm \varepsilon} \approx \sqrt{a}\left(1 \pm \frac{\varepsilon}{2a}\right)$$

$$\frac{1}{1 \pm \varepsilon} \approx 1 \mp \varepsilon \qquad \frac{1}{a \pm \varepsilon} \approx \frac{1}{a}\left(1 \mp \frac{\varepsilon}{a}\right)$$

$$\frac{1}{\sqrt{1 \pm \varepsilon}} \approx 1 \mp \frac{1}{2}\varepsilon \qquad \frac{1}{\sqrt{a \pm \varepsilon}} \approx \frac{1}{\sqrt{a}}\left(1 \mp \frac{\varepsilon}{2a}\right)$$

$$\sqrt[q]{(1 \pm \varepsilon)^p} \approx 1 \pm \frac{p}{q}\varepsilon$$

$$\frac{1}{\sqrt[q]{(1 \pm \varepsilon)^p}} \approx 1 \mp \frac{p}{q} \varepsilon$$

$$e^{\varepsilon} \approx 1 + \varepsilon \qquad a^{\varepsilon} \approx 1 + \varepsilon \ln a$$

$$\ln (1 + \varepsilon) \approx \varepsilon$$

$$\ln \frac{1 + \varepsilon}{1 - \varepsilon} \approx 2\varepsilon \qquad \ln \left(\varepsilon + \sqrt{\varepsilon^2 + 1}\right) \approx \varepsilon$$

$$\sin \varepsilon \approx \varepsilon \qquad \tan \varepsilon \approx \varepsilon$$

$$\cos \varepsilon \approx 1 - \frac{1}{2} \varepsilon^2 \qquad \cot \varepsilon \approx \frac{1}{\varepsilon}$$

$$\operatorname{arc} \sin \varepsilon \approx \varepsilon \qquad \operatorname{arc} \tan \varepsilon \approx \varepsilon$$

$$\sinh \varepsilon \approx \varepsilon \qquad \tanh \varepsilon \approx \varepsilon$$

$$\cosh \varepsilon \approx 1 + \frac{\varepsilon^2}{2} \qquad \coth \varepsilon \approx \frac{1}{\varepsilon}$$

$$\sinh^{-1} \varepsilon \approx \varepsilon \qquad \tanh^{-1} \varepsilon \approx \varepsilon$$

Generally valid:

$$f(\varepsilon) \approx f(0) + f'(0)\varepsilon$$

9.2. General statements on Fourier series, Fourier integrals, and Laplace transforms

Every unique periodic function $f(x) = f(x + kT_0)$ which is partially monotonous and continuous can be uniquely represented as a FOURIER series with a decomposition into the spectrum of $f(x)$ according to discrete frequencies kf_0.

$$f(x) = \frac{a_0}{2} + \sum_{k=1}^{\infty} [a_k \cos k\omega_0 x + b_k \sin k\omega_0 x] \qquad k \in N$$

with

$$a_k = \frac{2}{T_0}\int_0^{T_0} f(x)\cos k\omega_0 x\,dx \qquad \omega_0 = \frac{2\pi}{T_0} = 2\pi f_0$$

$$b_k = \frac{2}{T_0}\int_0^{T_0} f(x)\sin k\omega_0 x\,dx$$

$\frac{a_0}{2}$ mean value of the function $f(x)$; equal term

Specially valid for the period $T_0 = 2\pi$:

$$f(x) = \frac{a_0}{2} + \sum_{k=1}^{\infty} (a_k \cos kx + b_k \sin kx) \qquad k \in N$$

with $a_k = \frac{1}{\pi}\int_0^{2\pi} f(x)\cos kx\,dx$

$$b_k = \frac{1}{\pi}\int_0^{2\pi} f(x)\sin kx\,dx$$

With $D_k = \sqrt{a_k^2 + b_k^2}$; $\varphi_k = \arctan\frac{a_k}{b_k}$ the *spectral representation* is obtained

$$f(x) = \frac{a_0}{2} + \sum_{k=1}^{\infty} D_k \sin(k\omega_0 x + \varphi_k)$$

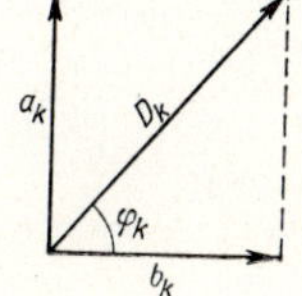

Complex representation:

$$f(x) = \sum_{k=-\infty}^{\infty} c_{\pm k}\, e^{jk\omega_0 x} \qquad k \in N$$

with $c_{\pm k} = \frac{1}{T_0}\int_0^{T_0} f(x)\, e^{\mp jk\omega_0 x}\,dx \qquad T_0 c_{\pm k} =$ *spectrum* for $f(x)$

The position of the integration interval is of no importance and can be stretched from $-\frac{T_0}{2}$ to $+\frac{T_0}{2}$.

Specially valid for the period $T_0 = 2\pi$:

$$f(x) = \sum_{k=-\infty}^{\infty} c_{\pm k}\, e^{jkx} \quad \text{with} \quad c_{\pm k} = \frac{1}{2\pi}\int_0^{2\pi} f(x)e^{\mp jkx}\,dx$$

The following is valid at the discontinuities x_0:

$$f(x_0) = \frac{f(x_0+0) + f(x_0-0)}{2}$$

with $\lim\limits_{x\to x_0+0} f(x) = f(x_0+0)$

$$\lim_{x\to x_0\ 0} f(x) = f(x_0 - 0)$$

Correlation

$$a_k = c_{+k} + c_{-k} \qquad k \in N$$

$$b_k = j(c_{+k} - c_{-k})$$

Every unique function $F(t)$, even if it is not periodic (process occurring only once), which is partially monotone and continuous can be uniquely represented as a FOURIER integral (FOURIER transformation) with a decomposition into a continuous spectrum of frequencies y in the infinite interval $T_0 \to \infty$; $t \in (-\infty, +\infty)$.

$$F(t) = \frac{1}{2\pi}\int_{-\infty}^{+\infty} f(y)e^{jyt}\,dy \qquad \text{condition:} \int_{-\infty}^{\infty} |F(t)\,dt| < \infty$$

with $f(y) = \int\limits_{-\infty}^{+\infty} F(t)e^{-jyt}\,dt$ *spectral function of* $F(t)$ yields

Restriction of the interval to $t \in [0, +\infty]$

$$\frac{1}{2\pi}\int_{-\infty}^{+\infty} f(y)e^{jyt}\,dy = \begin{cases} F(t) & \text{for } t \geqq 0 \\ 0 & \text{for } t < 0 \end{cases}$$

The interval $t \in [0, +\infty]$ permits the damping of the function with

e^{-xt} (reaching of convergence).

$$F(t) \to e^{-xt}\, F(t)$$

$$\frac{1}{2\pi}\int_{-\infty}^{+\infty} f_x(y) e^{jyt}\, dy = \begin{cases} e^{-xt}\, F(t) & \text{for} \quad t \geqq 0 \\ 0 & \text{for} \quad t < 0 \end{cases}$$

with $f_x(y) = \int_0^\infty e^{-xt}\, F(t) e^{-jyt}\, dt = \int_0^\infty e^{-(x+jy)t}\, F(t)\, dt$

Putting $s = x + jy$, we obtain

$$\frac{1}{2\pi j}\int_{x-j\omega}^{x+j\omega} e^{st}\, f(s)\, ds = \begin{cases} F(t) & \text{for} \quad t \geqq 0 \\ 0 & \text{for} \quad t < 0 \end{cases}$$

complex inversion formula

with $f(s) = \int_0^\infty e^{-st}\, F(t)\, dt$ — Laplace integral, Laplace transformation

In control engineering s is frequently replaced by $p = \sigma + j\omega$. $f(s)$ is the spectral function of the damped time function $e^{-xt}\, F(x)$.

9.3. Fourier series

Symmetry conditions for Fourier series

Case 1:

Function symmetrical to ordinate axis, even function

$$f(x) = f(-x)$$

$b_k = 0$, the series only contains cosine terms.

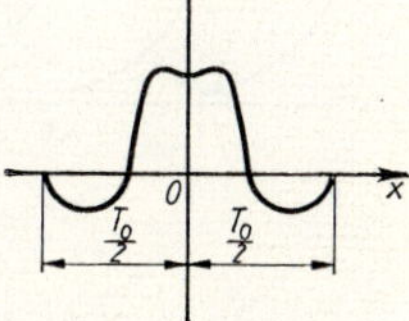

Case 2:

Function centrally symmetrical to origin of coordinates, odd function

$$f(x) = -f(-x)$$

$a_k = 0$, the series only contains sine terms.

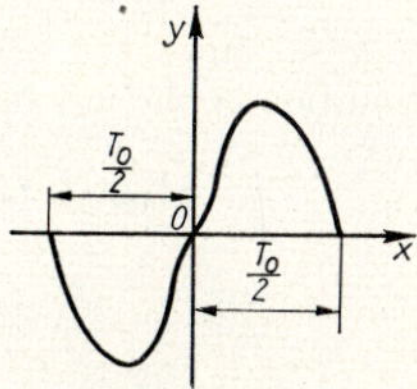

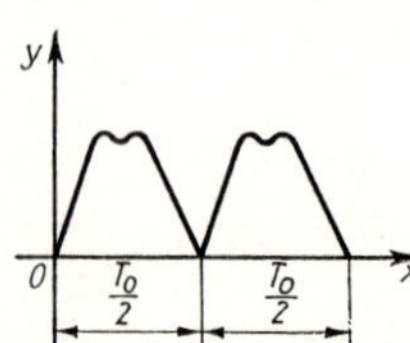

Case 3:

Same shape and same position of the half-cycles with respect to the x-axis

$$f(x) = f\left(x + \frac{T_0}{2}\right)$$

$$a_{2k+1} = 0; \quad b_{2k+1} = 0$$

The series only contains sine and cosine terms with even arguments.

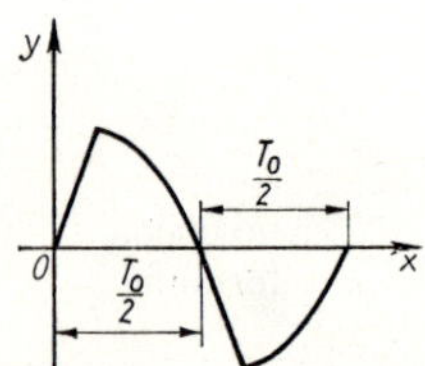

Case 4:

Same shape but different position of the half-cycles with respect to the x-axis

$$f(x) = -f\left(x + \frac{T_0}{2}\right)$$

$$a_{2k} = 0; \quad b_{2k} = 0$$

The series only contains sine and cosine terms with odd arguments.

Calculation of a Fourier series

The following rhythmically proceeding compensation process is to be represented by a Fourier series:

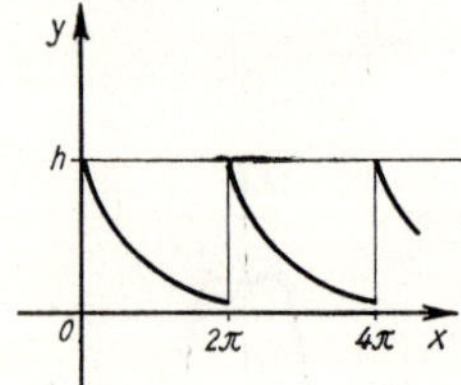

$$f(x) = he^{-x}$$

$$x \in [0, 2\pi],$$

$$T_0 = 2\pi$$

1st procedure of solution (calculation of the coefficients via the trigonometrical form)

$$a_k = \frac{1}{\pi}\int_0^{2\pi} he^{-x}\cos kx\,dx = \frac{h}{\pi}\int_0^{2\pi} e^{-x}\cos kx\,dx \qquad k \in N$$

Solution of the indefinite integral by partial integration

$$\int e^{-x}\cos kx\,dx = \frac{e^{-x}}{k}\sin kx + \frac{1}{k}\int e^{-x}\sin kx\,dx$$

$$= \frac{e^{-x}}{k}\sin kx - \frac{e^{-x}}{k^2}\cos kx - \frac{1}{k^2}\int e^{-x}\cos kx\,dx$$

The same integral occurs on the right and the left side. Collected:

$$\left(1+\frac{1}{k^2}\right)\int e^{-x}\cos kx\,dx = \frac{e^{-x}}{k}\sin kx - \frac{e^{-x}}{k^2}\cos kx;$$

$$\int e^{-x}\cos kx\,dx = \frac{k^2}{1+k^2}\left(\frac{e^{-x}}{k}\sin kx - \frac{e^{-x}}{k^2}\cos kx\right)$$

Considering the limits, $a_k = \dfrac{h(1-e^{-2\pi})}{\pi(1+k^2)}$

Respective calculation yields $b_k = \dfrac{hk(1-e^{-2\pi})}{\pi(1+k^2)}$

Hence $a_0 = \dfrac{h(1-e^{-2\pi})}{\pi}$; $\quad a_1 = \dfrac{h(1-e^{-2\pi})}{2\pi}$;

$$a_2 = \frac{h(1-e^{-2\pi})}{5\pi}; \ldots$$

$$b_0 = 0; \quad b_1 = \frac{h(1-e^{-2\pi})}{2\pi}; \quad b_2 = \frac{2h(1-e^{-2\pi})}{5\pi}; \ldots$$

Thus the FOURIER series will be

$$f(x) = \frac{h(1-e^{-2\pi})}{\pi}\left(\frac{1}{2}+\frac{1}{2}\cos x+\frac{1}{5}\cos 2x+\cdots\right.$$
$$\left.+\frac{1}{2}\sin x+\frac{2}{5}\sin 2x+\cdots\right)$$

Second procedure of solution (calculation of the coefficients via the complex form)

$$c_{\pm k} = \frac{1}{2\pi}\int_0^{2\pi} he^{-x}e^{\mp jk}\,dx = \frac{h}{2\pi}\int_0^{2\pi} e^{-(1\pm jk)x}\,dx$$

Integration immediately yields

$$c_{\pm k} = \left.\frac{-he^{-(1\pm kj)x}}{2\pi(1\pm kj)}\right|_0^{2\pi} = \frac{-h}{2\pi(1\pm kj)}(e^{-2}\underbrace{e^{\mp 2\pi kj}}_{1}-1)$$

$$c_{\pm k} = \frac{h(1-e^{-2\pi})}{2\pi(1\pm kj)}$$

From $c_{\pm k}$ the coefficients a_k and b_k are calculated:

$$a_k = c_{+k} + c_{-k} = \frac{h(1 - e^{-2\pi})}{2\pi} \left(\frac{1}{1 + kj} + \frac{1}{1 - kj} \right)$$

$$= \frac{h(1 - e^{-2\pi})}{2\pi} \cdot \frac{1 - kj + 1 + kj}{1 + k^2} = \frac{h(1 - e^{-2\pi})}{\pi(1 + k^2)}$$

$$b_k = j(c_{+k} - c_{-k}) = \frac{jh(1 - e^{-2\pi})}{2\pi} \left(\frac{1}{1 + kj} - \frac{1}{1 - kj} \right)$$

$$= \frac{jh(1 - e^{-2\pi})}{2\pi} \cdot \frac{1 - kj - 1 - kj}{1 + k^2} = \frac{hk(1 - e^{-2\pi})}{\pi(1 + k^2)}$$

The coefficients obtained in the two calculations are of course the same. However, it will be seen that the calculation using the complex form yields much simpler integrals; for this reason, this method requires partly a smaller amount of calculation, especially when $f(x)$ is an e-function.
The line spectrum of $f(x)$ will be

$$2\pi c_{\pm k} = \frac{h(1 - e^{-2\pi})}{1 \pm kj} = \frac{h(1 - e^{-2\pi})(1 \mp kj)}{1 + k^2}$$

$$= \frac{h(1 - e^{-2\pi})}{1 + k^2} + j \frac{\mp hk(1 - e^{-2\pi})}{1 + k^2} \approx \frac{h}{1 + k^2}$$

$$+ j \frac{\mp hk}{1 + k^2} \quad \text{because} \quad e^{-2\pi} \ll 1.$$

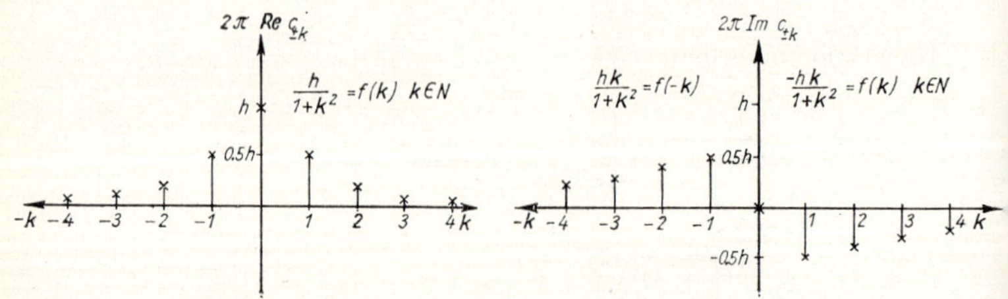

Special Fourier series

1. *Rectangular curve*

$$f(x) = \frac{4h}{\pi}\left(\sin x + \frac{1}{3}\sin 3x + \frac{1}{5}\sin 5x + \cdots\right)$$

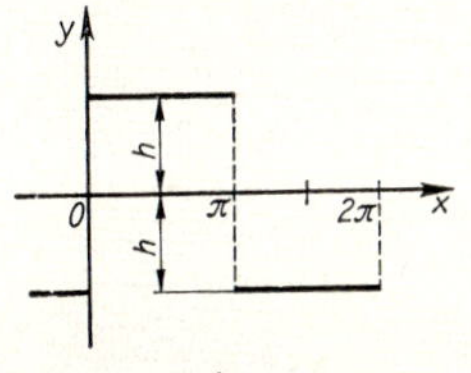

1.

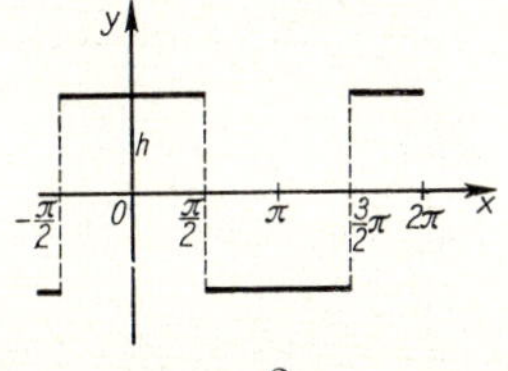

2.

2. *Rectangular curve*

$$f(x) = \frac{4h}{\pi}\left(\cos x - \frac{1}{3}\cos 3x + \frac{1}{5}\cos 5x - + \cdots\right)$$

3. *Rectangular curve* (shifted in the direction of the y-axis)

$$f(x) = \frac{h_1 + h_2}{2} + \frac{2(h_1 - h_2)}{\pi}$$

$$\times\left(\sin x + \frac{1}{3}\sin 3x + \frac{1}{5}\sin 5x + \cdots\right)$$

$h_2 = 0$ leads to rectangular pulse (right-hand figure)

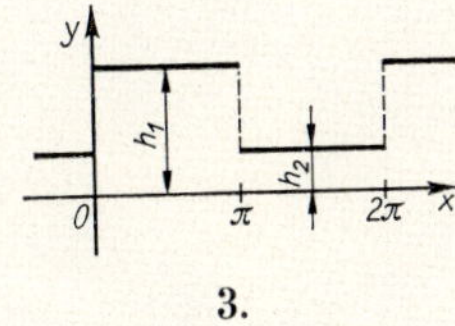

3.

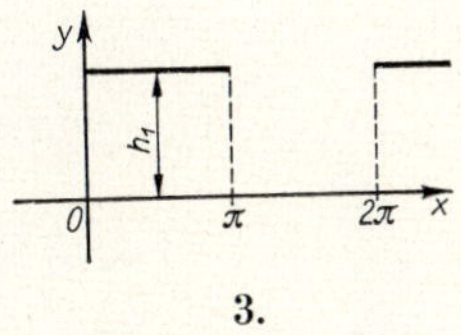

3.

4. *Rectangular curve* (shifted in the direction of the y-axis)

$$f(x) = \frac{h_1 + h_2}{2} + \frac{2(h_1 - h_2)}{\pi}$$

$$\times\left(\cos x - \frac{1}{3}\cos 3x + \frac{1}{5}\cos 5x - + \cdots\right)$$

$h_2 = 0$ leads to rectangular pulse (right-hand figure)

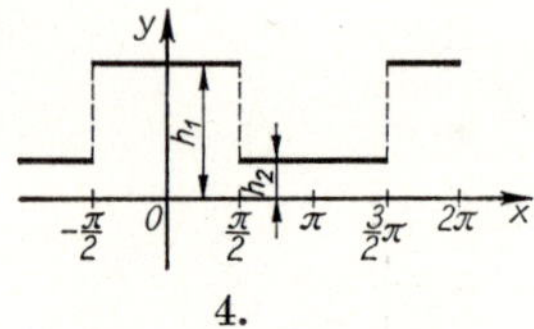

4.

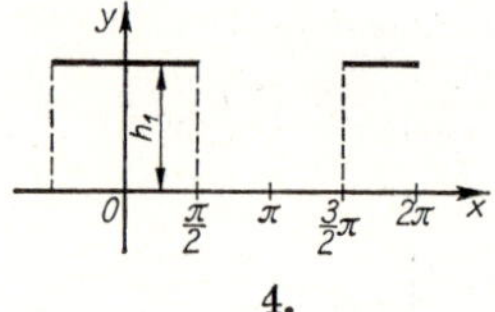

4.

5. *Rectangular pulse*

$$f(x) = \frac{2h}{\pi}\left(\frac{\varphi}{2} + \frac{\sin\varphi}{1}\cos x + \frac{\sin 2\varphi}{2}\cos 2x + \frac{\sin 3\varphi}{2}\cos 3x + \cdots\right)$$

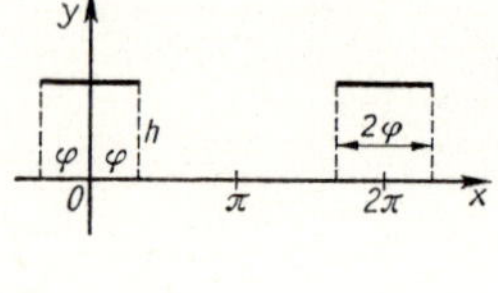

5.

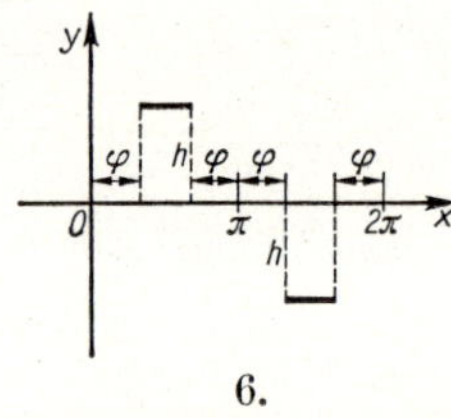

6.

6. *Rectangular pulse*

$$f(x) = \frac{4h}{\pi}\left(\frac{\cos\varphi}{1}\sin x + \frac{\cos 3\varphi}{3}\sin 3x + \frac{\cos 5\varphi}{5}\sin 5x + \cdots\right)$$

7. *Trapezoidal curve* (isoceles trapezoid)

$$f(x) = \frac{4h}{\pi\varphi}\left(\frac{1}{1^2}\sin\varphi\sin x + \frac{1}{3^2}\sin 3\varphi\sin 3x + \frac{1}{5^2}\sin 5\varphi\sin 5x + \cdots\right)$$

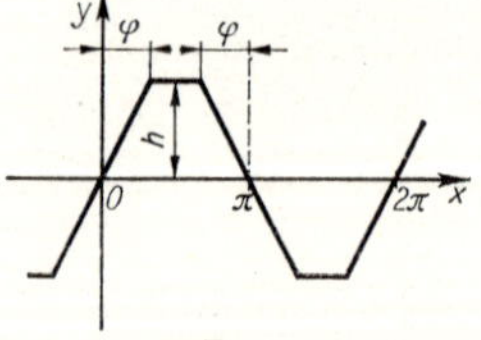

7.

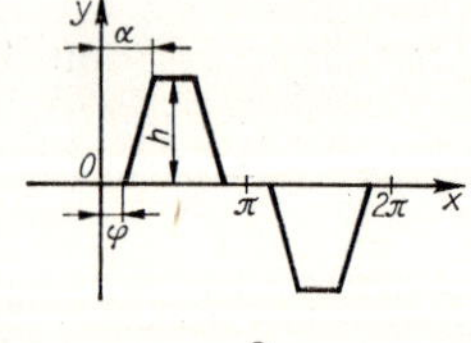

8.

8. *Trapezoidal pulse* (isoceles trapezoid)

$$f(x) = \frac{4h}{\pi(\alpha - \varphi)} \left(\frac{\sin \alpha - \sin \varphi}{1^2} \sin x + \frac{\sin 3\alpha - \sin 3\varphi}{3^2} \sin 3x + \frac{\sin 5\alpha - \sin 5\varphi}{5^2} \sin 5x + \cdots \right)$$

9. *Triangular curve* (isoceles triangle)

$$f(x) = \frac{8h}{\pi^2} \left(\frac{1}{1^2} \sin x - \frac{1}{3^2} \sin 3x + \frac{1}{5^2} \sin 5x - + \cdots \right)$$

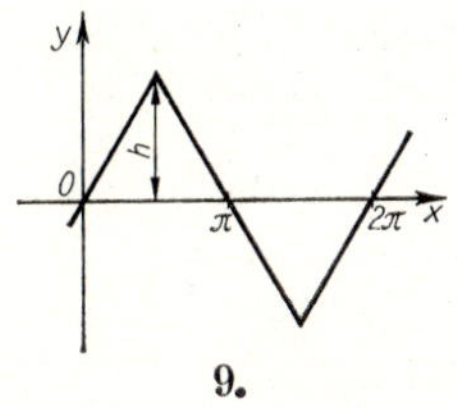

9.

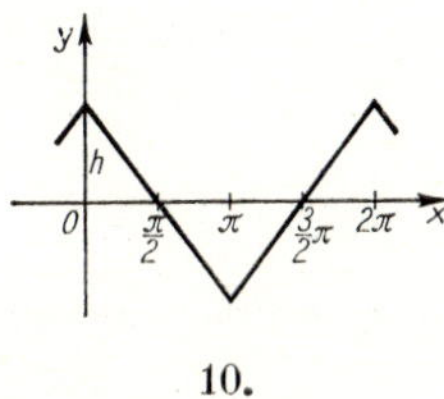

10.

10. *Triangular curve* (isoceles triangle)

$$f(x) = \frac{8h}{\pi^2} \left(\frac{1}{1^2} \cos x + \frac{1}{3^2} \cos 3x + \frac{1}{5^2} \cos 5x + \cdots \right)$$

11. *Triangular curve* (isoceles triangle)

$$f(x) = \frac{h}{2} + \frac{4h}{\pi^2} \left(\frac{1}{1^2} \cos x + \frac{1}{3^2} \cos 3x + \frac{1}{5^2} \cos 5x + \cdots \right)$$

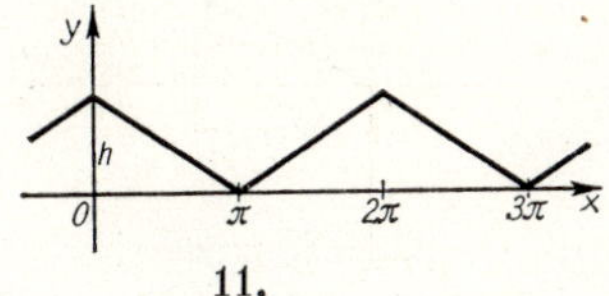

11.

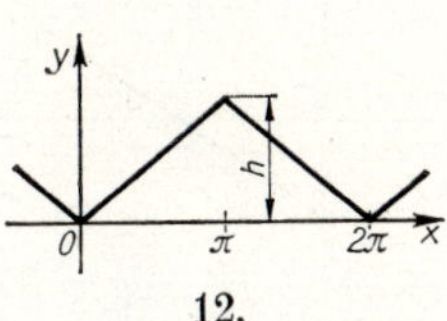

12.

12. *Triangular curve* (isoceles triangle)

$$f(x) = \frac{h}{2} - \frac{4h}{\pi^2} \left(\frac{1}{1^2} \cos x + \frac{1}{3^2} \cos 3x + \frac{1}{5^2} \cos 5x + \cdots \right)$$

13. *Triangular pulse* (isoceles triangle)

$$f(x) = \frac{h\varphi}{2\pi} + \frac{2h}{\pi\varphi}\left(\frac{1-\cos\varphi}{1^2}\cos x + \frac{1-\cos 2\varphi}{2^2}\cos 2x + \frac{1-\cos 3\varphi}{3^2}\cos 3x + \cdots\right)$$

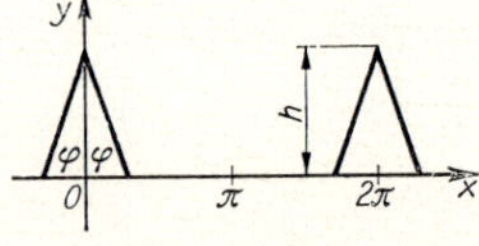

13.

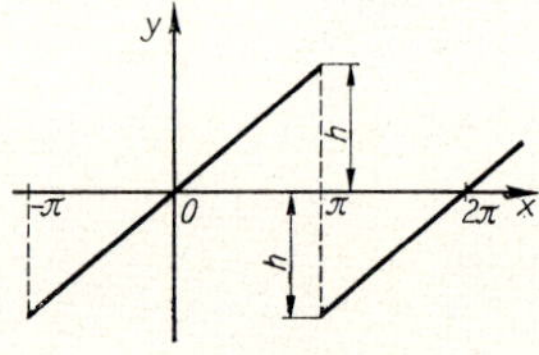

14.

14. *Saw-tooth curve* (rising)

$$f(x) = \frac{2h}{\pi}\left(\sin x - \frac{1}{2}\sin 2x + \frac{1}{3}\sin 3x - + \cdots\right)$$

15. *Saw-tooth curve* (rising)

$$f(x) = -\frac{2h}{\pi}\left(\sin x + \frac{1}{2}\sin 2x + \frac{1}{3}\sin 3x + \cdots\right)$$

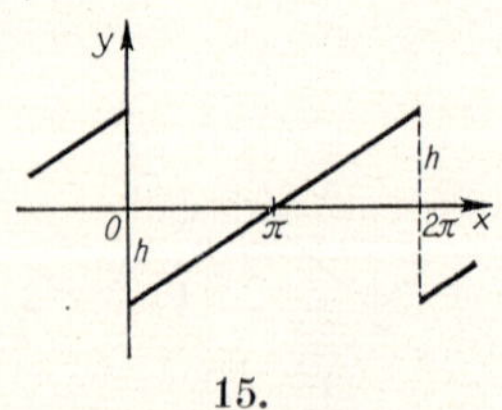

15.

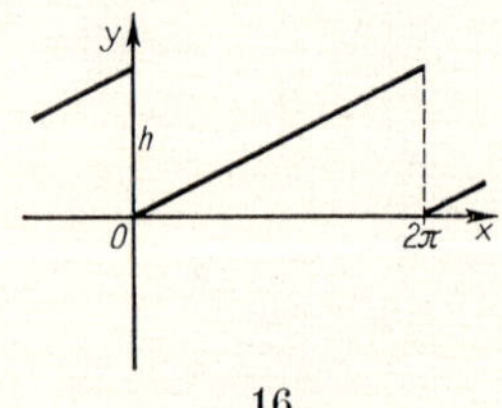

16.

16. *Saw-tooth curve* (rising)

$$f(x) = \frac{h}{2} - \frac{h}{\pi}\left(\sin x + \frac{1}{2}\sin 2x + \frac{1}{3}\sin 3x + \cdots\right)$$

17. *Saw-tooth curve* (falling)

$$f(x) = \frac{2h}{\pi}\left(\sin x + \frac{1}{2}\sin 2x + \frac{1}{3}\sin 3x + \cdots\right)$$

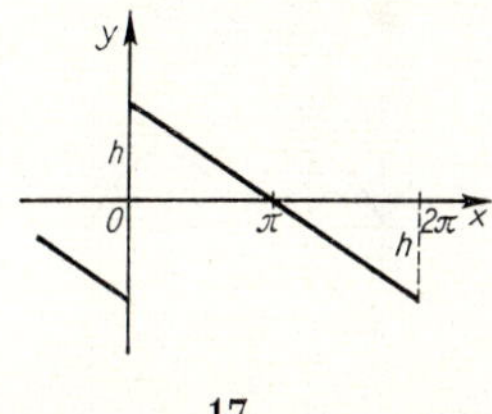

17.

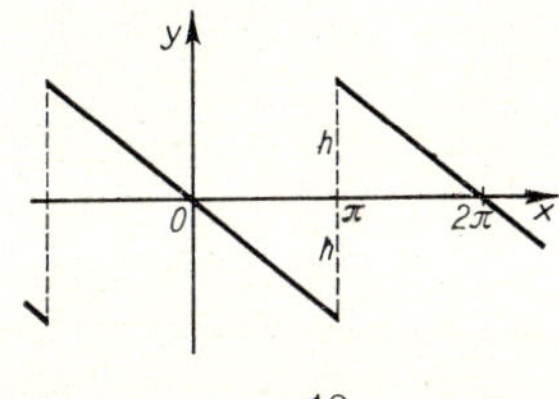

18.

18. *Saw-tooth curve* (falling)

$$f(x) = \frac{2h}{\pi}\left(-\sin x + \frac{1}{2}\sin 2x - \frac{1}{3}\sin 3x + - \cdots\right)$$

19. *Saw-tooth curve* (falling)

$$f(x) = \frac{h}{2} + \frac{h}{\pi}\left(\sin x + \frac{1}{2}\sin 2x + \frac{1}{3}\sin 3x + \cdots\right)$$

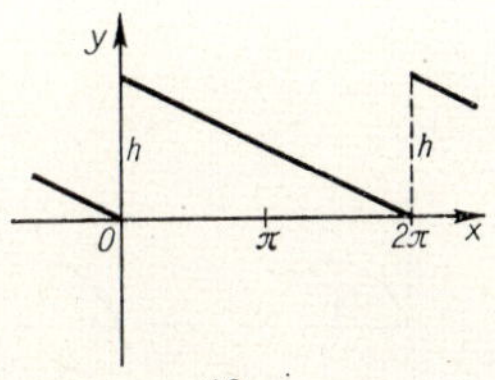

19.

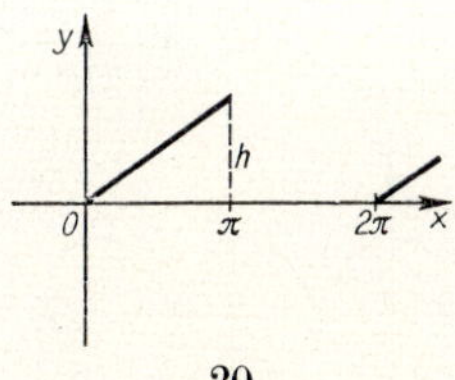

20.

20. *Saw-tooth pulse* (rising)

$$f(x) = \frac{h}{4} + \frac{h}{\pi}\left(\sin x - \frac{1}{2}\sin 2x + \frac{1}{3}\sin 3x - + \cdots\right)$$

$$- \frac{2h}{\pi^2}\left(\cos x + \frac{1}{3^2}\cos 3x + \frac{1}{5^2}\cos 5x + \cdots\right)$$

21. *Saw-tooth pulse* (falling)

$$f(x) = \frac{h}{4} + \frac{h}{\pi}\left(\sin x + \frac{1}{2}\sin 2x + \frac{1}{3}\sin 3x + \cdots\right)$$
$$+ \frac{2h}{\pi^2}\left(\cos x + \frac{1}{3^2}\cos 3x + \frac{1}{5^2}\cos 5x + \cdots\right)$$

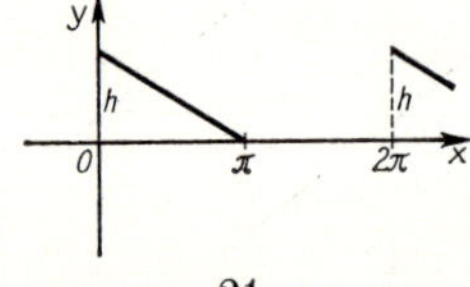

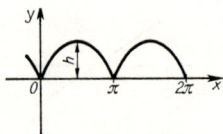

21.

22.

22. *Rectified sine curve* (full-wave rectification)

$$f(x) = \frac{4h}{\pi}\left(\frac{1}{2} - \frac{1}{1\cdot 3}\cos 2x - \frac{1}{3\cdot 5}\cos 4x\right.$$
$$\left. - \frac{1}{5\cdot 7}\cos 6x - \cdots\right)$$

23. *Rectified cosine curve* (full-wave rectification)

$$f(x) = \frac{4h}{\pi}\left(\frac{1}{2} + \frac{1}{1\cdot 3}\cos 2x - \frac{1}{3\cdot 5}\cos 4x\right.$$
$$\left. + \frac{1}{5\cdot 7}\cos 6x - + \cdots\right)$$

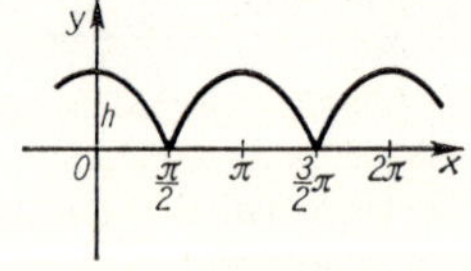

23.

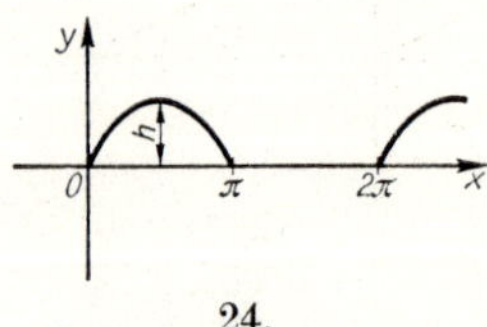

24.

24. *Sine pulse* (half-wave rectification)

$$f(x) = \frac{h}{\pi} + \frac{h}{2}\sin x - \frac{2h}{\pi}\left(\frac{1}{1\cdot 3}\cos 2x\right.$$
$$\left. + \frac{1}{3\cdot 5}\cos 4x + \frac{1}{5\cdot 7}\cos 6x + \cdots\right)$$

25. *Cosine pulse* (half-wave rectification)

$$f(x) = \frac{h}{\pi} + \frac{h}{2}\cos x + \frac{2h}{\pi}\left(\frac{1}{1\cdot 3}\cos 2x - \frac{1}{3\cdot 5}\cos 4x + \frac{1}{5\cdot 7}\cos 6x - + \cdots\right)$$

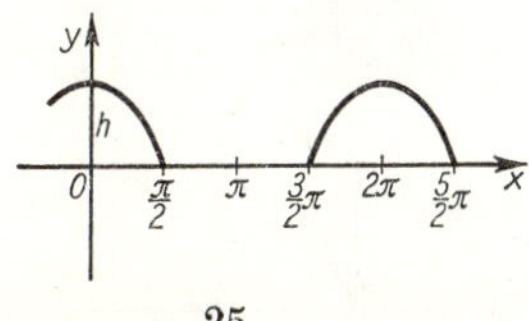

25.

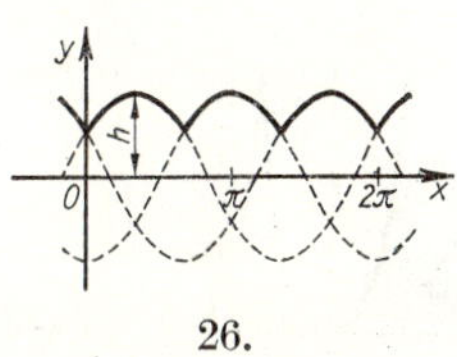

26.

26. *Rectified three-phase current*

$$f(x) = \frac{3h\sqrt{3}}{\pi}\left(\frac{1}{2} - \frac{1}{2\cdot 4}\cos 3x - \frac{1}{5\cdot 7}\cos 6x - \frac{1}{8\cdot 10}\cos 9x - \cdots\right)$$

27. *Parabolic arcs*

$$\left(\text{parabola equation} \quad y = \frac{h}{\pi^2}x^2 \text{ for } [-\pi, \pi]\right)$$

$$f(x) = \frac{h}{3} - \frac{4h}{\pi^2}\left(\cos x - \frac{1}{2^2}\cos 2x + \frac{1}{3^2}\cos 3x - + \cdots\right)$$

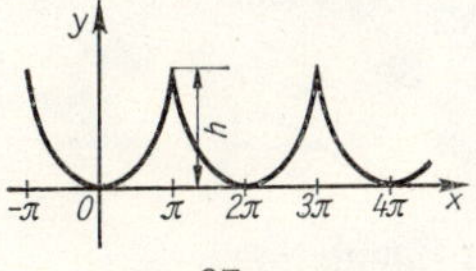

27.

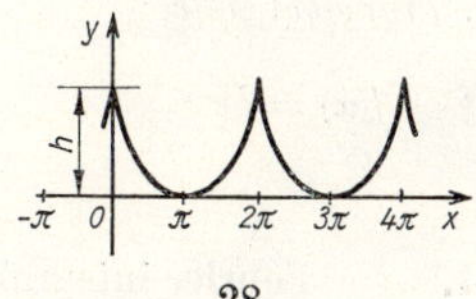

28.

28. *Parabolic arcs*

$$\left(\text{parabola equation} \quad y = \frac{h}{\pi^2}(x - \pi)^2 \text{ for } [0, 2\pi]\right)$$

$$f(x) = \frac{h}{3} + \frac{4h}{\pi^2}\left(\cos x + \frac{1}{2^2}\cos 2x + \frac{1}{3^2}\cos 3x + \cdots\right)$$

29. *Parabolic arcs*

(parabola equation $y = x^2$ for $[-\pi, \pi)]$

$$f(x) = \frac{\pi^2}{3} - 4\left(\cos x - \frac{1}{2^2}\cos 2x + \frac{1}{3^2}\cos 3x - + \cdots\right)$$

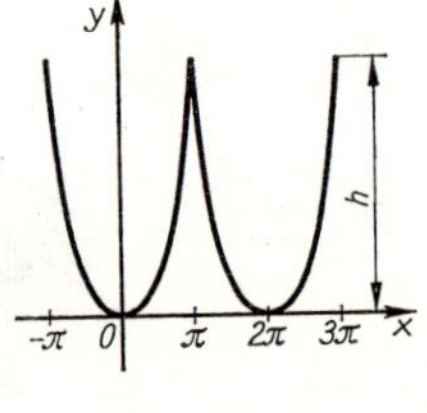

29.

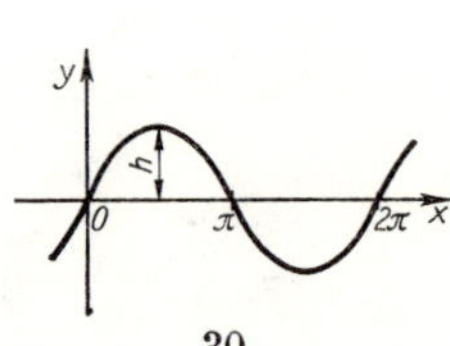

30.

30. *Parabolic arcs*

$$\left(\text{parabola equation } y = \frac{4h}{\pi^2}x(\pi - x) \text{ for } [0, \pi]\right.$$

$$\left.\text{and } y = \frac{4h}{\pi^2}(x^2 - 3\pi x + 2\pi^2) \quad \text{for} \quad [\pi, 2\pi]\right)$$

$$f(x) = \frac{32h}{\pi^3}\left(\sin x + \frac{1}{3^3}\sin 3x + \frac{1}{5^3}\sin 5x + \cdots\right)$$

31. *Transient state*

$$f(x) = he^{-x}$$

9.4. Fourier integral, example of calculation

The above discussed transient state $f(x) = he^{-x}$ (page 440) with $T_0 = 2\pi$ is assumed to be a nonperiodic process occurring only once:

$$he^{-t} = \begin{cases} F(t) & \text{for} \quad t \in (0, \infty] \\ 0 & \text{for} \quad t < 0 \end{cases}$$

The spectral function is

$$f(y) = \int_0^\infty F(t)\, e^{-jyt}\, dt = h \int_0^\infty e^{-t} e^{-jyt}\, dt = h \int_0^\infty e^{-(1+jy)t}\, dt$$

Integration yields

$$f(y) = h \frac{-1}{1+jy} e^{-(1+jy)t} \Big|_0^\infty = \frac{h}{1+jy} = \frac{h}{1+y^2} + j \frac{-hy}{1+y^2}$$

The FOURIER integral is

$$F(t) = he^{-t} = \frac{1}{2\pi} \int_{-\infty}^{\infty} \frac{h}{1+jy} e^{jyt}\, dy = \frac{h}{2\pi} \int_{-\infty}^{+\infty} \frac{1-jy}{1+y^2} e^{jyt}\, dy$$

The graphical representation of $f(y)$ gives the continuous spectrum:

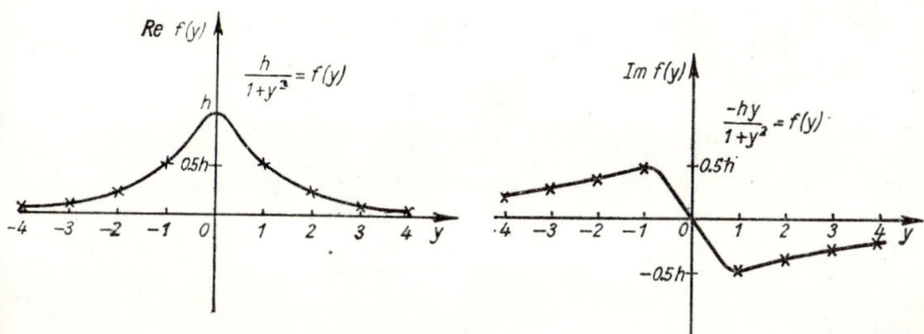

9.5. Laplace transforms

Written:

$$f(s) = \int_0^\infty e^{-st} F(t)\, dt = \mathbf{L}\{F(t)\} \qquad s \in K; \quad s = x + jy$$

with superfunction, original function $F(t)$ in the range $t \in (0, \infty)$, subfunction, image function, LAPLACE transform $f(s)$.

LAPLACE transformation $\quad \mathbf{L}\{F(t)\} = f(s)$

$F(t) \circ\!\!-\!\!\bullet f(s) \qquad$ (correspondence)

Inversion transformation $\mathbf{L}^{[-1]}\{f(s)\} = F(t)$

$$f(s) \bullet\!\!-\!\!\circ\; F(t) \quad \text{(correspondence)}$$

$$\mathbf{L}^{[-1]}\{\mathbf{L}\{f(s)\}\} = f(s); \quad \mathbf{L}^{[-1]}\{\mathbf{L}\{F(t)\}\} = F(t)$$

Within the range of partially continuous and monotonic functions $f(s)$ there exists, if at all, only a superfunction $F(t)$.

Criterion for the existence of the Laplace transform $f(s)$ in the range $t > 0$

1. $\int\limits_{-\infty}^{+\infty} |F(t)|\, dt < \infty;$ $F(t)$ partially continuous

2. $\int\limits_{0}^{T_1} |F(t)|\, dt = \lim\limits_{\delta \to 0} \int\limits_{\delta}^{T_1} |F(t)|\, dt$

3. $T_2 > T_1$; for at least one $s_0 \in K$

$$\lim_{T_2 \to \infty} \left| \int\limits_{T_1}^{T_2} e^{-s_0 t}\, F(t)\, dt \right| = 0$$

Functions that satisfy 1 to 3 are referred to as **L**-funtions.

4. If the convergence of $\mathbf{L}\{F(t)\}$ is given for x_0, it will also converge for all $x > x_0$ or $R(s) > R(s_0)$. The smallest possible value for x_0 is called convergence abscissa β (C. A.) When $\mathbf{L}\{F(t)\}$ converges in an absolute manner, one writes: $\mathbf{L}_a$-function.

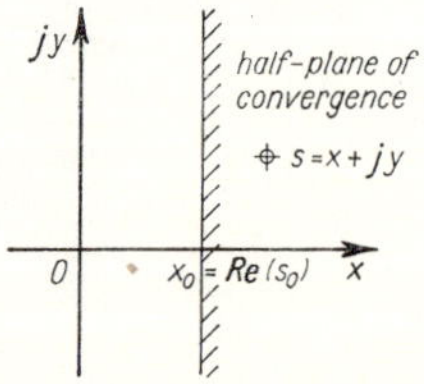

$$\int\limits_{0}^{\infty} |e^{-st}\, F(t)|\, dt < \infty \quad \text{for} \quad x > \beta$$

$\mathbf{L}_a$-functions are also **L**-functions.

Calculating rules

Linearity:

$$\mathbf{L}\{aF_1(t) + bF_2(t)\} = a\mathbf{L}\{F_1(t)\} + b\mathbf{L}\{F_2(t)\}$$

C. A. max. (β_1, β_2) means that the convergence abscissa is the greater of the two values β_1 and β_2.

Inversion $\mathbf{L}^{[-1]}\{af_1(s) + bf_2(s)\} = a\mathbf{L}^{[-1]}\{f_1(s)\} + b\mathbf{L}^{[-1]}\{f_2(s)\}$

Theorem of attenuation: $\mathbf{L}\{e^{-\alpha t}F(t)\} = f(s + \alpha)$ C. A. $= \beta - R(\alpha)$

$e^{-\alpha t}$ attenuation factor for $\alpha > 0$ $\alpha \in R$

inversion $\mathbf{L}^{[-1]}\{f(s + \alpha)\} = e^{-\alpha t}\mathbf{L}^{[-1]}\{f(s)\}$

Theorem of similarity: $\mathbf{L}\{F(\alpha t)\} = \dfrac{1}{\alpha} f\left(\dfrac{s}{\alpha}\right)$ C. A. $\alpha\beta$

$$\mathbf{L}\left\{\frac{1}{\gamma} F\left(\frac{1}{\gamma} t\right)\right\} = f(s\gamma) \qquad \gamma > 0$$

inversion $\mathbf{L}^{[-1]}\left\{f\left(\dfrac{s}{\alpha}\right)\right\} = \alpha F(\alpha t)$

$$\mathbf{L}^{[-1]} f(\gamma s) = \frac{1}{\gamma} F\left(\frac{t}{\gamma}\right) \qquad \gamma > 0$$

Theorem of displacement:

$$\mathbf{L}\{F_1(t)\} = e^{-st_0}\,\mathbf{L}\{F(t)\} \quad \text{with} \quad F_1(t) = \begin{cases} 0 \text{ for } t \in [0, t_0] \\ F(t - t_0) \\ \text{for } \ t \in [t_0, \infty) \end{cases}$$

or $\mathbf{L}\{F(t - t_0)\} = e^{-st_0} f(s)$ $\qquad t \in (t_0, \infty)$

$$\mathbf{L}\{F(t + \alpha)\} = e^{\alpha s}[f(s) - \int_0^\alpha e^{-st}\, F(t)\, dt] \qquad \alpha > 0$$

$$\text{inversion } \mathbf{L}^{[-1]}\{e^{-st_0}f(s)\} = \begin{cases} 0 & \text{for} \quad t \in [0, t_0) \\ F(t - t_0) & \text{for} \quad t > t_0 \end{cases}$$

Theorem of differentiation:

The following is valid for the superfunction:

$$\mathbf{L}\{F^{(n)}(t)\} = s^n f(s) - s^{n-1}F(+0) - s^{n-2}\dot{F}(+0)$$
$$- s^{n-3}\ddot{F}(+0) - \cdots - F^{n-1}(+0)$$

consequently first derivative: $\mathbf{L}\left\{\frac{dF(t)}{dt}\right\} = sf(s) - F(+0)$

second derivative: $\mathbf{L}\left\{\frac{d^2F(t)}{dt^2}\right\} = s^2f(s) - sF(+0) - \dot{F}(+0)$

inversion: $\mathbf{L}^{[-1]}\{sf(s) - F(+0)\} = \frac{dF(t)}{dt}$

The following is valid for the subfunction:

$$\frac{d^nf(s)}{ds^n} = \frac{d^n}{ds^n}\mathbf{L}\{F(t)\} = (-1)^n\,\mathbf{L}\{t^nF(t)\} = (-1)^nt^nf(s)$$

thus $f'(s) = -t\mathbf{L}\{F(t)\}$

$f''(s) = +t^2\mathbf{L}\{F(t)\}$

inversion: $\mathbf{L}^{[-1]}\left\{\frac{d^nf(s)}{ds^n}\right\} = (-1)^nt^nF(t)$

Theorem of integration:

The following is valid for the superfunction:

$$\mathbf{L}\left\{\int_0^t F(\tau)\,d\tau\right\} = \frac{1}{s}f(s) \qquad x > x_0$$

inversion: $\mathbf{L}^{[-1]}\left\{\frac{1}{s}f(s)\right\} = \int_0^t F(\tau)\,d\tau$

The following is valid for the subfunction:

$$\mathbf{L}\left\{\frac{F(t)}{t}\right\} = \int_s^\infty f(\sigma)\,d\sigma \qquad \sigma \in K$$

inversion: $\mathbf{L}^{[-1]}\left\{\int_s^\infty f(\sigma)\,d\sigma\right\} = \frac{F(t)}{t}$

Convolution integral: $F_1(t) * F_2(t) = \int_0^t F_1(\tau)\,F_2(t-\tau)\,d\tau$

Theorem of convolution: $f_1(s)\, f_2(s) = \mathbf{L}\{F_1(t)\}\, \mathbf{L}\{F_2(t)\}$

$$= \mathbf{L}\{F_1(t) * F_2(t)\} = \mathbf{L}\left\{\int_0^t F_1(\tau)\, F_2(T - \tau)\, d\tau\right\}$$

Commutative law: $F_1(t) * F_2(t) = F_2(t) * F_1(t)$

Associative law: $[F_1(t) * F_2(t)] * F_3(t)$

$$= F_1(t) * [F_2(t) * F_3(t)]$$

Inversion: $\mathbf{L}^{[-1]}\{f_1(s)\, f_2(s)\} = \mathbf{L}^{[-1]}\{f_1(s)\} * \mathbf{L}^{[-1]}\{f_2(s)\}$

Theorems of limit values:

$$\lim_{t\to+0} F(t) = \lim_{s\to\infty} sf(s)$$

$$\lim_{t\to+\infty} F(t) = \lim_{s\to0} sf(s)$$

9.6. Employment of Laplace transforms; solution of differential equations

Schematic mathematical procedure:

Differential equations + initial conditions → solution y | original space

↓ ↑

Laplace transformation (algorithms, tables) — $\mathbf{L}^{[-1]}$-transformation (algorithms, tables)

↓ ↑

Linear algebraic equation → solution for $\mathbf{L}\{y\}$ | image space

Instead of the direct solution, one chooses the procedure via the image space using the **L**-transformation whereby the solution of the differential equation will be reduced to a solution of a linear algebraic equation. It will be of advantage to take the initial conditions into account from the beginning. Moreover, the method of the solution of inhomogeneous differential equations becomes as simple as that of homogeneous differential equations. (Condition: disturbance term must have an **L**-transform).

Example 1:

Solve the differential equation

$$\ddot{y} + 5\dot{y} + 4y = t \quad \text{with} \quad y(0) = 0;\ \dot{y}(0) = 0$$

Applying the theorem of differentiation,

$$s^2\mathbf{L}\{y\} - sy(0) - \dot{y}(0) + 5s\mathbf{L}\{y\} - 5y(0) + 4\mathbf{L}\{y\} = \mathbf{L}\{t\}$$

Considering the initial conditions,

$$s^2\mathbf{L}\{y\} + 5s\mathbf{L}\{y\} + 4\mathbf{L}\{y\} = \mathbf{L}\{t\}$$

$$\mathbf{L}\{y\} = \frac{1}{s^2}\cdot\frac{1}{s^2+5s+4} = \frac{1}{s^2}\cdot\frac{1}{s+1}\cdot\frac{1}{s+4}$$

according to table 9.7./2

Retransformation

$$y = \mathbf{L}^{[-1]}\left\{\frac{1}{s^2}\cdot\frac{1}{s+1}\cdot\frac{1}{s+4}\right\}$$

Transformation into a sum by decomposition into partial fractions

$$\frac{1}{s^2}\cdot\frac{1}{s+1}\cdot\frac{1}{s+4} = \frac{A}{s^2} + \frac{B}{s} + \frac{C}{s+1} + \frac{D}{s+4}$$

$$\frac{1}{s^2(s+1)(s+4)} = \frac{A(s+1)(s+4) + Bs(s+1)(s+4)}{s^2(s+1)(s+4)} + \frac{Cs^2(s+4) + Ds^2(s+1)}{s^2(s+1)(s+4)}$$

Comparison of coefficients yields

$$A = \frac{1}{4};\quad B = -\frac{5}{16};\quad C = \frac{1}{3};\quad D = -\frac{1}{48}$$

Hence

$$y = \frac{1}{4}\mathbf{L}^{[-1]}\left\{\frac{1}{s^2}\right\} - \frac{5}{16}\mathbf{L}^{[-1]}\left\{\frac{1}{s}\right\} + \frac{1}{3}\mathbf{L}^{[-1]}\left\{\frac{1}{s+1}\right\} - \frac{1}{48}\mathbf{L}^{[-1]}\left\{\frac{1}{s+4}\right\}$$

$$y = \frac{t}{4} - \frac{5}{16} + \frac{1}{3}e^{-t} - \frac{1}{48}e^{-4t}$$

according to table 9.7./2;/1;/4

Example 2:

Solve the differential equation

$$\ddot{y} - 4y = 2 \sinh t \quad \text{with } y(0) = 0,\ \dot{y}(0) = 0.$$

Applying the theorem of differentiation,

$$s^2\mathbf{L}\{y\} - sy(0) - \dot{y}(0) - 4\mathbf{L}\{y\} = 2\mathbf{L}\{\sinh t\}$$

Considering the initial conditions,

$$s^2\mathbf{L}\{y\} - 4\,\mathbf{L}\{y\} = 2\,\mathbf{L}\{\sinh t\}$$

$$\mathbf{L}\{y\} = \frac{2}{s^2 - 4}\,\mathbf{L}\,\{\sinh t\}$$

Procedure of solution 1:

$$\frac{2}{s^2 - 4} = \mathbf{L}\{\sinh 2t\}$$

$$\mathbf{L}\{y\} = \mathbf{L}\{\sinh 2t\}\,\mathbf{L}\{\sinh t\}$$

Applying the theorem of convolution,

$$\mathbf{L}\{y\} = \mathbf{L}\{\sinh 2t * \sinh t\}$$

$$y = \mathbf{L}^{[-1]}\,\mathbf{L}\{\sinh 2t * \sinh t\} = \sinh 2t * \sinh t$$

(convolution integral)

$$y = \int_0^t \sinh(t - \tau) \cdot \sinh 2\tau\, d\tau$$

Solution by partial integration done twice

$$y = \left[\frac{1}{2} \sinh(t - \tau) \cosh 2\tau\right]_{\tau=0}^{t} + \frac{1}{2}\int_0^t \cosh(t - \tau) \cosh 2\tau\, d\tau$$

$$y = -\frac{1}{2} \sinh t + \left[\frac{1}{2}\left(\frac{1}{2} \cosh(t - \tau)\right) \sinh 2\tau\right]_0^t$$

$$+ \frac{1}{4}\int_0^t \sinh 2\tau \sinh(t - \tau)\, d\tau$$

$$y = -\frac{1}{2}\sinh t + \frac{1}{4}\sinh 2t + \frac{1}{4}y$$

$$y = -\frac{2}{3}\sinh t + \frac{1}{3}\sinh 2t$$

Procedure of solution 2:

$$\mathbf{L}\{\sinh t\} = \frac{1}{s^2 - 1} \quad \text{according to table 9.7./8}$$

$$\mathbf{L}\{y\} = \frac{2}{s^2 - 4} \cdot \frac{1}{s^2 - 1}$$

Retransformation

$$y = \mathbf{L}^{[-1]}\left\{\frac{2}{s^2 - 4} \cdot \frac{1}{s^2 - 1}\right\}$$

Transformation into a sum by decomposition into partial fractions

$$\frac{2}{(s^2 - 4)(s^2 - 1)} = \frac{A}{s^2 - 4} + \frac{B}{s^2 - 1}$$

(A decomposition into the denominators $(s - 2)(s + 2)(s - 1)(s + 1)$ is not expedient since $\frac{a}{s^2 - a^2}$ is **L**-transformable).

$$A = \frac{2}{3}; \quad B = -\frac{2}{3}$$

$$y = \frac{1}{3}\mathbf{L}^{[-1]}\left\{\frac{2}{s^2 - 4}\right\} - \frac{2}{3}\mathbf{L}^{[-1]}\left\{\frac{1}{s^2 - 1}\right\}$$

$$y = \frac{1}{3}\sinh 2t - \frac{2}{3}\sinh t \quad \text{(as above) according to table 9.7./8}$$

Example 3:

Solve the differential equation of the uniformly accelerated motion $\ddot{y} = b$ with $y(0) = y_0$; $\dot{y}(0) = v_0$.
Applying the theorem of differentiation,

$$s^2\mathbf{L}\{y\} - sy(0) - \dot{y}(0) = \mathbf{L}\{b\}$$

Considering the initial conditions,

$$s^2\mathbf{L}\{y\} - sy_0 - v_0 = \frac{b}{s}; \quad s^2\mathbf{L}\{y\} = \frac{b}{s} + sy_0 + v_0$$

$$\mathbf{L}\{y\} = \frac{b}{s^3} + \frac{y_0}{s} + \frac{v_0}{s^2}$$

Applying the theorem of convolution to

$$\frac{b}{s^3} = \frac{b}{s^2} \cdot \frac{1}{s} = \mathbf{L}\{bt\}\,\mathbf{L}\{1\} \quad \text{according to table 9.7/1;/2}$$

$$y = \mathbf{L}^{[-1]}\,\mathbf{L}\{bt\}\,\mathbf{L}\{1\} + \mathbf{L}^{[-1]}\left\{\frac{y_0}{s}\right\} + \mathbf{L}^{[-1]}\left\{\frac{v_0}{s^2}\right\}$$

$$= \mathbf{L}^{[-1]}\,\mathbf{L}\{bt * 1\} + y_0 + v_0t = \int_0^t b\tau\, 1\, d\tau$$

$$+ y_0 + v_0t \quad \text{(convolution integral)}$$

$$y = \frac{b}{2}t^2 + v_0t + y_0$$

It can be seen that the method of solving differential equations by using the **L**-transformation will only simplify the procedure of solution in the case of complicated equations.

Example 4:

Solve the differential equation of the harmonic oscillation

$$m \cdot \frac{d^2y}{dt^2} = -mg + k(a - y) \quad \text{with} \quad y(0) = 0; \quad \dot{y}(0) = v_0$$

(k spring constant)

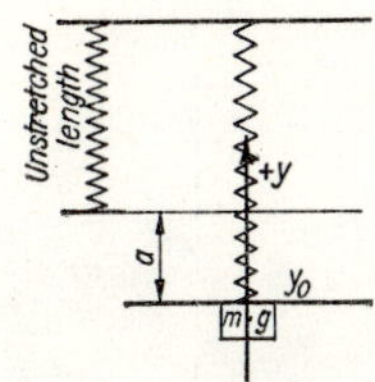

When $-mg = ka$, then $m\ddot{y} = -ky$; $\ddot{y} + \frac{k}{m} y = 0$

$$s^2\mathbf{L}\{y\} - sy(0) - \dot{y}(0) + \frac{k}{m}\mathbf{L}\{y\} = 0$$

Considering the initial conditions,

$$s^2\mathbf{L}\{y\} - v_0 + \frac{k}{m}\mathbf{L}\{y\} = 0; \quad \left(s^2 + \frac{k}{m}\right)\mathbf{L}\{y\} = v_0$$

$$\mathbf{L}\{y\} = \frac{v_0}{s^2 + \frac{k}{m}}$$

Retransformation:

$$y = v_0\mathbf{L}^{[-1]}\left\{\frac{1}{s^2 + \frac{k}{m}}\right\} = v_0\sqrt{\frac{m}{k}}\,\mathbf{L}^{[-1]}\left\{\frac{\sqrt{\frac{k}{m}}}{s^2 + \frac{k}{m}}\right\}$$

$$y = v_0\sqrt{\frac{m}{k}}\sin\left(\sqrt{\frac{k}{m}}\,t\right)$$ according to table 9.7./16

9.7. Table of correspondences of some rational Laplace integrals

Superfunction $F(t)$	Subfunction $f(s) = \mathbf{L}\{F(t)\}$
1. h	$\frac{h}{s}$
2. t	$\frac{1}{s^2}$
3. $\frac{t^n}{n!}$	$\frac{1}{s^{n+1}}$ $n \in I$
4. $e^{\pm at}$	$\frac{1}{s \mp a}$ Re $s >$ Re a

Superfunction $F(t)$	Subfunction $f(s) = \mathbf{L}F(t)$
5. a^t	$\frac{1}{s - \ln \lvert a \rvert}$
6. $(a + e^{-t})^n$	$\sum_{k=0}^{n} \binom{n}{k} \frac{a^{n-k}}{s + k}$
7. $-ae^{-at}$	$\frac{s}{s + a}$
8. $\sinh at$	$\frac{a}{s^2 - a^2}$
9. $\cosh at$	$\frac{s}{s^2 - a^2}$
10. $\frac{1}{a}(e^{at} - 1)$	$\frac{1}{s(s - a)}$
11. $te^{\pm at}$	$\frac{1}{(s \mp a)^2}$
12. $\frac{e^{bt} - e^{at}}{b - a}$	$\frac{1}{(s - a)(s - b)}$
13. $\frac{be^{bt} - ae^{at}}{b - a}$	$\frac{s}{(s - a)(s - b)}$
14. $e^{-bt} \sin at$	$\frac{a}{(s + b)^2 + a^2}$
15. $e^{-bt} \cos at$	$\frac{s + b}{(s + b)^2 + a^2}$
16. $\sin at$	$\frac{a}{s^2 + a^2} \quad a \neq 0$
17. $\cos at$	$\frac{s}{s^2 + a^2} \quad a \neq 0$
18. $\sin(\omega t + \varphi)$	$\cos \varphi \cdot \frac{\omega}{s^2 + \omega^2} + \sin \varphi \times \frac{s}{s^2 + \omega^2}$

Superfunction $F(t)$	Subfunction $f(s) = \mathbf{L}F(t)$	
19. $1 - 2\sin at$	$\dfrac{(s-a)^2}{s(s^2+a^2)}$	$a \neq 0$
20. $\cos^2 at$	$\dfrac{s^2+2a^2}{s(s^2+4a^2)}$	$a \neq 0$
21. $\cosh^2 at$	$\dfrac{s^2-2a^2}{s(s^2-4a^2)}$	$a \neq 0$
22. $\dfrac{t^{n-1}}{(n-1)!}$	$\dfrac{1}{s^n}$	$n > 0$

For more exhaustive tables see special literature, for example DOETSCH: „Anleitung zum praktischen Gebrauch der LAPLACE-Transformation", R. Oldenbourg, Munich.

10. Theory of probability; statistics; error calculation; mathematical analysis of observations

10.1. Theory of probability

Definitions

Every *event* E, i.e. every possible outcome in a trial, is assigned a number $P(E)$, the probability of the event E.

$$0 \leqq P(E) \leqq 1$$

Elementary definition of probability

$$P(E) = \frac{g}{m}$$

Among the m possible outcomes in an event E there are g favorable outcomes (successes).

Example:

The probability of throwing a 6 with a die is $P = \frac{1}{6}$.

Statistical definition of probability

$$P(E) = \lim_{n \to \infty} \frac{h}{n}$$

When a trial is performed n times, the event E will occur h times. The probability Q of the nonoccurence of the event E (*opposite event* $\bar{E}$) is

$$Q = 1 - P$$

Theorem of summation of probabilities — Probability of "Either—Or"

The probability of the occurrence of one of the *incompatible* events $E_1, E_2, \ldots, E_k$ is

$$P = P(E_1) + P(E_2) + \cdots + P(E_k)$$

Example:

The probability of throwing a 3 or a 6 is $P = \frac{1}{6} + \frac{1}{6} = \frac{1}{3}$.

Conditional probability

The probability of the occurrence of the event E_2 under the condition that the event E_1 has occurred previously is called the conditional probability $P(E_2/E_1)$ of the event E_2.

Theorem of multiplication of probabilities — probability of "Both—And"

The probability of the events E_1 and E_2 occurring simultaneously is

$$P = P(E_1)\,P(E_2/E_1) = P(E_2)\,P(E_1/E_2)$$

Example:

When taking a card out of a pack of cards (32 cards), the probability of taking a king will be $P(E_1) = \frac{4}{32} = \frac{1}{8}$. When the card taken is a king, the probability of taking again a king out of the remaining cards will be $P(E_2/E_1) = \frac{3}{31}$. Thus, when taking two cards from the complete pack of 32, the probability of taking two kings will be

$$P = P(E_1)\,P(E_2/E_1) = \frac{1}{8} \cdot \frac{3}{31} \approx 0.012$$

Two events E_1 and E_2 are said to be *independent* if

$$P(E_2/E_1) = P(E_2) \quad \text{or} \quad P(E_1/E_2) = P(E_1)$$

For k independent events $E_1, E_2, \ldots, E_k$ the probability of

the occurrence of *all events simultaneously* is:

$$P = P(E_1)\,P(E_2) \times \cdots \times P(E_k)$$

the occurrence of *none of the events* is:

$$P = [1 - P(E_1)]\,[1 - P(E_2)] \times \cdots \times [1 - P(E_k)]$$

the occurrence of *at least one event* is:

$$P = 1 - [1 - P(E_1)] \times [1 - P(E_2)] \times \cdots \times [1 - P(E_k)]$$

Formula of total probability

The event E is to occur together with one and only one of the k incompatible events $E_1, E_2, \ldots, E_k$. In this case,

$$P(E) = P(E_1)\,P(E/E_1) + P(E_2)\,P(E/E_2) + \cdots + P(E_k)P(E/E_k) = \sum_{i=1}^{k} P(E_i)\,P(E/E_i)$$

Bayes' formula (formula of the probability of hypotheses)

The event E is always to occur together with one and only one of the k incompatible events $E_1, E_2, \ldots, E_k$. The probability of the event E_i occurring under the condition that E has occurred previously will then be

$$P(E_i/E) = \frac{P(E_i)\,P(E/E_i)}{\sum_{j=1}^{k} P(E_j)P(E/E_j)}$$

Series of independent trials

Two trials are said to be *independent* when the result of one trial has no influence on the outcome of the other trial. A series of n independent trials is performed with one and only one of the k incompatible events with the probabilities $P_1 = P(E_1), P_2 = P(E_2), \ldots, P_k = P(E_k)$ occurring in each trial. Denoting the probability that in n trials the event E_1 will occur m_1 times, $E_2 m_2$ times, etc. $(m_1 + m_2 + \cdots + m_k = n)$, we have

$$P_n(m_1, m_2, \ldots, m_k) = \frac{n!}{m_1!\,m_2! \cdots m_k!} P_1^{m_1} P_2^{m_2} \ldots P_k^{m_k}$$

For the special case of $k = 2$, we have **Bernoulli's scheme:**

$$P_1 = P; \quad P_2 = 1 - P = Q, \quad m_1 = m, \quad m_2 = n - m$$

$$P_n(m) = \frac{n!}{m!(n-m)!} P^m Q^{n-m}$$

Properties of $P_n(m)$

$$\sum_{m=0}^{n} P_n(m) = 1$$

$P_n(m)$ is the coefficient of x^m in the expansion of the polynomial $(Q + Px)^n$ in powers of x (*binomial law of probability distribution*).

$$P_n(m) < P_n(m+1) \quad \text{for} \quad m < nP - Q$$

$$P_n(m) > P_n(m+1) \quad \text{for} \quad m > nP - Q$$

$$P_n(m) = P_n(m+1) \quad \text{for} \quad m = nP - Q$$

The maximum of $P(m)$ is at $\mu = \overline{m}_0 m_0 = nP - Q$.

$$\overline{m}_0 = \left\{ \begin{array}{l} m_0 \\ \text{of the smallest} \\ \text{integer containing } m_0 \end{array} \right\} \text{if } m_0 \left\{ \begin{array}{l} \text{is an integer} \\ \text{is not an} \\ \text{integer} \end{array} \right.$$

Random quantities

A quantity ξ whose values depend on chance is called a random variable if a *distribution function* of the probabilities of ξ exists.

$$F(x) = P(\xi < x)$$

Properties of $F(x)$

$F(x)$ is a monotonic nondecreasing function, i.e. from $x_1 < x_2$ follows $F(x_1) \leqq F(x_2)$.

$F(x)$ has, at the most, countably many jump points.

$F(x)$ is continuous on the left side.

$$F(-\infty) = 0, \quad F(+\infty) = 1$$

Particularly important types of distribution functions

Discrete distribution functions (*step functions*)

A random variable that can take only a finite number of values x_i $(i = 1, 2, \ldots, n)$ determines a discrete distribution function. The values taken are the *points of jumps* and the respective probabilities are the *heights of the steps* of the distribution function.

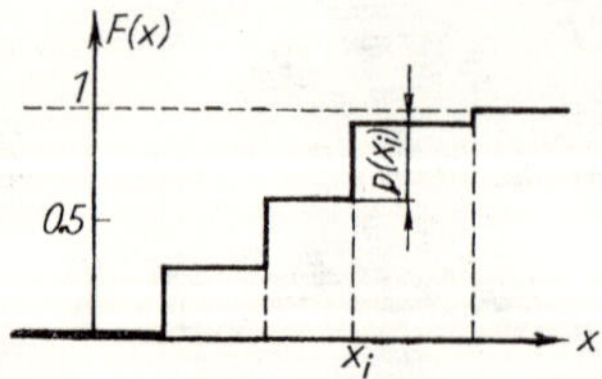

$$p_i = P(\xi = x_i)$$

$$\sum_{i=1}^{n} p_i = 1$$

$$F(x) = \sum_{x_i < x} p_i$$

Continuous distribution functions

A random variable ξ possesses a continuous distribution function $F(x)$ if a nonnegative function $f(x)$ exists so that

$$F(x) = \int_{-\infty}^{x} f(z)\, dz;$$

$$\frac{dF(x)}{dx} = f(x)$$

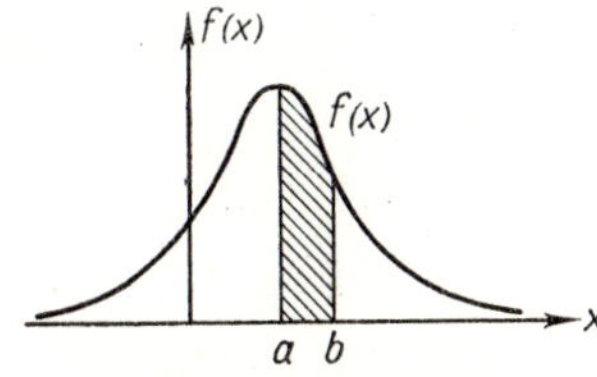

The hatched area represents the probability

$P(a \leqq \xi \leqq b) = F(b) - F(a)$

$f(x)$ is called the *density function* of ξ; $f(x) \geqq 0$

$$\int_{-\infty}^{+\infty} f(x)\, dx = 1 \qquad P(a \leqq \xi \leqq b) = F(b) - F(a) = \int_{a}^{b} f(x)\, dx$$

Moments

$$m(\xi - x)^k = \sum_{i=1}^{n} x_i{}^k P_i$$

Moments of the order of k

$$m(\xi - c)^k = \int_{-\infty}^{+\infty} (x - c)^k f(x)\, dx$$

For $c = 0$, the $m(\xi)$ are called *initial moments*. The initial moment of the first order is called the **mean value** μ of the distribution. When $c = \mu$, one speaks of *central moments*. The central moment of the second order $m(\xi - \mu)^2 = \sigma^2$ is called the **dispersion (variance)** of the distribution. σ is referred to as the **standard deviation (mean square deviation).**

Arithmetical rules

(C constant; ξ, η various random quantities)

$\mu(C) = C$	$\sigma^2(C) = 0$
$\mu(\xi + \eta) = \mu(\xi) + \mu(\eta)$	$\sigma^2(\xi + \eta) = \sigma^2(\xi) + \sigma^2(\eta)$
$\mu(\xi\eta) = \mu(\xi)\,\mu(\eta)$	
$\mu(C\xi) = C\mu(\xi)$	$\sigma^2(C\xi) = C^2\sigma^2(\xi)$
$\mu(\xi + C) = \mu(\xi) + C$	$\sigma^2(\xi + C) = \sigma^2(\xi)$

Coefficient of variation: $v = \dfrac{\sigma^2}{\mu}$

Obliquity: $\gamma_1 = \dfrac{m(\xi - \mu)^3}{\sigma^3}$

Kurtosis: $\gamma_2 = \dfrac{m(\xi - \mu)^4}{\sigma^4} - 3$

Obliquity is a measure of the *symmetry* and kurtosis is a measure of the *slope* of the curve.

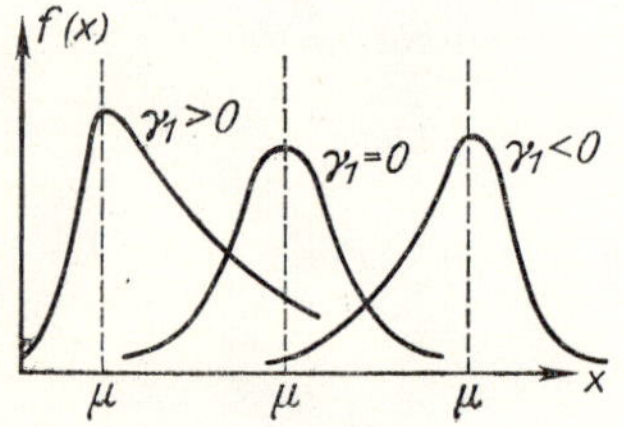

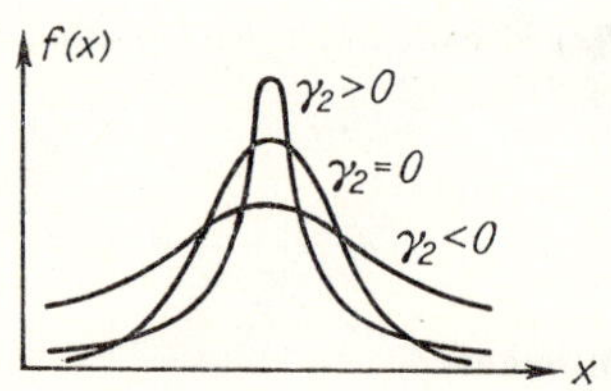

10.2. Statistics

Statistical coefficients of measure of a sample taken at random

Mean value $\bar{x}$ **Dispersion (variance)** s^2

For the n individual values x_i

$$\bar{x} = \frac{1}{n} \sum_{i=1}^{n} x_i \qquad s^2 = \frac{1}{n-1} \sum_{i=1}^{n} (x_i - \bar{x})^2 = \frac{1}{n-1} \left[\sum_{i=1}^{n} x_i^2 - \bar{x} \sum_{i=1}^{n} x_i \right]$$

For n individual values in M classes with the frequencies

$$f_i \left(\sum_{i=1}^{M} f_i = n \right)$$

$$\bar{x} = \frac{1}{n} \sum_{i=1}^{M} f_i x_i \qquad s^2 = \frac{1}{n-1} \sum_{i=1}^{M} f_i (x_i - \bar{x})^2 = \frac{1}{n-1} \left[\sum_{i=1}^{M} f_i x_i^2 - \bar{x} \sum_{i=1}^{M} f_i x_i \right]$$

For large values of x_i it is recommendable to introduce an auxiliary mean value D: $z_i = x_i - D$.

$$\bar{x} = D + \frac{1}{n} \sum_{i=1}^{n} z_i \qquad s^2 = \frac{1}{n-1} \sum_{i=1}^{n} (x_i - \bar{x})^2$$

$$= D + \bar{z} \qquad = \frac{1}{n-1} \sum_{i=1}^{n} (z_i - \bar{z})^2$$

$$\bar{x} = D + \bar{z} \qquad s^2 = \frac{1}{n-1} \sum_{i=1}^{M} f_i(z_i - \bar{z})^2$$

$$= D + \frac{1}{n} \sum_{i=1}^{M} f_i x_i \qquad = \frac{1}{n-1} \left[\sum_{i=1}^{M} f_i z_i^2 - \bar{z} \sum_{i=1}^{M} f_i z_i \right]$$

Range of variation: $R = x_{\max} - x_{\min}$

Coefficient of variation: $V = \frac{s}{x} 100\%$

Outlier problem

In a random sample of $n + 1$ measured values one value x_{n+1} is strikingly large. Let the basic totality show a normal distribution, $\bar{x}$ and s be the mean value and standard deviation of the sample *without* the outlier. The outlier is disregarded when

$$x_{n+1} > \bar{x} + ks$$

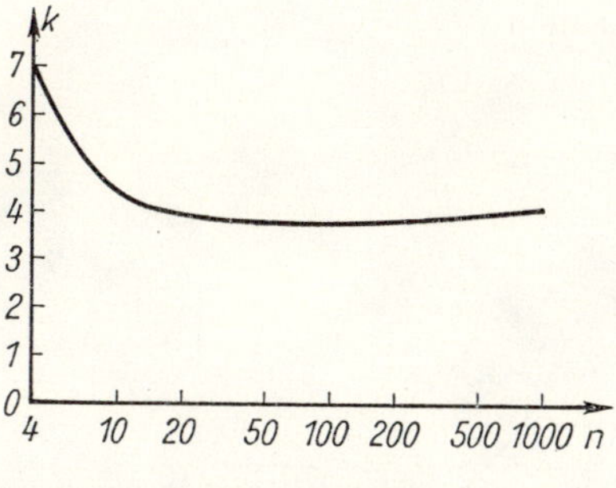

The value of k can be read from the diagram. In practical work, it is sufficient for $10 \leqq n \leqq 1000$ to calculate with $k = 4$.

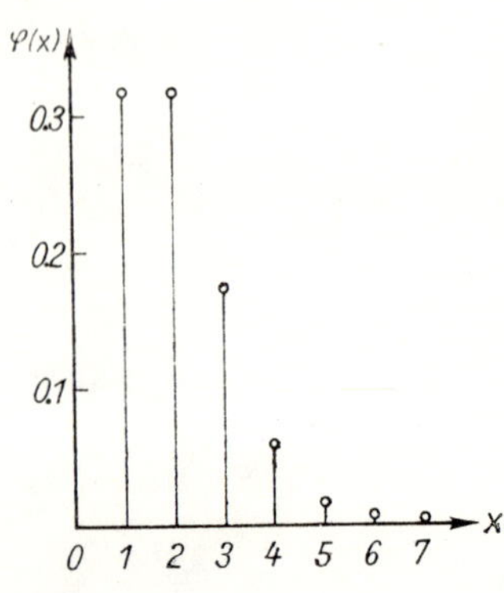

Distribution functions in statistics

Mean value μ,

Dispersion σ_2,

Obliquity γ_1,

Kurtosis γ_2

(For definitions cf. page 467.)

Binomial distribution

$$\varphi(x) = \binom{n}{x} p^x q^{n-x}$$

$$p + q = 1$$

$$n \in N;\ x = 1, 2, \ldots, n$$

$$\mu = np \qquad \sigma^2 = npq$$

$$\gamma_1 = \frac{q - p}{\sigma} \qquad \gamma_2 = \frac{1 - 6pq}{\sigma^2}$$

Poisson distribution

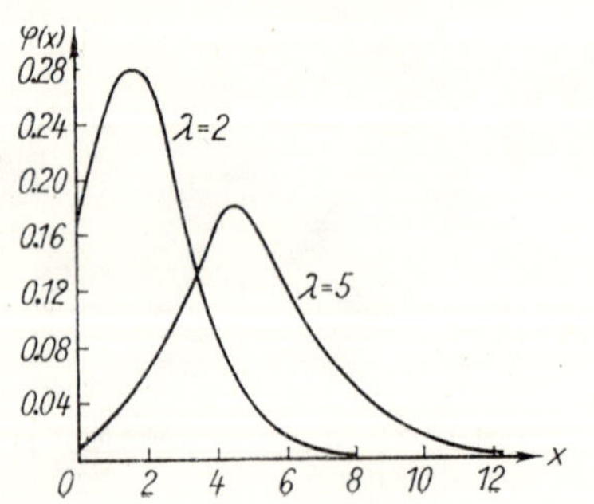

$$\varphi(x) = \frac{\lambda^x}{x!} e^{-\lambda}$$

$$\mu = \lambda \qquad \sigma^2 = \lambda$$

$$\gamma_1 = \frac{1}{\sqrt{\mu}}$$

$$\gamma_2 = \frac{1}{\mu}$$

This distribution is practically applicable only for small $p\left(<\frac{1}{2}\right)$. For large values of p, calculation is performed with $q = 1 - p$.

Normal distribution (Gauss-Laplace distribution)

$$\varphi(x) = \frac{1}{\sigma\sqrt{2\pi}} e^{-\frac{(x-\mu)^2}{2\sigma^2}} \qquad \text{(bell-shaped curve)}$$

Approximate values for drawing the bell-shaped curve

$x =$	$\mu \pm \frac{1}{2}\sigma$	$\mu \pm \sigma$	$\mu \pm \frac{3}{2}\sigma$	$\mu \pm 2\sigma$	$\mu \pm 3\sigma$
$y =$	$\frac{7}{8} y_{\max}$	$\frac{5}{8} y_{\max}$	$\frac{2.5}{8} y_{\max}$	$\frac{1}{8} y_{\max}$	$\frac{1}{80} y_{\max}$

Normal form of the normal distribution: ($\sigma = 1$, $\mu = 0$)

$$\varphi(\lambda) = \frac{1}{\sqrt{2\pi}} e^{-\frac{\lambda^2}{2}}$$

Every normal distribution $\varphi(x)$ can be reduced to the normal form by the transformation $\lambda = \frac{x - \mu}{\sigma}$ (*normalization*).

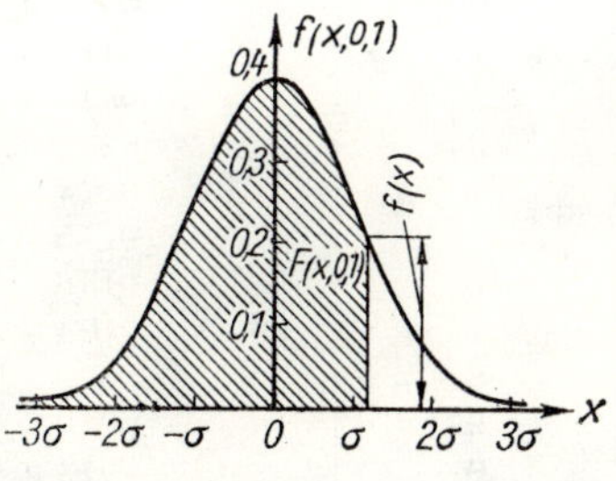

Integral form of the normal form of normal distribution:

$$\Phi(\lambda) = \int_{-\infty}^{\lambda} \varphi(t)\, dt$$

Table of values for the normal distribution

λ	$\varphi(\lambda)$	$\Phi(\lambda)$
0.0	0.3989	0.5000
1	3970	5398
2	3910	5793
3	3814	6179
4	3683	6554
5	3521	6915
6	3332	7257
7	3123	7580
8	2897	7881
9	2661	8159
1.0	2420	8413
1	2179	8643
2	1942	8849
3	1714	9032
4	1497	9192
5	1295	9332
6	1109	9452
7	0940	9554
8	0790	9641
9	0656	9713
2.0	0540	9772
1	0440	9821
2	0355	9861
3	0283	9893
4	0224	9918
5	0175	9938
6	0136	9953
7	0104	9965
8	0079	9974
9	0060	9981
3.0	0044	99865
2	0024	99931
4	0012	99966
6	00061	99984
8	00029	99993
4.0	000134	999968
4.5	000016	999997
5.0	000002	99999997

10.3. Error calculations

Fundamental definitions

x true value, x' measured value of a quantity

absolute error: $\Delta x = x' - x$ *correction:* $-\Delta x$

relative error: $\delta = \left|\frac{\Delta x}{x}\right| \approx \left|\frac{\Delta x}{x'}\right|$

percentage error: $\varepsilon = 100\delta\% \approx 100\left|\frac{\Delta x}{x'}\right|\%$

Input error

Error in the result of a calculation due to inaccuracy of the entered data.
If $\Delta x_1, \Delta x_2, \ldots, \Delta x_n$ are the errors of the quantities $x_1, x_2, \ldots, x_n$, the quantity $y = f(x_1, x_2, \ldots, x_n)$ will have the input error

$$\Delta y \approx \sum_{\nu=1}^{n} \Delta x_\nu \frac{\partial f(x_1, x_2, \ldots, x_n)}{\partial x_\nu} \approx dy$$

Maximum absolute input error

$$|\Delta y| \leqq \sum_{\nu=1}^{n} |\Delta x_\nu| \left|\frac{\partial f}{\partial x_\nu}\right|$$

Maximum relative input error

$$\delta = \left|\frac{\Delta y}{y}\right| \leqq \sum_{\nu=1}^{n} |\Delta x_\nu| \left|\frac{\frac{\partial f}{\partial x_\nu}}{f}\right|$$

Special cases

a) $y = x_1 + x_2 + \cdots + x_n$

$$|\Delta y| \leqq |\Delta x_1| + |\Delta x_2| + \cdots + |\Delta x_n|$$

$$\delta \leqq \left|\frac{\Delta x_1}{x_1}\right| + \left|\frac{\Delta x_2}{x_2}\right| + \cdots + \left|\frac{\Delta x_n}{x_n}\right|$$

b) $y = x_1 - x_2$

$$|\Delta y| \leqq |\Delta x_1| + |\Delta x_2|; \qquad \delta \leqq \frac{|\Delta x_1| + |\Delta x_2|}{|x_1 - x_2|} \ (x_1 \neq x_2)$$

c) $y = Cx_1x_2 \cdots x_n$

$$|\Delta y| \leqq |C| \, [|\Delta x_1 x_2 \cdots x_n| + |x_1 \Delta x_2 x_3 \cdots x_n| + \cdots + |x_1 x_2 \cdots x_{n-1} \Delta x_n|]$$

$$\delta \leqq |C| \left[\left|\frac{\Delta x_1}{x_1}\right| + \left|\frac{\Delta x_2}{x_2}\right| + \cdots + \left|\frac{\Delta x_n}{x_n}\right| \right]$$

d) $y = \dfrac{x_1}{x_2} \quad (x_2 \neq 0)$

$$|\Delta y| \leqq \frac{|x_2| \, |\Delta x_1| + |x_1| \, |\Delta x_2|}{{x_2}^2}$$

$$\delta \leqq \frac{|x_2| \, |\Delta x_1| + |x_1| \, |\Delta x_2|}{|x_1 x_2|} \qquad (x_1 \neq 0)$$

e) $y = x^n \quad (x > 0)$

$$|\Delta y| \leqq |\Delta x| \, |n x^{n-1}| \qquad \delta \leqq \left| n \frac{\Delta x}{x} \right|$$

f) $y = C \log_a x \quad (x > 0)$

$$|\Delta y| \leqq \left|\frac{C}{\ln a}\right| \left|\frac{\Delta x}{x}\right|; \qquad \delta \leqq \left|\frac{1}{\ln a}\right| \left|\frac{\Delta x}{x \log_a x}\right|$$

10.4. Calculus of observations

Observations of equal accuracy

True value of the measured quantity: X
Measurement results of a series of n measurements: $x_1, x_2, \ldots, x_n$
Sum convention (Gauss): $x_1 + x_2 + \cdots + x_n = [x]$

Mean value $\overline{x}$: $\overline{x} = \frac{[x]}{n}$

Apparent error: $v_i = \overline{x} - x_i$,

true error: $\varepsilon_i = X - x_i$

Mean error of the individual value:

$$m = \sqrt{\frac{[\varepsilon\varepsilon]}{n}}$$

$$\approx \sqrt{\frac{[vv]}{n-1}}$$

Mean error of the mean value $\overline{x}$:

$$m_{\overline{x}} = \frac{m}{\sqrt{n}} = \sqrt{\frac{[\varepsilon\varepsilon]}{n^2}}$$

$$\approx \sqrt{\frac{[vv]}{n(n-1)}}$$

The result of the series of measurements will then be: $\overline{x} \pm m_{\overline{x}}$.

Gaussian law of error propagation

When the quantities $\overline{x}_i \pm m_{\overline{x}i}$ $(i = 1, 2, \ldots, r)$ enter into the calculation, then $y = f(\overline{x}_1, \overline{x}_2, \ldots, \overline{x}_r)$ will have mean error

$$m_y = \sqrt{\sum_{\nu=1}^{r} \left(\frac{\partial f}{\partial x_\nu}\right)^2 m_{\overline{x}_\nu}{}^2}$$

Example:

Determination of the wall thickness of a hollow cylinder; outside diameter D, inside diameter d

wall thickness $w = \frac{1}{2}(D - d)\ \frac{\partial w}{\partial D} = \frac{1}{2}\ \frac{\partial w}{\partial d} = -\frac{1}{2}$

Five measurements each for D and d were carried out.

D	v_i	v_iv_i	d	v_i	v_iv_i
9.98	−0.01	0.0001	9.51	+0.01	0.0001
9.97	−0.02	0.0004	9.47	−0.03	0.0009
10.01	+0.02	0.0004	9.50	±0.00	0.0000
9.98	−0.01	0.0001	9.49	−0.01	0.0001
10.02	+0.03	0.0009	9.52	+0.02	0.0004
49.96	0.00	0.0019	47.49	−0.01	0.0015

$$\bar{D} = \frac{49.96}{5} = 9.99 \qquad \bar{d} = \frac{47.49}{5} = 9.50$$

$$m = \sqrt{\frac{0.0019}{4}} = 0.02 \qquad m = \sqrt{\frac{0.0015}{4}} = 0.02$$

$$m_{\bar{D}} = 0.01 \qquad m_{\bar{d}} = 0.01$$

According to the law of error propagation, we obtain for the wall thickness

$$\bar{w} = \frac{1}{2}(\bar{D} - \bar{d}) = \frac{0.49}{2} = 0.245$$

$$m_w = \sqrt{\frac{1}{4}\, m_{\bar{D}}^2 + \frac{1}{4}\, m_{\bar{d}}^2} = \sqrt{0.25 \cdot 0.0001 + 0.25 \cdot 0.0001}$$

$$m_w = \frac{0.01}{\sqrt{2}} = 0.00707 \approx 0.007$$

Observations of different accuracy

When k series of measurements are carried out with the nth series consisting of p_n measurements, we obtain the *weighted mean* of all series of measurements.

$$\bar{x} = \frac{[p\bar{x}]}{[p]} \qquad \bar{x}_n \text{ mean value of the } n\text{th series of measurements}$$

Apparent error of the mean value $\bar{x}_n$: $v_n = \bar{x} - \bar{x}_n$

Mean error of the unit of weight: $\mu = \sqrt{\dfrac{[pvv]}{n-1}}$

Mean error of the weighted mean:

$$m = \frac{\mu}{\sqrt{[p]}} = \sqrt{\frac{[pvv]}{[p]\,(n-1)}}$$

Intermediate observations

a) *Linear relationship*

n pairs of values (x_i, y_i) $(i = 1, 2, \ldots, n)$ were measured with the same accuracy $(n > 2)$ which are to satisfy the linear equation $y = a_0 + a_1 x$ (*fitted straight line*).

Set-up: $a_0 + a_1 x_i - y_i = v_i \qquad [vv] = \min$

a_0, a_1 are determined from the *normal equations:*

$$na_0 + a_1[x] = [y]$$

$$a_0[x] + a_1[xx] = [yx]$$

If the values $(y_i, x_{1i}, x_{2i}, \ldots, x_{ri})$ $(i = 1, 2, \ldots, n;\ n > r + 1)$ are to satisfy a linear equation, the coefficients a_i of the linear equation $y = a_0 + a_1 x_1 + a_2 x_2 + \cdots + a_r x_r$ have to be determined from the following *normal equations*:

$$na_0 + a_1[x_1] + a_2[x_2] + \cdots + a_r[x_r] = [y]$$

$$a_0[x_i] + a_1[x_1 x_i] + a_2[x_2 x_i] + \cdots + a_r[x_r x_i] = [yx_i]$$

$$(i = 1, 2, \ldots, r)$$

Example:

$n = 5$ pairs of values were measured:

(4, 3), (7, 4.5), (8, 5), (9, 6.1), (10, 6.4).

The fitted straight line is to be ascertained.

$$[x] = 4 + 7 + 8 + 9 + 10 = 38$$

$$[y] = 3 + 4.5 + 5 + 6.1 + 6.4 = 25$$

$$[xx] = 16 + 49 + 64 + 81 + 100 = 310$$

$$[yx] = 12 + 31.5 + 40 + 54.9 + 64 = 202.4$$

Normal equations: $5a_0 + 38a_1 = 25$

$$38a_0 + 310a_1 = 202.4$$

$$a_0 = \frac{25 \cdot 310 - 38 \cdot 202.4}{5 \cdot 310 - 38 \cdot 38} = 0.555$$

$$a_1 = \frac{5 \cdot 202.4 - 25 \cdot 38}{5 \cdot 310 - 38 \cdot 38} = 0.585$$

Fitted straight line: $y = 0.555 + 0.585x$

With unequal accuracy of the individual values, i.e. if $(y_i, x_{1i}, \dots, x_{ri})$ has the weight p_i, the a_j values are determined from the following normal equations:

$$a_0[p] + a_1[px_1] + \cdots + a_r[px_r] = [py]$$

$$a_0[px_i] + a_1[px_1x_i] + \cdots + a_r[px_rx_i] = [pyx_i] \quad (i = 1, 2, \dots, r)$$

b) *Nonlinear relationship*

The n measured pairs of values (x_i, y_i) $(i = 1, 2, \dots, n)$ are so satisfy an integral rational function $y = a_0 + a_1x + \cdots + a_rx^r$ $(r + 1 < n)$. The problem is solved with the normal equations of case (a) replacing x_i by x^i.

Example:

In the case of the *fitted parabola* the following normal equations are obtained:

$$na_0 + a_1[x] + a_2[x^2] = [y]$$

$$a_0[x] + a_1[xx] + a_2[x^2x] = [yx]$$

$$a_0[x^2] + a_1[xx^2] + a_2[x^2x^2] = [yx^2]$$

Conditional observations

In a series of measurements, the values $x_1', x_2', \dots, x_r'$ were determined with the same accuracy. The values are to satisfy t strictly linear equations of condition

$$[a_ix] = a_{i1}x_1 + a_{i2}x_2 + \cdots + a_{ir}x_r = c_i \quad (i = 1, 2, \dots, t)$$

The *corrections* v_n ($n = 1, 2, \ldots, r$) of the measured values are calculated from the *correlates* k_j ($j = 1, 2, \ldots, t$):

$$v_n = \sum_{i=1}^{t} a_{in} k_i \qquad (n = 1, 2, \ldots, r)$$

The correlates are calculated from the *normal equations:*

$$[a_i a_1]\, k_1 + [a_i a_2]\, k_2 + \cdots + [a_i a_t]\, k_t = w_i \qquad (i = 1, 2, \ldots, t)$$

$$w_i = c_i - [a_i x']$$

If the values $x_1', x_2', \ldots, x_r'$ were determined with different accuracies, i.e. if x_n was determined with the weight p_n, we have the following equations of determination for v and k:

Corrections: $$v_n = \sum_{i=1}^{t} \frac{a_{in} k_i}{p_n} \qquad (n = 1, 2, \ldots, r)$$

Normal equations for the correlates:

$$\left[\frac{a_i a_1}{p}\right] k_1 + \left[\frac{a_i a_2}{p}\right] k_2 + \cdots + \left[\frac{a_i a_t}{p}\right] k_t = w_i \qquad (i = 1, 2, \ldots, t)$$

$$w_i = c_i - [a_i x']$$

Example:

The values $x_1' = -1.2$ and $x_2' = 1.9$ are to satisfy strictly the equations

$$3x_1 + 4x_2 = 5$$
$$-7x_1 + 2x_2 = 11$$

Then we have

$$w_1 = 5 - 3x_1' - 4x_2' = 5 + 3.6 - 7.6 = 1$$
$$w_2 = 11 + 7x_1' - 2x_2' = 11 - 8.4 - 3.8 = -1.2$$
$$[a_1 a_1] = a_{11} a_{11} + a_{12} a_{12} = 3 \cdot 3 + 4 \cdot 4 = 25$$
$$[a_1 a_2] = a_{11} a_{21} + a_{12} a_{22} = 3 \cdot (-7) + 4 \cdot 2 = -13$$
$$[a_2 a_1] = a_{21} a_{11} + a_{22} a_{12} = (-7) \cdot 3 + 4 \cdot 2 = -13$$
$$[a_2 a_2] = a_{21} a_{21} + a_{22} a_{22} = (-7)\,(-7) + 2 \cdot 2 = 53$$

Normal equations:

$$25k_1 - 13k_2 = 1$$

$$-13k_1 + 53k_2 = -1.2$$

From these we obtain $k_1 = 0.032$ and $k_2 = -0.015$.
Now the corrections can be calculated. We have

$$v_1 = a_{11}k_1 + a_{21}k_2 = 3 \cdot 0.032 + (-7) \cdot (-0.015) = 0.2$$

$$v_2 = a_{12}k_1 + a_{22}k_2 = 4 \cdot 0.032 + 2 \cdot (-0.015) = 0.1$$

When these corrections of the measured values x_1', x_2' are made, we obtain as final values $x_1 = -1$ and $x_2 = 2$ which strictly satisfy the conditional equations.

11. Linear Optimization

11.1. General

Optimization is defined as procedures by which certain target conditions can be optimized by the economical use of given possibilities. The problem is either a problem of a maximum value or a problem of a minimum value which, in most cases, contains a great number of variables.

Linear optimization is a mathematical procedure for the determination of an optimum by means of a mathematical model of linear equations or inequations in which the *objective function*, which is also linear, is brought to an optimum under certain restricting conditions. The mathematical model is the mathematical formulation of the problem.

The solution of the optimization problem requires the compliance with the problem-specific optimum criterion.

Linear optimization is mainly employed for solving economic or technical-economic problems, of which only relatively simple ones (with only a few variables) can be manually solved in an economical manner. Complicated problems require the use of electronic data processing equipment partly involving high demands on the storage capacity of the plant. The procedure of the determination of extreme values using differential calculus will fail since the variables will disappear from the linear equations in differentiation.

The maximum-value problem in which the restricting conditions show only positive absolute values is called the *normal form of optimization.*

The algorithm of the mathematical model used represents the collection of all rules for the calculation of the solution. From the multitude of existing procedures the graphical method and the simplex algorithm will be discussed.

Setting up the mathematical model

Target or objective function:

$$z(x) = c_1x_1 + c_2x_2 + c_3x_3 + \cdots + c_nx_n = \sum_{k=1}^{n} c_kx_k \rightarrow \text{optimum}$$

written in matrix form with

row vector $\mathbf{c} = (c_1, c_2, c_3, \ldots, c_n)$

column vector $\mathbf{x} = \begin{pmatrix} x_1 \\ x_2 \\ \vdots \\ x_n \end{pmatrix}$ a_{ik} = constant

b_i = constant

$\mathbf{cx} \to$ optimum c_k = constant

x_k = variable

Restricting conditions (constraints)

(for maximum problem $\leqq b_i$; for minimum problem $\geqq b_i$)

$$\begin{matrix} a_{11}x_1 + a_{12}x_2 + \cdots + a_{1n}x_n \gtreqless b_1 \\ a_{21}x_1 + a_{22}x_2 + \cdots + a_{2n}x_n \gtreqless b_2 \\ \vdots \quad \vdots \quad \vdots \quad \vdots \\ a_{m1}x_1 + a_{m2}x_2 + \cdots + a_{mn}x_n \gtreqless b_m \end{matrix}$$

or

$$\sum_{k=1}^{n} a_{ik}x_k \gtreqless b_i \qquad i, k \in N$$

written in the form of a matrix with

coefficient matrix $\mathbf{A} = \begin{pmatrix} a_{11} & a_{12} & \dots & a_{1n} \\ a_{21} & a_{22} & \dots & a_{2n} \\ \vdots & \vdots & \cdot & \vdots \\ a_{m1} & a_{m2} & \dots & a_{mn} \end{pmatrix}$; $\mathbf{b} = \begin{pmatrix} b_1 \\ b_2 \\ \vdots \\ b_m \end{pmatrix}$

$\mathbf{Ax} \gtreqless b$ normal form $b_i \geqq 0$

Condition of nonnegativity:

$x_k \geqq 0 \quad k = 1; 2; \ldots; n$

The restricting conditions and the nonnegativity condition determine the range of definition.
The conditions $\mathbf{A} \cdot \mathbf{x} \leqq b$ and $\mathbf{A} \cdot \mathbf{x} \geqq b$ respectively are always obtainable by employing the inequation $a \geqq b \Leftrightarrow -a \leqq -b$.
The set of all permissible solutions is formed by all vectors $\mathbf{x}$ for which the restricting conditions are valid.
The solution $\mathbf{x}_0$ for which the objective function becomes an optimum is called *optimum programme*.

11.2. Graphical procedure

This procedure can be applied only to objective functions with two variables.

In a rectangular system of coordinates with the axes x_1 and x_2, the range of all ordered pairs of values (x_1, x_2) that satisfy the restricting conditions and the condition of nonnegativity is graphically ascertained; this is done by graphically representing the straight lines

$$a_{11}x_1 + a_{12}x_2 = b_1$$

$$a_{21}x_1 + a_{22}x_2 = b_2$$

etc.

If the restricting conditions (constraints) are given in the form of inequalities, each of these straight lines will divide the area into a range possible for these conditions and a range impossible for these conditions. The permissible conditions are located in the range that is permissible for all conditions (the region of feasible solution).

The optimal solution is found by drawing the picture of the modified objective function $z(x_1, x_2) = c$ (c an arbitrary constant) and parallel shifting of this $c \to c_{\max} \Rightarrow$ optimum = maximum. The optimal solution is unique if the straight line for $c_{\max}$ runs through a corner point of the possible range (*corner solution*); it is ambiguous if the graphically represented objective function runs parallel to one of the above indicated straight lines of the restricting conditions.

In minimization, the straight line must be shifted in the direction of $c \to c_{\min}$.

Example:

Find the optimum programme of the objective function $z(x_1, x_2) = 10x_1 + 15x_2 \to$ maximum.

Restricting conditions:

$$x_1 + 2x_2 \leqq 102$$

$$15x_1 + 3x_2 \leqq 450$$

$$x_1 \leqq 25$$

$$x_2 \leqq 45$$

Nonnegativity condition:

$$x_1 \geqq 0$$

$$x_2 \geqq 0$$

Solution:

$$x_1, x_1 \geqq 0 \Rightarrow \text{first quadrant}$$

$$\left.\begin{aligned} x_1 &\leqq 25 \Rightarrow \text{straight line} \mid\mid x_1\text{-axis} \\ x_2 &\leqq 45 \Rightarrow \text{straight line} \mid\mid x_2\text{-axis} \end{aligned}\right\} \Rightarrow \text{rectangle separated}$$

$$x_1 + 2x_2 = 102 \wedge 15x_1 + 9x_2 = 450 \text{ two straight lines}$$

Range of permissible solutions: Hexagon $O, P_1, P_2, P_3, P_4, P_5$
With $c = 300 \Rightarrow z(x_1, x_2) \equiv 10x_1 + 15x_2 = 300$ straight line, a set of permissible solutions, no optimum.
Parallel displacement of the straight line through $P_3 \Rightarrow$ optimum = maximum.

$$\mathbf{x}_0 = (22, 40)$$

Value of the objective function for optimum:

$$z_{\text{max}} = 10 \cdot 22 + 15 \cdot 40 = 820$$

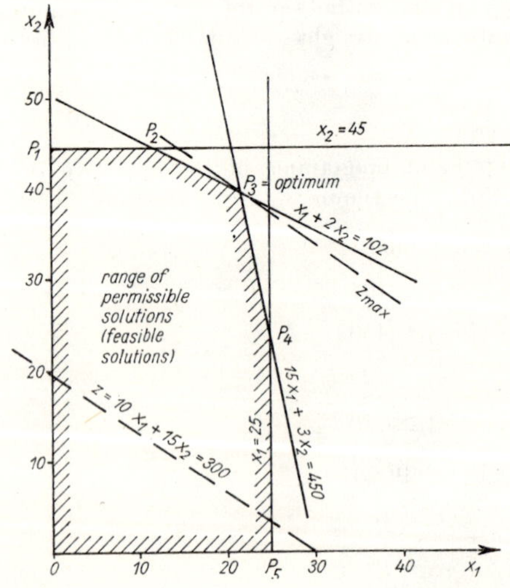

11.3. Simplex procedure (simplex algorithm)

The simplex procedure is an iterative process for approaching the optimum stepwise and is applicable to problems with two or more variables. It can be simplified for the application to transport optimization problems (minimization of the cost matrix).

By *solution basis* we understand a solution satisfying the restricting conditions that has, at most, as many variables (*basis variables*) with a value $\neq 0$ as restricting conditions that are independent of each other exist. The set of the basis variables is called *basis*; the variables that do not form part of the base and are put equal to zero are called *nonbasis variables*.

Each solution basis corresponds to a corner point of the permissible range in the graphical representation (cf. 11.2.).

Procedure of solution

1. *Normal form of linear optimization*

The restricting conditions given in the form of inequations $\mathbf{Ax} \leqq \mathbf{b}$ are transformed into equations by introducing fictious variables (slack variables).

$$\begin{array}{lllllll} a_{11}x_1 & + a_{12}x_2 & + \cdots + a_{1n}x_n & + u_1 & & & = b_1 \\ a_{21}x_1 & + a_{22}x_2 & + \cdots + a_{2n}x_n & & + u_2 & & = b_2 \\ \vdots & \vdots & \vdots & & \vdots & & \vdots \\ a_{m1}x_1 & + a_{m2}x_2 & + \cdots + a_{mn}x_n & & & + u_m & = b_m \end{array}$$

In the base solutions, the slack variables give a picture of the reserves which are still unused.

The objective function will be

$$\begin{aligned} & z(x_1, x_2, \ldots, x_n, u_1, u_2, \ldots, u_m) \\ & = c_1x_1 + c_2x_2 + \cdots + c_nx_n + 0u_1 + 0u_2 + \cdots \to \max \end{aligned}$$

Objective functions $z \to \min$ are transformed into $z' \to \max$ by multiplication by -1.

The restricting conditions given in the form of equations $\mathbf{Ax} = \mathbf{b}$ are transformed into the normal form as explained above by the introduction of artificial variables with the objective function changing as follows:

$$z = c_1x_1 + c_2x_2 + \cdots + c_nx_n - N(u_1 + u_2 + u_3 + \cdots)$$

with $N \gg c_1; c_2; \ldots$

The restricting conditions given in the form of inequations $\mathbf{Ax} \geqq \mathbf{b}$ are transformed into equations by introducing negative slack variables.

$$a_{11}x_1 + a_{12}x_2 + \cdots + a_{1n}x_n - u_1 = b_1$$

etc.

To obtain the normal form, the slack variables must become positive; this is achieved by subtracting each line from that with the largest b_i. The differences and the line with the largest b_i form a system of equations of normal form.

2. *First solution basis (initial solution)*

Put all $x_k = 0$, all $u_1 \neq 0 \Rightarrow$ all u_1 belong to the base of the first base solution.
Representation of the basis of the first solution basis:

$$\begin{aligned} u_1 &= b_1 - a_{11}x_1 - a_{12}x_2 - \cdots - a_{1n}x_n \\ u_2 &= b_2 - a_{21}x_1 - a_{22}x_2 - \cdots - a_{2n}x_n \\ &\vdots \\ u_m &= b_m - a_{m1}x_1 - a_{m2}x_2 - \cdots - a_{mn}x_n \end{aligned}$$

Representation of the basis of the objective function of the first solution basis:

$$z_1 = c_1x_1 + c_2c_2 + \cdots$$

Solution basis: $(0, 0, \ldots, b_1, b_2, \ldots, b_m)$
[given in the sequence of the special values for $(x_1, x_2, \ldots, u_1, u_2, \ldots)$]

3. *Checking the objective function of the first solution basis for optimality*

Criterion: The result obtained in a maximum problem can be improved as long as coefficients $c_i > 0$ occur in the basis representation of the objective function.

4. *Selection of a nonbasis variable for replacement by a basis variable*

Introduction of that nonbasis variable from z_1 that has the largest coefficient c_i, since this will bring the greatest increase in value of z.

5. *Selection of the basis variable to be exchanged*

Form all quotients $\dfrac{b_i}{a_{ij}}$, x_j being the new basis variable to be introduced.
The smallest of these quotients determines the basis variable to be exchanged belonging to the same equation $(\mathbf{u}_l)$: $\mathbf{a}_{ij} > 0$.

6. *Exchange of the variable*

By transformation of the respective equation from the basis representation of the first solution basis an equation for the new basis variable $x_j = b_j/a_{ij} - a_{i1}/a_{ij} \cdot x_1 - \cdots - u_l$ is obtained whose right side is substituted in the other equations for x_j.

7. *Second basis solution*

$$\begin{aligned}
u_1 &= b_1 - a_{11}x_1 - a_{12}x_2 - \cdots \\
&\qquad - a_{1j}\left(\frac{b_j}{a_{ij}} - \frac{a_{i1}}{a_{ij}}x_1 - \cdots - u_l\right) - \cdots - a_{1n}x_n \\
u_2 &= b_2 - a_{21}x_1 - \cdots \\
&\vdots \qquad \vdots \\
u_{j-1} &= \cdots \\
u_{j+1} &= \cdots \\
&\vdots \qquad \vdots \\
x_j &= \frac{b_j}{a_{ij}} - \frac{a_{i1}}{a_{ij}}x_1 - \cdots - u_l
\end{aligned}$$

On the left side there is the new basis in which x_j and u_l are exchanged as compared to the first basis. The procedure is then continued as described from 2 onward until all coefficients of the base representation of the objective function of the pth solution basis $= 0$. Now the optimum will have been reached.

Example:

The example solved graphically in 11.2. will now be repeated using the simplex algorithm.

$$z(x_1, x_2) = 10x_1 + 15x_2 \rightarrow \text{maximum}$$

Restricting conditions:

$$\begin{aligned}
x_1 + 2x_2 &= 102 \\
15x_1 + 3x_2 &= 450 \\
x_1 \qquad &= 25 \\
x_2 &= 45
\end{aligned}$$

Nonnegativity condition:

$$x_1 \geqq 0$$
$$x_2 \geqq 0$$

Note: When there are only two variables involved in the problem, the simplex procedure requires much more work than the graphical method.

1. *Normal form*

$$\begin{aligned} x_1 + 2x_2 + u_1 &= 102 \\ 15x_1 + 3x_2 + u_2 &= 450 \\ x_1 \qquad + u_3 &= 25 \\ x_2 + u_4 &= 45 \end{aligned}$$

2. *First solution basis*

Basis representation:

$$\begin{aligned} u_1 &= 102 - x_1 - 2x_2 \\ u_2 &= 450 - 15x_1 - 3x_2 \\ u_3 &= 25 - x_1 \\ u_4 &= 45 \qquad - x_2 \end{aligned}$$

Solution basis: (0, 0, 102, 450, 25, 45) corresponds to point P_0 of the graphical solution

Basis representation of the objective function:

$$z_1' = 10x_1 + 15x_2$$

Value of the objective function:

$$z_1 = 10 \cdot 0 + 15 \cdot 0 + 0 \cdot 102 + 0 \cdot 450 + 0 \cdot 25 + 0 \cdot 45 = 0$$

3. *Checking for optimality*

Not optimal as in the basis representation $c_1 > 0$; $c_2 > 0$.

4. *New basis variable* x_2 as in basis representation largest coefficient.

5. *Basis variable to be exchanged*

$$\frac{b_1}{a_{12}} = \frac{102}{2} = 51; \quad \frac{b_2}{a_{22}} = \frac{450}{3} = 150; \quad \frac{b_4}{a_{42}} = \frac{45}{1} = 45$$

smallest value

$\Rightarrow u_4$ is to be removed.

6. *Exchange of the variables*

$$x_2 = 45 - u_4$$

7. *Second solution basis*

$$\begin{aligned}
u_1 &= 102 - x_1 - 2(45 - u_4) = 12 - x_1 + 2u_4 \\
u_2 &= 450 - 15x_1 - 3(45 - u_4) = 315 - 15x_1 + 3u_4 \\
u_3 &= 25 - x_1 = 25 - x_1 \\
x_2 &= 45 - u_4 = 45 - u_4
\end{aligned}$$

Basis solution: $(0, 45, 102, 450, 25, 0) \Rightarrow$ point P_1

Basis representation: $z_2' = 10x_1 + 15(45 - u_4)$

$$= 10x_1 - 15u_4 + 675$$

Objective function: $z_2 = 10 \cdot 0 + 15 \cdot 45 = 675$

8. *Not optimal* as factor of $x_1 > 0 \Rightarrow x_1$ to be exchanged

9. $\dfrac{b_1'}{a_{11}'} = \dfrac{12}{1} = 12$ smallest value $\Rightarrow u_1$ to be removed

$$\frac{b_2'}{a_{21}'} = \frac{315}{15} = 21$$

$$\frac{b_3'}{a_{31}'} = \frac{25}{1} = 25$$

10. *Third solution basis*

$$\begin{aligned}
x_1 &= 12 - u_1 + 2u_4 &&= 12 - u_1 + 2u_4 \\
x_2 &= 45 - u_4 &&= 45 - u_4
\end{aligned}$$

$$u_2 = 315 - 15(12 - u_1 + 2u_4) + 3u_4 = 135 + 15u_1 - 27u_4$$
$$u_3 = 25 - (12 - u_1 + 2u_4) = 13 + u_1 - 2u_4$$

Basis solution: $(12, 45, 0, 135, 13, 0) \Rightarrow$ point P_2
Basis representation:

$$z_3' = 10(12 - u_1 + 2u_4) + 15(45 - u_4)$$
$$= -10u_1 + 5u_4 + 795$$

Objective function: $z_3 = 10 \cdot 12 + 15 \cdot 45 = 795$

11. *Not optimal* as factor $u_4 > 0 \Rightarrow u_4$ to be exchanged

12. $\left(\frac{b_1''}{a_{16}''} = \frac{12}{-2} = -6 \text{ to be omitted as } a_{16}'' < 0\right)$

$$\frac{b_2''}{a_{26}''} = \frac{45}{1} = 45; \quad \frac{b_3''}{a_{36}''} = \frac{135}{27} = 5 \Rightarrow u_2 \text{ to be removed}$$

$$\frac{b_4''}{a_{46}''} = \frac{13}{2} = 6.5$$

13. *Fourth solution basis*

$$x_1 = 12 - u_1 + 2\left(5 + \frac{15}{27}u_1 - \frac{1}{27}u_2\right) = 22 + \frac{1}{9}u_1 - \frac{2}{27}u_2$$

$$x_2 = 45 - \left(5 + \frac{15}{27}u_1 - \frac{1}{27}u_2\right) = 40 - \frac{5}{9}u_1 + \frac{1}{27}u_2$$

$$u_3 = 13 + u_1 - 2\left(5 + \frac{15}{27}u_1 - \frac{1}{27}u_2\right) = 3 - \frac{1}{9}u_1 + \frac{2}{27}u_2$$

$$u_4 = 5 + \frac{15}{27}u_1 - \frac{1}{27}u_2 = 5 + \frac{5}{9}u_1 - \frac{1}{27}u_2$$

Basis solution: $(22, 40, 0, 0, 3, 5) \Rightarrow$ point P_3
Basis representation:

$$z_4' = 10\left(22 + \frac{1}{9}u_1 - \frac{2}{27}u_2\right) + 15\left(40 - \frac{5}{9}u_1 + \frac{1}{27}u_2\right)$$

$$= 820 - 7\frac{2}{9}u_1 - \frac{5}{27}u_2$$

$$z_4 = 10 \cdot 22 + 15 \cdot 40 = \underline{\underline{820}}$$

11.4. Simplex table

The simplex table represents a clear tabular arrangement of the mathematical procedure described in 11.3.

Variables and slack variables			x_1	x_2	u_1	u_2	u_3	u_4	$\frac{b_i}{a_{ij}}$
Basis variable		0	10	15	0	0	0	0	
I	u_1	102	1	2	1	0	0	0	51
	u_2	450	15	3	0	1	0	0	150
	u_3	25	1	0	0	0	1	0	—
	u_4	45	0	$\boxed{1}$	0	0	0	1	45
II		−675	10	0	0	0	0	−15	
	u_1	12	$\boxed{1}$	0	1	0	0	−2	12
	u_2	315	15	0	0	1	0	−3	21
	u_3	25	1	0	0	0	1	0	25
	x_2	45	0	1	0	0	0	1	—
III		−795	0	0	−10	0	0	+5	
	x_1	12	1	0	1	0	0	−2	(−6) to be omitted
	u_2	135	0	0	15	1	0	$\boxed{27}$	5
	u_3	13	0	0	−1	0	1	2	6.5
	x_2	45	0	1	0	0	0	1	45
IV		−820	0	0	$\frac{-65}{9}$	$\frac{-5}{27}$	0	0	⇒ optimum
	x_1	22	1	0	$\frac{-3}{27}$	$\frac{2}{27}$	0	0	
	u_4	5	0	0	$\frac{-15}{27}$	$\frac{1}{27}$	0	1	
	u_3	3	0	0	$\frac{3}{27}$	$\frac{-2}{27}$	1	0	
	x_2	40	0	1	$\frac{15}{27}$	$\frac{-1}{27}$	0	0	

Explanations

The variables and slack variables are given in the head line.

Column 1: Basis variable of the respective base

Column 2: Value of the basis variable

Columns 3 to 8: Coefficients of the variables and slack variables in the basis representations

Column 9: Quotients from the absolute term and coefficient of the variable to be exchanged

First solution basis

Line 1: Objective function; largest coefficient 15 determined, x_2 to be exchanged

Lines 2 to 5: Normal form

The smallest quotient, 45, determines the exchange of u_4.

Main element: Coefficient of x_2 in the equation of u_4, $\boxed{1}$ bordered

Second solution basis

Line 5: Divide line 5 of I by the main element 1.

Line 1: Multiply line 5 by 15 and subtract from line 1 in I so as to obtain the coefficient 0 in the column of the variable x_2 to be exchanged $\Rightarrow$ basis representation of the 2nd basis solution.

Line 2: Multiply line 5 by 2 and subtract from line 2 in I.

Line 3: Multiply line 5 by 3 and subtract from line 3 in I.

Line 4: Multiply line 5 by zero and subtract from line 4 in I, i.e. take this line unchanged.

The smallest quotient 12 determines the replacement of u_1 by x_1 which has the largest coefficient in the objective function. Main element $\boxed{1}$

Third solution basis

Line 2: Divide line 2 of II by the main element 1.

Line 1: Multiply line 2 by 10 and subtract from line 1 in II $\Rightarrow$ basis representation of the third solution basis.

Line 3: Multiply line 2 by 15 and subtract from line 3 in II.

Continue the procedure until the line 1 (basis representation) no longer shows any coefficients larger than zero.

12. Algebra of logic (Boolean algebra)

12.1. General

The basis of this algebra is the Boolean algebra which was created as a calculus (*calculus of logic*) for the description of logical relations that mathematically symbolizes the statements as *true* or *false*. Its development into a two-value Boolean algebra as switching algebra is based on the fact that also in digital techniques with storage-free binary elements only the states "on" and "off" are possible:

$$x = \mathrm{L} \quad \text{when} \quad x \neq 0$$

$$x = 0 \quad \text{when} \quad x \neq \mathrm{L}$$

Notations:

Statements: true — false; yes — no; on — off
Symbols: L — O
Input variable: x_ν; number of input variables k
Output variable: $y = F(x)$
Constant connection: L; constant interruption: O

Basic logical interconnections:

Disjunction, logical addition, logical *OR*, alternative

Notation: $y = x_1 \vee x_0$ (read: x_1 or x_0) (other signs used are: $+$; $\cup$)

Switching technique: Parallel connection

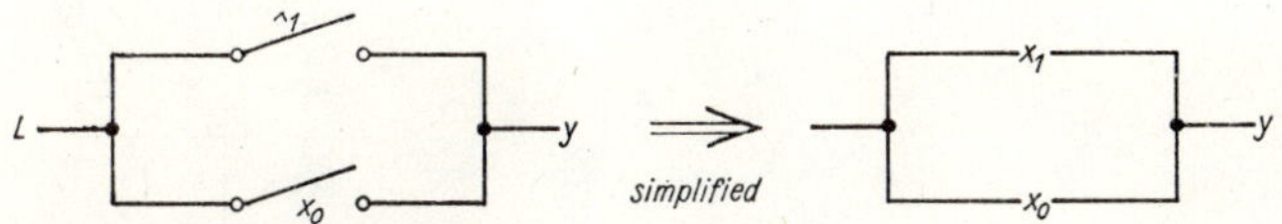

Table of switching availabilities

x_1	x_0	$y = x_1 \vee x_0$
O	O	O
O	L	L
L	O	L
L	L	L

Conjunction, logical product, logical *AND*

Notation: $y = x_1 x_2$ (read: x_1 and x_2) (other signs used are: $\wedge$; .; $\cap$; &)
No sign always means a conjunction.
Switching technique: Series connection

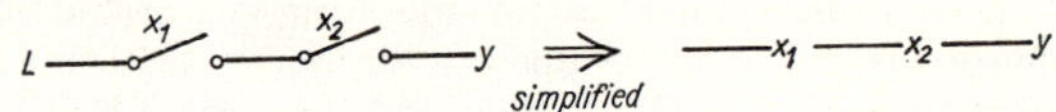

Table of switching availabilities

x_1	x_2	$y = x_1 x_2$
O	O	O
O	L	O
L	O	O
L	L	L

Negation, logical negation, inversion

Notation: $y = \bar{x}$ (read: x bar; not x)
Switching technique: Closed-circuit contact

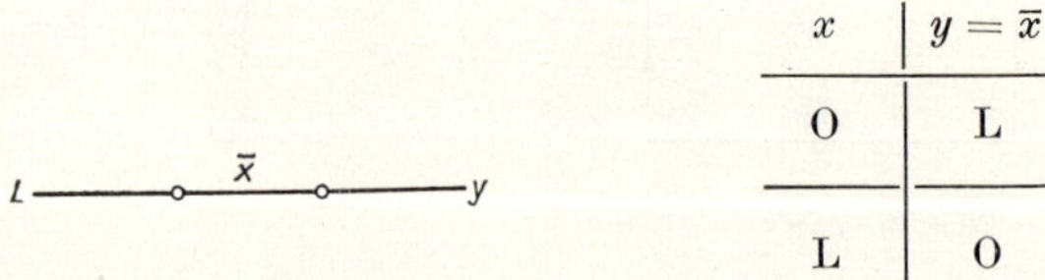

x	$y = \bar{x}$
O	L
L	O

The following configurations are obtained from the three basic interconnections:

$$\begin{array}{lll} O \vee O = O & O \wedge O = O & \overline{O} = L \\ O \vee L = L & O \wedge L = O & \overline{L} = O \\ L \vee O = L & L \wedge O = O & \\ L \vee L = L & L \wedge L = L & \end{array}$$

Number of possible groups of interconnections n:

e.g. $n = 2^{2^k}$ k number of input variables

$k = 2$ input variables: $n = 2^{2^k} = 2^{2^2} = 16$ possibilities (cf. table page 496)

$k = 3$ input variables yield $n = 2^{2^3} = 256$ possibilities

12.2. Arithmetical laws, arithmetical rules

Commutative law:

$$x_1 \vee x_0 = x_0 \vee x_1 \qquad x_1 x_0 = x_0 x_1$$

Associative law:

$$x_2 \vee (x_1 \vee x_0) = (x_2 \vee x_1) \vee x_0 = x_2 \vee x_1 \vee x_0 \qquad x_2(x_1 x_0) = (x_2 x_1)\, x_0 = x_2 x_1 x_0$$

Distributive law:

$$x_2(x_1 \vee x_0) = x_2 x_1 \vee x_2 x_0$$

$$x_2 \vee x_1 x_0 = (x_2 \vee x_1)\,(x_2 \vee x_0)$$

Note: There is no equivalent in conventional algebra for the latter relationship.

Rule: In combined calculations, disjunction is given preference to conjunction.

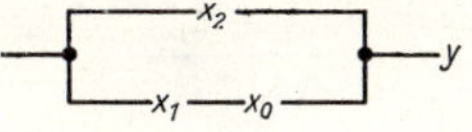

Example:

$$y = x_2 \vee x_1 \wedge x_0 = x_2 \vee (x_1 x_0)$$

but $\neq (x_2 \vee x_1)\, x_0$ (lower figure)

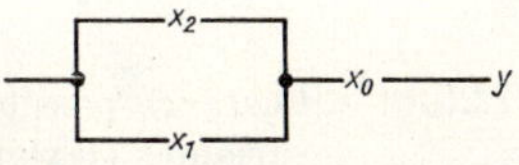

Arithmetical laws, arithmetical rules (table)

$O \vee O = O$	$O \wedge O = O \wedge L = L \wedge O = O$	$\bar{O} = L$
$O \vee L = L \vee O = L \vee L = L$	$L \wedge L = L$	$\bar{L} = O$
$x \vee O = x$	$x \wedge O = O$	$\bar{\bar{x}} = x$
$x \vee L = L$	$x \wedge L = x$	$(\bar{x}) = \bar{x}$
$x \vee x \vee x \cdots = x$	$xxx \cdots = x$	$(\bar{\bar{x}}) = \bar{\bar{x}} = x$
$x \vee \bar{x} = L$	$x\bar{x} = O$	

$x_1 \vee x_0 = x_0 \vee x_1$	$x_1 x_0 = x_0 x_1$ (commutative law)
$\overline{x_1 \vee x_0} = \bar{x}_1 \bar{x}_0$	$\overline{x_1 x_0} = \bar{x}_1 \vee \bar{x}_0$ (DE MORGAN's theorem)
$x_1 x_0 \vee x_1 \bar{x}_0 = x_1(x_0 \vee \bar{x}_0) = x_1$	
$x_1 \vee x_1 x_0 = x_1(\mathrm{L} \vee x_0) = x_1$	$x_1(x_1 \vee x_0) = x_1$
$x_1 \vee \bar{x}_1 x_0 = x_1 \vee x_0$	$x_1(\bar{x}_1 \vee x_0) = x_1 x_0$
$x_2 \vee (x_1 \vee x_0) = (x_2 \vee x_1) \vee x_0 = x_2 \vee x_1 \vee x_0$	$x_2(x_1 x_0) = (x_2 x_1)\, x_0 = x_2 x_1 x_0$ (associative law)
$x_2 \vee x_1 x_0 = (x_2 \vee x_1)\,(x_2 \vee x_0)$	$x_2(x_1 \vee x_0) = x_2 x_1 \vee x_2 x_0$ (distributive law)
$x_2 x_1 \vee x_2 x_0 = x_2(x_1 \vee x_0)$	$(x_2 \vee x_1)\,(x_2 \vee x_0) = x_2 \vee x_1 x_0$
$(x_2 \vee x_1)\,(\bar{x}_2 \vee x_0) = x_2 x_0 \vee \bar{x}_2 x_1$	
$x_2 x_1 \vee x_2 \bar{x}_0 \vee x_1 x_0 = x_2 x_0 \vee x_1 x_0$	$(x_2 \vee x_1)\,(x_2 \vee x_0)\,(x_1 \vee x_0) = x_1 \bar{x}_0 \vee x_2 x_0$
$\overline{x_{k-1} \vee x_{k-2} \vee \ldots \vee x_0} = \bar{x}_{k-1} \wedge \bar{x}_{k-2} \wedge \cdots \wedge \bar{x}_0$	$\overline{x_{k-1} x_{k-2} \wedge \cdots \wedge x_0} = \bar{x}_{k-1} \vee \bar{x}_{k-2} \vee \cdots \vee \bar{x}_0$

12.3. Further possibilities of interconnecting two input variables (lexigraphic order)

n \ x_1 / x_0	O / O	O / L	L / O	L / L	y_n	
0	O	O	O	O	$y_0 = 0$	Constancy; interruption
1	O	O	O	L	$y_1 = x_1 x_0$	Conjunction
2	O	O	L	O	$y_2 = x_1 \bar{x}_0$	Inhibition
3	O	O	L	L	$y_3 = x_1$	Identity
4	O	L	O	O	$y_4 = \bar{x}_1 x_0$	Inhibition
5	O	L	O	L	$y_5 = x_0$	Identity
6	O	L	L	O	$y_6 = \bar{x}_1 x_0 \vee x_1 x_0$	Antivalence; exclusive OR

Continuation

n \ x_1 / x_0	O / O	O / L	L / O	L / L	y_n	
7	O	L	L	L	$y_7 = x_1 \vee x_0$	Disjunction
8	L	O	O	O	$y_8 = \bar{x}_1\bar{x}_0$	NOR circuit; "NEITHER-NOR"
9	L	O	O	L	$y_9 = x_1x_0 \vee \bar{x}_1\bar{x}_0$	Equivalence
10	L	O	L	O	$y_{10} = \bar{x}_0$	Negation
11	L	O	L	L	$y_{11} = x_1 \vee \bar{x}_0$	Implication
12	L	L	O	O	$y_{12} = \bar{x}_1$	Negation
13	L	L	O	L	$y_{13} = \bar{x}_1 \vee x_0$	Implication
14	L	L	L	O	$y_{14} = \bar{x}_1 \vee \bar{x}_0$	Input-negated disjunction
15	L	L	L	L	$y_{15} = \mathrm{L}$	Constancy; short circuit

Six of the 16 possibilities are trivial ($n = 0; 3; 5; 12; 15$)

Antivalence (exclusive *OR*)

$x_1 \mp x_0 \quad y = \bar{x}_1x_0 \vee x_1\bar{x}_0$

x_1	x_0	$y = \bar{x}_1x_0 \vee x_1\bar{x}_0$
O	O	O
O	L	L
L	O	L
L	L	O

Equivalence

$x_1 \equiv x_0 \quad y = x_1x_0 \vee \bar{x}_1\bar{x}_0$

x_1	x_0	$y = x_1x_0 \vee \bar{x}_1\bar{x}_0$
O	O	L
O	L	O
L	O	O
L	L	L

Negation of the equivalence leads to antivalence and vice versa.

***NOR* circuit** (negated *OR*; *NEITHER-NOR*)

$$y = \overline{x_1 \vee x_0} = \bar{x}_1 \bar{x}_0$$

x_1	x_0	y
O	O	L
O	L	O
L	O	O
L	L	O

Sheffer function (negated *AND*; *NAND*)

$$y = \overline{x_1 x_0} = \bar{x}_1 \vee \bar{x}_0$$

x_1	x_0	y
O	O	L
O	L	L
L	O	L
L	L	O

Implication

$$y = x_1 \rightarrow x_0$$

Two definitions: $y_{\mathrm{I}} = \bar{x}_1 \vee x_0$

$y_{\mathrm{II}} = x_1 \vee \bar{x}_0$

x_1	x_0	y_{I}	y_{II}
O	O	L	L
O	L	L	O
L	O	O	L
L	L	L	L

12.4. Normal forms

Minterm (elementary conjunction)

$K_n{}^k = \bigwedge\limits_{\nu=0}^{k-1} x_\nu$ (read: sum of all conjunctions for $\nu = 0$ to $k - 1$)

Number of possible minterms 2^k

The term $K_n{}^k$ is called minterm if it contains the conjunctive connection of all k input variables (negated or nonnegated) weighted according to powers of 2.

For example: $K_{38}^6 = \bigwedge_{\nu=0}^{5} x_\nu = x_5\overline{x}_4\overline{x}_3x_2x_1\overline{x}_0$, weighted LOOLLO $\triangleq$ $38 = n$

Maxterm (elementary disjunction)

$D_n{}^k = \bigvee_{\nu=0}^{k-1} x_\nu$ (read: sum of all disjunctions for $\nu = 0$ to $k - 1$)

Number of possible maxterms 2^k

The term $D_n{}^k$ is called maxterm if it contains the disjunctive connection of all k input variables (negated or nonnegated) weighted according to powers of 2.

For example: $D_{11}^4 = \bigvee_{\nu=0}^{3} x_\nu = x_3 \vee \overline{x}_2 \vee x_1 \vee x_0$,
weighted LOLL $= 11 = n$

Disjunctive normal form, conjunctive normal form

Every disjunctive interconnection of conjunctions (*fundamental terms*) is called disjunctive normal form, e.g. $y = x_4x_2x_1 \vee \overline{x}_2x_1 \vee x_3$. Similarly, the conjunctive normal form, e.g. $y = (x_3 \vee \overline{x}_2)(x_3 \vee x_1 \vee \vee x_0)(\overline{x}_3 \vee \overline{x}_0)$.

Fundamental terms which can no longer be simplified are called *primeimplicants* of the function.
Canonical alternative (*disjunctive*) normal form (series-parallel connection).

Canonical conjunctive normal form (parallel-series connection)

Every disjunctive connection of minterms is called a canonical alternative normal form: $y^k = \bigvee_n K_n{}^k$ (chiefly used!).

Similarly, $y^k = \bigvee_n D_n{}^k$.

Number of possible canonical alternative normal forms

$n = (2)^{2^k}$ k number of input variables

Example:

In a technical problem, the three input variables x_2, x_1, x_0 are interconnected with the output variable y in the following way. Calculate a minimum form of the logic function.

Operating table with 2^k possibilities
k number of input variables

n	x_2	x_1	x_0	K_n
0	O	O	O	L
1	O	O	L	L
2	O	L	O	L
3	O	L	L	L
4	L	O	O	L
5	L	O	L	L
6	L	L	O	O
7	L	L	L	O

For $y = \mathrm{L}$, the canonical alternative normal form is valid:

$$y = \bigvee_n K_n \quad \text{for} \quad n = 0, 1, 2, 3, 4, 5$$

$$y = K_0 \vee K_1 \vee K_2 \vee K_3 \vee K_4 \vee K_5$$

$$= \bar{x}_2\bar{x}_1\bar{x}_0 \vee \bar{x}_2\bar{x}_1x_0 \vee \bar{x}_2x_1\bar{x}_0 \vee \bar{x}_2x_1x_0 \vee x_2\bar{x}_1\bar{x}_0 \vee x_2\bar{x}_1x_0$$

$$= \bar{x}_2\bar{x}_1(\bar{x}_0 \vee x_0) \vee \bar{x}_2x_1(\bar{x}_0 \vee x_0) \vee x_2\bar{x}_1(\bar{x}_0 \vee x_0)$$

$$= \bar{x}_2\bar{x}_1 \vee \bar{x}_2x_1 \vee x_2\bar{x}_1$$

$$= \bar{x}_2(\bar{x}_1 \vee x_1) \vee x_2\bar{x}_1 = \bar{x}_2 \vee x_2\bar{x}_1 = \underline{\underline{\bar{x}_2 \vee \bar{x}_1}}$$

Since $y = 0$ occurs only in two lines, it is better to choose the canonical alternative normal form of the O-decision and invert the result:

$$\bar{y} = K_6 \vee K_7$$

$$\bar{y} = x_2x_1\bar{x}_0 \vee x_2x_1x_0 = x_2x_1(\bar{x}_0 \vee x_0) = x_2x_1$$

$$y = \overline{x_2x_1} = \underline{\underline{\bar{x}_2 \vee \bar{x}_1}} \quad \text{as above}$$

12.5. Karnaugh tables

Every field represents the conjunction of the input variables given on the margin. KARNAUGH tables are established in the plane of up to five input variables (corresponding to 32 fields). If the number of variables is greater, the tables are no longer practical.
Only one variable changes from one column to another column and from one row to another row (Gray code!). The same applies to the margins, for example between the 1st and the 4th column or row.

Example:

$K^4_{13} = x_3 x_2 \bar{x}_1 x_0 \triangleq$ LLOL

	$\bar{x}_3\bar{x}_2$	$\bar{x}_3 x_2$	$x_3 x_2$	$x_3 \bar{x}_2$
$\bar{x}_1\bar{x}_0$	OOOO K_0	OLOO K_4	LLOO K_{12}	LOOO K_8
$\bar{x}_1 x_0$	OOOL K_1	OLOL K_5	LLOL K_{13}	LOOL K_9
$x_1 x_0$	OOLL K_3	OLLL K_7	LLLL K_{15}	LOLL K_{11}
$x_1 \bar{x}_0$	OOLO K_2	OLLO K_6	LLLO K_{14}	LOLO K_{10}

The desired output value L or O is entered in the intersection fields of rows and columns beside the respective elementary conjunctions K_n of the input variables. The fields are interconnected by *OR*.
Evaluation is performed by forming the largest possible blocks of two, four, or eight with $K_n = \mathrm{L}$ (or $K_n = 0$) which may also extend into the margins. An evaluation of the blocks, in which procedure all input variables are omitted whose values change within the blocks, yields the primeimplicants of the switching function.

Example:

Of four pumps x_0; x_1; x_2; x_3, no more than two pumps must operate together. The problem is to prevent that more than 2 pumps are switched on at the same time (operation of an interlocking device $K_n = \mathrm{L}$)

Operation of one pump: $x = \mathrm{L}$

Number of possible interconnections: $n = 2^k = 16$

n	x_3	x_2	x_1	x_0	K_n	n	x_3	x_2	x_1	x_0	K_n
0	O	O	O	O	O	8	L	O	O	O	O
1	O	O	O	L	O	9	L	O	O	L	L
2	O	O	L	O	O	10	L	O	L	O	L
3	O	O	L	L	L	11	L	O	L	L	L
4	O	L	O	O	O	12	L	L	O	O	L
5	O	L	O	L	L	13	L	L	O	L	L
6	O	L	L	O	L	14	L	L	L	O	L
7	O	L	L	L	L	15	L	L	L	L	L

The rows 7; 11; 13; 14; 15 could be omitted as according to the problem set these combinations of the variables must not occur. Taking them into consideration will often facilitate the calculation.

KARNAUGH Table

	$\overline{x}_3\overline{x}_2$	$\overline{x}_3x_2$	x_3x_2	$x_3\overline{x}_2$
$\overline{x}_1\overline{x}_0$	O	O	L	O
$\overline{x}_1x_0$	O	L	L	L
x_1x_0	L	L	L	L
$x_1\overline{x}_0$	O	L	L	L

3rd column	3rd row	block of four middle	also middle, bottom	also middle, right	also right bottom
$y = x_3x_2 \vee$	$x_1x_0 \vee$	$x_2x_0 \vee$	$x_2x_1 \vee$	$x_3x_0 \vee$	x_3x_1

$$= [x_3(x_2 \vee x_1 \vee x_0)] \vee [x_2(x_1 \vee x_0)] \vee x_1x_0$$

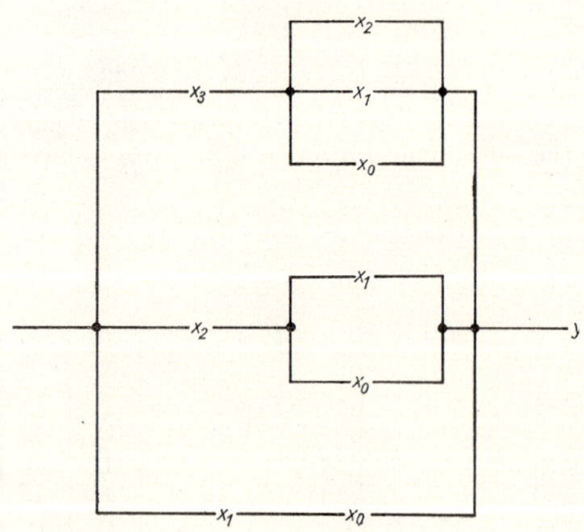

APPENDIX

The dual system (dyadic system)

Basic symbols: O, L
Place value: powers of 2

Decadic number	Associated dual code	Decadic number	Associated dual code
0	O	6	LLO
1	L	7	LLL
2	LO	8	LOOO
3	LL	9	LOOL
4	LOO	⋮	⋮
5	LOL	30	LLLLO

The Roman decimal system

Basic symbols: I = 1; V = 5; X = 10; L = 50; C = 100; D = 500; M = 1000

Manner of writing: Starting from the left with the symbol of the highest number; the symbols I, X, C are repeated, but not more than three times; the symbols V, L, D, are written only once.
When a symbol representing a smaller number is placed before that representing a higher number, then its value is subtracted from the following higher value.

Example:

1974 = MCMLXXIV

Greek alphabet

Letter capital	Letter small	Name	Letter capital	Letter small	Name
A	α	Alpha	N	ν	Nu
B	β	Beta	Ξ	ξ	Xi
Γ	γ	Gamma	O	o	Omicron
Δ	δ	Delta	Π	π	Pi
E	ε	Epsilon	P	ϱ	Rho
Z	ζ	Zeta	Σ	σ	Sigma
H	η	Eta	T	τ	Tau
Θ	ϑ	Theta	Y	υ	Upsilon
I	ι	Iota	Φ	φ	Phi
K	$\varkappa$	Kappa	X	χ	Chi
Λ	λ	Lambda	Ψ	ψ	Psi
M	μ	Mu	Ω	ω	Omega

Frequently used numbers and their common logarithms

(from MÜLLER, Tables of Logarithms)

	n	$\lg n$
π	3.1416	0.49715
2π	6.2832	0.79818
3π	9.4248	0.97427
4π	12.566	1.09921
$\frac{\pi}{2}$	1.5708	0.19612
$\frac{\pi}{3}$	1.0472	0.02003
$\frac{2\pi}{3}$	2.0944	0.32106
$\frac{4\pi}{3}$	4.1888	0.62209
$\frac{\pi}{4}$	0.78540	0.89509 − 1
$\frac{\pi}{6}$	0.52360	0.71900 − 1
π^2	9.8696	0.99430
$4\pi^2$	39.478	1.59636
$\frac{\pi^2}{4}$	2.4674	0.39224

	n	$\lg n$
π^3	31.006	1.49415
$\dfrac{\pi}{360}$	$8.7266 \cdot 10^{-3}$	$0.94085 - 3$
$\dfrac{2\pi}{360} = \dfrac{\pi}{180}$	$1.7453 \cdot 10^{-2}$	$0.24188 - 2$
$\text{arc } 1' = \dfrac{\pi}{180 \cdot 60}$	$2.9089 \cdot 10^{-4}$	$0.46373 - 4$
$\text{arc } 1'' = \dfrac{\pi}{180 \cdot 60 \cdot 60}$	$4.8481 \cdot 10^{-6}$	$0.68557 - 6$
1 rad; $\varrho^\circ = \dfrac{360}{2\pi} = \dfrac{180}{\pi}$	57.296	1.75812
$\varrho' = \dfrac{360 \cdot 60}{2\pi}$	3437.7	3.53627
$\varrho'' = \dfrac{360 \cdot 60 \cdot 60}{2\pi}$	$2.0626 \cdot 10^5$	5.31443
$\dfrac{1}{\pi}$	0.31831	$0.50285 - 1$
$\dfrac{1}{2\pi}$	0.15915	$0.20182 - 1$
$\dfrac{1}{4\pi}$	$7.9577 \cdot 10^{-2}$	$0.90079 - 2$
$\dfrac{3}{4\pi}$	0.23873	$0.37791 - 1$

	n	$\lg n$
$\frac{1}{\pi^2}$	0.101 32	0.005 70 − 1
$\frac{1}{4\pi^2}$	$2.5330 \cdot 10^{-2}$	0.403 64 − 2
$\sqrt{\pi}$	1.772 5	0.248 57
$2\sqrt{\pi}$	3.544 91	0.549 60
$\sqrt{2\pi}$	2.506 6	0.399 09
$\sqrt{\frac{\pi}{2}}$	1.253 3	0.098 06
$\frac{1}{\sqrt{\pi}}$	0.564 19	0.751 43 − 1
$c = \frac{2}{\sqrt{\pi}}$	1.128 4	0.052 46
$\frac{1}{c} = \frac{\sqrt{\pi}}{2}$	0.886 23	0.947 54 − 1
$c_1 = \sqrt{\frac{40}{\pi}}$	3.5682	0.552 46
$\sqrt{\frac{2}{\pi}}$	0.797 88	0.901 94 − 1
$\pi\sqrt{2}$	4.442 9	0.647 66
$\pi\sqrt{3}$	5.441 4	0.735 71

	n	$\lg n$
$\frac{\pi}{\sqrt{2}}$	2.2214	0.34663
$\frac{\pi}{\sqrt{3}}$	1.8138	0.25859
$\sqrt[3]{\pi}$	1.4646	0.16572
e	2.7183	0.43429
e^2	7.3891	0.86859
$M = \lg e$	0.43429	0.63778 − 1
$\frac{1}{M} = \ln 10$	2.3026	0.36222
$\frac{1}{e}$	0.36788	0.56571 − 1
$\frac{1}{e^2}$	0.13534	0.13141 − 1
e^π	23.141	1.36438
$\sqrt{e}$	1.6487	0.21715
g	9.80665	0.99152
g^2	96.170	1.98304
$\frac{1}{g}$	0.10197	0.00848 − 1

	n	$\lg n$
$\frac{1}{2g}$	$5.0986 \cdot 10^{-2}$	$0.70745 - 2$
$\frac{1}{g^2}$	$1.0398 \cdot 10^{-2}$	$0.01696 - 2$
$\sqrt{g}$	3.1316	0.49576
$\sqrt{2g}$	4.4287	0.64628
$\pi \sqrt{g}$	9.8381	0.99291
$\frac{1}{\sqrt{g}}$	0.31933	$0.50424 - 1$
$\frac{\pi}{\sqrt{g}}$	1.0032	0.00139
$\frac{2\pi}{\sqrt{g}}$	2.0064	0.30242
$\sqrt{2}$	1.4142	0.15051
$\sqrt{3}$	1.7321	0.23856
$\frac{1}{\sqrt{2}}$	0.70711	$0.84949 - 1$
$\frac{1}{\sqrt{3}}$	0.57735	$0.76144 - 1$

Index